GREAT BRITAIN

ESSAYS IN REGIONAL GEOGRAPHY

GREAT BRITAIN

ESSAYS IN REGIONAL GEOGRAPHY

BY

TWENTY-SIX AUTHORS

With an Introduction by

SIR E. J. RUSSELL, D.Sc., F.R.S.
DIRECTOR OF THE ROTHAMSTED EXPERIMENTAL STATION
EX-PRESIDENT OF THE GEOGRAPHICAL ASSOCIATION

Edited by

ALAN G. OGILVIE, O.B.E., M.A., B.Sc.
READER IN GEOGRAPHY IN THE UNIVERSITY OF EDINBURGH

CAMBRIDGE
AT THE UNIVERSITY PRESS
1958

CAMBRIDGE
UNIVERSITY PRESS

University Printing House, Cambridge CB2 8BS, United Kingdom

Published in the United States of America by Cambridge University Press, New York

Cambridge University Press is part of the University of Cambridge.

It furthers the University's mission by disseminating knowledge in the pursuit of education, learning and research at the highest international levels of excellence.

www.cambridge.org
Information on this title: www.cambridge.org/9781107626539

First edition 1928
Second edition 1930
First published 1930
Reprinted 1953
First paperback edition 2014

A catalogue record for this publication is available from the British Library

ISBN 978-1-107-62653-9 Paperback

CONTENTS

ILLUSTRATIONS

Certain of the figures and diagrams in this book have been based on maps of the Ordnance and Geological Surveys by permission of the Director General of the Ordnance Survey and the Controller of H.M. Stationery Office.

PREFACE TO THE FIRST EDITION

SINCE the publication, in 1902, of *Britain and the British Seas* by Mr (now Sir Halford) Mackinder, a steadily increasing band of geographers has been at work in these Islands, and a considerable number of papers have been published in the geographical periodicals, relating mostly to small parts of the kingdom. In the period, two comprehensive works of note have appeared: *The British Isles and Mediterranean Possessions*, the first volume, by twenty-two authors, of the *Oxford Survey of the Empire* (1914), in which the different chapters dealt with various aspects of the whole country; and *Les Iles Britanniques*, by Prof. A. Demangeon, a volume of the new *Géographie Universelle* (1927); while among smaller books mention should perhaps be made of Dr C. B. Fawcett's *The Provinces of England* (1919).

Between 1902 and 1927 the number of Departments of Geography in universities and university colleges in Great Britain has grown from three to twenty-one, and the network of these extends from Aberdeen to Exeter. Those responsible for geographical instruction in the universities have necessarily devoted much thought to the synthetic study of their own districts; hitherto, however, as a rule, none but their own pupils have been able to benefit from this. In addition, a good many local "regional surveys" have been made by individual effort and by co-operative work both within and outside the universities.

These considerations seemed to justify the preparation of a new composite volume, under the auspices of the British National Committee for Geography, dealing with Great Britain region by region, and thus permitting the expression of local characteristics, as it were from within.

To this end, the National Committee in August 1926 formed a "Joint Regional Studies Committee," with Sir John Russell as Chairman and consisting of the Authors of this book and several others, to arrange for the preparation of the volume and for its publication at the twelfth International Geographical Congress, Cambridge, 1928, as an appropriate means of presenting to the foreign Delegates to the Congress a synopsis by British geographers of the regional geography of Great Britain.

The Committee was well aware of the difficulty of dividing the country into regions for the purpose in view; and since many of the boundaries chosen are open to criticism, and some are quite arbitrary, no map has been inserted to illustrate them. It was felt that such a map might find its way into elementary text-books where it would mislead. It may be thought that an excessive subdivision of England has been made; but it was deemed desirable to secure a group of authors as representative as possible.

For the reasons above indicated, the contributors to this work mainly represent geography in the various university centres, but in some cases the collaboration of others has been obtained. Thus Mr Fagg and Mr Hutchings have long been associated with "regional survey" work, while Miss Kershaw, Mr Bygott, Mr Campbell, and Mr Gauld were regarded, by common consent, as the most competent authorities on the regions they undertook to describe. In the case of three regions joint authorship was held to be desirable, and in two of the essays special contributions have been made by botanists who are also ecologists: Prof. Tansley and Dr Smith.

The first essay alone deals with the entire Island; Dr Mill undertook to write a chapter on the climate of Great Britain as a whole, for the attempt of Authors to treat this subject piecemeal would clearly have led to much overlapping and would have been unsatisfactory in other ways.

The greatest possible freedom was given to the writers as to the manner of treating their subjects, the only important restriction being that of length. The book is therefore to be regarded as a collection of essays and not as a systematic handbook.

Space was allotted to the various regions by a sub-committee; and in this, weight was given to area, population, geographical variety and complexity. The imposition of space limits gives rise to the least agreeable task of an editor, and while recording the fact that but few serious reductions were required, I would express the hope that no Author has felt himself to be seriously hampered and that readers will appreciate the difficulty of getting the maximum amount of information into limited space without losing sight of the importance of "style."

In a geographical work there is no limit to the number of maps that could usefully be given. But here an average of only two could be allowed to each essay. The writers themselves chose the subjects for illustration and supplied the originals. Chapter II has more figures, the Authors having chosen to reduce their text and to provide maps in finished form. The drawing of all the other maps has been rendered possible by the contributions to a fund for the purpose made by the following: the respective Authorities of the Universities of Birmingham, Bristol, Cambridge, Leeds, Liverpool, London, Oxford, Sheffield, and Wales; those of the

University Colleges of Exeter, Leicester and Southampton; the Carnegie Trust for the Universities of Scotland; the Council of the Manchester Geographical Society; and the Organizing Committee of the International Geographical Congress of 1928.

To all of these contributors the National Committee for Geography is greatly indebted.

I have also to thank Col. H. St J. L. Winterbotham, C.M.G., D.S.O., and the Staff of the Geographical Section, War Office, for help they have given in connection with the drawing of the maps. I have to acknowledge, further, the courtesy of the Council of the Royal Meteorological Society in allowing a plate of the *Rainfall Atlas of the British Isles* to be copied as Fig. 1.

At the request of the National Committee for Geography, H.M. Ministry of Agriculture and the Scottish Board of Agriculture generously placed certain unpublished statistics at the disposal of the Authors of this work, and for this service they now express their thanks.

I have endeavoured to help readers unused to British measures by inserting metric and other equivalents of numerical data in the text or as footnotes at all important places; exact conversions are made where the data required them, otherwise they are given in rounded equivalents; *e.g.* the usual convention of isohyetal maps has been followed, by which the inch is taken as equal to 25 mm.

While very considerable advances have been made in our knowledge of the regional geography of Britain, much more work will have to be taken in hand or completed ere the material for a comprehensive treatise is available. Some of the present contributors do not feel that they have yet spent as long a time as they would have wished in the regions they were asked to discuss, before composing a synthetic description for final publication. However, the collection of essays is presented as indicating the points of view of those who are engaged in the subject and the general results at which they have already arrived. It is hoped that the volume will fill, for a time at least, a distinct gap in the geographical literature on Great Britain.

The time available for the actual editing and making of the book has been short, and I am obliged to the officials of the Cambridge University Press for the helpful consideration they have shown me in the circumstances.

A. G. O.

EDINBURGH UNIVERSITY
April 1928

PREFACE TO THE SECOND EDITION

The rapid exhaustion of the first edition may be taken as satisfactory evidence that there was a real need for such a work, and that the issue of a new edition is fully justified. This opportunity has been utilised to remove from the text certain minor blemishes due to the somewhat hurried preparation of the original book. Furthermore, many of the essays have been amplified in some measure, while one Author, Mr Fitzgerald, has largely rewritten his portions of the essay on Lancastria.

I have to record, with deep regret, the death of one Contributor, Dr W. G. Smith—before the present edition was prepared.

Five new maps have been added to the illustrations: Figs. 15, 16, 36, 51 and 52, and I have to thank the Council of the Royal Scottish Geographical Society for permission to reproduce the last two of these.

It is perhaps desirable to mention that the Authors, in planning *Great Britain*, did not contemplate its complete illustration by maps in the text. Readers of the book, to obtain the maximum advantage from it, are recommended to consult Ordnance Survey maps; and, for most of the topics discussed, those on the scale 1 : 253,440 (*i.e. c.* 4 miles to 1 inch) will be found to suffice.

A. G. O.

EDINBURGH UNIVERSITY
August 1930

INTRODUCTION

THE purpose of regional geography is to describe the regions of a country as they are and to discover the causes that have made them what they are. For Great Britain the causes go far back to very remote times; events that happened many millions of years ago have not only helped to shape the countryside of to-day but still profoundly influence the daily lives of its people. The principle of continuity affects not only the land but the people, and nowhere is it more necessary to know something of the history both natural and human of the country in order to understand and appreciate the things one sees.

As far back as geological records can be traced there was always some part of Britain existing as dry land. In the earliest ages it was part of the continent, now submerged, called Atlantis by the geologists: afterwards it was part of Europe: then in late times it became a separate island. Large parts of it, especially in the south and the east, have been several times submerged: first under the Palaeozoic sea, then under the Jurassic sea, then the Cretaceous sea, and some of it under the Wealden sea. Always, however, there stood out dry land that was not submerged; there was always something of Britain left. The submerged parts became covered with deposits of sand, silt and clay, which on emergence formed new land surfaces quite different from the old. In addition there have been from time to time gigantic volcanic movements.

The various land surfaces have had widely different climates. In Carboniferous times the climate was tropical, and there were vast areas of swamps in which grew immense trees: these by a peculiarity of their conditions never entirely disappeared, but became converted into the coal that has played and still plays a great part in the human history of the country.

In Triassic times the climate was hot and dry, even arid: the country was desert: the rocks very red, as indeed often happens in the deserts of to-day. Even in comparatively recent times, and since the country took somewhat of its present shape, the climate was arctic and most of the land north of London was covered with glaciers. These so much affected the aspect of the land that they may be regarded as among the most recent landscape-making agencies of Britain.

The first great division of the country, therefore, is into the uplands of the west and north formed of ancient rocks that have probably never been submerged, at any rate since Palaeozoic times, although now much eroded; and the newer formations of the east, midlands and south deposited more recently and now forming the English Plain.

The Plain, however, has not remained horizontal: great earth movements caused it to tilt towards the south-east. The result has been to expose the old formations in the west and north, but to keep the newer formations on the surface in the east and south. There has been some folding but no important mountain building in this region, and so the land slopes gently to the sea, with no cliffs of any size excepting at a few places in the north-east, and where the Chalk bridge was broken at Dover and Beachy Head.

A section across the country from the Thames estuary to the west covers almost the whole range of England's geological history. In the estuary the land is recent alluvium still in process of accumulation: it is but little above sea-level. As it gradually slopes upwards toward the west the older formations become exposed, and they in turn thin out as the rise continues till, still farther west, the underlying Chalk comes to the surface covered, however, with a layer of residual material. Around London the evenness of the slope is broken by the saucer-shaped depression known as the London Basin, but, once beyond this, the general rise continues to 600 or 800 ft. when the Chalk abruptly ends in a steep escarpment, one of the dominating features in the south of England.

This high land forms the Chalk Downs: open, wind-swept, and treeless in the exposed places, well wooded in its hollows with beech and many flowering shrubs and plants, almost free from hedges even when cultivated. To the north and west stretches a great plain, closely cultivated in the foreground, grassy and well wooded farther away, rising to another ridge in the distance. It is not quite a plain, there being one or two minor escarpments, and places where the rivers have cut down into the underlying clay, but toward the west it becomes a well-wooded plateau with a stony soil, very red in places, much arable land, stone walls in sharp contrast with the hedges elsewhere, and attractive stone houses. Finally the second great escarpment is reached, also 600–800 ft. high, formed on the Inferior Oolite, a limestone that can be cut in any direction and therefore is much valued for building. This forms a district among the most interesting in England. The escarpment, like that of the Chalk, commands extensive views to the west and north, but with this difference: while the view from the Chalk is that of a plain the view from here embraces the uplands of the west. On descending the escarpment a second difference is seen: the road winds steeply down a coombe, instead of descending gently on a long slope.

The bottom land next reached is the broken and undulating outcrop of Middle Lias, but this soon gives place to the clay vale of the Lower Lias, a sticky, impervious clay which readily becomes water-logged; it is, however, lightened in places by certain calcareous bands and by surface deposits of glacial drift. Passing across the greenish Lias soil one reaches the red Trias soil, all part of the same lowlying ground, traversed by two of the most famous rivers of England, the Avon and the Severn.

So far the sequence is the same whether one travels westwards as here, northwards, southwards or south-westwards, but there are certain differences.

In the south the Chalk, while tending to slope to the south-east, has been pressed into ridges and furrows by late-Miocene earth movements which were specially intense about Dorset. These ridges run east and west, presenting steep escarpments to the north: they terminate in the west against another fold system. The most remarkable feature of the Chalk country, however, is its wonderful break in the south-east, like a section cut out of an orange, forming two Chalk escarpments facing each other, and exposing high ridges of the underlying Greensand and Lower Wealden beds. This is the famous Wealden area, which inspired Topley's classical memoir*, Davis's studies of river systems, and in another direction some of Kipling's most intimate poetry.

In the north the comparatively recent London Clay and the boulder clays give rise to slight undulations inappreciable on the map and yet profoundly affecting the scenery in ways not easily described by the layman, giving the impression of a hilly country with an unusually vast expanse of sky. It was in this country that English landscape painting arose; on the Essex-Suffolk borders, Gainsborough and Constable learned to weld sky and land into one harmonious composition, as farther north in Norfolk did Crome and his great disciple John Sell Cotman. There is a break round the Wash, where on a great patch 70 miles long and 30 miles broad (110 by 48 km.) extending to the south and west, a former marsh or shallow lake became a deep layer of mild calcareous peat, called "fen" in England, similar to the *Niederungsmoor* of South Germany. Till the 17th century this was undrained and liable to much flooding, a safe refuge for the lawless and the outlawed, and associated in literature with the exploits of Hereward the Wake. Even now it is

* Memoirs Geol. Survey, *Jour. Roy. Ag. Soc.* vol. VIII, p. 241, 1872.

difficult country to live in, partly because of the danger of flood, partly because of the lack of good drinking water. In south Lincolnshire is a very flat area of silt recently deposited by the sea in that remarkable transfer of soil from north to south which occurs all along the east coast rounding off the north-eastern exposures and forming the long gravel noses or spits stretching towards the south. This deposition is still continuing, and it has been calculated that the Wash will become completely silted up in another 10,000 years. Farther north the chalk arises once more as a rounded mass called the Wolds running slightly north-north-westward up to the Humber, flanked to the west by the Oolite escarpment, here trending north and south overlooking the plain of the Trent. The Wolds reappear north of the Humber in east Yorkshire, fronting the great alluvial Vale of York but curving to end at the Vale of Pickering and Flamborough Head, beyond which the Oolite arises as the North York Moors.

Reverting to the section across the country, the regularity of the sequence ends with the Trias formation. Westwards and northwards lie the uplands, the ancient mountains which stood out of the waters in which the Oolite and the Lias were deposited, but now much worn down to a fragment of what they were. They are, however, not continuous: between them comes the Cheshire gap, a lowlying area covered with red Trias soil—the "Midland Gate" which, as Mackinder* reminds us, has played its part in English history. The edge of the Trias is one of the great boundary lines in England. Beyond it the landscape changes, and with it the utilisation of the land and to some extent the type of the people.

Of the older formations the Coal Measures come first, forming patches on a belt of country running from South Wales slightly to the east of north up to Northumberland. The largest coal resources are those of South Wales and of the eastern fields: Northumberland and Durham and especially that of South Yorkshire, Nottingham and Derbyshire, where the workings are pushed progressively eastward at ever greater depths. The other English coalfields lie in several areas of the Midlands, in Lancashire, and in Cumberland; while, as a new development, coal is now being won in Kent at a depth of some 2000 ft. Further, there are large coalfields in the Lowlands of Scotland.

Beyond the Coal Measures come the high lands, formed of much older rocks, some of which are igneous. In the south-western peninsula are two raised blocks, Dartmoor in the south, an open moor, and Exmoor in the north, somewhat more cultivated: in

* H. J. Mackinder, *Britain and the British Seas*.

between lies an undulating region of grey soil of the Culm Measures. In the west, towards Wales, is the rampart of hills just over the Severn formed by the Forest of Dean, the Malvern Hills, the Clee Hills, Wenlock Edge, the very ancient pre-Cambrian Longmynd, and, perhaps oldest of all the mountains in England, the Wrekin, which was already ancient and worn down before the Himalayas or the Rockies arose. These features nearly enclose the undulating country formed by the Devonian or Old Red Sandstone in which the Wye and the Usk have carved two of the most attractive valleys in the country. Beyond this lie the hills of Wales, rising for most of their length rapidly to their highest level, then falling away gradually to the coast; in the south they are rounded, but in the north they form the rocky heights of the Snowdon range.

The northern uplands of England, the Pennines, form a mass of high moorland, traversed even yet by few roads, but opening out on the east side in a series of pleasant dales to the lowlands of Yorkshire, and ending northwards in the Cheviots and the magnificent hill land of Scotland. Near the Border the highland runs almost from sea to sea, but on each side there is left a narrow strip of lowland, forming two corridors along which for nearly 2000 years there was much coming and going of armed men, often only cattle raiders but sometimes in great array moving on to great events. To this day only three important lines of railway and three important roads cross from England to Scotland. To the east, and separated at one part only by the Northallerton Gate, there is another high mass, the Yorkshire Moors, which almost cut off the north-eastern lowland from the south. On these moors the villages were formerly very isolated and some interesting old customs long survived*; now, however, they are being drawn into the industrial area. To the west is the mass of volcanic rocks forming the crags and fells of the Lake District, one of the most beautiful regions of Great Britain.

In Scotland the trend of the formations is, as in south-east England, from south-west to north-east, but in the main the rocks are old, and there is much igneous material. The formations divide the country into two masses of upland, the Highlands to the north and the Uplands to the south, separated by the Midland Valley: there is also a fourth area—a lowland—in the north-east.

The Highlands form a region of rounded massive country, with an even skyline and scored by deep narrow valleys or glens, often stepped in their upper parts, with lakes or "lochs," the bottoms

* For an account of the region round Danby see *Forty Years in a Moorland Parish*, by J. C. Atkinson.

of which may be far below sea-level; some are as much as 1000 ft. in depth. Many on the western side terminate in the sea, making this a coast of fjords.

The Southern Uplands are lower: they form rounded grass- and heather-covered hills, with much dale land in the west.

The Midland Valley is about 50 miles wide: it is terminated to the north by the sharp massive wall of the Highlands, but on the south it merges less sharply into the Uplands. It contains a variety of formations, including Coal Measures, oil-bearing shales, and others, distributed, however, unequally, partly because of the unevenness of the floor of older rock on which they were deposited, partly because of differences in the subsequent foldings and denudation. Here the hills are mainly of igneous rock.

Vegetation Regions

The vegetation regions of Britain are determined in part by the climate and in part by the soil. The four chief soil factors are the calcium carbonate, the clay, the organic matter, and the reaction, the last named, however, being a function of the other three. The climate, as shown later, is determined in its general features by the position of Great Britain in relation to the Gulf Stream drift, but modified by the anticyclones of the Continent: in any given locality it is profoundly affected by the contour of the land. In the western uplands the rainfall may exceed 100 inches (2500 mm.) per annum; it becomes less and less in passing towards the centre and the east: the line of 30 inches (750 mm.) annual fall, critical for agriculture, runs sinuously from Berwick to the Isle of Wight: eastward of this line the annual rainfall may be as low as 20 inches (500 mm.). Long-continued frost and drought are equally rare. England is essentially a country of grass and trees: land left to itself rapidly becomes covered first with grass and then with trees:

"England's green and pleasant land"

is the result of its moist, mild climate.

The Utilisation of the Land

So much for the country itself: we now turn to the use man has made of it. The lowlying shore facing Europe made invasion easy in the days (2000–500 B.C.) when people were moving in all parts of Europe and Asia. From about 1000 B.C. onwards there occurred, at intervals of about 500 years, invasions of such magnitude and importance as completely to alter the life of the country: first the

Bronze Age people, then the Celts of the Iron Age, then the Romans, then about A.D. 500 the Saxons, and finally, in 1066, the Normans*. This was the last invasion that succeeded although others have been threatened at various times.

Much of the best life of Britain has always been spent in the country, and each generation has added to it something attractive or useful. The long unbroken years of peace have tended to conserve the relics of the past: old customs, old houses and buildings, and especially old churches, are found in every district, many of them still showing signs of the damage done in the 16th century by Henry VIII and in the 17th by the Puritans, the last destructive force to march through the land. The vestiges of the pre-Celtic peoples are now almost obliterated: possibly something survives in the curious monosyllabic hill- and river-names of unknown derivation. The Celtic peoples have however left clear traces on the Chalk Downs of the south and east of England where they settled and farmed 2500 years ago—traces which are being unexpectedly revealed in the interesting air-surveys made by Mr O. G. S. Crawford: farther west their remains increase and in Wales they survive vigorously to this day.

The Romans laid down the main lines of the plan of the country, making the first trunk roads and founding the cities, but the Saxons, who followed about 500 A.D., began the filling in of the details. The counties represent the ancient kingdoms or the "shires" into which they were divided, while many of the villages and towns are the direct descendants of their settlements. The development continued under Norman guidance, and after the fusion, when the Normans and Saxons became one race, the countryside began to assume something of its present aspect: indeed in many villages the church, the manor-house, the road to town, and even the lanes about the villages remain where the Anglo-Normans first put them. Their three-field system of agriculture (which probably contained elements of the earlier agriculture of the country) survived with some changes till the end of the 18th century, and can still be traced in a modernised dress in a few places, *e.g.* at Haxey in Axholme, at Laxton in Nottingham and elsewhere†: many of the fields still bear their old names: even the twists and turns of the lanes, needed at first to avoid some Saxon ploughland or morass, have persisted though their original cause has long since gone. The enclosures, that began in the 14th and 15th centuries and led to the development of a great wool industry and therefore to sheep-raising, have

* See H. J. Fleure, *The Races of England and Wales*.

† Till recently: Soham, Clothall, Bygrave and others.

left their mark in the magnificent 15th century churches and 16th and 17th century manor- and farmhouses scattered over the Oolite formation and in the villages fed from the Chalk areas, where the industry developed as well. It is impossible to understand the geography, still less the life, of the English village, without some knowledge of its history.

Of the total land area of England and Wales eight-ninths is wild or planted and one-ninth is used for cities, towns, villages, buildings, roads, railways, public parks, open spaces, etc. The figures* are:

	Million acres	(ha.)
Arable land and permanent grass: holdings above 1 acre	25·75	(10·42)
Arable land and permanent grass: holdings of 1 acre or less	0·31	(0·13)
Rough grazing and heath land	5·02	(2·02)
Woods, forests and plantations	1·90	(0·77)
Towns, villages, houses and all other buildings, roads, railways, private gardens and parks not grazed, public parks, open spaces, etc. ...	4·15	(1·68)
Total land area	37·136	(15·02)

In Scotland the proportion of land occupied by towns, villages, houses, roads, etc. is much less than one-ninth; of the remainder, more is wild and less is cultivated than in England.

The division of the land between wild and cultivated, and of the latter between arable and grass, is primarily a matter of rainfall, and secondarily of altitude. In the eastern part of England and Scotland, where less than 30 inches (750 mm.) of rain falls annually, much of the land is in arable cultivation, with one year of grass in a rotation of four or five years; while in the western part, where the rainfall is higher, a smaller proportion is in arable cultivation, and of this the grass may remain two, three or more years in a rotation extending over six to nine years or even longer. The main agricultural division of the country, therefore, is into these two regions, the eastern being lowlying, relatively dry, and largely under arable cultivation, carrying also most of the pigs of the country; the western, including the highlands, where the wet areas are in grass and only the valleys and plains of lower rainfall are arable, and which carries most of the cattle and the sheep of the country.

* 1925 returns. Given in *The Agricultural Output of England and Wales, 1925*. Cmd. 2815, 1927, H.M. Stationery Office; probably the best account of the present position of agriculture in England and Wales. See also the *Atlas of Agriculture*, Ministry of Agriculture.

Within each of these regions there are certain differentiations in the agricultural systems. In the eastern part the dry winters and warm, somewhat dry summers suit both wheat and barley, but the barley crop tends to hug the eastern side and the wheat to push more out to the centre, while oats, being tolerant of wet weather, are grown all over the country. There is also a soil distinction: barley and sheep are associated with light soils, wheat and bullocks with heavier ones, while dairying cattle can be kept under a variety of systems adapted to different types of soil.

The soils of the English Plain from the east coast to the Oolite scarp fall into four great groups: (1) fen; (2) heavy soils; (3) loams; (4) light soils, (*a*) sands and gravels, (*b*) Chalk and Oolitic limestone.

Of these the *fen* is the lowlying area of about one million acres (404,000 ha.), south of the Wash, reclaimed by drainage and embankment from the 17th century onwards, perhaps the greatest achievement in land reclamation in England*. It has a black fertile soil, producing heavy crops of corn in the old days, then later of potatoes and, more recently, of sugar beet, with certain minor crops, celery and mustard, the latter forming the basis of an important industry in Norwich.

The *heavy soils* have always presented difficulty to the agriculturist. In the old days they were used for wheat and beans, and lay fallow for at least one year out of three: they gave poor returns and were occupied by only poor farmers and labourers who could not get on to better soils: the villages were and indeed often still are unattractive. Much of this land has always been too heavy to cultivate and so is left in wood: some formerly cultivated is now laid down to grass and used for dairying, but difficulties often arise because of the low rainfall. These soils shade off into the loams.

The *loams* are very fertile and suitable for a wide range of crops: there is a growing tendency to displace the ordinary farm crops, swedes, cereals, etc., by market gardening near the towns; by potatoes, of which large quantities are grown in the Holland division of Lincolnshire; by fruit, always a great industry in Kent, also increasing considerably in Norfolk, Cambridge and elsewhere, and carrying in its train the subsidiary industries of cider- and jam-making†. Hops are confined to Kent, Sussex and Hampshire.

The lighter loams shade off into the sands and Chalk soils. The

* For an account, see Lord Ernle, *English Farming Past and Present*; also J. Koethals-Altes, *Vermuyden, Sir Cornelius: The life work of a great Anglo-Dutchman in land reclamation and drainage*, 1925.

† Jam takes the excess of strawberries, raspberries and black currants: but there is at present no way of using the excess of plums, the public having lost the taste for plum jam.

light soils cover a considerable area in the eastern part of England and their peculiarities have given the chief character to the agriculture of the whole region. The lightest sands cannot be profitably cultivated by any known method and so are left as wild heaths, but some are being planted with conifers. They are now highly esteemed for residence and for sport. To an older generation, however, they were hateful: Hindhead itself, perhaps most favoured of all the sandy districts, being to Cobbett "certainly the most villanous spot that God ever made*." Market gardening, small fruit, and above all, flowers and shrubs for the great modern industry of ornamental gardening, occupy much land formerly waste. The more loamy sands can be used for agriculture, but are better suited for market-garden or arable cultivation than for grass; they are liable to dry out during the spring months of low rainfall, thus causing serious loss of crops. The East Anglian farmer has learned to mitigate spring droughts in two ways: good cultivation and the use of farmyard manure. The good cultivation necessitated suitable agricultural machinery, the manufacture of which has become an important Eastern Counties industry. The necessity for farmyard manure led to the winter feeding of bullocks, and so to settlement in villages in order that the labourers could easily attend to the animals: this kept the country people together and led to an organised village life, and to the many social and political consequences resulting therefrom. Norfolk has always been a fertile source of new ideas in agriculture and indeed in many other directions, while Suffolk has produced some of the finest and most beautiful animals in Britain.

On the lighter sands and on the Chalk and Oolitic limestone soils extensive use is made of sheep for the manuring and treading of the land: they are by far the best agents for overcoming drought and maintaining fertility and have been the basis of chalkland farming from time immemorial. In the various regions of the Chalk area special breeds of sheep have been evolved to suit the local conditions, elaborate rotations have been worked out to supply the necessary food, and the shepherd has attained high rank in the countryside†. To-day there is a tendency for dairy cows to displace sheep, but this leads to a considerable reduction in total output of food from the farm.

* *Rural Rides in Kent, Surrey...*, Wm Cobbett; one of the raciest accounts of southern England. 1st edition, 1830: many subsequent editions.

† See W. H. Hudson, *A Shepherd's Life: Impressions of the South Wiltshire Downs*, 1910; also A. D. Hall and E. J. Russell, *Agriculture and Soils of Kent, Surrey and Sussex*, 1911.

On the western side there is also considerable differentiation of the agriculture. In the main, grass is the predominating crop. Cattle are spread over the whole area, but least in the south central counties, Dorset, Wilts, Gloucester, Worcester, Monmouth and Brecon. Dairy cattle are more restricted, being concentrated in and about the Staffordshire-Cheshire and the Somerset-Dorset regions, while within these fresh milk is still further restricted to the parts easily accessible by rail.

Sheep tend to concentrate still farther west and north, being associated with the great open spaces of the uplands.

The western type of agriculture begins with the Lias vale, which lies north and west of the Oolite and bears a heavy dark greenish-brown soil, too heavy for arable cultivation in present economic conditions, and therefore left in grass: it stretches from Somerset, across Gloucester, Rutland, Leicester, and Northampton. Meat production is the chief husbandry, and this involves the buying of cattle in spring, leaving them to graze all the summer, and selling them in autumn. There is little to do in the winter, and fox-hunting has developed as a great sport which has profoundly influenced British life. In certain famous regions—round Oakham, Melton Mowbray, in the Pytchley, Quorn, Belvoir country—hunting has become an important industry, bringing much money to the countryside, encouraging the breeding of horses, and various minor occupations.

The old houses were made of timber frames filled in with brick or daub and wattle: the modern ones are of an ugly yellow brick.

Beyond the Lias comes the Trias, the marls of which form a bright red soil extending through Warwickshire, north Shropshire and Cheshire as a fertile plain. The arable land produces potatoes and vegetables for the large manufacturing towns: the grass area produces milk, the famous Cheshire cheese, and pigs; but the sandy pebble beds—Bunter beds—are not generally in cultivation: they are used as recreation grounds.

Beyond these lie the Coal Measures, which, when used for agriculture, form poor grass or arable land, much troubled by subsidences: the surface is poor, as if to compensate for the wealth below.

Farther west and north the agriculture is determined by elevation and climatic conditions; it is much affected also by the formation of acid in the soil and of peat—conditions set up by high rainfall.

The lowlying regions each have a characteristic system of agriculture. In the south-western peninsula the system is largely dependent on dairying and cider in the eastern parts, while farther westward there are many small general farms and an increasing area under early fruit and vegetables, culminating in the interesting early potato and broccoli culture about Penzance and the cut-flower industry of the Scilly Islands. In the Gloucester-Hereford area cider apples are important: the actual orchard area is not increasing much as yet, but old orchards are being improved and renewed under the guidance of the National Fruit and Cider Research Institute at Long Ashton and of some of the more enterprising cider makers. Here also are found the famous Hereford cattle, bred for beef but not for milk, more tolerant of the high-lying moist country than the dairy shorthorns. Farther north and east stretches the wonderful fruit and vegetable area of Evesham and Pershore which owes its existence to the light glacial drift overlying the Lias clay, and forming a belt of undulating ground traversed by the Avon, free from excessive frost, and well served by the railways. The Trias plain of Shropshire and Cheshire has already been mentioned as producing milk, potatoes, pigs and (near its towns) vegetables. Passing into Lancashire, and crossing the belt of chemical and other factories in the Mersey and Ship Canal zones into an agricultural area, we meet the important potato-growing regions extending north of Liverpool, producing also much milk and far more poultry than any other county. Farther north still in Ayrshire we are again in a rich dairy and potato country with an important fruit industry also. Passing along the same formations to the other side of Scotland we reach, near Dunbar, the fertile red soils which produce potatoes well known for high quality as "cookers." Farther westward there is an early potato industry on the south bank of the Firth of Forth, while throughout the Lothians, one of the best farmed districts in Britain, much attention is directed to seed potatoes. Climate and soil are both favourable to spring-sown crops: the winters are cold, preventing the growth of the winter weeds that are so troublesome in the mild southern parts of England: the spring and summer may be dry, but evaporation is relatively low, and repeatedly it happens that clover and rotation grasses are flourishing here, while they are parched and dying in the south.

Still farther north in Perthshire is the interesting raspberry industry of Blairgowrie, while farther again comes the good farming of Aberdeenshire. Here general farming is very successful: in the cool moist climate oats and swedes grow well, as does temporary grass; this is commonly left for periods of from two to six or more years, so that the distinction between arable land and permanent grassland no longer exists.

Turning now to the higher-lying land: through the west from

south to north this is mainly left for open grazing. Sheep are more successful than cattle: not all sheep, but usually only those having long wool. Each district has evolved its own breed; some are strikingly handsome, especially the Scottish Black-faced rams. They are brought down to the Lowlands to be fattened, and the October collections and sales are remarkable sights. Dogs are trained to collect the sheep in the most wonderful way; it is indeed a form of sport, and the sheep-dog trial is as popular in the uplands as the League football match in the lower levels. In the north and south, miles of stone wall restrain the wanderings of the sheep: the Welsh shepherds round Snowdon and those of the Scottish hills with great ingenuity save themselves this trouble by putting with each flock a certain number of old sheep who know how far the flock may wander and who also will not permit another flock to intrude on their ground.

Lower down, or in drier places, more grass can grow and more cattle be kept. The most hardy are the black cattle: the Welsh, the Galloways and others; they are not used for dairying but only for breeding; the cows suckle their calves. At still lower levels the conditions are less severe; the more sensitive but more productive red, white and roan cows can now be kept. The land is divided into small fields some of which are mown for hay while others are used for grazing; even a certain amount of cultivation is possible, though only two crops can be grown to advantage: swedes and oats, these being the only ones to stand the wetness. Milk is produced for sale or made into butter (not always good). In a few special regions, the Shetlands, Exmoor and Dartmoor, ponies are bred. The line between the cultivated and the uncultivated land is never sharp; enterprising men of all ages have endeavoured to push it higher and to take in more of the moorland; usually they have failed, being beaten by the high rainfall and the very acid soil; the heather and whortleberries (*Vaccinium*) remain in possession. There have been many tragedies in the moorland parishes*.

These features are common to all the uplands. As between the different regions, however, there are great differences. In the north-west of Scotland many of the people still retain their old Gaelic language and certain old characteristics. Much of the land is left as deer-forest and the rivers are preserved for fishing: the area is the recreation ground of the rich. The northern uplands of England are inhabited by people with a strong Scandinavian strain, as shown by their features, their place-names and their dialect; the men

* For the story of the reclamation of Exmoor see C. S. Orwin, *The Reclamation of Exmoor Forest*, 1929.

indulge in sports some of which are almost unknown elsewhere, while others, as the fell hunting, have their own peculiarities; the women are highly efficient housewives, making wonderful cakes and strange delicacies such as rum butter. Where the uplands and the lowlands meet on the eastern side, with its low rainfall, there is perhaps the most delectable region in the north of England, the Yorkshire dales centring round Richmond and Leyburn: a country held in high repute in olden days, as testified by the castles and ruined monasteries still remaining.

The south-western uplands breed deer and ponies and so a very distinct type of sport has been evolved: stag-hunting with hounds. In Scotland the deer are stalked and shot.

The Welsh uplands differ wholly from both. The inhabitants are the old race driven westwards by the invasions of 1500 years ago; they have retained their ancient characteristics: their love of song,

"Cymry lan, gwlad y gan,"

and their reverence for the bard and the student. But, with the extinction of the wolf, they ceased to hunt; there is nothing to correspond with the fox-hunts of England. Football is now the great sport in Wales. A still more striking difference is their strong Puritanism which prevented them from building beautiful churches. Craftsmanship and religion were never linked as in England from the 13th to 16th centuries; and craftsmanship alone could never tackle the unpromising stone of Wales. They made no attempt to adorn their churches or their chapels, or even to give them good proportions, nor did they adorn their houses. The villages consist of dreary rows of cottages and are almost devoid of artistic interest. It is only when one is invited inside the cottage, to sit round the fire and listen to the songs and tales of older days, that one begins to appreciate the meaning of the village life; only in the village Eisteddfodau, where singers and bards compete with poems in strange metres, does one realise how much of the life of an older time still remains in the hills of Wales.

The Highlands of Scotland present a completely different picture. The settlements are largely restricted to the fertile (or less infertile) patches in a vast infertile region and the people crowd on to small-holdings or crofts insufficient to afford any high standard of living. Fishing and agriculture are both followed by the crofters, but the population constantly tends to exceed the carrying power of the land, and so there is a constant migration to the south or to other parts of the Empire. The Lowlands, on the other hand, form a much richer country: agriculture was always more prosperous, and the great city of Edinburgh arose early, along with others of less size but hardly less note: Aberdeen, Dundee, etc. In this region the Scottish nation grew up, protected from invasion from the south by the mass of the Southern Uplands: in the east were seats of learning, and, in the west, when foreign trade developed, there arose the great industrial and commercial area of Glasgow and the Clyde towns. A curious change is coming over this western region, however: a large foreign population, chiefly Irish, is taking possession, ousting the Scotsmen, and doing by peaceful penetration what no previous invaders were ever able to do by force. What the end may be, no one can tell, but the process shows that the movements of population in Britain are not yet over, and the problems of human geography are by no means all solved.

E. JOHN RUSSELL

I

THE CLIMATE OF GREAT BRITAIN

Hugh Robert Mill

CLIMATE, though often described as average weather, should rather be viewed as the sequence and the sum of weather over a long period. It is produced by energy radiated from the Sun working in the Earth's atmosphere, but this process involves so many simultaneous and successive actions and reactions that it can only be made clear by a series of approximations.

The five paragraphs which follow set out with the utmost brevity five stages of approach to the description of British Climate.

(1) If the Earth consisted only of a homogeneous lithosphere with a smooth and rigid surface the temperature of that surface would depend on the latitude only. Temperature would vary from the equator to the poles in strict accord with the astronomical climate produced by the rotation and revolution of the globe and the inclination of its axis to the plane of its orbit. The hot tropical zone with nearly equal day and night would show a great diurnal range of temperature and a small annual range. It would be separated by a gradual transition on each side, the temperate zones, from the two cold polar zones. In the polar zones there would be a great seasonal difference in the length of day and night, allowing feeble heating in summer from a low sun and intense cooling at night in winter by terrestrial radiation; the range of temperature in 24 hours would be slight but the annual range would be great. The heat or cold acquired by the surface in any latitude would be exactly the same from east to west and would not affect contiguous regions to north or south.

(2) The lithosphere's surface is not homogeneous but composed of different rocks differing in thermal conductivity, and so parts of the solid surface in the same latitude are more rapidly heated and cooled than others, and the temperature of the surface is not exactly the same along the same parallel of latitude. The nature of the surface has to be considered.

(3) The surface of the lithosphere is not smooth but ridged into upheavals and depressions. All places in the same latitude are therefore not equally exposed to the sun's rays; the ridges catch the sunlight on one side at a favourable angle for absorption and so are more heated by day on the exposed than on the sheltered

side. Aspect has to be taken into account, and the detailed configuration of mountain and plain exerts a dominant influence on the distribution of local climates.

(4) The Earth does not consist of a lithosphere only; there is a mobile hydrosphere or water envelope, and the chief result of the irregular configuration of the lithosphere is that the water gathers into the deeper hollows and so gives rise to the broad geographical division into land and sea. If the Earth consisted of land and sea alone the astronomical zones of soil-temperature would not be changed on the land; but similar zones would not be established on the sea although solar radiation strikes impartially on both. The heating of the water in the Tropics and the chilling of the water in the polar zones alter the density and set up vertical and horizontal movements resulting in a circulation of warm water towards the polar and of cold water towards the tropical seas. In the northern hemisphere the rotation of the Earth causes these currents constantly to deviate towards the right as they flow on their way, and the edges of the continents guide them in their onward course, so that a warm stream sets eastward and northward towards the coast of northern Europe and a cold current flows southward and westward along the coast of North America. Thus the temperature of the water in the North Atlantic does not increase uniformly from north to south like that of the land surface, but rather from west to east. Heat from the Sun penetrates to a far greater depth in water than in solid land and thus raises the temperature of a far thicker layer to a far less degree. Similarly water cools by conduction and radiation much more slowly than does the land, giving a much smaller diurnal and annual range in the same latitude. Water when it receives heat turns into vapour which rises and accumulates until it is saturated and exerts a certain pressure which is greater as the temperature is higher; but on being cooled saturated vapour returns by degrees to the liquid state. The heat used up in evaporation is restored when the vapour, which may have been carried some distance from its source, condenses as mist or rain. Water, when cooled, solidifies to form ice which tends to accumulate in the polar zones, and the effects of evaporation and freezing greatly complicate the circulation of the sea.

(5) The physical structure of the Earth is completed by an all-embracing atmosphere of air resting in uneasy equilibrium on land and sea and permeated by the vapour of water. Clear dry air allows solar radiation to pass through to the land or sea below it, taking a very small toll of the heat with only a slight rise of temperature, and similarly at night air cools down very little by its own radiation

while allowing radiation from the land or sea to pass through. The presence of water vapour reduces the diathermancy of air, allowing more heat to be absorbed and reducing the heating of the land by day and its cooling by night, the net result being to retain heat in the air near the Earth's surface and to modify extremes. The temperature of the air is raised or lowered mainly by contact with the surface beneath it and always to a greater extent by land than by sea. Being much more mobile than water and expanding more with an equal rise of temperature, air is set into circulation by the gain or loss of heat more actively than the water of the sea, but it follows nearly the same direction at the surface. Under the direct influence of radiation a regular interchange of air takes place between the hot Tropics and the cold polar regions across the temperate zones. Over the North Atlantic the warm air currents from the Tropics flow towards the coasts of Europe from the south-west, while the cold air currents from the polar area flow towards the shores of America. The currents of the air are much less stable than those of the ocean, and in the unceasing conflict of tropical and polar winds Great Britain, though usually in the warm current, is often invaded by polar air and is sometimes in the front where the two streams meet. When polar air meets an opposing current of tropical air the two do not mix except superficially, but the denser air from the north flows under the lighter air from the south which may thus be raised to a considerable height in the atmosphere. This gives rise to a whirling or cyclonic movement affecting the distribution of pressure in the atmosphere.

Gain or loss of heat in air takes place not only by radiation and by contact with things outside; it may result from mechanical changes; thus when air is compressed heat is liberated, and when air is rarefied heat is absorbed. When moist sea winds are forced to ascend by any cause the air expands and is consequently cooled, the water vapour contained in it condenses in clouds which may form a screen nearly stopping radiation of heat to or from the surface below.

The little particles of water in clouds coalesce as they fall to form rain drops, and when a violent upward current is set up through rain the larger drops are shattered into droplets with the separation of electricity which may lead to disruptive discharges in the form of lightning.

The contrast between the thermal characters of land and sea and their influence on the air give rise to the broad distinction between oceanic climates of high rainfall and small range of temperature and continental climates which are very dry and have a great range between the temperature of day and night and of summer and winter. Great Britain enjoys a blend of both climates.

The great scheme of circulation between the Tropics and the polar area is complicated by diurnal and seasonal alternations due to the unequal heating and cooling of land and sea, producing regular breezes blowing from sea to land by day and from land to sea by night, and the system of monsoon winds blowing from sea to land in summer and from land to sea in winter. In the Tropics these effects are powerful enough periodically to slacken or reverse the main circulation; but in Great Britain they only appear feebly in spells of very calm weather. On steep slopes the air chilled by contact with cold land or snow at night may flow downward and displace the warmer air on valley floors or plains. This effect is a powerful agent of air circulation in the polar zones, and in Great Britain it occasionally produces damage to vegetation on low-lying ground in winter. As a rule the temperature of the air at the surface diminishes regularly as the height of the land increases, the rate of fall being about 1° F. in 300 ft. When isothermal maps are made to show the distribution of temperature, the figures are corrected to their value at sea-level before they are plotted. Thus the isotherm of 40° drawn across a mountain 3000 ft. high indicates an actual temperature of 30° on the summit.

Configuration of Great Britain

When dealing with British climate it is desirable to bear in mind the broad lines of the form and relief of the land. Great Britain, an island held in a mesh of the 10-degree net between the parallels of 50° and 60° N. and the meridians of 0° and 10° W., is broad in the south and narrows in a form approaching a triangle towards the north-west as far as the Firth of Forth in 56° N., thence it widens and trends due north in a broken rectangle to the Pentland Firth in 59° N. England is nearly bisected by the meridian of 2° W. which only touches the extreme eastern tips of Scotland, and the mainland of Scotland is nearly bisected by the meridian of 4° W. which traverses only the western peninsulas of Wales and England.

The orographical structure can be understood by dividing the island into two main divisions, the Lower Lands or English Plain, mainly in the south and east but branching westward through Cheshire and Gloucestershire, and the Higher Lands in the north and west, the dividing line running from Axmouth in the south northwards to the Severn estuary and thence in a bold curve to the Humber. In the eastern division the hills are mainly the scarped edges of chalk and limestone rocks rarely reaching much above 500 ft. (150 m.)* and nowhere touching 1000 ft. These low hills and plateaux have little practical effect on temperature but they exercise a distinct influence on rainfall. The nature of the soil has a marked effect on local climate. The permeable covering of the Downs and gravel hills allows rain to soak away and the dry ground responds to solar radiation and warms the overlying air. On the other hand, the flat impermeable claylands which alternate with them hold moisture which keeps the air damp and chilly and conduces to the formation of fog. The regions dealt with in ten of the eleven chapters that follow lie wholly or mainly in this division and their climate corresponds to their situation. The division of the Higher Lands contains the South-Western Peninsula (Ch. v) dominated inland by Dartmoor, Exmoor and Bodmin Moor, and enjoying on the coast the most thoroughly oceanic climate of any part of the mainland. On the west there is Wales rising in parts to over 2000 and even 3000 ft. (610 and 910 m.) with steep fronts to the sea on south, west and north, and a less steep face to the lower valleys of the Severn and Dee: a typically mountainous region on a small scale. In the north the backbone of the Pennine Highland is linked westward with the Lake District or Cumbria and includes most of Derbyshire, Yorkshire, and Durham, blending in Northumberland with the Cheviots. In Scotland the Southern Uplands fill the south, merging with the Cheviots: the Midland Valley is a low belt uniting the estuaries of the Clyde, Forth and Tay and seamed with lines of low abrupt hills, and the Highlands fill the north with a great dissected plateau rising over large areas above 3000 ft. and in places touching 4000 (1220 m.). This mass is divided by the Great Glen running north-eastward from the Firth of Lorne to the Moray Firth into two great areas, that to the south in the broadest part of Scotland being largely remote from the influence of ocean air, that in the north less lofty and much more open to sea winds. As a whole Britain faces the Atlantic with a bold front sloping sharply to the ocean and on the other sides it declines much more gradually towards the North Sea into the low plains of the east and south which here and there reach back to the west between the orographic nuclei of the Higher Lands. The contrast between the east and west of Britain is marked by the low level plains which usually border the North Sea in a slightly indented shore and the narrow mountain valleys of the west which front the archipelago of the Hebrides and the Atlantic.

* An explanation of the metric and other numerical equivalents given in this work will be found in the Preface.

The broad lines of configuration explain the distribution of climate by their action on solar radiation and on the wind which carries moderate warmth and abundant vapour from the ocean or extreme temperatures and drought from the continent.

The four conditions of climate, *viz.* latitude, nature of soil, configuration of surface and distribution of land and water, govern the weather-producing power of solar and terrestrial radiation as exemplified in direct sunshine, moving air in the form of wind, temperature changes due to both agencies, evaporation of water and the condensation of water vapour as cloud and its precipitation as dew, rain, snow or hail. These activities form the main elements of climate.

Sunshine

In the latitude of Great Britain, the possible duration of sunshine, averaging 12 hours per day, is divided very unequally over the circle of the year. At 50° N. in southern Cornwall the Sun at noon at the summer solstice attains an altitude of 63½°, that is as high as at the winter solstice in 3° N., 200 miles from the equator. It is above the horizon for 16 hours 20 minutes, rising in the NE. by E. and setting in the NW. by W. At the winter solstice the meridian altitude of the Sun in Cornwall is 16½° which is lower than at the North Pole in summer. It is above the horizon for only 8 hours, rising in the SE. by E. and setting in the SW. by W. The altitude of the Sun diminishes exactly as the latitude increases, so at 60° N., north of Orkney, the altitude at the summer solstice is 53½°, but the Sun is above the horizon for 18 hours 40 minutes, rising in the NE. by N. and setting in the NW. by N.; on the other hand, at the winter solstice the noonday altitude of the Sun is only 6½° or 13 times its own diameter above the horizon, an angle so low that radiation is of no account, and the Sun is visible only for 5 hours 40 minutes, rising in the SE. by S. and setting in the SW. by S. Thus the longest summer day is 2¼ hours longer in the north of Scotland than in the south of England; and the shortest winter day is 2¼ hours shorter. The decrease of 10° in the Sun's altitude at all times of the year means a great reduction in the heating power of solar radiation at the surface of the land.

Observations show that, for the country as a whole, cloud, fog and haze allow the Sun to be visible for only about one-third of the time it is above the horizon, that is to say for about 1500 hours in the year out of 4380. The proportion of recorded sunshine diminishes from an average of about 40 per cent. of the possible under the high Sun of Cornwall to about 26 per cent. under the low Sun of

Orkney, and the duration of sunshine is always greater on the coast than in the interior of the country in spite of the wider horizon of the higher land, the effect being assignable to the greater cloudiness over the hills.

In the map of average sunshine duration for the year the brightest region includes the South-Western Peninsula and a belt of about 35 miles (60 km.) wide along the whole south coast and the east coast as far north as the Wash in which the average duration is more than 4½ hours per day; the darkest month, December, has only about 1½ hours of sunshine and the brightest, June, from 7 to 7½ hours per day. Wales, the whole west coast of England and the English Plain enjoy an annual average of 4 hours per day of sunshine, varying from between 1 and 1½ hours in December (when the east coast as far north as Aberdeenshire is equally favoured) to more than 6½ hours in June. England, north of 52° N. and more than 10 miles inland, has an average annual sunshine duration of between 3½ and 4 hours daily, ranging from less than 1 hour per day in December to just under 6 hours per day in June. The Midland Valley of Scotland has about 3½ hours per day of bright sunshine on the average for the year, the Southern Uplands about ¼ hour more, the Highlands about ¼ hour less. In December the only parts of Scotland with over 1 hour a day of sunshine are the extreme south-west and a strip of the east coast extending north to the Moray Firth. In June all Scotland receives on the average more than 5½ hours of sunshine per day while on the Solway Firth and along the east coast south of Peterhead there are over 6 hours per day.

The average duration in December is twice as long on the English Channel as on the Pentland Firth, whereas in June the duration in the south is only 25 per cent. longer than in the north thanks to the difference in the length of the day. The greater altitude of the Sun and the lower level of the land in the English Plain south of the Pennines make that the only part of Great Britain in which local solar radiation is a powerful and sometimes a dominant climatic factor. In other parts of the country near sea-level considerable heating by the Sun may be experienced in clear weather. but it is insignificant everywhere above 1000 ft. (300 m.) although there is much intense cooling by terrestrial radiation on cloudless nights on the higher lands.

Winds and Atmospheric Pressure

The weather and therefore the climate of Great Britain is usually dictated by the vast eddies set up in the air over the North Atlantic somewhere to the south of Iceland. In this centre of action the barometric pressure is low, cloudiness excessive and fog very frequent. Another centre of action situated near the Azores is characterised by clear skies, light winds and high barometric pressure, and as both centres of action swing at times far on either side of their mean positions the conditions which affect the British Isles from the west are infinitely varied, the more so because in winter there is another high-pressure centre and in summer a low-pressure area over the continent, either of which may extend its influence across the North Sea. The Icelandic centre of action gives rise at times to families of secondary depressions which move forward one after another and carry with them a regular sequence of winds, cloud and rain, producing according to the intensity of the barometric gradient gentle breezes and showers or gales and devastating rains.

The most frequent condition of weather is the steady flow of warm moist air from the south-west, and while it is in force the chief agent in shaping regional climates is the configuration of the land. When, however, a depression or one of the active small secondaries crosses Great Britain it brings almost the whole weather-making power with it and distributes it in wind or rain with no apparent regard to local configuration. The distribution of weather in such a case depends on the track of the centre of the depression which is the point of lowest barometer. The most active secondaries seldom carry their rain area farther than 50 (80 km.) or 100 miles from their track, and so the same kind of weather very rarely prevails in the north and south of Britain. The most common track for main Atlantic depressions is north-eastward, parallel to the coasts of Ireland and Scotland. Secondaries cross the country at any point, run in any direction, and the track may curve to right or left, or double on itself. They usually proceed from a westerly to an easterly quarter following a geographical feature, such as that of the English Channel, or the English Plain from the Bristol Channel to the Wash or the Humber, or the Midland Valley of Scotland from the Solway or the Clyde to the Forth, or the Great Glen from the Firth of Lorne to the Moray Firth, or the Pentland Firth. So much of the weather is produced by rapidly moving depressions that the direction and force of the wind at any place change from day to day and often from hour to hour; in every month many instances occur of wind blowing from each point of the compass. For the country as a whole, and there is little variation in the different regions, it is a bold but fair approximation to say that the wind blows on the average from a westerly quarter on half the days of the year, from an easterly quarter on one-quarter of the days of the year, and from north and from south each on one-tenth of the days; on the remaining one-twentieth of the days there is no wind at all. The fact that easterly winds are usually cold and are most felt on the east coast, probably accounts for the common error that such winds predominate on that side of Britain.

The force of the wind at sea is usually steady and varies gradually, whether it is light or rises to a gale, and its direction is steady also, the veering or backing from one direction to another being easy and progressive. The same is true of the free air high above the inequalities of the land surface; but on the ground itself friction against the obstacles round or over which the wind must pass confuses the direction and breaks the steady stream of air into turbulent eddies and irregular gusts. Gales (in which the velocity of the wind exceeds 40 miles an hour) occur twice as frequently on the west, north and north-east coasts as they do on the east and south coasts of England, and strong winds are everywhere most frequent in the winter months, November to February, and least common in the summer months, especially June and July.

The varied configuration of the land gives a bewildering variety of shelter from the wind, the effect being local rather than regional. However, along the whole coast, especially in the west and on wide flat plains such as those of Caithness and the Fenland, the force of the unobstructed wind makes the growth of trees practically impossible without artificial shelter. This is equally true of the moors of the Higher Lands in the north and west, and of the broad ridges of the limestone and chalk downs in the Lower Lands of the south.

The best representation of the true distribution of wind is given by maps showing atmospheric pressure. As the pressure diminishes by about one-tenth of an inch of mercury for every hundred feet of elevation the barometer readings have to be corrected to their value at sea-level before being plotted on maps on which isobars are drawn. The flow of the wind is nearly parallel to the isobars and the direction such that a person standing with his left hand towards the lower pressure and his right towards the higher has the wind on his back. When the average pressure for a month, or a year, is mapped the isobars show which wind-direction prevails on the balance. The average annual isobars run across Great Britain from south-west to north-east with the highest pressure in the south. The gradient of pressure and therefore the mean force of the wind is greatest in January when the barometer oscillates about 1018 millibars in the south of England and 1008 in the north of Scotland. In May and June the gradient, though still for westerly winds, is only one-fifth as steep, the pressure varying from 1016 millibars in the south of England to 1014 millibars in the north of Scotland. As the winds in the midwinter months, December and January, are on the average five times as strong as the winds of the early summer months, May and June, it is obvious that the oceanic influence on the climate of Great Britain is five times greater in winter, when the result is to make the climate warmer and damper, than it is in summer, when the result is merely to make the climate a little cooler and moister than it would otherwise be.

Temperature

While solar radiation is important in raising the temperature of the air in the south of Great Britain in summer and terrestrial radiation plays a great part in reducing the air temperature in the north in winter the prevailing south-westerly wind from the ocean is always acting as a modifying agent, and produces for the whole year the warmest and most equable climate which exists in any part of the world between the latitudes of 50° and 60°.

The mean monthly isothermal maps show the seasonal swing of temperature from the coldest period at the winter solstice to the warmest at the summer solstice and back again. It is only possible to refer here to the conditions in the two extreme months and to the average for the whole year. In January the general temperature of Great Britain averages 40° F. (4°·4 C.)* which is the same as that of places 20° farther south on the east coast of North America. On the west side the mean temperature is higher than 40° (with mean daily maximum temperatures above 43° (6°·1) and mean minima about 35° (1°·7)) from Sutherland to Wales, and also over the whole of south-western England from Gloucester to the Isle of Wight. The mean temperature reaches 42° (5°·6) in the Outer Hebrides and the west of Wales and 44° (6°·7) in the south of Cornwall where the mean daily maximum reaches 48° (8°·9) and the night minimum 41° (5°·0): here the winter is mildest, frost is rare and is very seldom severe. The eastern half of Great Britain has a mean January temperature below 38° (3°·3) and often below freezing point in the higher districts, while above 1800 ft. (550 m.) a snow covering is usually found. The mean day maximum in the eastern half is about 43° (6°·1) and the mean night minimum 33° (0°·6) or less, the figures being correspondingly lower on the higher ground. Speaking generally, temperature in January diminishes from the south-west to the north-east in England and

* Temperatures below are given in degrees F., with degrees C. in brackets.

from west to east in Scotland, but the night minima are not so low along the coast as in the interior.

Since all the warmth in winter comes from the sea the maximum temperature may occur at any hour of the day or night.

As the year goes on temperature varies irregularly as cold winds blow from north and east breaking the usual warm flow of air from the south-west; but on the whole it rises and the interior, especially in the English Plain, warms up more rapidly than the coasts. In July, when the Sun in the summer solstice has produced its full effect, the general mean temperature in Great Britain is 60° (15°·6) (as compared with 40° (4°·4) in January), and this temperature is the same as that of places 10° farther north in western North America and Siberia, so greatly does the cooling influence of the ocean winds moderate the heat due to the summer Sun. The warmest area is in the Thames Basin around London with a mean July temperature of 64° (17°·8) or more, the mean daily maximum being 72° (22°·2), the minimum 53° (11°·7), and the highest temperature of the day usually occurs in the afternoon. The coasts of the Bristol Channel and of the Wash are a few degrees cooler. Towards the north mean temperature falls off steadily and apparently more rapidly on the coasts than in the interior, if one follows an isothermal map, but this inequality is nullified in fact by the greater altitude of the land. A mean temperature of 59° (15°·0) is found on the Scottish Border with mean daily maxima of 67° (19°·4), and mean minima of 51° (10°·6), and thence northward through Scotland July temperature diminishes quickly from south-east to north-west, the mean for the north coast and the Outer Hebrides being only 55° (12°·8), with mean day maxima of 60° (15°·6) and mean night minima little under 50° (10°·0). These figures show an approach to the continental type of climate in the warm English Plain with its mean daily range of nearly 20° (11°), and the consistently oceanic climate of the cool northern coast of Scotland with its mean daily range of only 10° F.

The actual extremes, that is the maximum of the hottest day and the minimum of the coldest night on record, are even more striking in enforcing this climate contrast. For London the extremes are 100° (37°·8) and 4° (– 15°·6), a range of 96° (53°·4), while for Stornoway they are 78° (25°·6) and 11° (– 11°·7), a total range of 67° (37°). Maximum temperatures over 90° (32°·2) on any day in summer are practically confined to the interior of the south of England, and maxima exceeding 80° (26°·7) scarcely ever occur north of Aberdeen. At the other extreme temperatures on winter nights falling under 10° (– 12°·2) never occur on the south coast and are very rare on the west, but they occur frequently on the east coast and more often still in inland places, where temperatures below zero (Fahr.) are recorded in exceptional winters. Thus in Britain days may occur that would be accounted hot in the Tropics and nights that would rank as cold in the Polar regions, but these rare excesses do not contradict the general mild equability of the normal climate.

A fair criterion of the severity of winter is afforded by the number of days on which snow falls. It is very rare for a year to pass without snow being recorded at least once, though this has sometimes occurred at stations on the south and west coasts. Equally rare, except in high districts like the Peak of Derbyshire or the Central Highlands, are winters with over 40 days on which snow falls. Above 4000 ft. on the highest summits of the Highlands snow may persist in sheltered northern ravines all the year round, and in all parts of the island over 1000 ft. the winter climate is severe.

Cloud and Fog

In Great Britain the average evaporation from a free water surface is about 16 in. (400 mm.) in a year, and as the annual rainfall is everywhere greater than this, in some places ten times as great, most of the rain that falls must be carried in from the sea.

The relative humidity of the air, as measured by the percentage of saturation, increases at any given temperature as the amount of water vapour increases, also for any degree of saturation it increases as the temperature falls; therefore relative humidity is the most variable of all the elements of climate.

Mist, formed when moist air rises in the atmosphere and is cooled by expansion, is known as cloud, and clouds by day obscure the light of the Sun so that the difference between the possible and the observed hours of sunshine expresses the amount of cloud, and the sunshine map may thus be viewed as the negative of a nebulosity map. The amount of cloud is least around the coast and in the south, greatest in the interior of the country and in the north. Mist in contact with the surface may result from warm moist air drifting over cold ground or a cold water surface: this is the usual cause of the winter and spring fogs in Great Britain. Mist may also be produced by cold air spreading over a warm water surface or wet warm ground: this is the usual cause of the summer fogs at sea and the autumn fogs on land. Fog is most common in still weather when there is little or no wind. Impressive weather phenomena are always supposed to occur much more frequently than statistics indicate, and the reputation for fogginess of the British climate is probably due to this illusion. As a fact fogs occur on the average on only about twenty days in a year at any place in Great Britain and rarely last for many hours. The black fogs of large towns can hardly be classed as natural phenomena; they are merely mists intensified and coloured by the stupid production of smoke.

Rainfall

The regular dry and rainy seasons which divide the year in the Tropics do not extend to Great Britain where rain may fall on any day in the year. It is unusual for more than 14 days to pass without rain and almost equally rare for rain to fall on more than 14 days consecutively. The dry spells or absolute droughts are most common in the English Plain where they occur on the average about once in a year and they may last for as much as a month on very rare occasions. Rain spells are most frequent in the Higher Lands in the west where they sometimes last for a month or more, especially in winter. The occurrence of droughts and rain spells is very irregular both in time and space.

It is possible to speak more definitely of the average number of rain-days in a year and here the geographical relationship is very clear. In the east of Essex the number is only 150, but the number increases towards the west and north and reaches 250 days in the extreme north-west of Scotland. Speaking generally the average number of rain-days in the English Plain is 175 or rather less than every other day in the year. In the Higher Lands of the west and north rain falls on the average on more than 200 days in the year, and in the western half of Scotland on more than 225 days, whilst the general figure for the whole of Great Britain is 204.

The distribution of the amount of rainfall throughout the year varies from place to place. In the east of England the month of highest rainfall is usually July or August when summer thunderstorms are frequent in that area, and October comes next, the driest month being February. Throughout southern and western England October is very definitely the month of greatest rainfall, and March or April that of least. In Wales and throughout the west of Scotland December and January have by far the most rain, and the least occurs there in May or June. On the average the wettest month at any place has twice the rainfall of the driest month, but the variation from year to year is so great that the average proportions are often completely upset.

The amount of rain which may fall in one day varies very greatly; for the English Plain it averages about 0·15 in. (3·8 mm.) and for

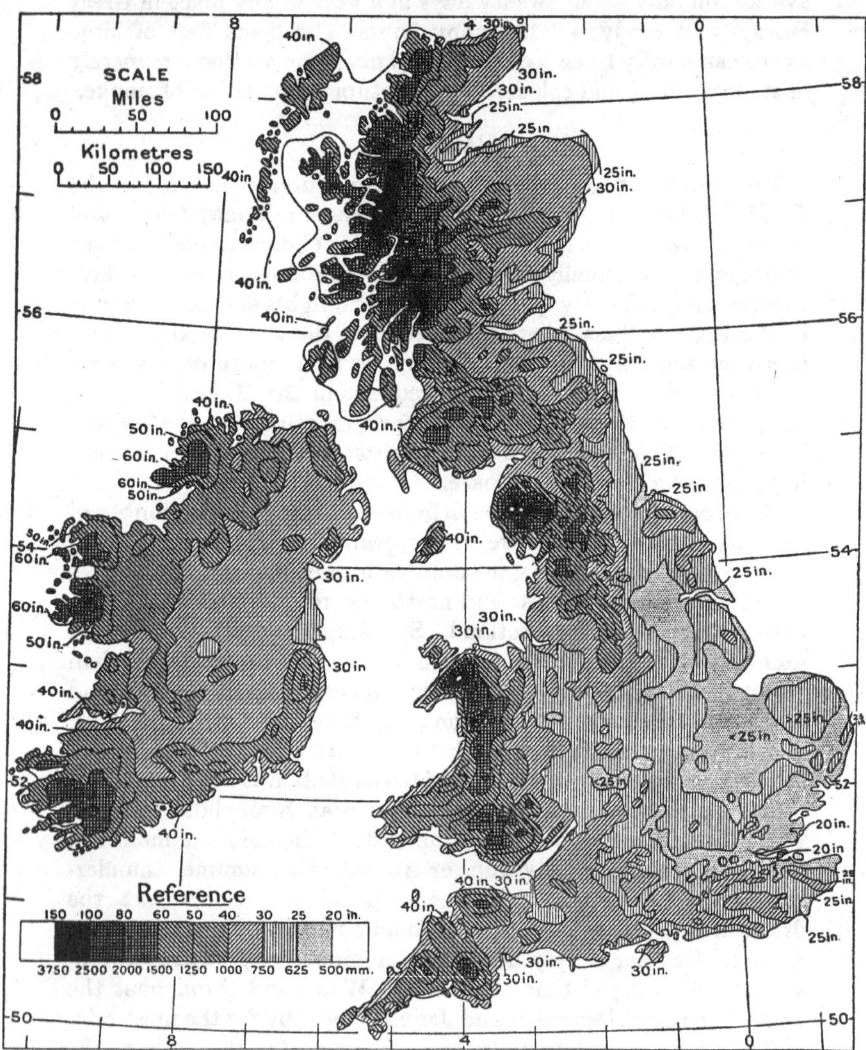

Fig. 1. Average annual rainfall of the British Isles. (Reduced from Pl. I of the *Rainfall Atlas of the British Isles*, 1926, by permission of the Council of the Royal Meteorological Society.)

the Lake District 0·60 in. (15·2 mm.), but in any part of the country it may exceed 9 in. (230 mm.) once in a century or two.

When the uprising of moist air in which clouds are being formed is vigorous the showers which fall are heavy and when the uprising is continuous the showers are long. The heaviest showers, often with hail, occur in summer thunderstorms due to local heating, when 3 in. of rain or more may fall in one hour on any small area in the English Plain. The wettest days, on the other hand, when 4 in. or more fall in 24 hours, occur during the passage of a cyclonic storm. Electrical disturbances in powerful cyclones account for winter thunderstorms which are common incidents of the weather in the west and north.

When rain occurs with a steady wind blowing horizontally from the sea it is distributed in an orderly manner contrasting with the sporadic splashes of great intensity produced by thunderstorms and strong cyclones. Even when the vapour they contain is nearly saturated such winds deposit little rain when crossing low ground, but on meeting the slope of high land the air is forced upwards, expands in adjustment to the diminishing pressure, cools and lets the rain fall in increasing volume as the slope grows steeper, the heaviest fall occurring just beyond the summit when the air pauses at its greatest height before beginning to descend the leeward slope. When the temperature of descending air is increased by compression, the relative humidity diminishes and rain ceases, not to recommence unless the air meets and mounts a still higher ridge.

A map of annual rainfall shows by its isohyetal lines the control which the orographic relief of the country exercises on the fall of rain brought by the prevailing wind which in Great Britain is in the main from a westerly quarter, and in a map showing the average rainfall of 35 years the influence of prevailing winds blowing in accordance with the isobars of a map of average atmospheric pressure is unmistakable. No other instance of the control of a mobile distribution by the fixed forms of the land is so perfect or so invariable as this. At any one place the annual rainfall may vary from two-thirds of the average in the driest year to one-and-a-half times the average in the wettest year: these ratios hold good alike for the eastern plains with their moderate average rainfall and for the western mountains where the annual fall may on the average be ten times as great. The wettest year rarely occurs simultaneously over any considerable area, and it sometimes happens that in a particular year one station experiences its maximum annual rainfall, while another station experiences its minimum.

The general average rainfall of the whole island of Great Britain is very nearly 40 in. (1000 mm.). In round figures 55,000 square miles (say 142,000 km.^{2}) of the surface in the Lower Lands receive less than 40 in., and 33,000 square miles (79,000 km.^{2}) in the Higher Lands receive more than 40 in. High rainfall (above 40 in.) occurs on one-fifth of the surface of England, on three-quarters of the surface of Wales and on two-thirds of the surface of Scotland. On a map of average rainfall (Fig. 1) there are five clearly-marked areas of high rainfall distributed as follows:

(1) The South-Western Peninsula of which only the low plain on the east and a very narrow strip along the coast have less than 40 in. On Exmoor, Dartmoor and Bodmin Moor large areas have over 50 in. (1270 mm.) culminating in each case with more than 60 in. (1520 mm.) and on Dartmoor 80 in. (2030 mm.) on the highest tors.

(2) Wales with only a narrow and discontinuous strip on the coast where a little less than 40 in. falls. The central mountains have practically everywhere more than 50 in., and more than 60 in. with many patches of 80 in. on the higher parts, while in the north-west the rainfall on 120 sq. miles (310 km.^{2}) of Snowdonia exceeds 100 in. (2500 mm.) and some of the higher valleys, notably Llyn Llydaw, have the tremendous average rainfall of over 200 in. (5000 mm.). The combination of high level and high rainfall gives great economic importance as a source of water power to these barren and cheerless wastes.

(3) The Pennine Highland carries a long and wide band of rainfall over 40 in. from Derbyshire to the Tyne Gap, with more than 50 in. in the northern half, and considerable areas with more than 60 in. in the Forests of Rossendale and Bowland and on the heights of Crossfell. The north-western extension of this high land in Cumbria carries a rainfall almost equal to that of Snowdonia on the group of fells from which the long valleys of the Lake District radiate outwards, 120 sq. miles having a fall of more than 100 in.

(4) The Southern Uplands of Scotland and the Cheviots receive more than 40 in. of rain everywhere except on the south-western peninsulas, the borders of Ayr Bay and the lower Tweed valley, and bring this high fall to within 10 miles of the east coast in Northumberland. The great rounded hills at the headwaters of the Annan, Tweed and Clyde receive more than 60 in. over large areas.

(5) The Highlands and the Western Islands have 40 in. or more, though this figure is not exceeded in the flat Outer Hebrides or at one or two points on the coast of the mainland. On the landward side the 40 in. line is drawn about two-thirds of the way to the east coast, except in the Grampians (where the rainfall is still little known) which carry tongues of high rainfall into Banffshire and

Kincardineshire. The west of the mainland and the Isles of Skye and Mull have more than 50 in. for the most part, and the rainfall exceeds 100 in. over about 800 sq. miles (say 2100 km.^{2}) in the east of Argyll and the west of Inverness-shire and Ross.

Very little of the land with over 40 in. of rain is of agricultural value, though often bearing excellent pasture, but the less remote parts furnish gathering grounds for the supply of water by gravitation to large towns, and the more remote have potential value for the development of water power.

The English Plain and most of the east of Scotland have less than 40 in. of rain save for one small patch on the North York Moors. Rainfall exceeding 30 in. (760 mm.) borders all the Higher Lands in a broad band in the west and in a narrower band in the east where the Pennines rise from the English Plain. The lower valley of the Severn and the Cheshire Plain have a rainfall less than 30 in. right through to the west coast; the Midlands and the east of England have on the whole less than 30 in. also. The hill ridges of the oolitic and chalk escarpments, however, carry more than 30 in. of rain north-eastward to the borders of Oxfordshire and Worcestershire and eastward south of the Thames over Salisbury Plain and along the North Downs and South Downs flanking the Weald and on the isolated elevation of the Forest Ridges. A rainfall exceeding 30 in. is recorded on the Yorkshire Wolds and the North York Moors. Nearly one-quarter of the surface of England and the same proportion of Scotland has rainfall averaging between 30 and 40 in. The greater part of the broad valleys through which the rivers of eastern England flow, the Thames below Oxford, the Trent below Newark, the Ouse below York and the rivers of the Fenland along most of their course have a rainfall less than 25 in. (630 mm.), falls exceeding that amount being confined to the broad ridges of the Chilterns, the East Anglian Heights, and the Wolds of Lincolnshire and Yorkshire. Less than 25 in. of rainfall is normal on a narrow shore strip bordering the Moray Firth, the Firth of Forth and at the mouths of the Tweed and Tees, but the whole of this dry area north of the Humber does not extend to 300 sq. miles (780 km.^{2}), while farther south there are 5000 sq. miles (say 13,000 km.^{2}) equally favoured. The driest strip in England is a patch of some 50 sq. miles (130 km.^{2}) on the extreme easterly border of the Thames estuary, where the average annual total rainfall just falls short of 20 in. (500 mm.). One-half of the area of England including its richest agricultural land, but only one-fifteenth of the area of Scotland, has less than 30 in. of rainfall.

Climate and Plants

Various elements of the climate combine to affect the distribution of vegetation on the land and to influence the life-processes of plants and animals. It has been shown by the Phenological Committee of the Royal Meteorological Society as a result of 35 years' observation that the average date of the flowering of a group of selected plants is earliest on the coast of the South-Western Peninsula (April 27), is later on the higher ground of the interior and towards the north, so that in northern England the date is May 11, along a line from the mouth of the Tees on the east coast southward on the plain surrounding the Pennine hills and northward again through Shropshire and Lancashire to the Lune. More remarkable is the fact that a week later (May 18) the same group flowers simultaneously all round the coast of Scotland from the Moray Firth southward to the Tweed and thence around the slopes of the Pennines as far south as Derbyshire and north again through Cumberland and the west coast of Scotland as far as Ross-shire. Nothing could show more distinctly how in spring the oceanic influence remains superior to the solar power in the northern half of Great Britain.

Books of Reference

The most satisfactory collection of climatic data will be found in *The Book of Normals* published by the Meteorological Office of the Air Ministry in five sections, giving average values of Temperature, Rainfall and Sunshine for each month, and small-scale maps of the distribution of mean annual and mean monthly maximum and minimum temperature, duration of sunshine and amount of rainfall. The data in the *Book of Normals* are set out in a series of ingenious and interesting graphs and discussed from the point of view of health in Dr Edgar Hawkins's *Medical Climatology of England and Wales*, London, Lewis & Co., 1923. Rainfall is treated in great detail in M. de C. Salter's *Rainfall of the British Isles*, London, The University of London Press, 1921, and maps of high accuracy are given in the *Rainfall Atlas of the British Isles* published by the Royal Meteorological Society in 1926.

II

THE SOUTH-EAST

C. C. Fagg and G. E. Hutchings
with a contribution by A. G. Tansley

"The Straits of Dover fill the history of this island because they have afforded our principal gate upon a full life." (H. Belloc, *The Old Road.*)

The south-eastern corner of England comprising the counties of Kent and Sussex and parts of Surrey and Hampshire constitutes a clearly defined region whose physiography, varied vegetation, and history as a human environment are very definitely related to its geological structure*. To demonstrate these relationships as fully as possible within the space available we have resorted largely to the use of maps and diagrams instead of verbal description. While we hope these will convey at a glance a general impression of the region we believe they will repay more careful study.

The Sub-regions, Geology and Land Forms

The first map shows the counties, the physiographical subdivisions of the region and its position relatively to the London

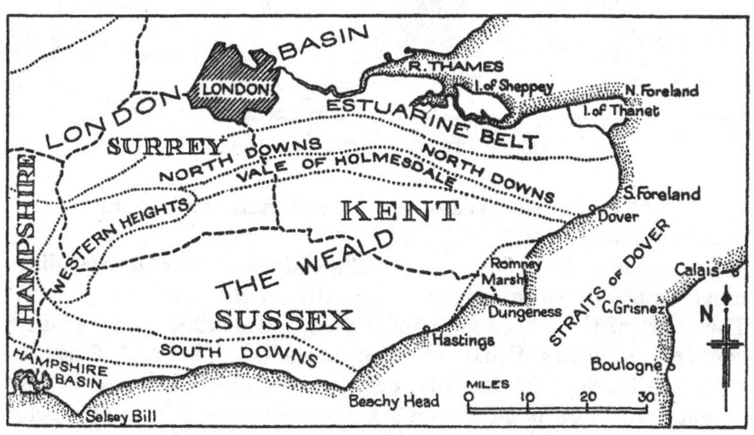

Fig. 2. The South-East: showing physiographical sub-regions.

Basin and the Continent. Nearly half the area is occupied by the Weald proper. This is bordered on the north by the Greensand Ridge and Gault Vale (Holmesdale). The Western Heights, approaching

* In broad outline the geology is relatively simple. The many problems presented by its minor details are beyond the scope of this work.

1000 ft. (305 m.), form the highest part of the region. They owe their prominence to the abnormal thickness of the Lower Greensand chert beds. The Gault and Greensand outcrops south of the Weald are physiographically unimportant. The Western Heights and Holmesdale are often included in the Weald, the whole being enclosed by the North and South Downs and the Chalk Uplands of Hampshire. The North Downs are flanked in Surrey by the Tertiaries of the London Basin and in Kent by the "Estuarine Belt" of low Tertiary hills and marshes. A minor fold brings the Chalk again to the surface in Thanet. The South Downs dip to the sea in their eastern half and are flanked by the Tertiaries and alluvium of the Hampshire Basin in the west. Romney Marsh, at

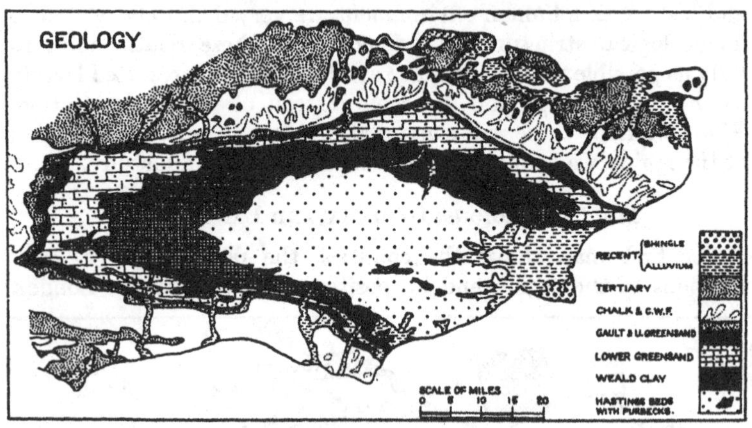

Fig. 3. The South-East: Geology.

the eastern end of the Weald, is a wide stretch of alluvium formed behind a shingle bar.

Geologically the region is an anticlinorium with its axis extending ninety miles eastwards from Hampshire to the Straits of Dover. The heart of the Weald is about 850 sq. miles (2200 km.^{2}) of hilly country (Hastings Beds). Except for small inliers of Purbeck Beds these lowest Cretaceous rocks are the oldest exposed in the region. This core is surrounded by successive outcrops of newer formations, the strata dipping gently to north and south, giving rise to a series of scarped ridges and alternate vales roughly parallel to the anticlinal axis.

The table (Fig. 4) represents the geological strata and their characteristics, their disposition being indicated by the geological map and various sections.

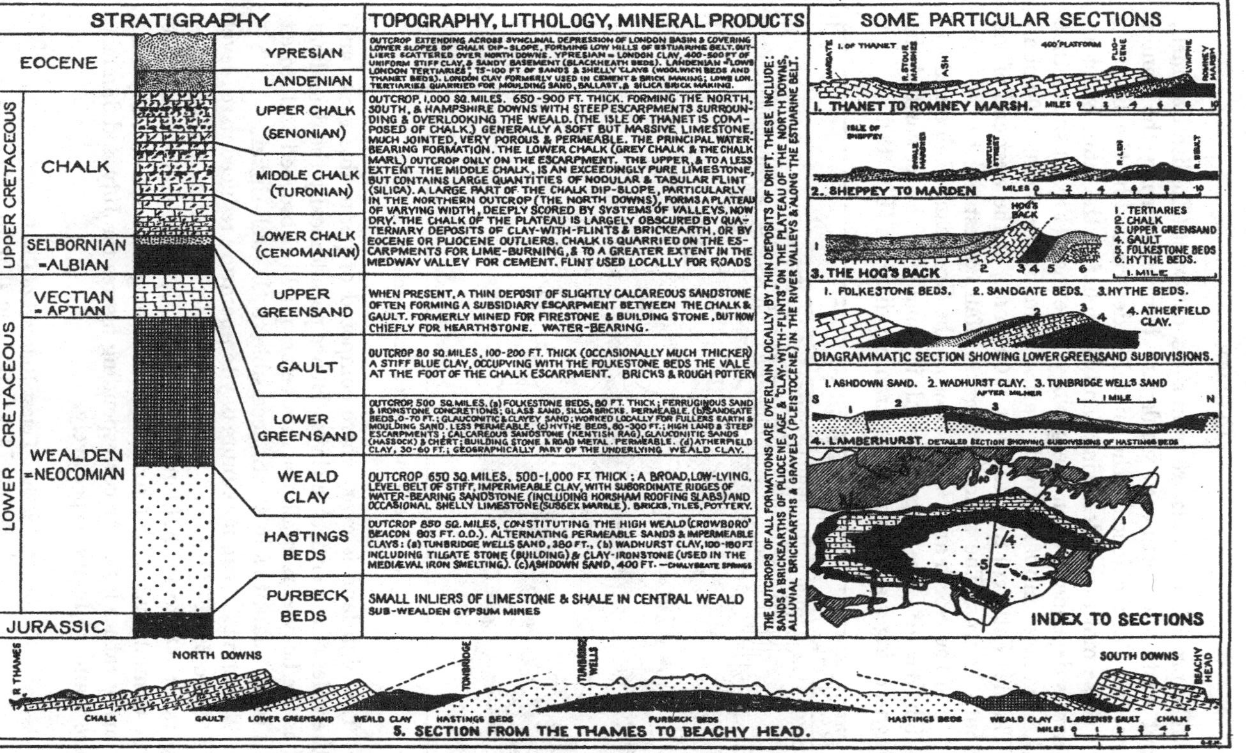

Fig. 4. A tabular summary of the Geology of the South-East.

Climate

The prevailing winds are from the south-west (English Channel), but east and north-east winds are frequent during the winter months. The incidence of the rainfall is determined by the direction of these moisture-laden winds and the relief of the land, which in turn is dependent upon the geology. The map (Fig. 5) shows that the average annual rainfall varies from under 20 in. (500 mm.) (Thames Estuary) to over 40 in. (1000 mm.) at the west end of the South Downs. The zones of high rainfall correspond to the high land of the North and South Downs, the Greensand Ridge and the High Weald*. The maximum fall is usually to the leeward of the highest ridges as shown in the transect diagram (Fig. 12).

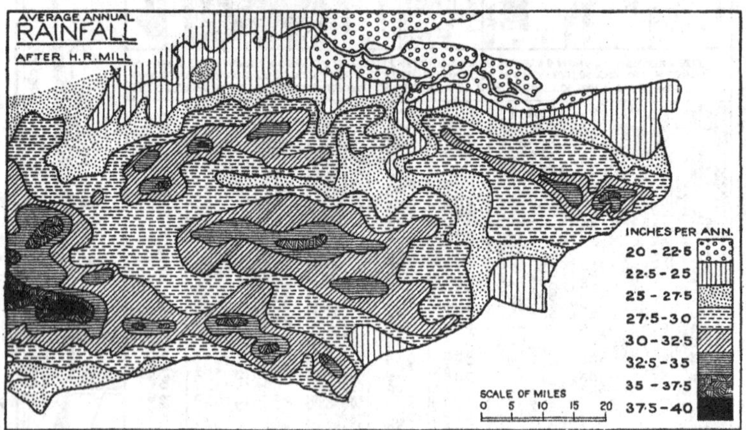

Fig. 5. The South-East: average annual rainfall.

The high land is everywhere composed of permeable rocks and the vales of impermeable clays. Hence paradoxically the drier land has the heavier rainfall, the rain being quickly absorbed and stored in large quantities in strata accessible to the well-sinker.

The annual mean maximum day temperature reduced to sea-level is 55° to 57° F. (July 68° to 71° F.)† and the minimum night temperature 43° to 45° F. (Jan. 34° to 36° F.)‡. Kent and Sussex have an annual average of over 4½ hours bright sunshine daily and Surrey rather less. The corresponding figures for July and January are 7 and 1½ hours respectively.

* *I.e.* the "Forest Ridges," of Ch. I.
† 12°·8 to 13°·9 C. (20°·0 to 21°·7 C.).
‡ 6°·1 to 7°·2 C. (1°·1 to 2°·2 C.).

Natural Drainage

The river system of the region is of special interest, Prof. W. M. Davis having used it to illustrate the classic thesis* in which he gave us the now well-understood terms "consequent," "subsequent" and "obsequent," as applied to streams. Our dominant rivers of to-day (Medway, Mole, Arun, Adur, Ouse) have their sources in numerous springs and streamlets in the much faulted Hastings Beds. These, uniting into streams, feed meandering subsequents on the Weald Clay, which in turn join the main consequents. The Medway and Mole breach the Greensand Ridge, and in the Holmesdale receive fresh subsequents, fed by secondary consequents rising in the springs on the dip slope of the Greensand, and obsequents from springs at the foot of the Chalk escarpment. The Stour and Darenth, having lost their Wealden headwaters to the Medway, now have their sources in the Holmesdale, while the Wey rises in the Western Heights. All these rivers penetrate by narrow gaps the North or South Downs and so reach the Thames or English Channel. The relief sketch (Fig. 6) and transect (Fig. 12) diagrammatically illustrate these phenomena.

The Chalk areas are now entirely devoid of streams except for the passage of the primary consequents through gaps and occasional bourne flows. They are, however, deeply scored by dry valley systems connected with the existing rivers. Many theories have been framed by geologists to explain the formation of these valleys without river action. It has been demonstrated†, however, that rivers have been the denuding agents in these as in other valleys. They are now dry because erosion of the Gault outcrop has progressively lowered the water-level in the Chalk (see Fig. 7). There is good evidence that in early Palaeolithic times the Gault Vale, and consequently the Chalk water-table, were about 100 ft. (30 m.) higher than at present. Vigorous streams must then have occupied the now dry valleys. Even now, after very wet seasons, temporary streams known as bournes or nailbournes flow in some of the valleys in question.

The region has passed through one cycle of denudation, and is in an advanced stage of the second. The initial uplift of the Weald brought into being a series of consequent streams on either side of the central crest. By mid-Tertiary times these and their tributaries had reduced the area to a peneplane which is still represented by the accordant summit plains of the existing higher

* "The Development of Certain English Rivers." *Geog. Jour.* 1895.

† C. C. Fagg, *Trans. Croydon Nat. Hist. Soc.* 1923.

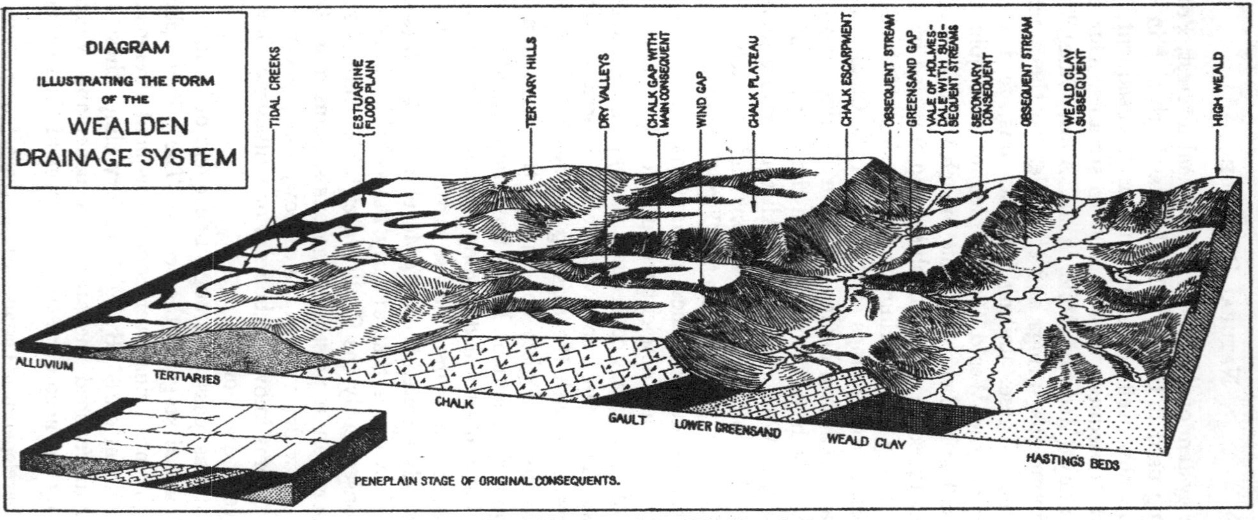

Fig. 6. Block-diagram of the Wealden drainage.

hills. Subsequent uplift gave new vigour to the drainage system. The streams have been deepening and widening their valleys for a prolonged period, and little denudation is now taking place except in the High Weald and Greensand areas. The Medway, forty miles (64 km.) from its mouth, is only 50 ft. (15 m.) above O.D.; the Mole in its upper reaches is a sluggish meandering river. Considerable areas of alluvium have accumulated about the estuarine tracts of the rivers in consequence of the comparatively recent (post-Neolithic) submergence.

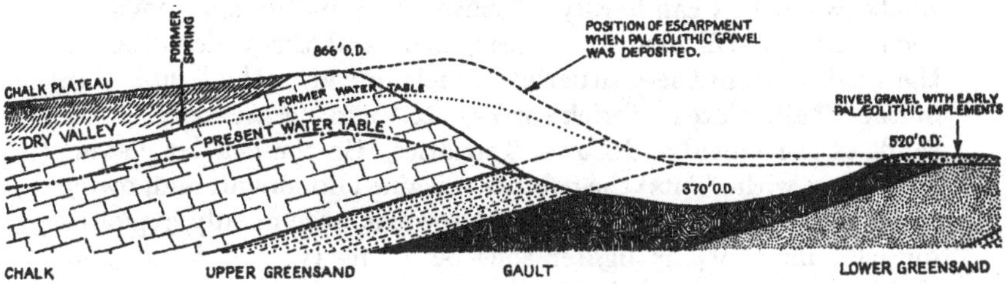

Fig. 7. A dry-valley profile.

Vegetation

(Contributed by A. G. Tansley)

Though apparently never themselves covered by ice the south-east counties of England probably supported nothing more than tundra vegetation during the maximum extension of the Quaternary ice sheet. We have good evidence that as the ice receded the first tree invaders of Britain were birch and pine. The continental early post-glacial pine period ("boreal") was dominated in the British Isles by birch, though pine, now no longer native in England, was not absent. Later came oak, with lime and elm (late "boreal" and "Atlantic"), and later still beech and hornbeam, probably in the "sub-boreal" period though the possibility that they were introduced by the Romans cannot be entirely excluded. During these last two periods the climate was decidedly warmer than to-day, and the deciduous trees (certainly oak and its associates, probably beech and hornbeam) ranged farther north than now. To-day indigenous beech and hornbeam extend no farther than the Cotswolds and Chilterns respectively. The south-east is still the richest part of England in natural or semi-natural forest growth and marks the farthest existing north-westward extension of at all fully developed central and west European *summer deciduous forest*, dominated by beech and secondarily by oak. Of these two the oak alone now extends to Wales, Ireland and central Scotland.

Beech forest, the climax type, does not, however, and probably never did, occur on all the soils of the south-eastern counties. Beech, intolerant of soils which remain water-logged for considerable periods, avoids the heavy clays which cover great tracts in the Weald and London Basin. On these soils oak is the natural climax dominant. The great forest which once covered practically the whole area between the North and South Downs and the Hampshire uplands was of oak. In its migrations beech kept on the whole to the light shallow soils derived immediately from the Chalk, where oak can hardly establish itself, and its most constant occurrence is still along the escarpment and the valley sides of the Chalk. From these situations beech spread to the lighter soils of the Chalk plateau (brick-earths, etc.) and occasionally to the sands of the Weald and Lower Tertiaries. On the heavier plateau soils (Clay-with-Flints) the oak has remained dominant, and, owing to the fact that in historic times the woods were mainly exploited for oak, most of the lighter soils bear this type of forest also, though beech can and does colonise them when it gets a chance. Ash and birch are not climax dominants in the south-east but respectively enter into the succession of vegetation to beechwood on chalk and to oakwood on sandy soils. Hornbeam is in England associated with oakwood and is nearly always coppiced. It occurs especially on sandy clays where hornbeam coppice locally displaces hazel over wide areas. Yew is especially associated with chalk scrub, locally forming pure yewwood, but also occurs, though far more sparingly, on other soils. Alder and willow are the dominant trees on very wet soils.

Of the semi-natural, that is non-arable, vegetation other than forest we may distinguish broadly (1) grassland, stabilised by mowing or pasturing, the latter by far the commoner practice in England, (2) heathland, partially stabilised by burning and to some extent by pasturing also, and (3) scrub, *i.e.* bushland dominated by shrubs and stabilised by coppicing. Of grassland there are several types, depending partly upon soil and partly upon the treatment of the land. *Chalk grassland* with its characteristic assemblage of grasses and herbs is one of the most distinctive. It occupies the shallow soils derived immediately from the Chalk and has been typical sheep pasture from very early times. With pasturing excluded chalk grassland would be occupied by scrub and beech forest. Another well-marked type (often called *grass-heath*) occurs on sandy soils and shows many transitions to and mixtures with heath proper. Left unpastured it would develop into oak or sometimes into beech forest. Its pasture value is generally poor. A third type, *neutral grassland*, so-called because its soil type is neither characteristically alkaline like chalk grassland, nor characteristically acid like grass-heath, is developed on clays and loams, and forms the richest pasture bearing the most nutritious grasses for cattle. It corresponds with the heaviest oak forest, and occurs largely on Weald Clay, Gault, London Clay and to a less extent upon Clay-with-Flints. On alluvial soils where the water level is high neutral pasture passes over to *water-meadow*, with various moisture-loving grasses. This is commonly mown for hay, or forms a rich pasture. Where the summer water level is at or near the surface *fresh-water marshland*, corresponding with alder and willow woodland, is produced. It is useless for pasture unless artificially drained.

Heathland dominated by ericaceous undershrubs is developed mainly on sandy soils with distinctly acid reactions. It occupies much of the sandy deposits of the Hastings Beds and Lower Greensand, of the Chalk plateau and of the Lower Tertiaries. It is most readily colonised by birch and subspontaneous pine (together with various shrubs) but may develop oak or beech forest with a ground vegetation differing from that of other forest-bearing soils. The factors which check or prevent the development of climax forest from heathland are various, including in different cases extreme acidity (probably) or extreme dryness of soil, heath fires, pasturing, cutting, and paucity of seed parents of the climax trees. The result of varying incidence and different combinations of these factors has been to produce over widespread areas of sandy soil the varied type of vegetation which has been called *oak-birch-heath*, essentially a mixture of various stages of succession from heath to grass-heath and dry oak forest.

Scrubland is essentially a transition stage from grassland or heathland to forest. Scrub may maintain itself for long periods under certain conditions, and particularly in the absence of seed parents of forest trees, or where abundant ground vermin destroy the tree seeds or seedlings. Scrubland composed of shrubs which regenerate readily after cutting is maintained artificially as coppice. *Coppice-with-standards* is a mixture of shrub vegetation, cut at regular intervals (ten to fifteen years) with trees (generally oaks) which has been through mediaeval times and until recently the typical south English mode of culture of woody vegetation. It is still the commonest type of semi-natural woodland in the south and the midlands of England, but is now increasingly replaced by plantations of exotic conifers. Hazel is the commonest shrub of coppice, and is often planted. *Chalk scrub*, and the corresponding semi-natural calcicolous coppice, is a well-marked type distinguished by a variety of shrub species occurring but rarely on other soils. Heath scrub and neutral scrub are less distinctive though each has species less commonly found in the other.

The only other types of natural vegetation in the region are maritime. They may be classed as *saltmarsh*, occurring on not too mobile mud or sand of estuaries covered by high spring tides, and consisting of a characteristic collection of herbs of very distinctive character—the so-called "halophytes"; *sand-dune* vegetation showing a definite succession from marram grass to inland sand-loving vegetation with increasing stabilisation of soil; *shingle beach* vegetation closely allied to that of sand dunes, but with a few peculiar species. Drained salt marsh, from which the salt is quickly washed out by percolating rain water, forms excellent pasture grassland. Sand dune is generally used for rabbit warren, golf links or seaside building. There are also characteristic seaside grasslands used for pasture, whose origin and nature are inadequately known.

The following table shows the principal types of natural non-maritime vegetation in south-eastern England:

Shallow chalk soil	Lighter soils of chalk plateau	Heavier soils of chalk plateau and Wealden and London Clays
Beech forest	Beech forest (often oak-wood by culture)	Oak forest
↑	↑	↑
Ashwood	Ash-oakwood	Neutral scrub
↑	↑	↑
Chalk scrub (with yew)	Scrub	Neutral grassland
↑	↑	
Chalk grassland	Neutral grassland	

Sandy soils, often occupied by oak-birch heath	Wet sandy soils	Alluvial soils with high water table
Beech forest (local)	Wet heath	Alder-willow woodland
↑	↑	↑
Oak forest	Sphagnum bog	Marshland
↑		
Birchwood		
↑		
Heath scrub		
↑		
Heath or grass-heath		

As regards the geographical distribution of plant species, not only the beech and hornbeam, but also a number of herbaceous plants of central European or Mediterranean origin and even some from the steppes of south-eastern Europe, reach south-eastern England (and partly also East Anglia) but do not penetrate farther into the British Isles. Most of these are species of a drier continental climate and are therefore confined in Britain to the somewhat drier climate and to the dry chalk and sandy soils of southern and eastern England.

Prehistory

Our region has without doubt been the home of man from very remote antiquity and was probably his earliest habitat in Britain.

The earliest fossil remains that have been dignified by the name of man (*Eoanthropus*) were found in an ancient gravel at Piltdown in the Sussex Weald. If the "eoliths" of the North Downs are accepted as artefacts they were used by still earlier inhabitants than Piltdown man. The Pleistocene river gravels, especially in Kent, have yielded abundant Palaeolithic implements and the chief skeletal remains of Palaeolithic man discovered in this country are from north Kent.

In early Neolithic times England and the Channel Islands were still joined to the Continent by a land bridge enabling the "Mediterranean" peoples who were then spreading westwards to expand into England, where, with a few descendants of earlier Palaeolithic inhabitants, they held the ground until the close of the Neolithic phase. Even the South-East, however, was sparsely populated, the whole Weald and most of the low-lying land of the vales and the estuarine belt being densely forested. Settlement was confined mainly to the open high land of the chalk and Greensand ridges and the marshes of the Thames and Medway estuaries. Here Neolithic man founded simple pastoral communities which remained even after the lowland agricultural settlements were long established. Parts of the Old Road along the North Downs and other hill roads only loosely associated with the newer village communities survive from the days of upland settlement and migration along the ridges. Belloc speaks of the "natural causeway" of the North Downs as an invitation westwards to peoples entering this country in east Kent. It was indeed the only open route inland, a narrow strip, dry and level except for the river gaps, between the almost impenetrable oak forests of the Weald and the deeply scored chalk plateau, devoid to this day of good east and west roads. This oldest human trackway in Britain (the mediaeval "Pilgrims Way") led westward to Salisbury Plain, for long the focus of prehistoric civilisation. Too little is known of the life of Neolithic times in this country, and its differentiation from late Palaeolithic has not been clearly established. The close of the Neolithic period was marked by a widespread cultural advance and the use of bronze in many parts of Europe. A people of mixed Mediterranean and Alpine stock appeared in England trading in metal and bringing the familiar megalithic culture. In our region an interesting group of megalithic structures remains where the Old Road crosses the Medway valley.

The first attempts to establish communities in the wooded lowlands were by Nordic-Alpine visitors of the Bronze Age who brought the art of husbandry from central Europe. The use of iron and the coming of Belgic tribes from Gaul hastened the extension of settlement along the estuarine belt and the margins of the Weald. Further colonisation took place during the Roman occupation when a number of existing village sites were fortified, and military roads, including the coast roads of east Kent and the great Watling Street along the Tertiary tract to London, were constructed. The western Weald was penetrated and the Sussex and Hampshire coast stations connected with the Thames.

Historical and Economic

If we endeavour to estimate the factors that have made our region what it is to-day we find they are all related to four major facts: (1) Kent and east Sussex have been the chief gateway to and from the Continent, in peace and in war, from prehistoric times down to these days of aerial transport; (2) the fertility of the soil and great length of coast have made our region primarily the home of peasants and fisher-folk; (3) subordinately, but of some importance and antiquity, non-agricultural industries have developed, especially in connection with timber (shipbuilding, charcoal burning, iron smelting) and chalk (lime and cement); and (4) during the past hundred years the phenomenal growth of London and increasing transport facilities have combined to subordinate the regional life of the population to the domination of the metropolis.

Gateway to the Continent

"The little peninsula of Kent seems as it were a pivot whereupon Britain may be said to swing, inclining now to the ways across the North Sea and now again to the ways across the English Channel....Both by land and sea these links across our coastal waters are the last in long chains. These stretch along the northern and southern sides of the mountain zone which runs like a great scar across Central Europe."*

* H. J. Fleure, *Human Geog. in Western Europe.*

The historic entries of Romans and Germanic peoples are but recent episodes in this age-long drama of Kent, the first scenes of which were enacted by plants and animals. The economic and strategic importance of the Kent-Sussex coast is revealed by the history of its ports. As Belloc has pointed out, practical difficulties in crossing the Straits and the absence of any one superior harbour caused the development of several ports to be used as weather or tidal conditions determined (see Fig. 8). The Roman Ports were replaced by the Saxon Cinque Ports and their numerous "members" which occupied the coast from Seaford to Faversham. The multiplicity of ports, and the uncertainty of the point of landing on the English shore, necessitated an inland focus upon which commercial

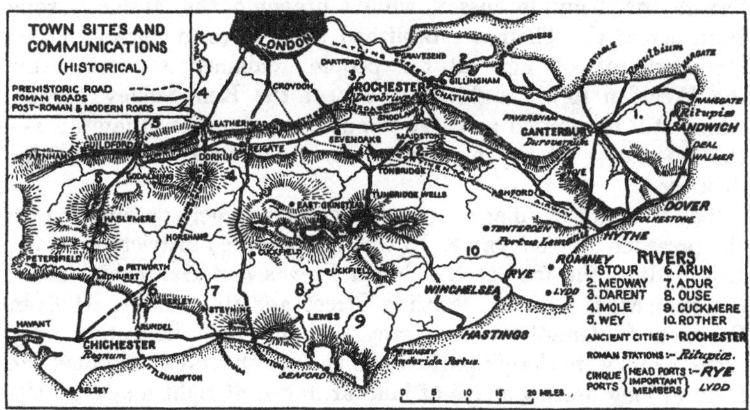

Fig. 8. The South-East: town sites and communications.

and military routes might converge. Hence the significance and early importance of Canterbury as the terminus of the Old Road and later of Watling Street.

The Cinque Ports were of Anglo-Saxon foundation. To the five original "Head Ports," Sandwich, Dover, Hythe, Romney and Hastings, were added Winchelsea and Rye and later numerous "Members" or "Limbs" which finally included fifteen corporate towns and twenty-four non-corporate villages, sharing the responsibilities and certain privileges of the Head Ports. From the time of the Saxon kings until the period of maritime expansion in the 16th century the Ports fulfilled the functions of a Royal Navy, providing ships and men for naval warfare and transport of armies; they also controlled the nation's fishing industry. The Ports and their members formed a powerful civic confederation which enjoyed a large measure of self-government. Their presence has not only affected the coastal region economically but has given a bold independence to its people. The decline of the Ports was caused not only by the rise of West-Country mariners but by the gradual silting up of their harbours. Dover alone remains as a modern port.

Rural Development

The post-Roman settlements of Nordic peoples though not discontinuous with the past may be regarded as marking the beginning of that phase of rural life which reached its climax in feudalism, persisted more or less unchanged until the dawn of industrialism and in large measure still survives. In our region the Jutes conquered and assimilated the Cantii in Kent, the South Saxons the Regnii in Sussex, while a branch of the Middle Saxons from across the Thames established themselves in Surrey. These three groups of agricultural people remained separated by the Wealden forest and the woods of the Kent-Surrey border, which still harboured a sparse population of earlier inhabitants—the *Nibelungs* of the mediaeval charcoal-burning and iron industries.

The sites of the village communities were largely determined by the presence of good water. In this the favoured districts were the Holmesdale, the foot of the South Downs and Hampshire chalk escarpment, the line of Watling Street and its counterpart from Croydon to Guildford in Surrey. Along these lines the early settlements were closely packed and the parishes often assumed the form of long narrow strips of land arranged at right angles to the geological strike. The tracks between these primitive settlements have become important roads in later times. In the case of Watling Street, straightened by the Romans, the minor ports and fishing villages of the Estuarine Belt have extended to it (*e.g.* Faversham). In Holmesdale the tracks between villages together make one long road from east Kent to west Surrey, which has for long replaced the old road of the chalk ridge. Spring water was abundant also in the High Weald, where iron ores were worked and forged from prehistoric times. Village communities of later origin and not so regularly disposed are fairly numerous. They are much more thinly scattered on the Weald clay. In the Saxon kingdom of Kent large areas of the great Forest of Anderida known as "denes" were set apart for the use of distant townships as "pannage for hogs" and sources of timber. Most of the Forest however, especially in Surrey and Sussex, remained untouched and later became the property of the Crown. The "drofways" or roads leading to the denes, and subsequently adopted as highways between the villages, were the first roads of the Weald. Under the feudal system numerous manors were cut out of the royal forests and new townships were formed from the denes. The demand for timber for the shipbuilding industry of the coasts, for building, and later for the iron industry, was the chief factor which accelerated the clearing and settlement of the Weald. The development of Romney Marsh and large areas of the eastern Weald as pasture land gave rise during late mediaeval times to extensive sheep-rearing and the now defunct wool industry.

Village communities were thinly scattered on the chalk downs where water supply has always been difficult. This defect was to some extent overcome, particularly on the South Downs, by the construction of so-called dewponds.

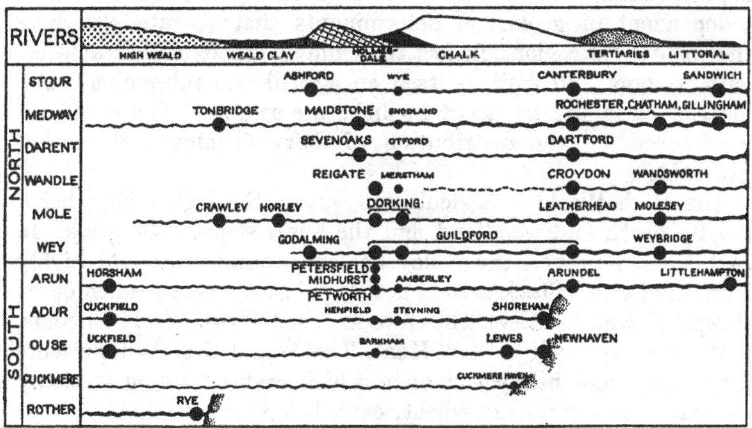

Fig. 9. The South-East: Diagram showing the relationship of town sites to rivers and geological outcrops.

The disposition of village sites relative to geological outcrops is indicated in the transect diagram (Fig. 12). Those members of the closely packed lines of villages in Holmesdale and along Watling Street and its Surrey equivalent, which happened to fall on a main consequent river, have developed into the market towns and cities of this essentially agricultural region. This phenomenon is strikingly shown in the diagram (Fig. 9), which significantly includes all the towns of importance, except the great resorts, of mushroom growth, shown on the map (Fig. 11). At Guildford the chalk outcrop becomes so narrow that the two principal lines of townships unite in the Wey gap. Guildford, exhibiting characters of both Holmesdale and Tertiary towns, is actually formed by the amalgamation of two earlier settlements. Similarly Rochester, owing to the Medway inlet, fulfils the functions of a port and a Watling Street town.

Agriculture

Of the 2¼ million acres (say 910,000 ha.) which compose our region roughly 60 per cent. are under crops or grass and of this the arable bears to the grassland about the proportion 5 to 8. Approximately one-sixth of the arable is wheat and one-sixth oats, barley accounting for one-sixteenth only. The present-day agriculture and soils have been fully dealt with by Hall and Russell*. We may conveniently review the main features by considering the sub-regions (Fig. 2), though in the fertile Medway valley widespread deposits of alluvial brick-earths have rendered farming largely independent of geological determinants that operate elsewhere throughout the region. In fact the highly cultivated Medway area, with its hops and fruit, is itself an agricultural sub-region. Also local markets and transport facilities are as potent factors as soil in determining the distribution of dairy farming and market gardening.

The High Weald is agriculturally poor, the higher land being mostly heath and woodland and the lower slopes poor grass. In east Kent, however, the valley bottoms contain rich and highly cultivated soils. Stock raising is the mainstay (cattle, milch cows, sheep and some poultry). The crops are chiefly wheat and mangolds, with hops and fruit in east Kent. The Weald Clay belt presents a very wet and heavy soil upon which arable farming is thinly scattered. The crops are wheat, oats, beans and mangolds. Most of the area is either in grass for hay or woodland. Dairy stock, sheep and cattle are kept.

In the Holmesdale the Gault is scarcely ever under the plough. It is either in grazing for cattle and milch cows or woodland. On either edge of its outcrop is a narrow strip of arable which persists with few breaks throughout the Weald. At the foot of the escarpment this is formed of Chalk Marl, Upper Greensand (when present) and soils (scarp drift) which creep from these on to the edge of the Gault. The other very narrow strip is along the junction of the Gault and Folkestone Beds, whose admixture makes a light loam in which potatoes, roots and oats are commonly grown. Of the Lower Greensand series the sandy Folkestone Beds are unfertile, being largely in heath or woodland. The Hythe Beds in Kent, particularly

* A. D. Hall and E. J. Russell, *Agriculture and Soils of Kent, Surrey and Sussex*, 1911.

in the Medway basin, give some of the most fertile and highly cultivated soils. Fruit and hops are the staple crops, and some barley is grown. The Western Heights, formed of Folkestone and Hythe Beds, are characterised by barren heaths and coniferous woodlands.

On the South Downs sheep farming is the basis of agriculture, the arable farming being directed to folding and winter feeding. The pasture is not rich enough for cattle, but the moisture-retaining properties of the porous chalk keep it fresh when better quality grassland is parched by drought. The soft springy turf makes chalk grassland a favourite training ground for race-horses and hunters. These remarks apply in less measure to the North Downs which are not so open and are largely covered by superficial deposits. Clay-with-Flints forms the chief covering and is mainly under woodland (coppice-with-standards) and grass, though good wheat is grown on the lower portions. The Tertiary outliers are given up to heath and woodland.

In the Estuarine Belt the Lower Tertiaries, particularly Thanet Sand, form rich agricultural land, fruit, hops, potatoes, wheat, barley, roots and lucerne being the chief crops. The London Clay when not woodland is typically sheep pasture as in "Sheppey." Romney Marsh and the alluvial areas of the Estuarine Belt are now mainly sheepwalks.

The coastal plain around Chichester in south-west Sussex is almost entirely under the plough and very largely in wheat.

One of the most distinctive features of the region is the cultivation of hops and fruit in mid-Kent and east Sussex. More than two-thirds of the hops produced in Great Britain are from this district, while a quarter of the whole area devoted to small fruit and orchards is within our region.

Industries

Minor industries using agricultural products (*e.g.* brewing, flour-milling) are distributed over the region. The manufacturing industries are concentrated mainly in the Estuarine Belt and Medway valley. The towns and villages of north Kent and the Medway estuary are maritime in character, though many are now separated from the sea by wide stretches of marshland. Transport by sailing barges and lighters is important in this zone, many industries being dependent upon navigable creeks. Rochester and Faversham are ancient ports associated with fishing and ship-building; Whitstable is celebrated for its oyster fisheries, and Sheerness and Chatham have naval dockyards.

The Medway, the largest of the Wealden rivers, reflects many features of the Thames, with its tidal inlet. Rochester, Chatham and Gillingham, known as the Medway Towns, form the largest industrial population centre of the whole region, and may appropriately be called the London of the Medway. Coal and iron, timber and wood-pulp for paper-making, an old Kentish industry, are the chief imports. Large quantities of cement are exported. The principal industries of the lower Medway are shipbuilding, cement-making, engineering for transport and agriculture and oil refining. Most of these have developed from the primitive industries of the region. The familiar steam rollers and tractors bearing the arms

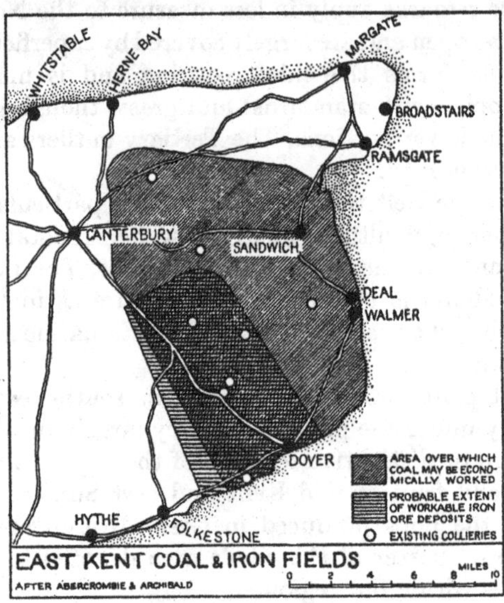

Fig. 10. The East Kent coal and iron fields.

of Kent (*Invicta*) are manufactured at Rochester in the works founded by their inventor, a Kentish agriculturalist. A large seaplane and motor-body works has replaced much of the older boat-building industry of the Medway. Motor-omnibuses are built at Maidstone and railway locomotives and rolling-stock at Ashford on the Stour. The important cement industry of the Medway is a maritime development of the ancient lime-burning industry which is distributed along the chalk escarpment and the river gaps throughout the region. Other mineral products and industries —building-stone, road-metal, brick-making, rough pottery, etc.— directly dependent upon geological formations are distributed as shown in Fig. 4 and the transect diagram (p. 39).

The mediaeval iron industry of the Weald deserves special notice. Its former extent is indicated by the frequent occurrence of such place-names as Furnace Farm, Hammer Pond, Cinder Hill, etc. The Romans extracted iron at several places in Sussex, probably adopting already existing furnaces. Domesday Book does not mention the Wealden ironworks, but it is probable that so important an industry was continuous from Roman times. Wealden iron was abundantly used from the 13th century onwards. Smelting as distinct from primitive extraction by forging was probably employed as early as the 14th century. Ordnance and domestic articles were produced and the industry continued into the early part of last century.

The chief ore used was the clay-ironstone from the Wadhurst Clay but other ores from the Hastings Beds were also used. Most of the furnaces were in Sussex, where ancient differences in land tenure probably permitted less restricted use of timber for fuel. The introduction of blast furnaces threatened to cause a serious depletion of the forest timber which was increasingly demanded for naval and other shipbuilding. Legislation for the preservation of forests caused the industry to decline, its final disappearance being due to the introduction of pit coal in other parts of the country. The development of the East Kent Coalfield is likely to revive iron smelting in our area. This development is still in its infancy. Coal is obtained from deep workings beneath the iron-yielding secondary rocks. The opening up of this coalfield was made the subject of investigation by a Joint Regional Committee. A thorough survey was made by Prof. Patrick Abercrombie and John Archibald in whose published report a scheme of regional development has been laid down*. How far the industrial and other interests concerned will execute this scheme remains to be seen, but it stands for much and it is a measure of the progress of regional geography in Britain that such a survey and plan have been made with a view to achieving this great industrial development while preserving the rich historic interests and great natural beauty of the region.

Metropolitan Expansion and Dominance

The area of our region is roughly 3500 sq. miles (say 9000 km.^{2}), and the population in April 1911 was 3,000,000, an average density of less than two persons per acre. Approximately 70 per cent. were natives of the region, and of the 30 per cent. immigrants one-third were Londoners.

* *East Kent Regional Planning Scheme. Preliminary Survey* (Hodder and Stoughton), 1925.

These figures give an inadequate indication of the extent to which the surviving regional life had at that time been overlain by "metropolitan" development. A census taken in August when the resorts are packed with visitors would give a very different result. Further, while few of the permanent immigrants will participate in the rural life of the region, a large proportion of the native population is now directly dependent upon the daily, week-end, seasonal and permanent influx of Londoners. Since the War this metropolitan overflow has enormously increased. It is of a twofold nature, residential and pleasure-seeking. From all parts of the region, business and professional men travel by rail or motor to London daily. Except near the metropolis however they have not yet greatly interfered with the agricultural land. The most desirable building sites—the ridge of the North Downs, the Greensand hills and the High Weald—are among the less fertile parts of the region. These residents and the seaside resorts, by

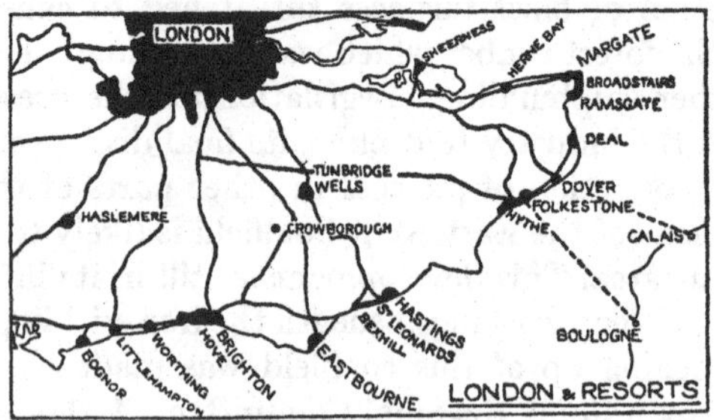

Fig. 11. The South-East: resorts and main railways.

creating local markets, have even stimulated dairy and poultry farming and market gardening in some parts. With the great army of holiday-making invaders, however, they are exerting a considerable and increasing influence upon the rural life of the region.

Brighton, the point on the south coast nearest and most easily accessible to Londoners, was the first and is now the largest of the resorts. It is virtually a detached portion of London, as in slightly less degrees are they all. In the 18th century the fishing village of Brighthelmstone contained a few hundred inhabitants. It became a fashionable resort under the patronage of the Prince of Wales, afterwards George IV, from 1782 onwards. In 1801 the population of Brighton and Hove had reached 7500, by 1851 this number had increased tenfold and in 1911 it had reached 173,000.

The importance of water supply and track-ways for the early

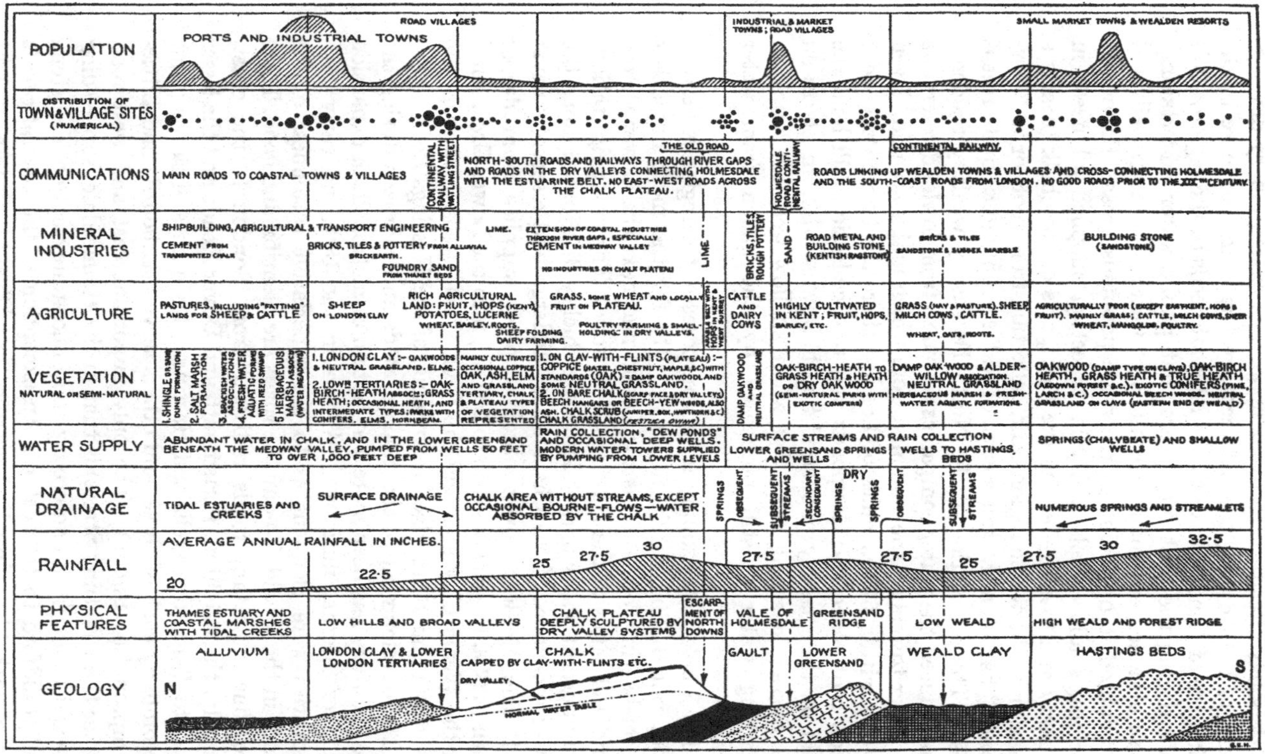

Fig. 12. The South-East: a Diagram showing the relationship of certain aspects of the Region to one another and to the geological outcrops.

settlers has already been dwelt upon. The extraordinary development of both during the past century has been due more to the influence of London than to the internal needs of the region. It is only during the past fifty years that public water undertakings have obtained enormous quantities of water from deep wells and distributed it by mains in the more populous areas. Previously springs, shallow wells and rain collection supplied all needs in country and town alike. On the high parts of the chalk (nearly a quarter of the whole region) rain collection from roofs or in "dew-ponds" has been the sole source of water supply until quite recent years. The occupation of the North Downs by wealthy residents has induced the water companies to pump from lowland reservoirs to water towers on the summits, from which mains are laid. Even to-day within fifteen miles of London there are farms dependent upon roof collection; dewponds and roof collection still hold the field on the sparsely populated South Downs, while in places on the Sussex coast, *e.g.* near Rye where shallow wells are brackish, drinking water is still purchased by the gallon from water bearers.

The whole water supply for all purposes in the region is derived from rain falling within its limits. By far the most important water-bearing formation is the Chalk, and next the Lower Greensand and sands of the Hastings Beds series. The pumping stations of important water undertakings are thickly distributed along the lower dip-slopes of the Chalk whose underground reservoirs never fail. The Medway towns obtain copious supplies from deeper borings (700 to 1000 ft.—213 to 305 m.) into the Lower Greensand. In the Weald Clay area and the High Weald wells draw from the Ashdown and Tunbridge Wells Sands of the Hastings Beds.

The development of the modern roads and railroads during the past century has been more a symptom of metropolitan expansion than of internal initiative. Writers in the 18th and early 19th centuries were eloquent as to the superlative rottenness of the roads in the Weald. To-day the roads are plentiful and good. There are four important roads radiating from London to the coast, namely, the Brighton, Portsmouth and Dover Roads and the lesser road to Tunbridge Wells and Hastings. Main lines of railway now accompany each of these, and others serve Eastbourne and Littlehampton (see Fig. 11). The only important east and west road (other than Watling Street) is that in the Holmesdale, superseding the Old Road, from Canterbury and Folkestone through Maidstone and Guildford to Hampshire. An exception to the general rule is the remarkable railroad which runs over Weald Clay in a straight line for 50 miles (80 km.) from Redhill in Surrey to Ashford in Kent. This was constructed at the invitation and with the co-operation of a group of far-seeing Kentish landlords and farmers.

The transect diagram (Fig. 12) is a summary and amplification of the material we have been able to deal with in this brief essay. It perhaps looks rather formidable at first sight, but its simplicity will emerge if the reader will start at the foot with any particular geological zone and read upwards. For instance, the Tertiary belt is seen to be composed of London Clay and Lower London Tertiaries forming low hills and broad valleys with an average annual rainfall of 22·5 in. (560 mm.). It is characterised by surface drainage and an absence of springs. The water supply is obtained from deep wells in Chalk and Greensand. Vegetation—oakwoods, neutral grassland, etc.; agriculture—sheep farming, lucerne, potatoes, etc.; and so on.

The geological section and most of the data given represent a rectangular sector of the region between the Thames and the Central Weald roughly corresponding with the Medway basin. This area affords the completest epitome of our region, and though local variations would sometimes be considerable it is fairly typical of the results that would be obtained by similar treatment of any sector. Limitations of scale and space have prevented us from using this mode of presentation as fully as we should wish. Our region is particularly adapted to it, but we believe that for most regions it is one of the best methods of concisely exhibiting regional phenomena.

We are conscious of having given a very inadequate account of our most fascinating region, of dealing disproportionately with some aspects and totally omitting to deal with others. This we fear has been inevitable in trying to compress so much into so short a space, and on this ground we would ask the indulgence of the reader.

III

THE LONDON BASIN

Hilda Ormsby

THE LONDON BASIN may be defined as the V-shaped trough of land lying between the North Downs and the Chilterns and drained by the River Thames and its tributaries below the Goring Gap. It tapers westward in the narrowing Vale of Kennet and opens wide to the east to embrace an arm of the sea between the Essex and Kentish coasts. It is both a geological and a geographical unit, for the physical basin coincides with the Chalk syncline. It does not coincide, however, with the boundaries of the Lower Thames Basin, which extends beyond it considerably in the south, where the watershed between the Thames and Channel drainage is formed, not by the North Downs, but by the central ridges of the Weald. In the north-east the boundary is less well defined than in the rest of the Basin, for the Chilterns drop gently to the low chalk plateau of East Anglia, and, north of Ipswich, the drainage is no longer to the Thames nor to its estuary, but direct to the North Sea.

London occupies the centre of the Basin, but the tentacles of the metropolis stretch to the rims and beyond; and it is obvious, as indeed the name implies, that London dominates the London Basin and must be the main theme in any synthetic geographical description.

GEOLOGY AND RELIEF

The relation which the Chalk syncline that determines the Basin bears to underlying rocks does not greatly concern the geographer and may be ignored in a brief study of this character; mention must be made, however, of the continuous layer of Gault clay that underlies the Chalk, for it has an important bearing on the water-supply of London. The Chalk, rising to heights of over eight hundred feet (240 m.) on the crests of the escarpments that mark its limits to north and south, dips slightly more steeply towards the centre of the Basin in the south than it does in the north. The Thames hugs the southern steep side of the Basin from the neighbourhood of Staines eastward, probably following a minor eroded anticline within the major syncline. A number of minor folds and faults affecting the floor of the Basin are reflected faintly in the

Tertiary and late deposits which carpet that floor* (see Figs. 13 and 14).

The higher levels of the North Downs are capped with Clay-with-Flints, but on the lower slopes much of the Chalk lies bare. Westwards, owing to the increasing steepness of the dip, the Downs taper to a ridge less than a quarter of a mile wide in the Hog's Back, but broaden rapidly, east of Guildford, to a maximum of ten miles near Dartford. The comparative narrowness of the exposure of the Chalk in the North Downs has given little opportunity for river development, such rivers as do traverse them having their origin to the south and crossing them in deep trenches. There is, nevertheless, a somewhat intricate development of dry valleys to the south of Croydon, where the exposure widens to about ten miles.

North of the Thames the Chalk is only exposed in the river valleys that score the dip-slopes of the Chilterns, and in the major heights rising above the escarpment; for, even when it emerges from the Tertiary deposits that line the bottom of the Basin, it is covered on the lower slopes by ancient drift gravels, and higher up by clays or loam. The Chilterns differ, then, from the Downs, not only in their gentler and longer slopes, with a consequently better developed drainage system, but in being much more completely covered with residual formations and drift deposits—a condition which has had the effect of preserving the general plateau-level and imparting a general flatness to the summits of the spurs.

Overlying the Chalk, and filling the bottom of the trough to a depth amounting to some 550 ft. (170 m.) in places, are sands, gravels and clays belonging to the Tertiary period. Among these the most extensively developed, if not the most important from the point of view of human geography, is the London Clay, which is exposed over three-fifths of the Basin. It is the predominance of this clay formation in the London Basin that makes the latter contrast so markedly with the Paris Basin; for, whereas the former is basin-like both in form and structure, the enormous development of limestone among the Tertiary rocks has imparted a plateau-character to the centre of the Basin of Paris. The London Clay, being easily eroded, forms the surface of most of the low ground in the Brent valley, in the lower Lee basin, and over a strip of country from four to eight miles wide south of the Thames. Its flat, ill-drained surface is dotted here and there with remnants of later Eocene sandy and gravelly deposits and glacial gravels, which appear to have been preserved in minor north-east to

* Wooldridge, "The Minor Structures of the London Basin," *Proc. Geol. Assn.* vol. XXXIV, p. 175.

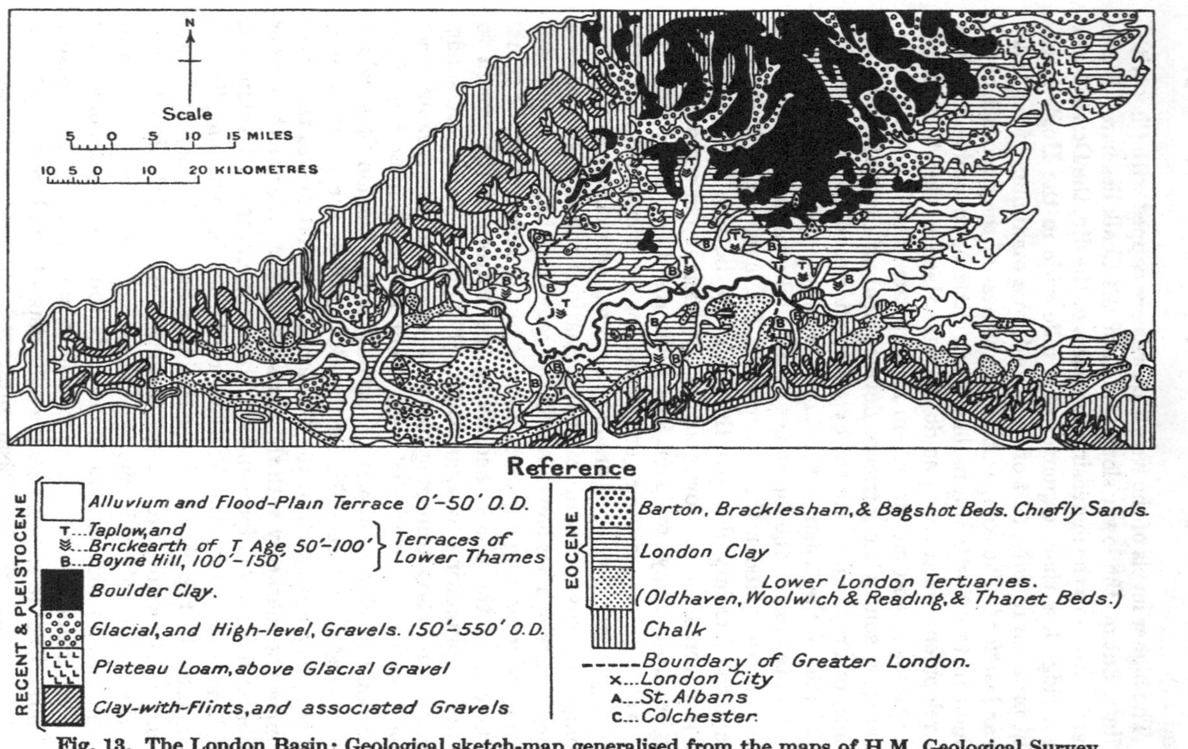

Fig. 13. The London Basin: Geological sketch-map generalised from the maps of H.M. Geological Survey.

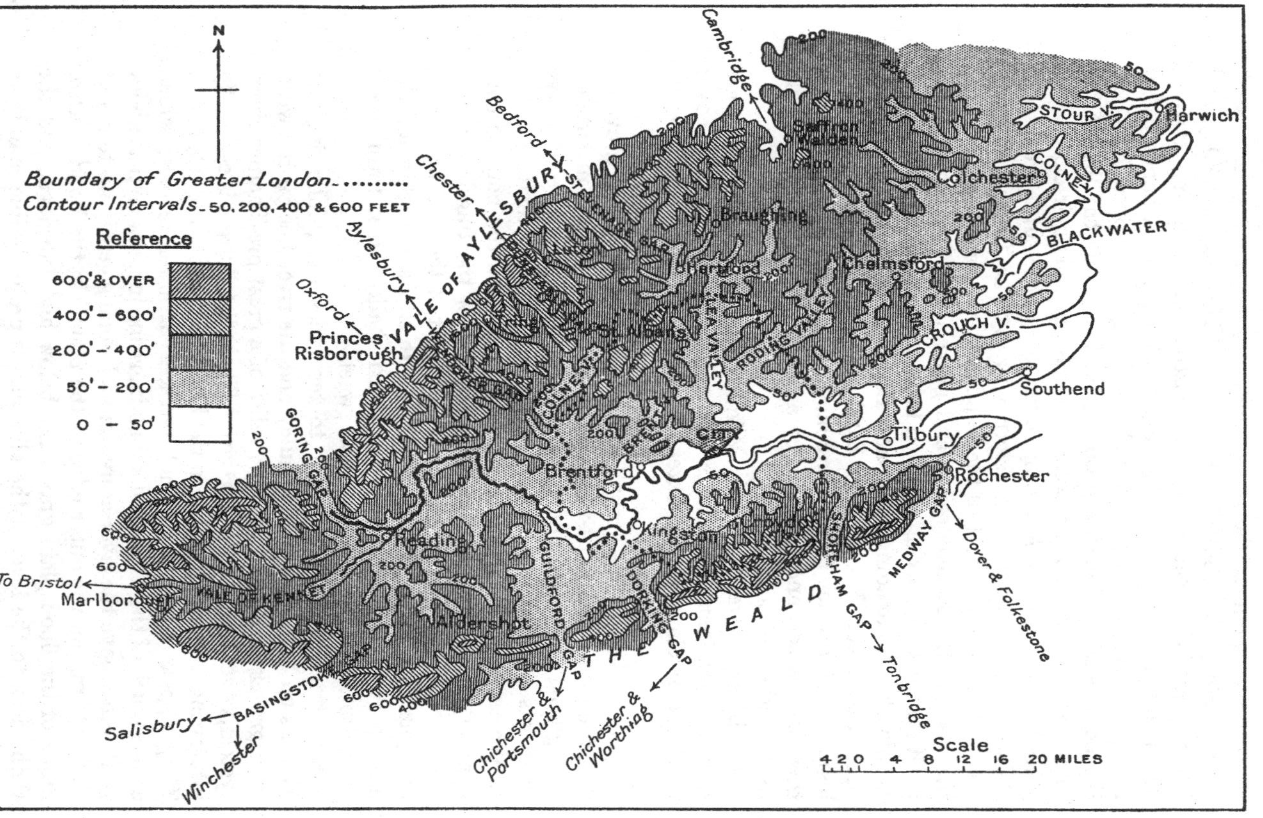

Fig. 14. The London Basin: generalised relief map.

south-west synclines. These remnants cap low hills of clay, such as Stanmore, Hampstead, Harrow and Highgate, Finchley and Winchmore Hill, Bushey Heath, Elstree and Chipping Barnet. The London Clay is exposed at levels varying from a hundred feet in the valleys to three hundred on the gravel-capped hills. In the north-west, between the Colne and Lee, along the line of the Chalk outcrop the Lower London Tertiaries and their capping of glacial gravels form a low ridge overlooking the subsequent Lee-Colne trough. This is often referred to as "the Tertiary Escarpment." In the south-east of the London Basin, particularly east of the Darent, the Eocene sands and gravels and the London Clay itself are missing, leaving exposed large stretches of Thanet Sands, Woolwich and Reading Beds and Blackheath Pebble Beds lying upon the Chalk. The Thanet Sands east of the Darent become loamy and form rich agricultural land. South of London, however, they are coarser, and, where they are not built over, form chiefly heath and common.

In the south-west, between Windsor and Guildford, sands are again predominant—this time of the Bagshot series. They cover the London Clay over an extensive area, forming a dry, infertile soil.

Boulder clay, left as ground moraine by the ice which drove an extension into the north-west of the Basin, still exists in patches with glacial gravels west of the Lee valley, one of the more extensive remnants covering the Hornsey-Finchley ridge, extending north to Whetstone. In the Essex section of our area the boulder clay is a very important feature. It overlies now the Bagshot Sands, and now the London Clay, and covers most of the area north of a line joining Colchester and Romford. It is a sandy clay for the most part, stony and containing much chalk, and it forms good agricultural soil, well drained and sweet.

In a broad belt, some eight to ten miles wide, following the Thames on both sides of the valley and the river Lee on the west, river gravels of Pleistocene age, relics of a great pre-Thames river-system, cover the London Tertiaries. Their distribution is especially interesting to the geographer in the London area (see Fig. 15, p. 60). They appear to form distinct terraces: a lower terrace, which covers much of the Thames flood plain, rising in a low step at about 10 ft. O.D. above the river alluvium and sloping gently up to about 50 ft. (15 m.) on the outer edge of the plain. In the London area, much of the flood plain gravel lies below high-tide level, and the slight drop to the level of the alluvium is probably due mainly to shrinkage of the latter resulting from embanking. Much of the flood plain gravel, then, must be regarded as land not conducive to settlement. The Taplow Gravels of the middle terrace, on the other hand, which lie at levels of from 50 to 150 ft. above O.D., and drop steeply to the flood plain, have attracted settlement. They floor a great part of North London, form a long strip to the west of the Lee valley, and spread in a broad band from the Thames near Kingston eastward to Mitcham, covering Mitcham Common and filling the bottom of the Croydon valley. They occupy most of the area above the flood plain between the Thames and Colne, and attain in places a thickness of 30 to 40 ft. A light yellow loam, known as Brick-earth, covers much of the Taplow Gravel and probably at one time overlaid most of it. This friable, fertile, easily-worked loam, probably resulting from the disintegration of London Clay and Bagshot Sand, but by some considered as an eolian deposit of the loess type, never attains a thickness of more than a few feet, and owing to its utility in brick-making and to its value to market-gardeners, has disappeared from large surfaces which it once covered, leaving only a few traces. The Boyne Hill Gravels, covering a higher terrace of an earlier stage in the history of the Thames, are now represented chiefly on the south side, where they still cap the brink of the flood plain in Richmond Park and Roehampton, in Wandsworth and Clapham, and they form patches several square miles in area near Weybridge and between Croydon and Thornton Heath.

Finally, deposits classed as Alluvium, but varying much in composition, cover much of the Thames flood plain. These deposits are products of marsh, tidal saltings, and stagnant pools, where the accumulation of peat was possible. The two main expanses of alluvium are found, the one up-stream flooring the low-lying basin between Staines and Chertsey about the confluences of the Colne and Wey, and the other, at a much lower level, flooring the flood plain below Barking, in Wanstead Flats, and from Plaistow Marshes to West Thurrock, where the valley is traversed by an anticlyne which brings the Chalk to the surface and transforms the broad flood plain into a low-walled trench for a space. This contraction of the valley may have some connection with the lake-bottom-like expansion of the valley above it and the vast deposits of alluvium. Alluvium accompanies the Wey, Mole and Wandle valleys in broad stretches, and the Thames itself at Kew, Barnes and Hammersmith, and again from Westminster to Greenwich. Farther east it follows the Lee and Roding valleys, and forms all the Essex coast land, except where islands of harder rock approach the estuary, as at Southend.

Alluvium may vary from a fine silt, which is its normal character, to a coarse sand or gravel, and may consist largely of clay or peat. The more recent alluvium of the lower Basin was deposited mainly during a period of gradual and intermittent subsidence during which deposits of marsh clay were laid down under tidal conditions, and layers of peat marked periods of slight elevation during which a rank vegetation covered the marsh.

River System

We see then, that the London Basin, although it possesses little of the dissected plateau character that distinguishes the Basin of Paris, is by no means without variety of subsoil and relief. The influence of the surface geological formations on river development, vegetation, agriculture and settlement, is as striking as their influence on relief. The streams draining into the London Basin to the Thames may be divided into four groups. First, those that rise outside the Basin altogether, drawing their headwaters from the sandstone ridges of the Weald and maintaining their original course consequent upon the Wealden uplift. The largest of these streams is the Medway; others are the Wey, Mole and Darent*. Then there are the rivers which rise within the Basin, but whose valleys, for the most part now dry, have been cut back almost, or quite, to the escarpment. Such valleys are specially noticeable in the long dip-slope of the Chalk in the Chilterns, but are also exemplified in the North Downs by the valleys converging on Croydon and by the rivers Wandle and Cray. Thirdly, still smaller streams draw their waters from springs at the line of junction of the Lower London Tertiaries and the Chalk. Such are the Hogsmill Brook, the Beverley Brook, and the Ravensbourne, in the south. In the north, a north to south ridge which forms a water-shed from the Thames to the chalk escarpment and may prove to be a deep-seated feature, throws the drainage right and left to the Colne and Lee. Hence there is no development of smaller streams like those of South London, draining direct to the Thames. Fourthly, quite small streams drain the gravels overlying the London Clay. Many of them, particularly those within London, are now covered in or converted into sewers, but the area they drained in common was considerable and they brought down a large volume of water, for the Valley Gravels were and are capable of holding large supplies, owing to the impervious London Clay beneath. Such streams are the now-buried Effra in the south; the Brent, which has a large development on the wide stretch of London Clay in the north; the

* See previous Essay.

defunct Westbourne, Tyburn, Turnmill Brook or Holborn, Walbrook and Hackney Brook of North London.

The rivers and streams of the London Basin present very different types of valley according to the geological formations they traverse. Deep, branching, steep-sided dry valleys characterise the upper chalk slopes. They are highly developed in the Downs behind Croydon and in the upper basins of the Lee and Colne in the Chilterns. Water does not begin to flow regularly in these valleys until the valley sides are trenched to the level of the water-table in the Chalk. In the Chilterns very few of the valleys carry surface water, although after a period of heavy rain water may flow for a short time. The larger valleys, even the dry upper sections, are carpeted by strips of gravel and alluvium washed down from the plateau surface at a time when, probably, there was a great flow of water from the ice sheet which lay to the north and from the snow fields of the Wealden heights. Solution of the Chalk has probably also played its part in hollowing out the coombes and valleys as well as the hollows in the plateau. The Chiltern valleys are much too large to have been carved out by the present insignificant streams that drain them. Their broad, flat-bottomed flood plains appear much too wide for their sluggish, meandering waters. They are probably a legacy of that torrential period that succeeded to the still grip of the ice. The valleys of the London Clay are wide and shallow. The streams flow swiftly in rainy weather, undercutting and carving away their banks rapidly, in spite of the willows and rank vegetation which clothe them. In dry weather they diminish rapidly in volume, frequently drying up altogether, or forming a chain of stagnant pools. The run-off from the clay would be much more rapid than it is, were the clay not to a very large extent covered with meadow and pasture, which, unless elaborately drained, has an enormous capacity for holding water. Thus a network of natural and artificial streams intersects the London Clay surface, in marked contrast to the waterless surface of the Chalk. In the gravels, the rivers cut steep-sided, V-shaped valleys. Streams collecting their waters on the flat London Clay bring them through the Valley Gravels to the Thames in deep, narrow cuttings. A noteworthy, if miniature example is the Holborn or Turnmill Brook which, after collecting a number of tributaries from the low levels of St Pancras, passed through the deep valley now below Holborn Viaduct, to form the Fleet.

We may subdivide the London Basin reasonably enough into five types of country, based on the physical features: the Kennet valley, the Lower Thames valley, the North Downs, the Chilterns and the Essex Plain and Marshlands. But, lying right across these natural divisions, the vast agglomeration of humanity known as Greater London forms a geographical fact that so masks and modifies natural features and at the same time forms in itself so definite a barrier, that we find ourselves forced to make physical divisions subsidiary to others which define units of a more complex origin. Hence our analysis will resolve itself into a study of the Western Basin, the Eastern Basin and Greater London, though these divisions can have no definite boundaries.

The Western Basin

The Thames enters the London Basin through a flat-bottomed, steep-sided north to south gap in the Chilterns, a mile wide and some half dozen miles long. The river swings from side to side, impinging on the chalk cliffs on either hand and leaving broad stretches of alluvial flats within the curves of its meanders. Throughout this part of its course it is fed by springs issuing from the chalk slopes. At Pangbourne it leaves the Goring Gap and receives from the south the Pangbourne or River Pang which rises, a perennial stream, in the Berkshire Downs and gives its name to the village at the confluence. The Goring Gap carries road, railway and navigable river (for the Thames is locked) from the London Basin to the west. Beyond the Chilterns the routes branch at Didcot Junction, passing south to Portsmouth, north to Oxford and west to Bath and Bristol. The old Bath Road, however, branching at Reading, utilised the Kennet valley. The valleys of the Thames, Kennet and Pang isolate a triangular hill composed of Chalk capped with Tertiary deposits and rising to 340 ft. (104 m.) above sea-level. Tilehurst Common occupies the summit. Reading, originally a bridge-town at the crossing of the Kennet, where a spur of high ground from the east approaches the river, now spreads westwards, up the slopes of the Loddon-Kennet divide and across the flat to Tilehurst Hill; via Caversham bridge it is also climbing up the slopes of the northern bank of the Thames. Its nodal position, controlling the diverging routes from London to Southampton, Bath and Oxford, could not fail to give it importance as a market. The diverse regions of cornland, downland, water-meadows and sandy heathland, which surround it, made it an exchange centre for products of various kinds. Local initiative and capital combined to give it its two main industries of biscuit-making and seed-growing, though no doubt the important railway communications and possibly at first local wheaten flour were contributory factors. The essentially agricultural nature of the western end of the London

Basin is emphasised by the presence of a University in which the Faculty of Agriculture plays an important part.

The Kennet takes its rise near the western extremity of the London Basin, in springs from the Lower Chalk, and flows almost due east, somewhat north of the synclinal axis of the Basin. Marlborough, Hungerford and Newbury mark the junction of converging valley routes. North and south of Marlborough the Upper Chalk rises steeply to an escarpment overlooking a broad terrace of Lower Chalk, which having a thin covering of loam can be cultivated and which, in its turn, drops steeply to the Oxford plain and to the Vale of Pewsey. The bottom of the western end of the Vale of Kennet is devoid of London Tertiaries, but is floored with wide though thin stretches of Clay-with-Flints, which are cultivated on a sheep-farming basis connected with the pastures of the Downs. As one approaches Newbury the Chalk is less and less exposed and the London Tertiaries appear, capped, for the most part, with hill gravels, heavily wooded, into which the Kennet and its tributaries trench deeply. The broad flood plain is floored with irrigated water-meadows. With the decline in agriculture the rural population has diminished, detracting from the market activities of the Vale towns. It is the great Bath Road and the crossing of the rails to London and Southampton that give Newbury its nodality to-day.

Below Reading, the Thames turns northwards to Henley, receiving the Loddon from the London Clay vale to the south, and winds back into the lower slopes of the Chilterns from which it had emerged below Pangbourne. Henley, Great Marlow and Maidenhead (pop. 16,700) are summer river-resorts, but they also provide homes for a population that travels daily to London for work. The riverside and the main-line railway service are the chief attractions to settlement in this section of the Thames valley, for while the river between Henley and Great Marlow cuts into the rising chalk slopes, elsewhere the flood plain is flat, waterlogged and liable to serious floods. At Windsor the dead-level of the plain is broken by an outcrop of Chalk, exposed as the result of the erosion of a north-south fold. The Thames, cutting into the exposure of Chalk, forms a steep bluff on which stands Windsor Castle. Windsor Great Park extends its lawns and woodlands of oak, ash and elm over the water-holding London Clay south of the river, but on the Bagshot sands and gravels the scene changes to wide stretches of heath and common, as at Bagshot and Chobham, with a warmer colouring contributed by bracken and heather. Here parks and water-meadows give place to open land largely devoted to military training camps. The military training camp at Aldershot, the rifle ranges of Bisley, the Staff College at Camberley, Wellington College and Sandhurst Military College are placed here because the land is of little or no agricultural value, is healthy and offers excellent sites for camps and ample space for manœuvring. Settlement grew up round the edges of the exposure of Bagshot Sands where the water seeps out at the junction with the London Clay. The beautiful and health-giving qualities of this heath and pine country have attracted many residents in recent years, but the absorbing occupation of the countryside appears to be catering for the military. In the south and west the Bagshot Sands are covered by a loamy deposit known as Bracklesham Beds. These yield a more fertile soil than the sands, but, owing to their impermeable nature, often give rise to marshy conditions. It is somewhat surprising to find, on the surface of these sands, areas of swamp and peat-bog, as for instance at Chavy Down. A thin layer of clay or of pan, such as occurs not infrequently in this formation, or a bed of extremely fine sand, is sufficient to hold the water, locally, at the surface, and leads to the development of bog moss; and this, once established, tends to increase the area of marsh by its rapid growth and its remarkable capacity for holding water.

In the great bend of the Thames between Staines and Kingston the broad stretches of Thames gravels, covered to a great extent with brick-earth, or down-wash from the brick-earth, furnish light, warm, easily-worked soil suitable for market gardening and horticulture.

Between Farnham and Guildford the North Downs are represented by the Hog's Back, where the steeply-dipping strata expose the Chalk in a narrow ridge. It carries the high road between the two market towns. Guildford, that "happy-looking town" as Cobbett terms it, has a fortunate position as a market town. It lies on the steep bluff of chalk, at the northern entrance to the gap that the Wey, or a precursor of the Wey, has carved—a half-mile wide—in the Chalk in passing from the Weald toward the Thames. It commands the Portsmouth road from Kingston via Esher and Ripley, now being linked more directly to London by the Kingston By-Pass, and the main road from London to Chichester which skirts the edge of the London Tertiaries from Leatherhead. Five first-class roads and three lines of rail converge on Guildford from the north; and the Wey is navigable for small craft through the gap. A busy and prosperous town in coaching days, Guildford passed through a period of coma with the advent of the railway, from which it has now been effectively aroused. For the Portsmouth Road hums with life again and the old coaching inns are basking in prosperity once more. In addition to its command of routes, Guildford is fortunate in a situation near the boundary between very diverse types of farming land, so that interchange of products would naturally take place there. East of Guildford the angle of dip of the Chalk becomes less acute and the outcrop consequently widens. The brink and the higher shoulders are capped with flinty clays and loams and are well wooded with pine and beech, but Merrow Downs and Clandon Downs on the lower slopes exhibit more typical downland, with gently contoured slopes, fine herbage dotted with twisted hawthorn bushes or studded with yew and juniper. Between Leatherhead and Dorking the Downs reach a breadth of four miles. Here the River Mole has cut a broad but winding gap followed by a main road and the railway from London to Horsham, which however takes a more direct route than the road by tunnelling through the spurs of Chalk that project into the meanders of the river. The Dorking-Leatherhead gap is the main natural gateway leading from the south coast to London and offers a comparatively easy route across the Downs. Between the Mole and the Darent gap all the roads across the Downs are marked by steep gradients; on two that are first-class roads the steepest gradients are respectively 1 in 11 and 1 in 6. The factor of slope, though it has decreased in importance, as far as road transport is concerned, with the advent of motor traction, still controls routes and is still a serious item in railway transport. Road and railway in this section utilise gentle slopes of the dry valleys which seam the dip-slope and the wind gaps forming the valley heads in the escarpment. But the railways, after following the dry valleys on the London side, are of course obliged to tunnel to get a sufficiently gentle gradient down the escarpment towards the Weald. The market town of Croydon has become one of London's largest suburbs, easy routes along the valley, and a direct route to the sea coast, affording it the benefits of a main line service to Town.

The Darent, rising within the Lower Greensand formation to the south of the Chalk, enters the Basin through a broad trench, in which the sluggish river meanders, like its sisters Wey and Mole, to the west. Its valley appears to bear some relation to the north and south line of flexure, which brought the Chalk to the surface between Dartford and Gravesend*. The plateau of Tertiary sands and gravels bearing Bexley Heath, Woolwich Common and Blackheath,

* Wooldridge, "Structural Evolution of the London Basin," *Proc. Geol. Asscn.* vol. XXXVII, p. 162.

ends abruptly along this line between Erith and Crayford where its scarp-like edge overlooks Dartford Marshes. The Cray and Darent both pour their waters into the great triangle of alluvium which borders Long Reach. Roads and villages hug the slopes between the fifty- and hundred-foot levels, and a little lower the railway follows the edge of the flats. Dartford, where Watling Street, crossing the Darent, meets the road from Sevenoaks and Tonbridge, developed at the cross-roads beside the ford. The town is expanding northwards, and if the project for a Thames tunnel to Purfleet is carried out, Dartford may become a very important focal point on the Kentish shore. Crayford, also on the Roman road, has a somewhat similar position but is growing less rapidly, and the villages of the Cray valley are becoming residential suburbs of London. Where the Thames impinges on the southern border of the flood plain, the steep-to shores have encouraged the growth of shipping settlements. Most of these are in process of becoming manufacturing towns. The growth of Woolwich has been stimulated by the establishment of the Royal Dockyard and Arsenal and recently by the great dock construction on the opposite bank of the Thames.

The population has increased by nearly fifty per cent. in the last thirty years. Plumstead, on the edge of the marsh, is a flourishing suburb of Woolwich. Erith and Greenhithe, pleasantly situated, the one on a gravel, the other on a Chalk bluff, have profited by the deep water to become yachting stations, but have grown rather as residential and week-end resorts. Northfleet on the Chalk outcrop is devoted to cement-making.

Gravesend marks the end of the shore-way which stretches along the Thames from Northfleet and the sudden widening of the Thames beyond the Chalk outcrop to the outer estuary. Here the Thames pilots are taken on board for the navigation of the lower reaches. The growing importance of Tilbury is reflected in the opposite ferry town; hundreds of men resident in Gravesend cross the Thames daily by ferry to their work in Tilbury. During the War a bridge of boats linked the two shores.

Now let us turn to the stream basins north of the Thames. It would be difficult to find a greater contrast than that offered by the Colne basin in the Chalk and the basin of the Brent in the London Clay. The upper Colne basin lies between 250 and 400 ft. (say 75–120 m.) above the sea. The Chalk is masked everywhere by hill gravels and loams except on the steep valley slopes where it is exposed. The valleys are deeply trenched, steep-sided, flat-bottomed and floored with gravel or alluvium. Erosion, owing to the spring origin of the streams and to the clear nature of the chalky water, is slight and there is a tendency for peat to form in the valley bottoms. These flats are utilised for the growing of water-cress which is irrigated from the river. Lines of pollard willows follow the streams, which flow often flush with the water meadows. The steep slopes are often wooded, while the flattish uplands are devoted to sheep pasture, scattered beech woods or cultivation, according to the nature of the drift and residual deposits that cover them. The beauty of the Chalk slopes and the variety and warmth of colour imparted to the vegetation by the sandy soils form a pleasant landscape, which, with the generally southern aspect and healthy elevation, made this an attractive area for modern settlement as soon as the electric railway brought it into close touch with the City and as land became available. But owing to difficulties of communication due to relief, the development is slow and is toward the creation of "better-class" suburbs.

The Colne is a subsequent stream in its upper course. Watford, where the river touches the "Tertiary Escarpment," depended on the clear water of the Colne and on the excellent barley of the clay-covered uplands for the brewing industry for which it was famous. Owing to an excellent rail service, it is now a London dormitory and a centre for small manufacture. Wheat used to be the other main crop of the plateau, the specially fine straw being the raw material for the straw-plaiting and hat-making industries of Luton, Dunstable and St Albans. Competition from the East and the vagaries of fashion have brought the industry into low water. On the other hand brewing and furniture-making still use local raw material. Modern industries dependent on imported material, such as light engineering, paper-making, rubber working, printing and book-binding, have made Watford, Luton* and St Albans busy centres. Harpenden is mainly residential.

St Albans lies just north of the great vale drained by the subsequent reaches of Colne and Lee, between the Chalk slope and the "Tertiary Escarpment." Verulamium, the British and Romano-British centre which preceded St Albans, occupied a defensive position on the right bank of the River Ver, whose consequent valley leads north-west from the Colne trench to the Dunstable gap. Watling Street, directed to the citadel straight across the London Clay plain and following the Silk Stream, passed through it to Chester. After the destruction of Verulamium, the founding of St Albans on the steep slope of the left bank necessitated a diversion of Watling Street. The ridgeway from London, following

* In the upper Lee valley.

the waterparting on the Tertiary plateau and passing through St Albans, rejoins the line of the ancient Watling Street about two miles north of the city and continues through the Dunstable gap as the Holyhead Road.

The Brent basin is floored by London Clay, dotted over with patches of sand and gravel, preserved in a series of minor synclines. We may cite Hampstead and Highgate capped with Bagshot Sand; the Finchley Ridge, where the hill gravels are overlain by boulder clay; Dollis Hill, overlain by hill gravel; Harrow-on-the-Hill, covered with Bagshot Sands; and farther north Pinner Hill, Oxhey Wood, Bushey Heath, Stanmore, Totteridge, Chipping Barnet Ridge, all capped with hill gravels. From all these hills what the French would describe as a *ruissellement* of water spreads in a network over the impermeable surface of the London Clay. The Bushey Heath-Stanmore mass, for instance, radiates in all directions streams which are picked up by the Colne and Brent. From the edge of the Hampstead and Highgate gravel-caps flowed the famous streams of Old London—the Westbourne, Tyburn and Holborn or Turnmill Brook.

The monotonous green water-meadows, the lines of willows and elms with their dull foliage, form a strong contrast to the warmer colouring of the heathland and beechwood on the gravel-covered Chiltern slopes. The shallow, meandering valleys in the clay, the often turbid streams, the soggy meadows, have little of the attraction of the clear, spring-fed chalk rivers with their cliff-like, wooded valley slopes, so that the relief afforded to the eye by the gravel-capped hillocks dotting the clay is most welcome. Early settlement in the Brent basin, as elsewhere, is closely related to local potable water supply, hence we find villages of the Brent basin clinging to the slopes of the gravel-capped hills where the water came to the surface. These villages are rapidly growing to be suburbs of London on account of their comparatively healthy and pleasant situation, while the broad area of intervening clay is even to-day but sparsely peopled. Half a dozen lines of railway traverse the Brent basin, diverging as they make for the wind gaps in the Chilterns. The intervening gravel-capped hillocks make necessary certain diversions and tunnelling, but the large expanses of flat, unoccupied and cheap land have encouraged the development of railway sidings (as, for example, at Willesden), and in consequence the railway and to a certain extent canal facilities have in their turn encouraged the establishment of industries.

The Lee valley offers a contrast in its lower stretch to both Colne and Brent. Its uncompromising straightness is due to a monoclinal fold with a downthrow to the east. Gravel deposits, once laid down in the course of the river but left high and dry as it deepened its bed and shifted its course down the dip-slope, have to a large extent protected the London Clay on the west. At the same time the Lee has been cutting steadily into its left bank, so that, with its steep edges and broad, flat bottom, the valley looks as if it had been trenched in limestone rather than in clay. The run-off from the clay is heavy and rapid, and the valley, as a result, is exceptionally marshy. On the right bank settlement has been encouraged by an extensive covering of brick-earth. This, overlying the gravel, is well drained and fertile, for which reason and for its use in brick-making it is rapidly disappearing. The brick-earth is sold to the market gardeners for use in their greenhouses, and the underlying gravel is dug out by contractors for building purposes. Building is checked on the eastern side of the Lee valley by the presence of the great Forest of Epping, once a royal preserve which stretched far into Essex. Here the gravel capping of the London Clay only exists in patches, but is sufficient to give a considerable variety to the trees.

The London and North-Eastern Railway and the Old North Road from London Bridge follow the right bank of the Lee, passing an almost continuous string of suburbs: Tottenham, Edmonton, Ponder's End, Green Street and Enfield.

The Lee probably formed a navigable highway of some importance in the early history of the area. From the Thames small boats could ascend to Hertford, whence communication with the river Colne and St Albans was short and easy. In modern times, save for the important part it plays in the London water supply, and the small amount of traffic that utilises the navigable channel, the Lee valley is rather a hindrance than a help to communication. In the lower reaches, roads across the valley have to be supported on costly causeways, because floods, especially in the winter half-year, are not infrequent and cover large areas. Parts of the marsh have been drained and brought under cultivation in the form of market-gardens mainly of the allotment type; large areas are utilised, water permitting, as sports grounds, but the most profitable use to which the flood plain has been put so far has been as a site for reservoirs. The Lee contribution ranks next to that of the Thames in the supply of water to London.

The Eastern Basin

Two main features distinguish the Eastern London Basin from the western—the broad coastal plain with its drowned valleys along whose intricate channels the tide penetrates some twenty miles into the land, and the great spread of boulder clay which blankets the low chalk slopes*. A minor distinction is the broadening of the strip of gravel which borders the Thames, and which, owing to a less generous covering of brick-earth, discouraged settlement along the coastal strip. The boulder clay and the brick-earth provided excellent arable land, and in the early days of our history, when this area was colonised by an agricultural people from the Belgian plains, the long, winding channels gave safe and easy access from the coast across the strip of marsh, across the infertile, heath-covered gravels, across the forests of the cold, wet London Clay, to some of the best farming-land the country produces. Colchester, the Camulodunum of Roman Britain, the stronghold on a steep bluff overlooking the tidal waters of the Colne, with its then busy hythe at the foot of the hill, its high-roads branching to Braughing and London, grew up as the natural centre of the rich agricultural region that lay behind it. The double trend of this eastern end of the London Basin, which is brought out vividly by the contours selected in the relief map, and which suggests a number of interesting problems with regard to the history of the drainage of the area that space will not allow us to deal with here, is reflected in the human geography. The NW.–SE. trend of the consequent valleys, and the sympathetic development of the obsequent escarpment valleys cutting back into the plateau as at Saffron Walden, link agricultural Essex with Cambridge and the rich Fenlands. The longitudinal development of the upper Lee basin along the edge of the Tertiaries, and the remarkable NE.–SW. trend of the Stort, the Roding and the Wid valleys (a trend very marked also in the lower basins of the Colne, Blackwater, Crouch and the Mar Dyke north of Tilbury), emphasise the geographical centrality of London. The main lines of settlement to-day lie along the Roman road from London to Colchester, where a series of ford towns mark the ancient crossings of the streams; Old Ford on the Lee, Ilford on the Roding, Romford on the Rom, Chelmsford on the Chelmer. Ilford and Romford lie on patches of brick-earth. Brentwood marks the crossing of a hill capped with Bagshot Sands and is typical of that kind of settlement met with already in the London Clay districts, where a supply of water from some gravel reservoir is the determining factor. Chelmsford is, in addition, the point where a Roman road linked the Stane Street, from Braughing to Colchester, with the road converging on it from London. In the Eastern Basin the alluvium becomes an important element.

* H. Ormsby, *London on the Thames*, p. 11, map, fig. 4.

The Thames flood plain widens, the meanders of the river become more broad and sweeping and great stretches of tidal marsh bound it to north and south: Barking Levels, Dagenham, Plumstead and Erith Marshes and others. Until quite recent times settlement on these embanked and partially drained marshes was almost non-existent. Here and there an isolated farm with a few haystacks and a small patch of cultivation emphasised the loneliness of these great alluvial flats. Settlement is still almost confined to the edge of the gravel terrace that overlooks the marsh, where a wash of brick-earth or a thin covering of loam makes cultivation possible. The polders, however, form excellent grazing land, and sheep, cattle and horses feed there in large numbers. Near London, an accessible market has encouraged the more elaborate draining and sweetening of the rich marsh soil, and the fine crops that one can see from the Southend Railway show that it is not lack of fertility that abandons the land to rough pasture and waste.

Greater London

The third sub-division of the London Basin consists of Greater London itself. Its focus is London Bridge and the City. The trend is from south-east to north-west, following the line of greatest traffic from the Continent and the Ferry Ports towards the Industrial North. Greater London does not coincide with any set of geographical features. It extends over chalk, gravel, clay and alluvium. It is not spread out along the Thames valley bottom, nor does it follow any one tributary valley. But it does occupy in a truly geographical sense the centre of the Basin. From the City natural routes and lines inviting settlement radiate in many directions. It would be impossible to find a site within the Basin with greater natural nodality.

The actual foundation of London has been influenced primarily by the Thames, as a barrier that must be crossed to reach the Romano-British centres, first of Verulamium and later of Camulodunum and eventually the Roman stations beyond the London Basin. Whether or no London was the lowest possible crossing place on the Thames it is difficult to say. If we admit a slight subsidence of the Basin during and since the Roman occupation to account for the depth at which the Roman floor now lies, we must also modify our estimate of the fordability of the lower Thames reaches and their suitability for bridging. The most definite statement we can make at present in this connection is first, that the City of London is the only place where the fifty-foot contour line approaches the Thames on the north side between Purfleet and Hampton;

secondly, that a crossing at Greenwich would have involved the crossing of the Lee and its flood plain as well as the Thames; and thirdly, that a crossing at Brentford would necessitate a considerable detour in a journey from Kent to Verulamium. Actually the shortest route across the flood plain would have been from Battersea

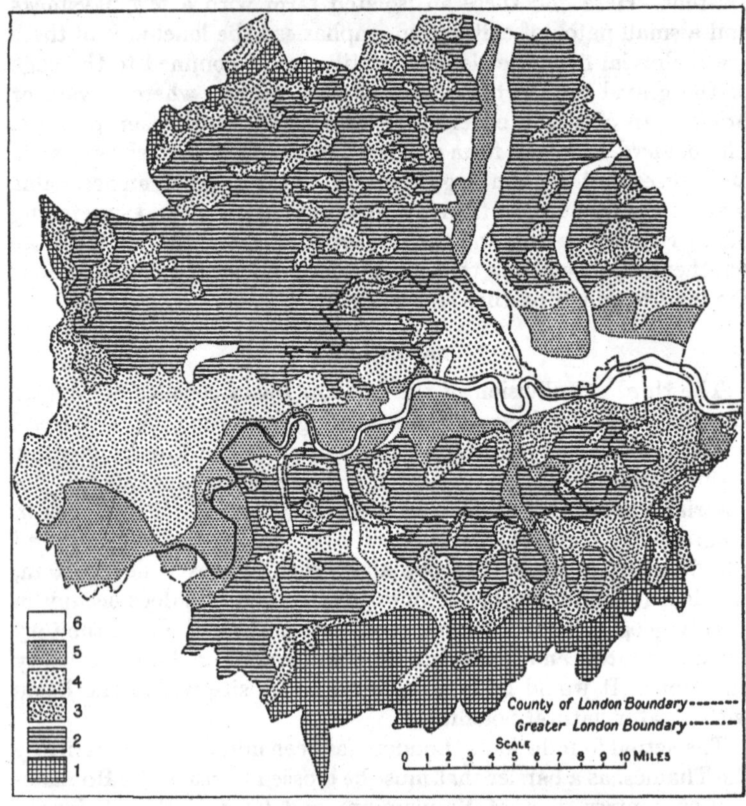

Fig. 15. Greater London: simplified map showing the surface deposits. Reference: 1, Chalk; 2, London Clay; 3, high level gravels of various periods; 4, Taplow Gravel; 5, flood plain gravel; 6, alluvium.

to Chelsea, involving very little detour, and the fact that this site was not chosen points to the selection of a ford for the original crossing. Westminster was the next place above Greenwich where the river, crossing the flood plain from south to north in a direct line, caused a broadening and shallowing of its channel and created fordable conditions at low water. Moreover at Westminster a low, sandy island rose slightly above the marsh, probably partly separated by a backwater or former river course from the mainland. The sections of Watling Street coming from north and south-east appear to converge on Westminster.

A second factor influencing the selection of the site of London was a defensible position for the protection of a bridge-head, for

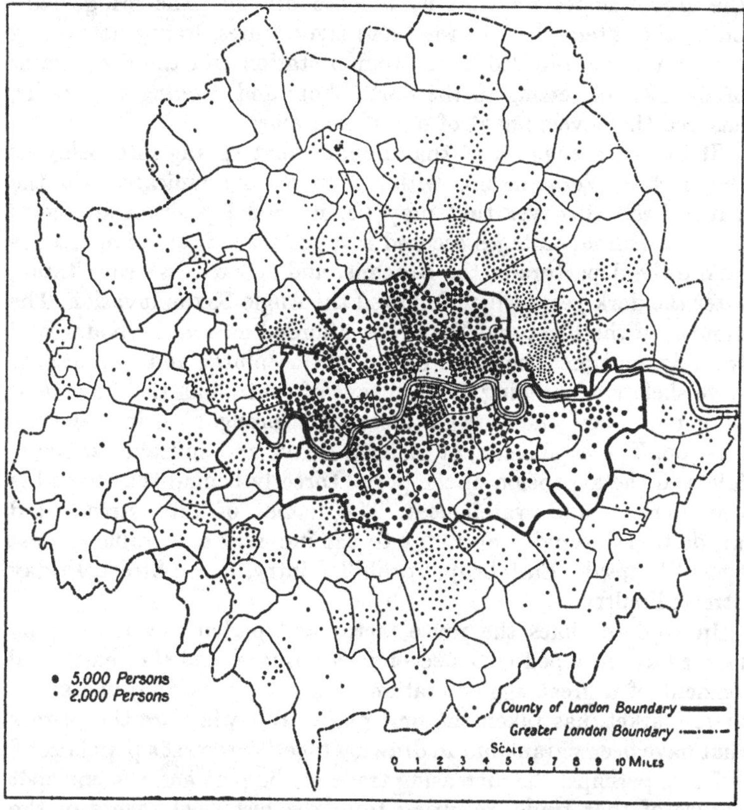

Fig. 16. Greater London: distribution of population according to the census of 1921.

obviously a ford crossing could not be regarded as permanent. Here again, if suitability for bridging had been the only consideration, one would imagine the Battersea crossing to have been more advantageous in that it secured comparatively firm ground on *both* shores. But the superior defensive position of the twin hills on which the City stands was in favour of the down-stream site. These hills, carved out of the Taplow Gravel terrace by a couple of deeply trenched rivulets, were flanked by navigable creeks, and the outer defences were provided by the flood plains of the Thames and Lee; moreover the main objective of the crossing had shifted from Verulamium to Camulodunum.

No doubt a third consideration was water supply, which the Walbrook and Turnmill Brook and the innumerable springs from the gravel slopes were amply able to provide. The bridge once built, the intersection of road and river routes, to say nothing of the protection afforded by a Roman station and the deep water of the current setting to the north shore and flooding the creeks, ensured the development of a port and mart.

It has been suggested* that London first sprang into being as the port of Verulamium, with which it communicated via the River Lee. However that may be, its nodal position and local geographical advantages secured it a position of some importance before the departure of the Romans, and ensured its resuscitation after the dark years which followed the Anglo-Saxon invasion. The mouths of the Fleet and Walbrook afforded natural harbours, and from Saxon times the small basins of Queenhithe and Billingsgate gave shelter to shipping and a means of controlling the movement of goods when the foreign merchants brought their wares to London. The establishment of legal quays and suffrance wharves followed the expansion of the port. Directly behind the quays lay the markets and warehouses. The names of City streets and wards to-day provide an index to the industry and traffic of past ages: Cheapside, Eastcheap, Cornhill, Vintry, Bread Street, Friday Street, Poultry.

In modern times the route factor and perhaps even the port factor have dropped into a secondary position, and the centripetal element of a great agglomeration of people, providing in itself a huge market, has taken the first place. But what are the factors that have been paramount in drawing together so vast a population?

First, perhaps, the increasing trade of the port and the entrepôt nature of that trade, resulting from the historical events of the 18th century and the artificial stimulus given by the Navigation Laws. These induced a multifarious activity and brought London into contact with the far corners of the earth and into intimate relations with western Continental Europe.

Secondly, a collateral financial development, dependent initially on London's geographical position and subsequently on accumulated capital, experience and prestige, which finally made London the leading money-market of the world, and resulted in the necessity for the representation in London of almost every

* H. Ormsby, *London on the Thames*, footnote p. 69.

leading trade and industry. London bankers finance the movement of goods between London and every quarter of the Globe, and of goods moving from one end of the world to the other that never see London at all.

Thirdly, the course of events in railway development, which led to the termination of the great railway lines on the outskirts of London, and the resulting enormous increase in the various functions of entrepôt trade already typical of the port. The convergence of traffic lines, accentuated greatly by the development of railways, entails a great transit movement across London both of passengers and goods and largely increases the need for warehouses and hotels.

A group of factors which have acquired great importance in the last quarter of a century, and whose effect has been centrifugal rather than centripetal, may be summarised as follows:

First, modern engineering developments, which have made possible the supply of enormous quantities of water to, and the discharge of equally vast amounts of sewage from, an ever-widening area. The collecting, purifying and discharging into the lower Thames estuary of London's sewage, from the inhabited districts of the Thames flood plain, is a colossal task. But much vaster is the storing, purifying, pumping and distributing an average of some 37 gallons (140 l.) of water per head per diem to 7 millions of people. London's earliest water supply was drawn from the Taplow Gravels of the Thames valley by the use of springs, wells and streams. By the 13th century, these supplies were being supplemented by waters brought in conduits from Tyburn or from fields where it was impounded, *e.g.* Lamb's Conduit Fields. Thames water was also distributed from house to house. An important step forward was taken when in 1580, by utilising the tidal race through the arches of Old London Bridge, water was pumped to the conduit heads of the City. But a much greater stride was made when 33 years later the New River Company tapped the pure water from springs in the chalk of the Lee basin in Hertfordshire, and brought it, in wooden aqueducts, 32 miles (51 km.) to London. Following a shortened course in a cement-lined canal, the New River still contributes to London's water supply*. During the 18th and 19th centuries a number of other companies sprang up, drawing, some on the flood plain gravels, some on the Thames or Lee and some on the subterranean water supply of the Chalk, held captive between the continuous impervious belts of Gault Clay and London Clay or obtained by very deep bores in the Downs and

* H. Ormsby, *London on the Thames*, fig. 34.

Chilterns. To-day this source of uncontaminated water is still important, especially to East and South-East London. At the present time several hundred private artesian wells, in addition to those belonging to the Metropolitan Water Board, are drawing on the chalk reservoir. It is not surprising that the water-table in the chalk is markedly lower than it was fifty years ago, as is evidenced by the levels at which springs put out and at which pumping has to be resorted to in the wells. At the end of the last century, serious threat of water shortage resulted in the appointment of two Royal Commissions, held in 1893 and 1899–1900, and the subsequent amalgamation of the various companies under the Metropolitan Water Board. Various schemes were brought before the Balfour Commission, including one for bringing water from Wales and another for damming up the Kennet to form a reservoir. Two great innovations mark the solution finally adopted for the problem of London's water supply. One is the construction of great reservoirs on the level of the flood plain, capable of storing thousands of millions of gallons; and the other is the bacteriological purification of water. By their means, London is now able to draw 60 per cent. of its supply of water from the Thames and the bulk of the remainder from the Lee. In 1913 the George V reservoir at Chingford in the Lee valley was opened. It occupies over 400 acres (160 ha.) of the Lee alluvial plain and necessitated the diversion of the river. In 1925 a much more grandiose undertaking, which had been interrupted by the War, was completed: the construction of the Staines and Littleton reservoirs. The latter covers 723 acres (293 ha.), has a storage capacity of 6750 million gallons* and is the largest completely artificial reservoir in existence. Both reservoirs are scooped out of the Thames gravels, for neither the Thames nor its tributaries near London offer facilities for water storage by the usual method of damming a valley. This method of storing means that the whole of the supply has to be pumped, first to the filtering beds and then to the service reservoirs. The fact that three-fifths of the normal flow of the Thames above Teddington can be impounded by the Metropolitan Water Board, emphasises the estuarine nature of the river at London.

Secondly, the electrification of railways already existing and the construction of a large number of new lines, together with the introduction of motor transport, have made it practicable to extend building operations for the housing of Londoners to a distance of from twenty to thirty miles from Charing Cross. Coincident with this outward movement has been the tendency to fill up hitherto

* = 255 million hectolitres.

unused and vacant spaces and ineligible building land. This occurs chiefly on the old marshlands and in the clay bottoms of the river valleys. A noteworthy example is the vale running east from Kingston, where rapid building extension, mainly concerned with the production of small houses and cottages, is taking place in the Malden, Raynes Park, Merton and Morden districts, as a result of land cheaply acquired and increased railway facilities. Another example is the expansion of London over the Lee marshland and along the London to Chelmsford line. The study of population maps based on the census returns indicates that there has been a definite geographical selection in the expansion of London (Figs. 15 and 16). Except in the centre and east, where the influence of the market and the growing port was strong, the marshlands have been avoided until recent times; insalubrity, coupled with difficulties of drainage and water supply, being the main contributing causes. The utilisation of the marshlands for habitation was only made possible in the first instance by the introduction of pumping and distributing systems for water and sewage. The spread was first along the Taplow Gravel terrace in the neighbourhood of the City and along the river banks, with extensions along the main roads. Later, the villages of the northern and southern heights, supplied with water from the high level water-holding strata such as Boyne Hill Gravels, Bagshot Sands, Glacial Gravels and Lower London Tertiaries, and linked with London by the high roads, were attached. At a later stage still, following the demand for open country and fresh air and a dry soil, and aided by deep well-borings, the outskirts of London spread to the chalk downs. The map for 1931 will show a prodigious expansion of building over the less desirable flats, where the problem is now one of drainage rather than of water supply, and where the provision of small houses, to accommodate the natural increase in population and that displaced by the reconstruction of large areas in central London, demanded land at a low cost. One result of this centrifugal movement is an easing of congestion, already sensible in the heart of London. The City has for some years past ceased to be a residential area, and the business quarter is now spreading with increasing volume into Holborn and Westminster, where large residences, too large for the modern cost of upkeep, are being converted into or replaced by banks, offices or hotels.

Again, the increased facilities for transport by rail and road, and the enormous recent expansion of light engineering, associated especially with the electrical and motor-car industries, is causing more and more factories to be established in and on the outskirts of London. They are springing up more especially along the lines of the trunk railways, the new arterial roads and along the Thames shores both up-stream and down. For the river with its tides offers cheap transport of bulky and heavy goods, notably coal and oil. It is an interesting and educative thing to stand on Battersea Bridge and watch the tugs with their convoy of six lighters, or the collier steamers hurrying up on the tide, with coal for the Chelsea Power Station or for the gas works or railway depôts that lie on both sides of the river up to Brentford, or along the Grand Junction Canal. There are some 9000 barges plying on the Thames, most of them having a carrying capacity of over 250 tons. It is estimated that 50,000 tons of goods a day pass under London Bridge, mostly for some up-stream destination*. This industrial development in Greater London is due in part to the moving out of industries from the congested central areas, and in part to the establishment of new industries. A factor of increasing moment in this connection will be the largely extended provision of electric power in and around London. The springing up of new workers' residential suburbs, as for instance at Becontree in Essex, encourages the establishment of factories by its supply of local labour.

Lastly, the gradual shifting of the Port down-stream has had the effect of causing a drift of population eastwards, though the rapidity of this movement is checked by the serious barrier offered to cross-river communication by the width of the estuary. The change has been a steady and gradual one and has resulted chiefly from the gradually increasing size of shipping, coupled with the fact that the Thames is a tidal river. At the end of the 18th century the congestion in the river was so serious and the loss of goods by theft, from craft lying in the stream or from the quays, so great, that the demand for enclosed basins became urgent. Unlike Hamburg and Rotterdam, London can only secure the advantages of basin accommodation for her shipping by the construction of locks, which retain the tidal water in the basins during the ebb. In the first fifty years of the 19th century the West India, London, St Katharine, East India and Surrey Commercial Docks were constructed, and functioned under the control of various companies. In 1882 the necessity for an outport made itself felt as the size of shipping continued to grow, and the building of Tilbury Dock, corresponding to Cuxhaven or Brunshausen, took place. Then came a period of inaction during which competing companies and divided authorities allowed the water-way to deteriorate and the

* *Royal Commission on Cross River Traffic in London, Minutes of Evidence*, 1926, p. 212.

accommodation to become antiquated and out of date. The Thames estuary, with its alternating currents and shifting shoals, requires costly dredging and careful supervision and has always necessitated skilled pilotage both above and below bridge. The construction of the great modern liner, with its demand for 40 feet (12 m.) of water, gave such deep-water ports as Liverpool and more especially Southampton an advantage over London. The strenuous efforts of Hamburg to create a North Sea port on German territory that should compete with the great river port of Rotterdam, and the increasing rivalry of Antwerp with its excellent rail and canal connections and its deep-water accommodation, had already provided London with a stimulus to improve her equipment. But London suffers, as all great ports of early growth must suffer, from the difficulty of increasing harbour accommodation within easy reach of her markets and warehouses and of acquiring land for elaborate systems of railway sidings. It is land-transport, especially the railway, that ensures the prosperity of the modern port; and a glance at the disposition of the London railway termini will show that the Port of London suffers grave disadvantages in this respect. Southampton Docks, owned and served by the Southern Railway, increases its traffic no doubt in part at the expense of London Docks. Yet, in a sense, Southampton, Harwich, Folkestone and Dover may be considered, like Tilbury, as outports of London, for the goods they receive come, for the most part, by rail or road to London and are in a large measure dealt with by the Port Authority in one or other of its multifarious capacities. Modern improvements of the Port of London—the great George V dock, the re-construction and re-equipment of the older docks, the deepening of the channel (to 30 ft. or 9 m. at l.w.s.t. at Woolwich), the construction of a great new passenger landing stage at Tilbury—improvements which have cost over sixteen millions sterling, have brought the Port of London into line with its foreign competitors and given it an equipment well ahead of its requirements*. The docks do not deal with the whole of the goods entering the port. Some 60% of the shipping uses the docks and some 40% the riverside wharves.

Improved policing of the river, modern equipment of wharves and warehouses, have made London River as important a factor in the modern port as the docks. The wharfingers companies conduct an enormous entrepôt traffic, and with their fleets of modern

* The value of the imports and exports of the Port of London, including Queenborough, in 1928 was £692,730,143, nearly 34 per cent. of that of the whole United Kingdom.

steel barges serve as distributors from the docks as well as from ships lying in the river: London, like New York, is essentially a lighter port. And the great market behind the port, including perhaps the largest market of skilled labour the world can offer, ensures it a future worthy of its great past. But the port can no longer cling to the market; the market must move to the port; and it is here that the question of cross-river communication becomes vital.

If, as is predicted, the trend of industry southward continues and becomes more emphatic, a certain decentralisation from the point of view of transport will become inevitable. Congestion of routes in and around London is a serious problem. The concentrating effect on traffic exerted by the Thames obstacle to land transport, an obstacle which has operated from the founding of the city, will be to some extent modified by the construction of more bridges and subways and circumventing routes. We have noted the extension of London eastwards into Essex. The development of the Kent coalfield and a consequent increase in the services offered by the Channel and estuarine ports may lead to a great industrial extension of South London pivoting upon a great railway terminus south of the Thames.

We have traced, perforce in a cursory manner, the main geographical factors that have influenced and do influence movement and settlement within the London Basin. We have noted the close relation existing between soil and drainage and human habitation and occupation, and we have seen that it is possible to subdivide the area into a number of geographical units. Yet as we proceed we are forced to realise the ever-tightening grip of London on the whole area, manifesting itself now in this way, now in that, but never slackening; and gradually welding, not only the London Basin, but the whole of south-eastern England into a geographic and economic whole. In the process the physical features tend to lose their sharpness of outline, change in relative importance, or exert a less direct control or influence, and the student of geography is forced as he works out the interrelations of human and physiographic facts to turn his eyes more and more frequently to the great pulsating organism, whose arteries and veins ramify throughout the countryside and whose intelligence controls the activities of half the world.

IV

CENTRAL SOUTH ENGLAND

O. H. T. Rishbeth

THE area which we have designated Central South England (*v*. Fig. 17) is included within fairly simple general lines. The northern half of a circle described from a centre near Lymington with a radius of 40 miles (64 km.) includes nearly 90 per cent. of the total 2950 sq. miles (7640 km.^{2}). The bulk of the remainder is covered by an isosceles triangle whose base runs approximately along Chesil Beach. The southern boundary follows, with notable deviations, an east-to-west line drawn from Selsey Bill to the head of Lyme Bay, and the inland boundary—about 160 miles (260 km.) in length—includes all of Dorset except the north-west, the south-eastern half of Wiltshire, all of Hampshire except the north-eastern corner, and small portions of Somerset and west Sussex.

The land and adjacent waters thus briefly delimited form a fairly definite physical region, viz. the "Hampshire" drainage basin together with two or three smaller basins farther west and so much of the boundary watersheds as pertains to them. Geographically the boundaries are not so easy to define. In certain directions in particular, considerations of geology, relief and drainage, and human settlement can lead to inconclusive or conflicting results, and for a decision the whole of the geographical circumstances have to be taken into account.

STRUCTURE AND LAND FORMS

For an area of low relief the influence of geological factors is marked. Land forms, soils, drainage, and in particular the characteristic coast-line are directly and obviously related to facts of physical history. Mid-Late Miocene earth movements, acting upon a surface already gently tilted to the south-east, called into being here a series of ridges and furrows. Aligned east-and-west roughly parallel with one another, these advance *en échelon* from the Dorset coast—where the disturbance was most intense—north-eastwards towards and into the London Basin. Three major features result, each running east and west and each comprising several minor undulations: two anticlinal upfolds on the north and south respectively, and between them a synclinal trough. All have a marked asymmetry, the northern limb of the anticlines being much steeper than the southern, so that the folds are almost monoclinal. This is partly why the northern anticline now appears much broader than the southern, and the trough lies somewhat towards the southern part of our area. Westwards these features terminate against another fold system and the synclinal trough is closed in that direction.

The materials in which these structures—to judge by present exposures—were created consist of rocks of Mid-Jurassic to Lower Tertiary (Oligocene) age. The Tertiaries, save for some residual

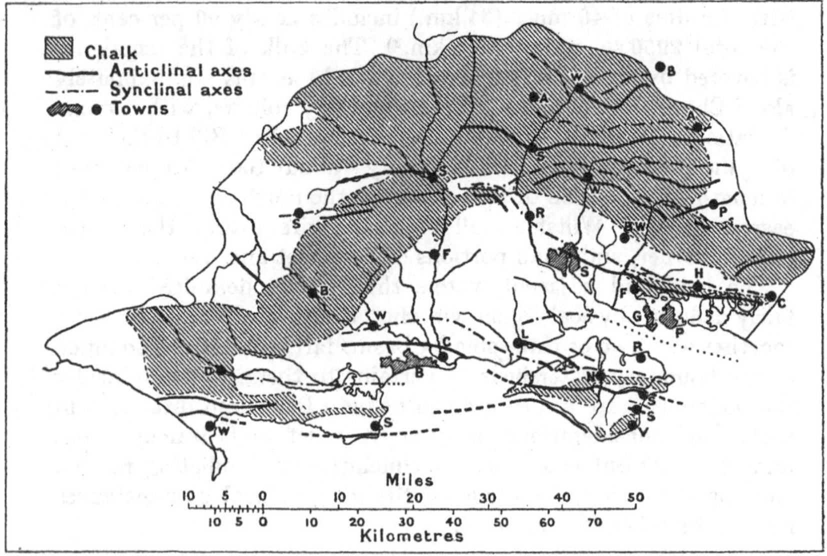

Fig. 17. Central South England: structural diagram, showing certain town sites.

sheets and remnants, have been stripped from the slopes of the synclines they once largely covered, and the Upper Cretaceous chalk now forms the surface of over one-half of the area, principally in the north, but also in the east, west and south. In the synclinal trough the weak Tertiary rocks were substantially preserved and in addition received vast accessions as downwash from the neighbouring slopes.

Neglecting provisionally such later accretions and also the effects of subsequent disintegration there resulted a geological basin, asymmetrical on the south, where it terminates in a sharply folded and nearly straight (east-to-west) edge, but forming a fairly regular semicircle on the north. The regional dip of the strata is towards a centre at or near the Solent. Conversely, from below this locality the beds slant upwards and outwards, cropping out in a series of concentric curving bands on the north, and in roughly straight lines along the south. The outcrops on the north are generally broad, those on the south are for the most part narrow.

This general structure is clearly expressed in the broad features of surface relief. On the chalk, the wide uplands on the north, and narrow ridges along the south, correspond to the gentle and steep anticlinal dip-slopes respectively and form the outer edges of the basin. The hollow or trough included between these uplands corresponds to the structural depression, though the axis of the syncline and that of the morphological hollow do not exactly coincide.

But to understand the existing relief a recognition of two further processes is necessary: the erosion of the anticlines, and the evolution of the coast-line and drainage system.

In our area the differential erosion of anticlines has produced a series of "secondary" vale, terrace and scarp landscapes chiefly around the outer margins. The east-to-west line of southern anticlines has degenerated into a series of parallel ridges and vales running through the Portland Peninsula and the Isles of Purbeck and Wight. The marked asymmetry of the folds has resulted in a line of sharp narrow ridges and scarps along the north, and in broad but bold tabular elevations on the south. In the northern part of our area the chalk uplands are being attacked from east, west and north, and represent, in fact, a remnant of a larger chalk area. The main line of attack is inwards along anticlinal axes—on the east along those of the Weald, in the west and north chiefly along those of Blackmore, Wardour and Pewsey. Here the dismantling of the outer edge of the Hampshire Basin is in process. Sapped by streams and worn down by sub-aerial agents, the chalk scarp is receding and a well-marked "cuesta" landscape is maintained. Along the northern rim of the chalk runs a long "Wealden" anticline, with characteristic periclinal swellings or domes. These domes, breached and in part excavated, form a series of marginal vales which help to bound our area in this direction. These, then, and other similar vales within the chalk area itself (*e.g.* Chilcombe, Bower Chalk, upper Meon valley) are the secondary lowlands referred to.

At the time when a line of overlapping fold-ridges formed a southern bulwark, the land extended farther southwards and the coast-line must have been longitudinal. At some undetermined, but comparatively recent, period this ridge line was breached by the sea. Along the transverse sections of the dissected domes very diverse rock-types were exposed. The result was a highly diversified coast-line along which numerous promontories, "noses," "bills," represent stronger materials, and inlets and bays, often of graceful curvature, normally represent the weaker. The Solent-Spithead channel has been styled a "drowned river valley," but marine erosion has largely determined its present form, and even such enclosed waters as Portsmouth and adjacent inlets are being actively modified by the same means. Some forms along this coast (*e.g.* Lulworth Cove) are classic examples of differential marine erosion, while others offer interesting examples of active and progressive deposition (Chesil Beach, Hurst Castle Spit). The prevailing up-Channel set of the currents, in both wearing away and masking this coast, have produced striking changes even in historical times. In this connection, too, may be mentioned the work of tidal scour and also the "double tides." These phenomena affect considerable stretches of this coast but are of most geographical consequence in Southampton Water.

Like the coast-line, the present drainage system is a fragment of a larger and simpler original. This seems to have originated with the formation of the Hampshire Basin and to have been consequent upon it. The main channel ran south-eastwards down the major syncline (River Frome—Solent—Spithead line) and gathered the drainage of the whole structural basin. Of the short southern affluents the Isle of Wight streams are the chief relics; those on the north are substantially represented by existing streams*. Oscillations of these coastal regions with regard to sea-level played a part in the disruption of this system. In Pleistocene times the land-surface apparently stood some 30 to 80 ft. (9–24 m.) higher than to-day and the valleys were correspondingly deeper and broader. Subsequent (relative) subsidence of the land, beginning perhaps in Neolithic and continuing into recent (historical) times, caused or aided sea encroachment and the breaching of the southern ridge near Bournemouth Bay. Thus the single system has been disrupted into a series of separate systems, the lower valleys throughout the area have been submerged and, with enfeebled outflow, are in process of silting up and becoming closed. The geographical significance of the foregoing events hardly needs emphasis: to them we owe Southampton Water and other harbours and also the difficulties with which certain of them contend†.

* Doubt has been cast upon the existence of the suggested upper Avon—lower Test ancient river line. (*Vide Brit. Asscn. Rpt.* 1925 (Southampton), p. 309.)

† *V.* pp. 81, 82.

CLIMATE

The general type of climate prevailing in southern England is dealt with in Chap. I. It is clear that in an area of less than 3000 sq. miles (7800 km.²), so situated, and with a coast-line of over 200 miles (320 km.), an area in which no point is more than 45 miles (72 km.) from the sea and very few as much as 900 ft. (270 m.) above it, extreme meteorological values will have little geographical significance. Such deviations from the mean as do matter geographically will be chiefly due to local irregularities. As regards temperatures, the mean monthly temperatures of January (39°–40° F.)* and of July (62°–63° F.)* may be taken as fairly representative for the area as a whole, though readings tend to show progressively greater extremes as one proceeds inland in an east-north-easterly direction. The coastal stations show a fairly uniform mean of 50°–51° F.† The prevailing winds have a westerly component. A twenty-five year record from the Isle of Wight shows: NW. 14; W. 21; SW. 18; S. 7; SE. 11; E. 7; NE. 13; N. 9 per cent. respectively. The preponderance of sea-winds will be noticed—and these shores have their share of Atlantic gales—and also the relative immunity, due partly to relief, from northerly and easterly winds. The average annual rainfall varies over the area from 40 ins. (1000 mm.) on the more exposed uplands of the south-west and north, to 27 ins. (675 mm.) or less in low-lying parts. The influence of relief is definitely marked, especially in such rain-shadow areas as Poole Harbour and certain coastal parts farther east and in the Isle of Wight. In view of its geographical consequences an important feature of the distribution of precipitation throughout the year is that, in spite of a steady rise to a maximum in October, some 40 per cent. of the total falls in the six driest months (February to July). Fogs and mist are fairly frequent over the Isle of Wight and adjacent waters (Isle of Wight about 29 days' fog per annum), but they are rarely persistent. The more southerly, and particularly the coastal portion of this area, on the other hand lies in the sunniest zone of the British Isles (1700–1800 hours and over, declining to 1400–1600 hours per annum farther inland).

Local irregularities play an important rôle in an area favoured for residential purposes. The coastal resorts proclaim their relative advantages and the well-known and almost Mediterranean mildness of climate permits southern gardens to display a variety of exotic shrubs and flowers. The higher parts of the chalk uplands (*e.g.* in

* 3°·9–4°·4 C. and 16°·7–17°·2 C. † 10°·0–10°·6 C.

Wiltshire), on the other hand, are apt to be bleak in winter and suffer from occasional heavy snowfalls. Economically more important are local variations due to the varying physical capacities—thermal and other—of rocks and soils in agricultural districts. *Vide* pp. 75, 82, 84 *infra*.

From the foregoing brief sketch of the physical constitution of the area, it may be possible to discern in it some degree of physical order or system. Regarding this order provisionally as a rudimentary and purely physical form of organisation, it remains to observe whether the area has evolved farther in the same direction. We have to examine whether the physical is a basis for biological organisation and whether this in turn leads to regional integration. Perhaps the best way to do this will be first to trace separately the evolution of major component parts, for this area is clearly divisible into such. The dominant factor differentiating these is rock composition, and we may thus divide our area directly and simply into: (1) Chalk Uplands, (2) Tertiary Lowlands, (3) Clay Vales.

The Chalk Uplands

Chalklands occupy some 1520 sq. miles* or about 52 per cent. of the total area. Their general distribution is indicated on the diagram (Fig. 17). From an undulating and irregularly disposed rim they slope downwards and inwards to sink below the Tertiary basin. At the outer margin the higher undulations mount commonly to 750–800 ft. (230–245 m.)—Walbury Hill, north Hampshire, 975 ft. (297 m.)—the intervening dips and hollows lying mostly 200–250 ft. lower. Along the inner margin where the Chalk dips beneath the Tertiaries, elevations of 180–150 ft. (55–45 m.) prevail. In the northern and larger chalk area the slope is not even. Here two "Wealden" anticlines, and others of like direction farther west, cause two main east-to-west belts of upland—a higher and broader in the north (north Hampshire and central Wilts) and a lower double ridge, much broken across, in the south. Between them they include an oblong depression, the upper basins of the Avon, Test and Itchen. The higher chalk uplands lie at an elevation of 450–550 ft. (137–168 m.), the lower at 250–350 ft. (76–107 m.), over wide areas.

The topography developed upon this surface has on the whole emphasised its uniformity. Here more youthful, elsewhere more mature, all forms combine to produce the effect of a "frozen undulating sea." Subtle dissolving curves, a certain witchery of

* *c.* 3940 km.²

lines, is half masked, half revealed by cultivation. Yet abrupt straight-edged slopes—up to an angle of 38° or 40°—are quite as characteristic of the Chalk, especially on northern exposures. Over large areas also, spreading over plateaux and down broad "hogs-back" spurs, sheets of residual clays, gravels, flints, form the surface and provide extensive downwash.

The drainage system was evolved at a time when the chalk area stood higher and was more extensive and when the climate was probably sub-Arctic. The vigorous erosion of streams then cut across these platforms a wide web of confluent drainage lines. The valleys, over-large for their present streams, sink well below, and often appear lost in, the rolling plateau surfaces (Salisbury Plain). Normally dry in their upper courses, these valleys with their broad gravel-lined floors form corridors through the uplands, leading gently up to notches and wind-gaps cut in the outer chalk rim. Most of these gaps serve as gates to and from the area, the most important being those of the rivers Stour (near Blandford), Nadder (Tisbury), Wylye (Warminster) and Test (Basingstoke). Beyond the chalk rim, streams, generated on chalk dip-slopes which have since disappeared, are struggling along a fluctuating margin with those flowing away from our area.

In attributing poverty to the chalk soils discrimination is necessary. The Upper Chalk soils are naturally dry and porous, qualities most in evidence on hill tops and higher slopes. Soils on the Middle Chalk are usually rubbly and intractable. But the superficial "Clay-with-Flints" deposits greatly modify conditions in many upland areas and, with suitable manipulation, large stretches of chalklands, light loams mostly, have been brought under the plough. The marly Lower Chalk, where it is laid bare on lower valley slopes, is almost everywhere noted for its fertility.

The index of productivity is apt to be water-supply. Chalk areas are liable to suffer from drought and hydrological conditions assume unusual importance. In earlier (Neolithic?) times, when a moister climate prevailed here and the water-table stood some 200 ft. (60 m.) higher than now, conditions on the uplands must have been severer but on the whole more propitious. Since then the water-table has perhaps progressively sunk and now often lies 250 ft. (76 m.) or more below the surface in upland parts. Moreover wells sunk into the water-bearing strata may show seasonal variations of 50–60 ft. (15–18 m.) and even of 120–140 ft. (37–43 m.) in the water-level. Hence the variation in spring levels in different seasons and periods and the prevalence of dry watercourses and intermittent streams ("bourne," "winterbourne," "lavant")*. But while in higher parts—except where clay catchments exist—wells, rain-water, road-drainage† and dew-ponds form the rather uncertain supplies of man and beast, the valleys have plentiful springs and, lower down, sparkling clear streams. As a whole the Chalk forms a great reservoir of useful water. Around its southern margin in this area a long line of towns and villages and an increasing population in the whole area is supplied from the dip-slope springs. At Havant, Portsmouth has the largest supply of spring water in the British Isles (average yield per 24 hours: 10–12 million gallons or 45–54·5 million litres), while from its wells at Otterbourne, Timsbury and Twyford, Southampton has available a daily supply of some 11·5–12·5 million gallons or 52·3–56·8 million litres.

An authority has recently stated, on the basis of ecological investigations, that "the vast bulk if not all of it [the English chalk grassland] would pass into woodland if pasturage were withdrawn‡." Such, too, is the inference to be drawn from even relatively casual observation. This fact, if it is such, must be borne in mind in speculating upon the origins of human settlement in these areas, especially as it seems unsafe to assume a post-glacial interval in which vegetation was absent or sparse. The residual clay cappings, once more extensive, favour forest growth. Tree roots fissure the chalk, tap, hold up and distribute evenly available water supplies. Add a moister climate and a higher water-level and it seems probable that these areas were originally forested. Apart, therefore, from the virtual impossibility of settlement in low-lying parts (*v.* p. 83 *infra*), early man probably found upon these uplands certain advantages of a positive kind. Foremost amongst these were perhaps their relative definiteness in respect of size, shape, and direction. Chalk spurs and hill-tops offered limited, and therefore measurable, problems of clearness, organisation and control while the ridges offered relatively obvious, if sinuous, lines for diverging, converging or continuous routes. Add the possibilities of defence suggested by the remarkable straight-edged slopes (*v. supra*)—natural bastions and glacis when cleared of tree growth—and the equally remarkable properties of chalk almost unique amongst

* *Vide* J. Howard Brown, "Bourne Flows" (*Geography*, Vol. xv, ii, No. 84, June 1929).

† The chalkland roads until recently were chiefly made with flints. Imported materials are now generally used for main roads.

‡ A. G. Tansley, "The Vegetation of the Southern English Chalk," in *Festschrift Carl Schröter* (Rascher. Zürich. 1925). (*Vide* also further references included in General Bibliography cited at end of this article.)

rocks, and there was probably sufficient of natural advantage here to attract and to lure on the early forest immigrants into this area. Moreover, if the beech and pine grew here as they do to-day, the forests were freer of undergrowth and also drier, and openings once made closed again less rapidly. To-day the "Clay-with-Flint" areas still support considerable woodlands (oak, hazel, firs, thorns)*, especially in the east. For the rest, conditions of slope, soil, exposure, but also of land tenure and historical tradition, account for the presence of most of the "hangars," hill-crest clumps, hedge-row strips, parklands and game preserves. The somewhat melancholy scrub (mostly yew and juniper) along the windswept ridges is probably degenerating or regenerating forest, and the famous downland grass itself is only maintained by the constant nibbling of rabbits and sheep.

Tenanted perhaps first by hunters and then by hunter-pastoralists, the chalk uplands later became the site of a Romano-British or Celtic civilisation†. Long occupation of restricted spaces seems by then to have removed much of the forest cover, and demarcation of clan, if not of private, property prevailed. Tracks, ditches, village "sites" are the chief surface vestigės of this well-humanised landscape, in which pasture was probably distinguished from arable and both from the still forested lowlands and valleys. The possibilities, too, of chalk had been more fully realised for the erection of fortifications which had advanced far beyond their probable origins in strengthened natural sites. Vast and intricate structures, impressive even to-day, were reared in this soft and workable material which, packed and grass-grown, retains almost indefinitely a very high angle of slope‡. In this and other ways it seems that early civilisation was materially advantaged here by the nature of the rocks and terrain.

The Saxon settlements which succeeded this civilisation kept strictly to the valleys. Adapted to marshy lowland forests§, this intrusive civilisation—whatever the actual historical facts of invasion—seems to have utilised the valley corridors which lead into the heart of the hills. The uplands they perhaps devastated and shunned, and never since have these uplands been much used for

* The ash is more common in the Isle of Wight and in the western part of the area.

† For this and some of the following points *vide* O. G. S. Crawford, *Air Survey and Archaeology*, 2nd edn. 1928.

‡ *Cf.* Maiden Castle (Mai Dun) near Dorchester and compare with the angle of its piled slopes the angle of road and railway cuttings in the chalk (*e.g.* at Butser Gap).

§ *Vide* R. G. Collingwood in *Antiquity*, September 1929, p. 265.

human habitation. To the Saxons, however, the chalklands appear ultimately to owe many characteristic and persistent features of organisation. The long rows of villages strung like beads up the branching stream lines; the double-ways running parallel up either side of the lower valley floors and coalescing higher up where the stream fails and the valley is dry; the characteristic herring-bone pattern of chalk ridge-road with side-tracks mounting the "hollow-ways" and spurs—these, and the sweeping lynchet-lines mantling lower valley slopes with graceful curves of terrace and balk, are signs still decipherable of the intimate partnership of these people with their natural surroundings. In the valleys they had water, wood and meadows; on the slopes soil, shelter and moisture; on the hill-tops the pasturage of the downs. Characteristic parish types, particularly that hill-and-valley strip form which includes water-meadow, arable and downland, were based upon this early type of landscape integration.

Since Saxon days the chalklands have passed through many vicissitudes. Prime wool-producers in the days when the woollen industry flourished in the South and when Winchester and Southampton had wool-staples and exported wool, they were progressively enclosed and put to the plough. Once—in some areas twice—in the 19th century much land was sown to grain when high prices for this prevailed. But the chalklands are not the best natural cornlands and this state could not endure. Now they are very largely enclosed or cultivated without enclosure. Very little "downland" is left and those who expect "wide open spaces" are liable to extensive disillusionment. Distinction must, however, be drawn between the eastern and northern (Hants) portions and the north-western and western (Wilts and Dorset) lands where more open land still exists. In five chalk areas selected as representative of the whole, 510 sq. miles (1320 km.^{2}) out of a total 710 sq. miles (1839 km.^{2}) were already enclosed in 1925, and most of the remainder is utilised.

Farming is extensive: farms of 500–1000 acres (200–400 ha.) and single fields of 100 acres (40 ha.) are common. Smaller farms prove unprofitable and there is a tendency to amalgamate. The density of the human population, though higher than sometimes represented, is low, 18·3 persons per 100 acres being the average for the districts referred to above. The population, moreover, is declining in places: the chalk areas of Dorset and western Wiltshire showed a progressive decline between 1901 and 1921 and NE. Hants appears to be going in the same direction* (Fig. 18). But there are some

* The use of certain areas of the chalklands as military training-grounds has an important, though fluctuating, influence on population.

notable exceptions and these will be referred to later. In some cases (NE. Hants) there is evidence that this decline is due to unfavourable conditions and failure to compete. The farmers are poor and can keep few hands. In other cases a decline may signify an improvement or a change in methods of farming—more extensive farming with machinery or a change to a more pastoral type. Perhaps it is all these things, together or singly, which still keep chalk-landscape unique. In spite of fields, plough and hedgerow, they retain an indefinable freedom. NE. Hants has an air of remoteness. Elsewhere the plough furrows hillside and hollow in silence and the last downlands seem lost in solitude.

Through the premature specialisations mentioned above the chalklands seem to have arrived at a more stable and balanced mixed farming, based upon their three natural major products: sheep, cattle, grain. The balance between these varies, but there is a general tendency now towards milk and meat production. This tendency is a response to a demand coming mainly from outside. It represents an attempt, as does the regime as a whole, to adjust the internal organisation of production in the area to the external demands made upon it.

Speaking broadly, the chalk has three types of terrain—valley bottom, hill-slope (or rather medium elevations), downland. The valley floor produces meadow grass or hay almost perennially; the downlands afford summer and winter (daytime) grazing; the intervening arable yields fodders and some grain. Sheep are the integrating factor. They are fed, according to time and season, on all three sections and so improve the land. Thus they are kept for their intrinsic value and also as a vital link in a somewhat intricate and variable rotation system. For this reason, though the centre of gravity in production may shift from one product to another, the sheep have remained, and probably will always remain.

Once or twice, as remarked above, grain production came to the fore. In the period between 1913 and 1925 direct (human) food crops (chiefly wheat and potatoes) declined from 10 per cent. to 8·6 per cent. of the total area under cultivation in the five districts referred to above. This decline of the already small proportion under these crops is characteristic of recent tendencies.

In the same area and period, sheep have also declined by over 30 per cent. (*i.e.* from a population of about 75 to one of 50 head per 100 acres or 40 ha.). The explanation lies in the increase of dairy and other cattle. The population density of the former (7·24 per 100 acres in 1925) has increased by 42 per cent. since 1913; that of grazing cattle by 46 per cent. (4·9 per 100 acres in 1925) in the same period. Causally connected with this change has been the partial replacement of the specialised local breed of downland sheep ("Hampshire Downs," "Dorsets" in Dorsetshire and Isle of Wight) by various imported types of grassland sheep. The result is a disturbance of the balance above noted; the fields are less well fertilised and there is again a move back to the "Hampshire Down" sheep. So geographical factors, implicit in every detail of the situation, are here slowly but effectively reasserting themselves. In other parts (N. Hants, Wilts, and Dorset) the flocks, 1000 head and more, graze over the downs under the care of shepherds who keep the sheep from the unfenced fields. Here shepherds and their flocks have still a flavour of the primitive, and sheep fairs, of chalkland fame, are held, some in historic towns and some in the still more ancient forts and camps whither gather all the trackways of the hills.

The developments here outlined have geographical significance so far as they indicate the evolution of the area. Within the somewhat rigid limits of their natural possibilities the chalklands, by single productions ("monocultures"), later by permutations and combinations of several, have sought to discover and to meet the demands made upon them. Wool, grain, milk, meat*, these have not been consumed locally but are produced for larger and more distant markets. In meeting those demands they have come to share in a larger life or "scheme of things" outside them, and in doing this they have survived and progressed. We must ultimately ask what "scheme of things" they are now existing to serve, for in this lies their ultimate regional meaning.

Something has been said of the distribution of population in the chalk areas. We need not enter here into a detailed description of that distribution nor into a description of individual towns. Our purpose will be served if we note the presence of the twin chalkland "capitals"—Winchester to the east, Salisbury to the west of the long northern depression, each at a gathering place of valleys and routes leading to the chalk rim, and, more especially, southwards. Andover and Whitchurch occupy similar but subsidiary positions farther north. Further, the long row of gate-towns around the outer rim—Maiden Newton, Blandford, Wilton, Warminster, Basingstoke and Alton—and the equally striking series of "transition" towns—Dorchester, Wimborne, Fordingbridge, Romsey, Bishop's Waltham and Chichester—which mark the junction with the Tertiary area to the south (*v.* p. 83 *infra*).

* Two minor species of production—poultry farming and water-cress growing—have peculiar geographical interest in connection with chalkland development.

The Tertiary Lowlands

In the trough which extends from Dorchester to Chichester and from Romsey to Newport (Isle of Wight) the original flooring of marine strata has been preserved. It has also received upon its surface, as the product of long-continued denudation, downwash from the neighbouring heights. These water-sorted remnants now cover large areas of the lowlands and lend them some of their most distinctive geographical characteristics. As broad gravel sheets, pervious, resistant, water-distributing, they mantle and partly occasion plateaus—notably that of the northern New Forest. As wide sloping belts or as terraces they accompany and mark the ways of ancient rivers. The outer margins of the Test and Avon valleys are spread with them, as are the broad seaward slopes from Poole to Portsmouth, once the valley flanks of the ancient Solent river. Along the floors of existing valleys they stretch as wide terraces and flats of gravels and loams. Topographically their function is twofold. Acting as caps to hills and ridges they accentuate and sustain inequalities and account for sharper lines in the landscape*. Elsewhere their effect is masking and muffling, leading, as in part of the New Forest, to uniformity and sense of space. Where they are thick and clean-washed, the result is barrenness, as of heaths. Where they are thin, lowlying, or loam-filled, they result in fertile plains.

The older rock materials of these lowlands are mainly estuarine sands and clays of very diversified composition. Their structural disposition as noted above has had little effect upon topography. Generally the clays result in wet lowlands, mostly in belts, except in the New Forest. Where, as very often, sands and clay are interbedded, an irregular hummocky hill-land has been evolved, the hills having often sandy upper layers surmounted by caps of gravel. Sudden declivities often mark geological junctions, and minor flexures in the underlying chalk account for some striking ridges, notably Dean Hill and the Portsdown anticline.

The coast-line, a succession of crumbling cliff and shingle sweeps, reflects faithfully its physical genesis. The subsidence and choking up of the lower valley stretches (p. 72) has resulted in abnormally wide floors and alluvial flood plains, in braided streams, and often also in land-locked lagoons and sand-barred mouths. Tidal mudflats spread wide in Poole and the eastern harbours, though

* *Vide* Heywood Sumner, "Natural Landmarks in the Bournemouth and New Forest Districts," *Proc. Bournemouth Nat. Sci. Soc.* 1929.

Southampton Water is kept clear by tidal scour and by dredging*.

In general the forms in this area, owing to the weakness of the rocks and to recent subsidence, show an early maturity of form. That the synclinal trough as a whole, in spite of the material accessions received (*v. supra*), has retained its form as a depression is due partly to these same causes, to the evacuation of much material into the sea, and to the greater relative resistancy of the chalk.

Water is not lacking in these lowlands: the clay floors, especially when forested, harbour marshes, and on peaty moors are bogs. From the sands and gravels of the hills numerous springs, oozing out at sedgy soaks, feed the gravel-floored streamlets. But few of these Tertiary sources are reliable and, though wells and springs serve for local use, the larger supplies derive from the chalk and systems of rural pipe-supply are extending (p. 76).

The soils reflect the sub-surface variety—though the relation is widely obscured by drift and rainwash—and they are characterised by extraordinary diversity in detail. The great bulk are loams of varying clay, sand and gravel content, and nearly all stand in need of lime. For purposes of ordinary mixed farming they are on the whole rather poor. The river alluvium and valley gravels (*e.g.* of the Avon) are fertile, and there exist some useful wheatlands. Apart from this, progress has been most apparent where special soil capacities have been discovered and turned to account.

The primitive vegetation covering was mainly dense tangled forest, wet where it stood on clays, and composed mainly of oak, ash, hazel and associated forms. On drier sandy slopes beech, pine, birch and bracken replace the former and the division lines are often remarkably distinct. The pine, which grew here in pre-glacial times, later disappeared. Reintroduced in the 18th century, it is now invading heath and hill and vigorously claiming supremacy in many former oak-clad localities. There are those who view with apprehension the encouragement of pines in the New Forest†. The rolling heathlands have probably never been much other than they

* For coastal, and also tidal, phenomena in the area *vide* "Bibliography of Hampshire Basin" (Sections III and VIII), *Geog. Teacher*, vol. XIII, 1926, p. 489; also, J. A. Rupert-Jones, *Tidal Research* together with a paper (in collaboration with G. H. J. Daysh) on *The Physical Characteristics of Southampton Water*, 1929 (privately printed).

† *Vide Geog. Jour.* vol. LXIII, i, 1924, p. 79. Considerable correspondence upon the matter has also appeared in recent years in the daily press (*e.g. The Times*). Cf. also Heywood Sumner, "New Forest and Old Woods," Pres. Address, *Bournemouth Nat. Sci. Soc.* 1926.

now are; surprisingly clear correlations between soils and vegetation can be traced in and round them and vegetation "contour lines" followed on the sides of their marsh-floored valleys. For long, settlement and cultivation made little impression on the forests*. Even now there are farms not four miles from Southampton that have the air of clearings in a newly settled country.

Primitive man probably chose the heathlands for settlement. There were also certain early trading stations along this coast—notably at Christchurch Harbour; but these and the valley and gravel ridgeways leading northwards from them were probably used for the chalklands behind†. As settlement gained ground a characteristic landscape, very different from that of the chalklands, was evolved. The heaths are mostly still commons and waste. On heavy clays woods still cover wide areas. There has always been, and still is, something sporadic and haphazard in the settled landscape. The cause lies in the patchy and variable physical background, in its lack of clear lines and of system, in the jumble of hill, marsh, and forest. Parish boundaries reflect the difficulties of division; roads and tracks seem to wander irrationally; main routes till recently led mostly *through* the area. Few settlements of any size lie inland apart from the line (p. 80 *supra*) along the edge of the chalk.

The striking change that has supervened is historically a thing of yesterday. It proceeded mainly from the geographical evolution of the coastlands. The first impetus seems to have come with the rise of seaside resorts, itself a phenomenon of wider import. Bournemouth (population in 1921, 91,760) was in 1856 a mere hamlet. In the 18th century Southampton was a watering-place. The presence of rapidly expanding towns stimulated rural activity and production. The immediate hinterlands sought to produce such necessities and luxuries as they could. This again reacted in other directions and strengthened tendencies already existing. The area—the New Forest in particular—was of old a favourite residence and hunting ground for the wealthy‡. Establishments

* It is of interest to note that shipbuilding, particularly the building of naval vessels, flourished in the estuaries and harbours of the Hampshire coast around and near Southampton from the end of the 17th to the early 19th century—apart, that is, from earlier activities. In its earlier phases, at any rate, this industry drew from the New Forest and other woods its supplies of oak, and interesting maps exist showing Admiralty reservations in the New Forest for this purpose.

† *Vide* "Bibliography of Hampshire Basin" (Section VI). Add: Heywood Sumner, "Geography and Prehistoric Earthworks in the New Forest District." *Geog. Jour.* vol. LXVII, 1926, pp. 244 f.

‡ The influence in more recent times of the Cowes Regatta in this direction should not be overlooked.

such as Portsmouth and the military camps inland added the retired service officer class to these. Later came a wider class of settler—retired from business or professions—and an army of "pensioners" of smaller means. All these sought quiet and pleasant surroundings, the nearer countryside with access to towns. Many had funds at their disposal and preferred to combine pleasure and production. Thus arose a type of resident-producer which has played no small part in general development. Last of all has come tourist traffic, providing fresh employment and business.

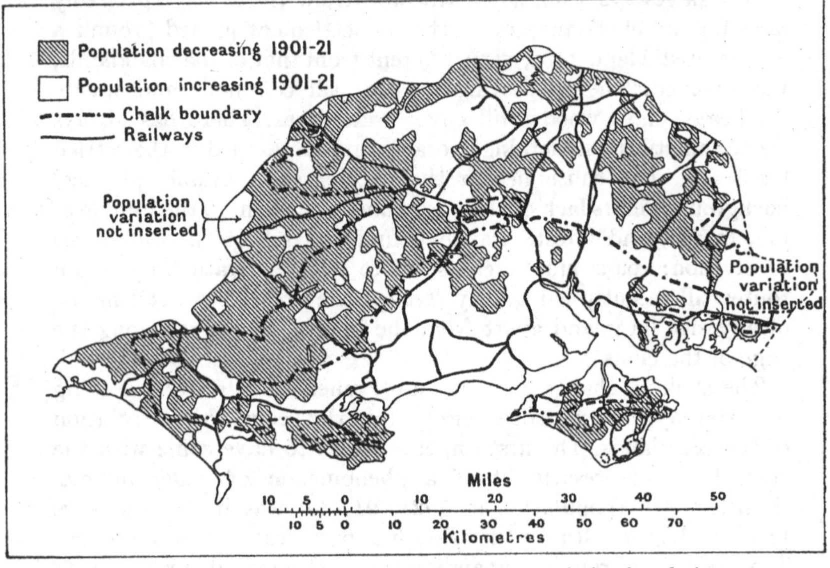

Fig. 18. Central South England: population variation in relation to geological features and main railway routes.

In this residential occupation geographical factors have manifestly operated. The very sites which are useless for agriculture are those which are chosen for residence. The gravel scarps, pine-clad hills, steep slopes; wide aspects from plateau rims and ridges; dry sands, gravel roads, plentiful water; a mild climate and especially the motor-car—all have played their part. It is not surprising that the Tertiary areas show a steady rise in population in recent years. Before very long a continuous belt of settlement will probably fringe the greater part of this coast (v. p. 87 *infra*).

So long as these lowlands were "backwoods" of London, poverty of soil and an easy climate offered little stimulus to agricultural development. Mixed farming here could not compete with that of better areas and there was a certain half-heartedness in the pursuit. But the same influences which have affected recent chalk-land evolution operated here also in their degree. The Tertiary area as a whole, including heaths, probably produces less, area for area, than the chalk. The proportion of land under cultivation is lower and the population density of live-stock, particularly of sheep, is less. There is the same marked rise in dairying, and particularly in milk production for the town populations*. Here the well-placed water-meadows of the Test and Itchen valleys have an advantage, and the lower Avon valley has also rich mixed agriculture.

The distinguishing feature, however, is the development of fruit growing and market gardening. Orchards, small fruit, vegetable and flower gardens have invaded wide areas and reclaimed much that was waste. The heaths frequently contain patches of loam. Here light and warm soils give early spring crops which motor transport carries quickly to markets. Thus along many of the roads former wildernesses are being covered with the bungalow of the small-holder. Specialisation in selection of soils and sites, aided by proximity of markets, is at work. Characteristic of this development is the Fareham district where the small-holders have developed the highly specialised strawberry-growing industry, the products of which are despatched by specially arranged transport to markets throughout the country.

Thus in the Tertiaries we have an area of recent colonisation and of rapid and somewhat surprising growth.

The Clay Vales

These are the valleys which lie scattered as separate fragments around the rim of our area, from the Isle of Wight, Purbeck and Portland in the south to the western end of the Weald in the east. The upland parts of these (*e.g.* the Purbeck and Portland heights) have rather the general character of the chalk. The lower portions, though largely developed in clays, contain lighter loamy soils as well. The term "Clay Vales" is therefore conventional. Mainly agricultural, comparatively rich and devoted to dairying, grain, and in the north-west to fruit growing, they are marginal both physically and economically to our area. Their chief interest to us lies in the reason for their inclusion within it.

* Milk has, in a sense, replaced hay as a "ready-money" crop since the advent of the motor-car. The dairyman often delivers his own milk in the towns, and from Southampton he can carry back oil-cake direct from the quay-side.

The Area as a Geographical Region

Physically, as we have seen, there is both unity and diversity within the area. The geological structure, relief and drainage system afford a basis for further integration. The geological evolution of these divisions, though convergent in some directions, has diverged in other and important ways. Is there, then, a true region here or are there only certain geographical units? To answer this we must glance at both the external and the internal influences that are at work.

The external influences come from two sides—from landwards and seawards. Taking the latter first: the area adjoins the English Channel and at a place where, opposite the Cotentin peninsula, it narrows and where the routes passing along it converge. It is thus a coastal area of this channel and also, in a wider sense, of north-western Europe. From the other, or land, side this coast-line is quite as much a possession of England as it is of the particular area under discussion. It is not Hampshire or Dorset alone, or even mainly, which populate the seaside towns and use them, nor have such towns as Portsmouth and Southampton a merely local, but rather a national significance.

Moreover, the shape of the area, with its extended coast-line, seems to operate against unity. The main lines of movement seem to run through, rather than within, the area. Each of the larger coastal towns has its route northwards, and main lines link these and the larger inland towns to London. Between themselves these towns have little basis for common cause: the coastal resorts compete with each other; Winchester and Salisbury stand side by side; Portsmouth and Weymouth as naval stations stand aloof. The road-system of the Romans seemed to treat this area as a side issue*, and at first sight the modern railway system appears to do the same.

The first factor making for regional integration lies in the development of the coastal areas. The coastal resorts, whatever their origin, have "taken root," as we saw, in their hinterlands. Each demanded, as it grew, products of necessity and luxury from a widening circle of country behind it. This process is continuing and the circles—or rather semicircles—of such tributary areas have now largely coalesced. This belt—for so it now is—tends to develop lateral communication within itself for purposes of greater mobility and freedom in matters of demand and supply†.

* *Vide* Map of Roman Britain (Ordnance Survey); 2nd edition (1928).

† The Bodleian map of Great Britain (*c.* 1300 A.D.; author unknown) shows, in addition to the great south-west route from London, via Winchester and

Again, immigration once started gathers momentum, and residential expansion is pushing farther inland. Portsmouth is colonising its hinterland; at Alton residents from the London area meet those from the Hampshire area, and many other examples might be given. The Tertiary areas (*v.* p. 85) have everywhere shown a continuous increase in population—the rate is in excess of that of England and Wales as a whole—and a wedge of increasing population is pushing northwards up through the centre of Hampshire.

Southampton

But in estimating the nature and extent of regional integration in this area we must examine the threefold function of Southampton in developing contacts both outwards (overseas), inwards (inland) and within the area.

Situated at a stream confluence, and approached by a sea-furrow which opens a double door SW. and SE. on a great maritime highway, Southampton is not far from the centre of the area as a whole and is almost the exact centre of the most habitable portion of it. Bearing in mind also the close economic nexus between north-western France and southern, central and northern England (*v. infra*), and also the peculiar "apex" position occupied by Southampton in relation to E.-W. communications along the south coast, this centrality is a factor of more than local significance and is in fact basic to the port's modern growth. Favoured by sheltered position, by deep and largely silt-free waters* and by double tides, the largest vessels can here approach open quay-sides and come and go almost at will. The natural channel, already roomy, is being continually improved and widened† while the triangular dock-area —in origin a south-pointing confluence spit—gives a maximum of water-frontage with minimum transhipment distances for persons and goods.

Behind this commercial "spear-head" the town, mounting gentle slopes, fills and overflows the larger triangle which lies between the Test and the Itchen. This triangle, however, is but one sector of the

Salisbury, to Devon and Cornwall, a south-eastern coastal through road from Canterbury to Southampton. It is interesting to witness the revival at the present time of the importance of this route or of certain large parts of it. (*Vide* last footnote on p. 92 *infra*. *Vide* also A. H. Schofield, *The West Sussex Coast and Downs* (Arundel, 1929), especially cap. xi (pp. 126 ff.).)

* *Vide* Rupert-Jones and Daysh, *op. cit.*, on p. 82, footnote.

† The ultimate aim is an uninterrupted fairway to the deep outer waters of 1000 ft. width and 35 ft.—finally, perhaps, 40 ft.—depth l.w.o.s.t. The large-scale dredging scheme recently approved represents a substantial step towards this end.

probable city of the future which will occupy a circle around its "trivium" of rectified and improved waterways. Into the heart of this city these waterways will admit the sea with its traffic and along the banks will grow and extend the still nascent industrial areas. Provided thus internally with natural radial lines, Southampton has also radiating from it outwards a system of natural routes leading by easy gradients in all directions of commercial importance.

It was the advent of the ocean liner with its requirements in the way of depth, speed and regularity of sailing which established the supremacy of Southampton as a passenger port. Its situation, neither too far seaward nor landward of the boundary-line between north-west Europe and the Atlantic, was another potent factor in the result. On the basis of these natural advantages has been built up an organisation—represented inwards by the Southern Railway, outwards by over thirty steamship companies—adequate for their commercial development. With 34 per cent. of the total United Kingdom ocean passenger traffic and a very large mail and trooping service, Southampton stands easily at the head of the passenger ports of the United Kingdom. Its connections are world-wide, 44 per cent. of its traffic is with other parts of the British Empire, and during the summer there is an average of one large vessel sailing every day to New York*.

The rise of Southampton as a cargo port has also been notable and involves considerations of some geographical interest. Though not comparable in this respect with London and Liverpool, Southampton in 1928 ranked third amongst United Kingdom ports in respect of value of cargo trade, while over the 8-year period 1921–8 it ranked fourth†. Its trade, moreover, during this same period has been much more stable than that of its competitors and exhibits a steady general upward tendency. The explanation is to be sought in the special nature of Southampton's trade, but behind this lie the geographical facts, notably those of position, indicated above.

* A corollary of this growth is the docks extension scheme which will ultimately add 15,500 ft. of quays to the existing 21,200 ft., involve the reclamation of about 2 miles of river-frontage (407 acres of mudlands), besides the extension and perfection of railway, storage and other accommodation at an estimated cost of some £13,000,000. The first instalment, providing for 3500 ft. of quay at a cost of £3,000,000, is expected to reach completion in 1931–2.

† This includes values of imports, exports (including re-exports of imported goods), and of cargo transhipped under bond, but it does not include the value of coastal trade. The ports with which Southampton has been compared are: London, Liverpool, Hull, Manchester, Glasgow. In respect of the total net coasting trade tonnage which used these ports in 1928, Southampton stands fourth (3·066 million tons) and close to Glasgow (3·330 million tons), which ranks third.

Southampton's cargo trade falls into three main categories: that connected with its regular shipping services—*i.e.* liner trade; secondly, seasonal traffic or trade in seasonal commodities; and thirdly, transit or transhipment trade. As a cargo carrier the liner has quite special characteristics. With speed, safety and regularity of sailing is coupled a rather more limited and specialised accommodation. Small "parcels" of freight—especially freight valuable in proportion to bulk (bullion, gems, ostrich feathers, etc.)—and also perishable goods found the liner adapted, or capable of adaptation (*e.g.* by the provision of refrigerating appliances, cooling chambers, etc.). To this new type of maritime trade the railway organisation, adapted to cope swiftly, smoothly and flexibly with miscellaneous and particularised cargoes, was the landward counterpart. Upon this basis, and using this machinery, there has been built up a considerable "secondary" liner-trade, more bulky and staple in kind —wool, grain, hides, etc.—while this, in its turn, is tending to attract other more truly cargo-boat trade*.

In the second place, Southampton's favourable position relative to NW. France, the Channel Islands—and even Spain and the Canaries—on the one side and the capacious markets of its hinterland on the other (*v.* p. 91 *infra*) has given an enormous stimulus to a seasonal trade (*e.g.* in fruit, flowers, vegetables, etc.). In 1927, some 35 per cent. of the total imports at the Docks came from France and the Channel Islands, and the full significance of this is apparent when it is realised that this trade has been retained, if not acquired, mainly in virtue of the special technical evolution indicated†.

Geographical position, again, has offered opportunities for the third characteristic trade of the port. Situation relative to NW. European ports and manufacturing centres, in conjunction with regular and fast shipping services, afford exceptional advantages for transit traffic. Goods from the Continent or the Channel Islands are delivered at Southampton to catch the outgoing liner and are often, in a surprisingly short space of time, "en route" thence to far parts of the world‡. Similarly continental liners call at Southampton

* In 1926 the average gross tonnage per vessel using the Southampton Docks was 4545; in 1927, 4508. This decrease was due to the increasing number of small cargo vessels.

† South African trade—in 1927 nearly 30 per cent. of the total (export and import) trade—shares to some extent in this seasonal character as does also the "North Pacific" trade in fruit, salmon, etc., opened up by the Panama Canal. But the trade of these countries is more diversified, and also more of a bulk nature, including wool, hides (S. Africa) and timber (North Pacific), etc. To this is now being added trade from New Zealand (meat, butter, etc.) and Australia.

‡ During the 8 years 1921–8 the value of Southampton's transit trade has

for cargoes from the Midlands and North while certain types of ocean freight—mainly perishable goods (meat, etc.)—destined for London markets, are habitually landed at Southampton, thus avoiding the circuitous route to Tilbury and effecting a saving of some 12 hours in delivery.

Thus it is that though much of the cargo trade is seasonal, it is also highly diversified. Whereas the northern ports suffer large-scale fluctuations, Southampton's trade, based on liner-traffic, preserves a more equable mien. Casual tramp trade is at a minimum, imports are largely those "necessary-luxury" commodities with which a modern community does not readily dispense*.

The influence of these developments upon the area can be summarised under several heads. In the first place, Southampton's rise as a port is giving birth to industries. So long as cheap power was synonymous with cheap coal, Southampton was relatively unfavourably placed. For many years, however, she has been developing her electric supply and her Undertaking has recently been named as one of the six selected generating stations under the Central Electricity Board's scheme for SW. England and S. Wales, while a further station will probably be sited on or near Southampton Water by about 1938. Add to this the now established use of oil for fuel and power† and it is evident that the position as regards power-supply is radically changed. Ship-building and repairing, engineering and air-craft manufacture are firmly rooted and flourishing‡. Electric cable, heavy electrical, tobacco, saw-milling, seed-crushing, margarine and many other industries are established and will probably increase. Many of these are small but in the aggregate they are attaining importance§.

In the second place, Southampton, the port, has to serve her own growth and that of the neighbouring areas. Her own population is expanding (1930: *c.* 175,000 estimated) as is that of her nearer hinterland (*v. infra*). Centrality, the possession of easy routes and

mounted to well above that of any other British port (1927, £33,426,000), and for the 5 years 1924–8 it was 42 per cent. of the average annual value of the total transhipment trade of the United Kingdom.

* Approximately 80 per cent. of the total entering Southampton docks is liner tonnage. Casual tramp trade is distinguished from regular cargo trade (*e.g.* in oil, timber, etc.), which is often specialised and carried in special vessels.

† In 1927 Southampton stood sixth amongst United Kingdom oil-importing ports with petroleum imports of *c.* 142,390,000 gallons valued at £2,634,000.

‡ The yacht industry alone, it is estimated, brings to the port an annual revenue of £1,000,000. In its floating dock, the property of the Southern Railway, the port possesses the largest of its kind in the world, with a length of 960 ft. and an accommodation capacity of 60,000 tons.

§ The Southampton Corporation Electricity Undertaking supplies (1930) over 450 factories from its mains.

a transport mechanism of growing scope and flexibility enable her to keep pace with her opportunities. She is steadily becoming more of a regional centre, a factor integrating a region. This is well illustrated in the growth of Southampton as a distributing centre. Within the last few years the town has been selected by several important firms for this specific purpose. From Southampton trade routes lead E., W., N., and also S.—for the Isle of Wight, despite the not inconsiderable trade of Cowes, is included in her commercial sphere. The same tendency reveals itself in the progress of the public utility services of the town: with regard to both electricity and water-supply in particular there has been a marked expansion and, at the same time, centralisation of distributive control*.

The hinterland falls into two clear divisions, the remoter and the more immediate. The remoter consists of the Midlands and of towns as far north as Glasgow, but by far the most important area is that of London. Of London, Southampton may be said, in this and in certain other ways, to be an "outport": trade with it, and with the "remoter" hinterland in general, is mainly specialised, *i.e.* trade in a few special commodities—meat, wool, timber, etc.—part of which is road-borne†. The nearer hinterland, on the other hand, lies mostly within road and rail distances of 40 miles, and is an area of general supply, though special "docks" trade is not excluded. Significantly, this area extends farther west and north-west than it does in the direction of London. Significantly, too, it is an area of considerable value from the point of view of commercial distribution. For though it is not uniformly or densely populated, it is developing and has in any case, because of its relative remoteness from other commercial centres, a high proportionate absorptive capacity. Hence its value to Southampton, its natural supplier; hence also Southampton's function with regard to it.

In these ways, therefore—in virtue of her own growth and the growth of her commerce and trade—Southampton stimulates production, and also consumption, within the area. In virtue of her system of transport, developed primarily for other ends, she is

* The Southampton Corporation Electricity Undertaking supplied in 1920 *c.* 15·7 sq. miles; in 1930, 256 sq. miles (population *c.* 230,000). The Southampton Corporation Waterworks supplied in 1907 *c.* 4 million gallons (average) daily; in 1927, 7·7 million gallons; in 1930, *c.* 8·95 million gallons to a population of nearly 275,000, through over 500 miles of pipe-line.

† Approximately 50 per cent. of the cargo landed at Southampton docks goes to London, a fluctuating quantity—sometimes as much as 30 per cent.—goes north (Glasgow, Manchester, Liverpool). Of wool imports, 50 per cent. is railed to Yorkshire (Bradford), the remaining 50 per cent., besides hides, skins, etc., are transhipped to continental ports. Cargoes destined for local consumption and distribution (Docks or Town Quay) are not included in the above.

establishing a position as its collecting and distributing centre. As such again she aids the general growth of population and residence, and is establishing a definite type of intercommunication*. These facts, and the facts regarding the area to which they apply, rest upon independent evidence. When, however, we examine the area affected, we find that it coincides closely with our region. Curtailed on the north-east where London reaches forward, this sphere of influence is expanding in the direction of south Wiltshire and Dorset. In the marginal areas—the "Clay Vales"—the process is in progress, but there are signs that the boundaries are expanding in those directions. To the coastal parts, therefore, and to Southampton as centre, virtually the whole of the area looks. From these directions and from Southampton as centre, lines of influence are developing outwards.

We may thus say that there is here a basis for regional integration; that this integration is progressive—in other words, that there is evolving here a definite geographical region.

Bibliographical Note: A bibliography of the Hampshire Basin area is published in *The Geographical Teacher*, vol. XIII, Pt. 6, No. 76, Autumn, 1926. References to some of the more recent literature are given above; to these may be added:

G. H. J. Daysh, *Southampton*, 1928. *Id.* in *Scot. Geog. Mag.* vol. XLV, July 1929.

H. Parsons: "Southampton" in *The Chamber of Commerce Journal.* Inter-Imperial Trade Number, Sept. 27, 1929.

Acknowledgment is also made of certain unpublished materials kindly supplied by individuals and of information and assistance given by representatives of various public Authorities and by the Docks and Marine Manager, Southampton Docks.

* There is some evidence for supposing that east-west communications along the south coast, both by rail and by road, will undergo development, a circumstance which, as indicated above cannot but greatly strengthen the commercial position of Southampton.

V

THE SOUTH-WEST

W. Stanley Lewis

REMOTENESS from the great centres of activity has preserved in the people of the South-West of England a strong sense of individuality, which is partly expressed by a highly developed spirit of regional consciousness. The origin of this must be sought in the far distant past, for it took shape in prehistoric time, and growing against a background of wild seas and wind-swept moors, acquired a strength and harshness that later found outlets in the fields of colonisation and discovery and in the less creditable pursuits of piracy and wrecking. Peninsular location and topography combined to foster an aloofness from the rest of the country, but the ease of making contacts by sea prevented the stagnation that might otherwise have resulted. The protected harbours of the gentler southern shore invited intercourse; and the possession of coveted metals encouraged the entry of Mediterranean influences that widened the horizon of the people at a formative period of their existence. This was but the first of a series of stimulating infusions from beyond the sea. With the growth of industrialism the geographical advantages declined in importance. Consequently the rate of development was retarded, and the region remained content to nurse the memories of past glories, and somewhat contemptuous of the advances in which it was unable actively to participate.

STRUCTURE AND LAND FORM

The setting of this old culture is appropriately one of ancient rock formations, in some cases so scarred and altered in the passage of time that it is difficult to determine the geological system or even period to which they belong. Weather-beaten fragments of a Pre-Cambrian continent are exposed in the plutonic complex of the Lizard district and in the low schistose plateau between Bolt Head and Start Point, but the main body of the peninsula is of Primary age. Devonian shales and limestones and Carboniferous grits, shales and limestones (Culm Measures) predominate amongst the stratified formations, and their resistant nature has helped to maintain the rugged relief that is characteristic of the region, but the framework of the modern scenery was constructed in Carbo-Permian time by the Armorican upheaval. The

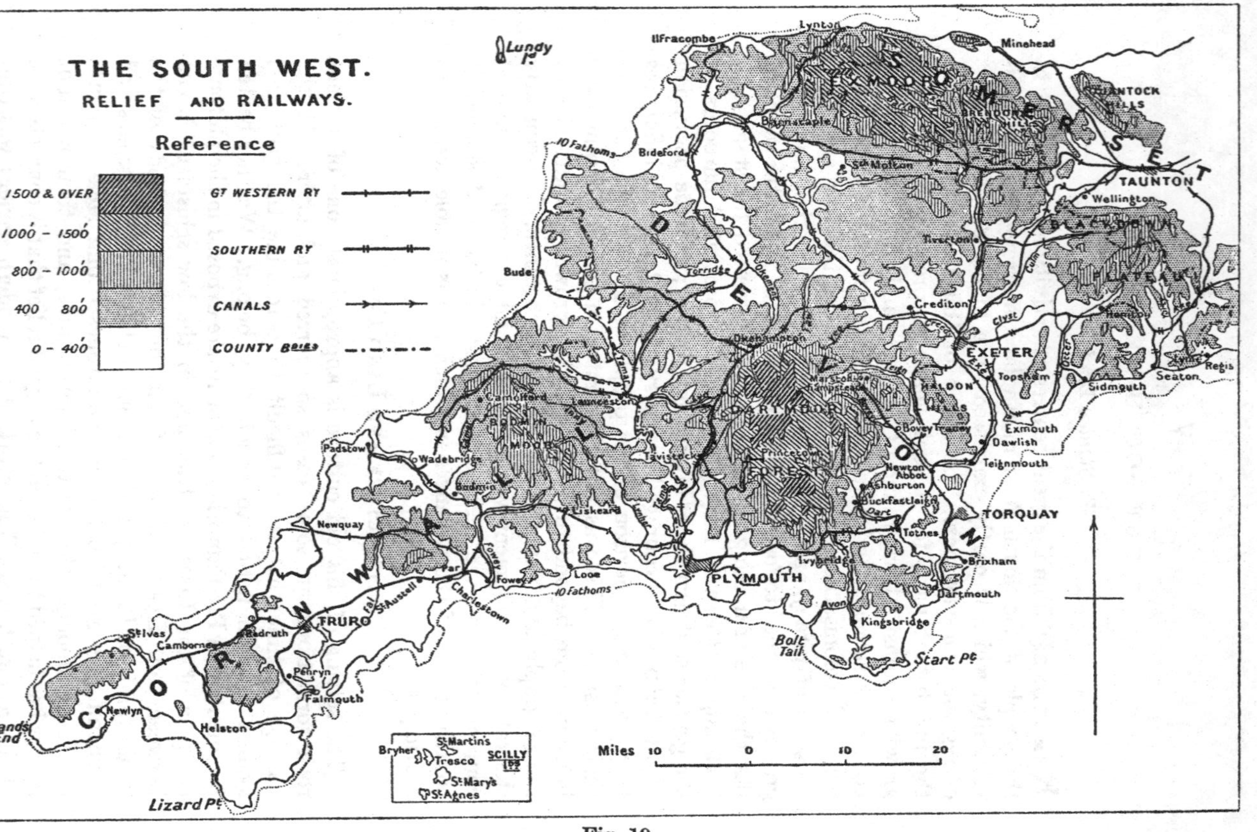
THE SOUTH WEST.
RELIEF AND RAILWAYS.
Reference
1500 & OVER
1000 – 1500
800 – 1000
400 800
0 – 400
GT WESTERN RY
SOUTHERN RY
CANALS
COUNTY BDIES
Lundy Iᵈ
Ilfracombe
Lynton
Minehead
EXMOOR
QUANTOCK HILLS
SOMERSET
TAUNTON
Wellington
BLACKDOWN
PLATEAU
Barnstaple
St Molton
10 Fathoms
Bideford
Bude
Torridge
DEVON
Tiverton
Crediton
EXETER
Topsham
Clyst
Seaton
Regis
Sidmouth
Okehampton
HALDON HILLS
DARTMOOR
FOREST
Camelford
Launceston
BODMIN MOOR
Tamar
Tavistock
Princetown
Bovey Tracey
Exmouth
Dawlish
Teignmouth
Newton Abbot
Ashburton
Buckfastleigh
Dart
TORQUAY
Totnes
Brixham
Dartmouth
Padstow
Wadebridge
Bodmin
Liskeard
Newquay
CORNWALL
Par
Fowey
Looe
PLYMOUTH
Ivybridge
Avon
Kingsbridge
Bolt Tail
Start Pᵗ
St Austell
Charlestown
10 Fathoms
St Ives
Camborne
Redruth
TRURO
Fal
Penryn
Falmouth
Newlyn
Lands End
Helston
Lizard Pᵗ
Bryher
St Martin's
Tresco
SCILLY
St Mary's
St Agnes
Miles 10 0 10 20

Fig. 19.

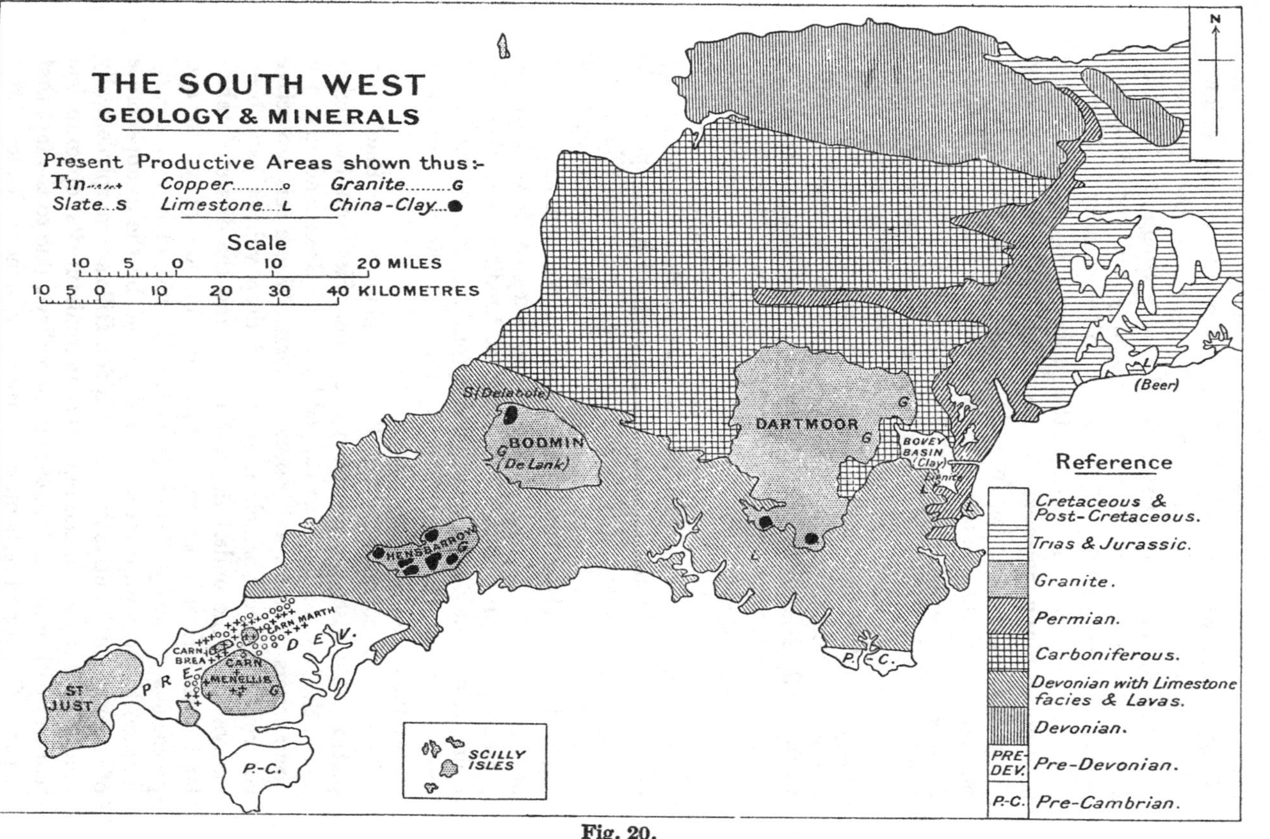
THE SOUTH WEST
GEOLOGY & MINERALS
Present Productive Areas shown thus:-
Tin.....+
Copper..........o
Granite.........G
Slate...S
Limestone...L
China-Clay...●
Scale
10 5 0 10 20 MILES
10 5 0 10 20 30 40 KILOMETRES
N
S (Delabole)
BODMIN
G
(De Lank)
HENSBARROW
G
DARTMOOR
G
BOVEY BASIN (Clay)
Lignite
L
(Beer)
P.-C.
ST JUST
P R E-
D E V.
CARN BREA
CARN MARTH
CARN MENELLIS
SCILLY ISLES
Reference
Cretaceous & Post-Cretaceous.
Trias & Jurassic.
Granite.
Permian.
Carboniferous.
Devonian with Limestone facies & Lavas.
Devonian.
PRE-DEV.
Pre-Devonian.
P.-C.
Pre-Cambrian.

Fig. 20.

trend of this disturbance is reproduced in the general W.-E. strike of the strata; while two of the fold axes can be traced without difficulty, the north Devon anticline on which lie the Quantock Hills, the Brendon Hills, Exmoor and Lundy Island and the Mid-Devon syncline. A second anticline probably runs from Bude Bay across the northern half of Dartmoor to the vicinity of Exeter; but here, as in the remainder of the peninsula, the complexity of the folds has been increased by massive igneous intrusions, mainly of granite, which, arguing from its appearance in Lundy Island, probably also forms the core of the north Devon anticline. In these great granite bosses is revealed the heart of the old Armorican highlands and, though shadows of their former selves, the five major outcrops of Dartmoor, Bodmin Moor, Hensbarrow, Carn Menellis, and St Just still dominate the surrounding landscape, but with a loss in height and grandeur as one moves westward. In the Scilly Islands the elevated portions of a sixth mass offer a last faint challenge to wind, weather and sea (*v.* Figs. 19 and 20).

The appearance of Secondary strata in the east is accompanied by a decided change in the character of the topography. Narrow outcrops of Bunter pebbles and Keuper sands strike across the neck of the peninsula and are in turn overlaid by Keuper marls that floor the lowlying Vale of Taunton. The change in strike almost coincides with the disappearance of the Armorican structural unity. Between the old and the new the Permian sands and breccias act as a connecting link, the dovetailing being emphasised by the projecting tongues of this formation at Crediton and Tiverton. Going east from the Permian outcrop one is reminded more and more in almost every way of the "plain" and "down" country of the Midlands and southern England.

In striking contrast to the Taunton plain is the Blackdown plateau that rises steeply from the vale. Here, on the higher levels, the Keuper marls are capped by Upper Greensand, in which occur layers of chert, and Eocene gravels. Where this protective covering has been breached, wide and deep valleys have been excavated in the underlying soft marls, and the dissected table-land thus presents an interesting combination of youth and precocious maturity. The base of the Greensand is very clearly marked by an abrupt increase in slope and by the disappearance of enclosure and cultivation. The Haldon Hills are morphologically related to this plateau and reproduce in miniature its topographical features with minor differences in the valleys, due to the fact that here the Greensand rests on the Permian. Further evidence of this Cretaceous overlap is supplied by the interposition of Jurassic beds between Keuper and Greensand along the eastern fringe of the Blackdown plateau.

The sequence of events during the Tertiary era is difficult to reconstruct, for a number of problems still await solution. Local earth movements on an extensive scale are proved by the NW. and NNW. systems of faults to which many of the valleys, particularly those on the eastern flank of Dartmoor, conform, by the synclinal basin of Bovey Tracey which continues across Dartmoor, and by the relics of peneplanes traceable at various levels. Much discussion has centred on the question of the origin of these interesting platforms and it is still uncertain whether they result from sub-aerial or marine action or from a combination of both; the absence of clues in their superficial deposits is one of the chief obstacles to a satisfactory solution*.

Three platforms can be distinctly recognised, at levels of approximately 1000, 800, and 430 ft., while there are slight traces of a fourth at a little below 2000 ft.† Of the certain occurrences the highest is actually somewhat less than 1000 ft. in altitude. It is best preserved on the eastern side of Bodmin Moor and on the south-eastern edge of Dartmoor, but fragments remain recognisable elsewhere in the surrounding upland. Above it rise detached heights, crowned by "tors," piles of granite blocks weathered along joints into fantastic shapes like masses of ruined masonry, and with steep yet rounded slopes strewn with "clitters," a chaos of huge boulders often tons in weight.

The middle platform, between 750 and 800 ft., is found nearer the sea, but shows to greatest advantage in the granite areas, *e.g.* near Camelford and west of Exeter, at Moretonhampstead. Economically it has been the most important of the three, for its surface deposits contained large quantities of stream tin and are rich with remains of the old tin-streamers' activities. It is possible that the Haldon Hills and Blackdown plateau have morphological affinities with this surface though the balance on present evidence is against this relationship.

The 430-foot platform has been established as Pliocene in age and, since it is bounded landward by an old cliff, is most likely of marine origin. It may be recognised at numerous places near the coast in Cornwall and south Devon; while, where the actual

* Cf. G. Barrow, "The High-Level Platforms of Bodmin Moor etc." *Quart. Jour. Geol. Soc.* 1908, and W. M. Davis, "A Geographical Pilgrimage from Ireland to Italy," *Annals Assoc. Amer. Geog.* vol. II (1912), pp. 80 f.

† (= 305, 244, 131, 609 m.) *V.* L. Sawicki, "Die Einebnungsflächen in Wales und Devon," *Sitz.-Ber. d. Warschauer Gesell. d. Wissensch.* 1912. Sawicki gives the altitudes as 300, 240, 150 and 600 m. respectively.

platform is no longer visible, the rejuvenated rivers testify to the elevation that gave it being. The rocky gorges and incised meanders of the Tavy, Dart, Teign and Tamar prepare one for the peculiarities of their beds in the higher courses, where they occupy V-shaped gashes sunk in older and wider valleys. At short distances from the banks these young ravines in the older surface are barely visible. Once again vigorous youth and calm maturity interlock with remarkable scenic effect.

Subsequent disturbances of the land level are revealed by the occasional raised-beaches and best of all by the many rias, often with submerged forests near their mouths, as at Falmouth and Torbay.

In its broad lines the modern scenery is a subdued copy of that which existed in Permian time and still reflects strongly the influence of rock content and structure. The upland areas coincide with actual outcrops of granite or suggest its existence at no great distance from the surface. The Devonian formation of Exmoor differs considerably from that of south Devon and Cornwall; for the latter shows a limestone facies and is associated with contemporaneous lavas whereby the topography is diversified and the agricultural possibilities are increased. Nevertheless both districts exemplify the characteristic Devonian scenery, alternations of steep, rounded hills and deep, narrow valleys. Everywhere the Culm weathers to bare and rugged heights, and the cold mid-Devon and north Cornish plateau, wind-swept, often water-logged and thinly-settled, is a true reflection of its difficulties. Similarly the Greensand and Eocene gravels on the plateaux of the south-east are tame from a scenic point of view and useless for cultivation; but the intermediate Permian and New Red formations offer a pleasing contrast of low undulating plain, warm to the eye and rich as pasture or arable land.

Rivers

The whole of the peninsula is well-watered. On the whole the drainage systems show the expected tendency to radiate from the granite moorlands but there are several puzzling anomalies. Thus there are the peculiar cases of the Torridge which, rising near the Tamar, doubles back upon itself to find outlet at Bideford, of the Tamar and Exe, both almost crossing the peninsula, and of the curious course of the Tone. These and frequent examples of capture, such as that of the Teign near Newton Abbot, indicate both the epigenetic character of the present drainage and the fact that it differs considerably from that of Eocene time. Though the radial distribution of streams from Dartmoor is probably not older than Eocene, parts of the Tamar and of the Cornish streams must be of great geological age and may have fed larger rivers flowing in the Armorican valleys. In early Tertiary time the streams of north, central and south-east Devon appear to have been tributaries of a river flowing eastward through the Burlescombe gap, now followed by the railway from Exeter to Taunton, a road and a canal. Later a southward and westward tilt was imparted to the southern half of the peninsula. To this the drainage adapted itself, and streams flowing south received an impetus which enabled them ultimately to capture the headwaters of many of the eastward-directed rivers. The relations of the upper Teign, lower Teign and Lemmon support this theory, with which must be associated the uplift of the 430-foot platform*.

Climate

The abundance of streams is the result of a combination of impervious rock and heavy rainfall. Free exposure to the Atlantic guarantees high humidity and imparts the characteristic "softness" to the climate, but the principal contributory factor to the reputation for mildness is the relatively high mean temperature during winter months, when east winds are frequently increased in temperature by passage over the warm Channel. Both the extraordinary climatic conditions of such places as Torquay, Penzance and Falmouth, where sub-tropical plants flourish in the open and sufferers from bronchial complaints flock, exemplify the effect of local shelter from north and east winds. In fact, the configuration of the land modifies the general climatic circumstances in such an intricate manner and conditions vary so from place to place that generalisations need critical interpretation. The mean annual temperature is high, from 48°·6 F. on the uplands to 50°·7 F. in the coastal zone (9°·2 to 10°·4 C.) and the yearly range of temperature is small, 18° F. in the west to 22° F. in the east (10° to 12°·2 C.). Frosts are less frequent than elsewhere and snow rarely lies long, even on the high ground; on the south coast deep falls are almost unknown. The rainfall is heavy, particularly along the vertebral ridge and on Exmoor, and is usually accompanied by strong winds inimical to tree growth except in the valleys. Despite the humid

* A. J. Jukes Browne, "The Valley of the Teign," *Quart. Jour. Geol. Soc.* 1904. Other important sources dealing with the problems on drainage are:

E. A. Newell Arber, *The Coast Scenery of North Devon*, London, 1911.

H. Dewey, "On the Origin of some River Gorges in Cornwall and Devon," *Quart. Jour. Geol. Soc.* 1916.

E. M. Hendriks, "The Physiography of South-West Cornwall," *Geol. Mag.* 1923.

atmosphere sunshine is abundant but curtailed in spring and autumn by sea mists. The most inclement conditions are experienced on the uplands and on the bleak Carboniferous plateau, but, like the north coast, they are bracing whereas the southern exposures are notoriously enervating, especially in summer, a fact that is not without its psychological reaction.

Agriculture

Agriculture in the peninsula bears few imprints of very early origin for reasons both geographical and historical. For centuries in west Devon and Cornwall it was subordinated to fishing and mining; when it began to occupy serious attention the old language was dying and the penetration of English methods and ideas was easy and assured. Nevertheless certain important characteristics of modern farming are rooted deeply in the past. Thus the predominance of small-holdings is one result of the early enclosure of common fields in the West Country; the tortuous sunken lanes and the dispersed habitations may be traced to the same cause. The average size of properties increases from west to east, but in Devon and Cornwall does not exceed 75 acres (30 ha.); 70 per cent. of the Cornish holdings are less than 50 acres (20 ha.), and almost 50 per cent. are under 20 acres (8 ha.) in extent. There are comparatively few extensive farms such as are found in the chalk country. One effect of this is that the problem of labour supply is less acute, the farm being almost a family concern and a unit of cultivation. This fosters intensive development and leads to a scattered distribution of population with small concentrations in market villages.

The hilly nature of the country and the high average rainfall are reflected in the utilisation of the land, for the proportion of grassland to arable is about two to one; partly for climatic and partly for economic reasons the amount of arable land is gradually decreasing. Except in certain favoured areas where a specialised form of cultivation is practised, the ploughland is largely for the benefit of stock, the proportion so utilised increasing rapidly from east to west. Generally speaking, the rock formations break down into soils that give good growths of grass and clover; but soil surveys are difficult owing to rapid and frequent variations in soil types. The high granite moorlands are covered with a thin peaty sand, useless for cultivation, but the intervening valleys show a good thickness of clay, suited to hardy cereals when drained and supplied with lime. The lesser granite heights weather to a light fertile loam rich in potash. The Devonian strata break down to a dark red stony loam, well suited to root crops, and provide rich pastures, which however deteriorate rapidly and need to be rested. Where calcareous formations occur, as in the South Hams, *i.e.* between Plymouth and Torquay, very productive light loams result, easily worked and responding freely to treatment. The Carboniferous shales and grits of mid-Devon and north Cornwall give poor buff-coloured soils only useful where freely drained and even then handicapped by climate. Their eastward extension is approximately delimited by the Exe, where they become overlaid by New Red sands and marls that form some of the richest pasture land in the kingdom, well-drained and admirable for hardy cereals, root crops and garden produce. The thin soil of the Greensand heights is a moderate pasture land in strong contrast to the fatness of the neighbouring fertile valleys.

Generally speaking, the grazing is good, especially in those districts where the annual rainfall exceeds 35 in. (875 mm.), for the wild white clover upon which it depends is indigenous to the region. Further, the humid atmosphere ensures grass all the year round so that farms are self-supporting, while in most cases young store cattle can remain out through the winter, and retain their coats, whereby their value is enhanced.

There are definite differences between the farming as practised in west Somerset, Devon and Cornwall respectively. All rank high in the country by proportion of cattle to acreage of crops and grass, with Cornwall leading and Devon taking third place. But the farmer of west Somerset is in the nature of a "cow-keeper" since he is near enough to the centres of dense population to concentrate upon milk production. Thus west Somerset leads in the proportion of dairy cattle to crop and grassland, Cornwall follows and Devon still occupies third place; for while Cornwall too is a dairying county, being one of the first six in England in this branch of farming, and producing cream, butter, cheese and dried milk products, Devon is not so interested in dairy work as is popularly supposed. True, in the neighbourhood of towns milch cattle are found in large numbers, but the basis of Devon farming is the raising of "store" cattle. The distribution of the breeds favoured in the peninsula confirms this and at the same time emphasises the underlying climatic and economic factors. Along the populous southern side and in west Somerset the South Devon and Guernsey cattle, rich in quantity and quality of milk, predominate, but on the more exposed and poorer pastures of north Cornwall, north and east Devon, the "Red" Devon, a smaller and hardier beef type, prevails. The ultimate destination of the latter is one of the more famous pastures of the Midlands, Pevensey Levels, etc.

Sheep farming is practised everywhere, especially on the rough grazings of the higher levels, but nowhere on the same scale as in Wales. The breeds are local and are maintained almost entirely on grass with little folding on arable crops. Similarly, pigs are turned out to grass but not to the same extent, for like calves they are closely associated with the districts where separated milk is available; for this reason they, with poultry, are increasing rapidly on the small-holdings in Cornwall*.

Mention must be made of the interesting and important custom of depasturing cattle and sheep on the high levels of Bodmin Moor and Dartmoor in the summer from May till October. Both local and imported cattle stock are concerned in this movement, Galloway cattle being largely represented. In dry seasons stock is sent from very far afield, but the greatest numbers come up from the lower levels in Devon and Cornwall. The custom is not followed entirely to take advantage of the summer pastures, for it is found that stock grazed on this comparatively poor land thrive better when returned to their native parts. This is attributed to change of air and greater freedom of movement. On the open "Forest" of Dartmoor, which belongs to the Duchy of Cornwall and comprises over 30,000 acres (12,000 ha.), the custom is rich in interest for it is closely bound up with rights and traditions that have persisted through centuries. Space forbids anything but a perfunctory treatment†. All moorland parishes have Venville rights over the moor, whereby, in return for payment of a nominal sum, they may graze freely large tracts of moorland, while certain other parishes have commonage rights of sending cattle, taking turf and stone, etc. The Duchy land is divided into four quarters, each agisted to tenants locally known as "Moormen," who eke out small farms by breeding ponies, cutting peat and taking charge of stock for the summer season, charging about *5s.* per head for cattle and *25s.* per score for sheep. Better pastures on large enclosures of the moor known as "Newtakes" command higher prices. Associated in origin with these customs is that of common stocking after August on certain parts of the moorland edge; this persists in fields that have long been grassland, but is dying out rapidly.

The subordination of arable land to the needs of stock has been mentioned. It follows therefore that fodder crops and cereals suitable as animal foods occupy the largest proportion of the

* For important data concerning this and other aspects see *The Agricultural Output of England and Wales*, 1925, Min. of Agric. and Fisheries, 1927.

† See C. Vallaux, "La 'Dartmoor Forest'," *Ann. de Géog.* no. 180, 1914.

cultivated area. One relic of old British farming, the growth of mixed or dredge corn, *i.e.* oats and barley, dominates production in Cornwall and is still important in Devon. It gives a heavy yield and is a good dry food for pigs and cattle. Otherwise oats, fodder roots, clover and rotation grasses are the rule. Wheat and barley are almost negligible except in east Devon and west Somerset, where wheat is grown in small amounts for working up in local mills with stronger foreign sorts, and sufficient barley is produced to meet the needs of breweries, using water from the New Red marls, and to leave a small surplus for export.

It is only to be expected that the advantages of warm "growing" temperatures during the winter and freedom from frost are exploited wherever possible, especially since early produce commands high prices and compensates for distance from the markets. The classic example of such adaptation is centred on the parish of Gulval, near Penzance, on the southern slope toward Mounts Bay. Here a combination of favourable circumstances, slope, shelter, light volcanic soil and shore manure in the form of sea-weed and sand, makes it possible to grow two crops in the year. Early potatoes are ready in May, and on the same ground broccoli are then planted for the November to February markets. Occasionally this rotation is interrupted by a course of wheat or rye for packing straw. But the growing of these early crops involves a heavy expenditure of manure and is not possible far from the shore where sea-weed and sand can be easily obtained in large quantities, and the competition of Channel Islands and French potatoes and French broccoli is making itself felt in the London markets, to the detriment of Cornish growers, in spite of a highly developed co-operative organisation. Fortunately for this and other branches of market gardening there is a good demand by visitors to the district.

Along both banks of the middle and lower Tamar a prosperous small fruit industry has arisen. Strawberries are the staple crop, but except in thin years they suffer in the markets of London and the North from the competition of the Hampshire and French beds. Other fruits are absorbed mainly by local consumption.

Market gardens attain a local importance near most of the larger towns, particularly where a tourist population guarantees an active demand, and there is a tendency towards a revival in the systematic growth of fruits such as apples. Most of the orchards of Devon and Somerset produce good cider fruit, but, like other crops, suffer from the prevalence of fungoid diseases that flourish in the humid warmth. This has caused them to be neglected and to present a sorry picture by comparison with the orchards of Kent. It is largely a question of organisation, and tentative efforts are being made to capture a market now held by France. Probably the most interesting human effort to meet economic pressure through the medium of a "specialist" cultivation seen in this region is the bulb and flower industry of the Scilly Isles. The marketing of an oceanic climate in the form of spring flowers of all kinds is the ultimate means of livelihood to most of the inhabitants. From December to June tons of autumn-planted flowers are sent to the London market. But the problem of competition from the Channel Isles is equally acute here and, added to this, the "potato" parishes along the Mounts Bay shore and other sheltered spots on the mainland are now active flower-growers. The declining prosperity of certain other occupations is an incentive to the development of the opportunities that exist in south Cornwall for the extensive and intensive cultivation of early flowers, fruits and vegetables. Attention is now being directed in an organised manner towards the supply of graded produce and this, with an increasing recognition of the market requirements, is aiding these industries to enlarge the range and size of their operations.

Fishing

Agriculture has to meet the attack of several formidable opponents. These vary in strength from point to point. Everywhere the attraction of naval service reacts unfavourably upon labour supply with an intensity that increases westward. Along the coasts, the fishing industry is the principal competitor, though many fishermen combine small-holdings with shares in a boat, while in certain areas mining and quarrying monopolise the majority of the male population.

The harvest of the sea is an important asset in economic wealth of the peninsula. Just as the ocean penetrates deeply into the land so has its influence permeated the history of the people. Its attraction is most insistent along the broken coastland of the south in ramifying gulfs where sea and river merge, in calm estuaries and in sheltered coves notched in the towering cliffs. Here every small inlet has its fishing town or village with a fleet of small motor-fitted sailing boats, while occasionally, as at Brixham, Plymouth and Newlyn, the steam trawlers of eastern England find depth and shelter for anchorage and facilities for handling their superior catches. The north coast, craggy, wilder and more regular, is less important. Its shore settlements are fewer, difficult of approach and open to the full force of the Atlantic gales. Where occasionally land passes gently to sea as on the dune coasts of

St Ives and in the estuaries of Padstow and Bideford, the fisherman comes into his own, but distance from the fishing grounds presses him hard.

The variety of the fishing is remarkable; it results from variety of foreshore and from a fortunate situation across the southern limit of the cold-water herring and the northern limit of the warm-water pilchard. The latter is never found east of Start Point and rarely east of Plymouth, being a characteristic Cornish catch and the basis of a flourishing trade in fish and oil with Italy. Unfortunately its annual migrations are uncertain and catches are unreliable. Further, the rise in Italian standards of living is reacting unfavourably upon the market.

Other than pilchards the most important fisheries are those of mackerel and herring in the west, sole, plaice and whiting in the east, a variety that assures occupation during the whole year.

Newlyn leads with heavy catches of herring, pilchard and mackerel. Plymouth and Brixham with sole, plaice and whiting and St Ives with herring and mackerel follow in importance. The south-eastern ports are chiefly occupied with inshore fishing.

Like most other industries in the South-West the present state is no measure of the past prosperity. Lack of capital to modernise boats and tackle, cost of transport, distance from markets and the competition of east coast trawlers that drive the shoals seawards are difficulties enhanced by the innate conservatism and independence of the fishermen, who are loath to relinquish the complete freedom of small ownership in favour of companies or co-operative undertakings. The tourist market is too small to absorb the local catches; in fact the presence of a visiting population tends to divert the interests of the fishermen and to transform them into boatmen. Nevertheless this coastland has fostered a people of remarkable initiative; changing values and the passing of the need for the old "sea dog" type restrict the possibilities of a folk whose tradition and upbringing make it difficult for them to turn to other activities.

Mining

A similar tale of decay invests the story of mining. Some would identify Cornwall as the famed Cassiterides; whether or no this be the case its metalliferous mining is certainly of great antiquity. Historical records prove it to have been highly organised in Norman days, and tin mining may justifiably be regarded as one of the oldest and most important industries in the kingdom.

The lodes occur along the margins of contact between the granite and the "killas*," mainly in three belts, from Cape Cornwall to St Agnes along the southern and western sides of the Hensbarrow granite and along the southern edge of Bodmin Moor through Calstock to Tavistock, but with decreasing numbers and richness as they extend eastward. The most important ores are those of tin and copper, but deposits containing arsenic, tungsten and wolfram occur in workable quantities; more than thirty mineral centres exist, the majority being found west of Tavistock† (Fig. 20).

Tin and copper, especially the former, have always predominated, above all around Redruth and Camborne‡. A steady production of tin was maintained from the 13th century onwards. The early centres were located in Devon but gradually moved westward to the toe of the peninsula, where the country is now a desolation of abandoned mines, ruined stacks of old stamps, diverted water-courses and pit-mounds. The maximum of prosperity was attained in the "sixties" and "seventies" of last century when an average of more than 14,000 tons was annually produced. Then a fall in the price of tin caused by growing competition from foreign sources initiated the decline. By 1914 the annual production was halved, but the final collapse was delayed by the war-time demand for wolfram. This was but a temporary relief, for, with the post-war slump in prices, the output in 1921 fell to 1078 tons, and in 1922 Cornwall practically disappeared from the list of tin producers. It must not be inferred that the supplies of tin were exhausted. On the contrary there were still many rich lodes, but the days of easily-worked stream deposits had long since disappeared, and the price of tin was too low to enable the veins to be properly worked in competition with stream-tin from the Malay States. Once again too, lack of capital, lack of fuel and an unwise conservatism, expressed in the reluctance of miners to depart from their old methods, contributed to the decline. The gradual exhaustion of alluvial supplies from abroad and a rise in the price of tin recently stimulated the industry to fresh efforts. For a short period there was promise of a return to the old order of things. In 1927 about a dozen mines were in active operation, houses for employees were built by the mining firms and an output of some 4000 tons of dressed tin ore awakened hopes of a full, if slow, recovery. These hopes have not been realised. Large sums of money expended upon deep (Dolcoath, 3150 ft. = 960 m.) and water-logged workings have

* A Cornish term for the contorted slaty rocks.

† J. H. Collins, *Observations on the West of England Mining Region*. Plymouth, 1907.

‡ Populations: 9,916 and 14,578 respectively.

not resulted in the tapping of new veins and the few active centres are in a precarious position, unable to obtain the financial support required for further exploration. A similar difficulty besets those who are anxious to undertake dredging operations in the numerous creeks fed by drainage from the tin grounds.

But there is another side to the picture. The old-time tinners enjoyed numerous privileges under the stannary system. They had free rights of prospecting on unenclosed ground, were excused tithes and military service and possessed their own Courts, one for Devon and one for Cornwall, and even a "Stannary" Parliament with powers over the tin trade superior to the decisions of the State Council*. These privileges were granted in return for heavy taxes upon tin, assessed in the "coinage" towns, Helston, Liskeard, Lostwithiel, Penzance, Truro, etc., and were carefully guarded by the miners, for their Courts met regularly until the middle of the 18th century.

The feeling of freedom and communal interest conferred by these ancient rights has doubtless done much to maintain the spirit of clanship amongst the Cornish. Even though thousands of miners have emigrated to mines in South Africa, Australia and elsewhere the desire and effort to return is always manifest, and as a rule the family remains in the county. Much of the pioneer mineral work in the British Empire has been effected by Cornish miners or by others trained in the School of Mines at Camborne; just as many of the pioneer advances in mining engineering were first conceived in the mine levels of Cornwall.

Most of the tin areas produced copper at one time or another, and at certain periods the output of copper surpassed that of tin. Where both ores occur together, tin is found nearer the surface, and as the difficulties of working the copper are altogether too great to permit competition with the United States, most copper mining concerns ultimately transferred their attention to the raising of arsenic ores.

Apart from their mineral contents many of the formations are valuable in themselves. Penryn, on the flank of Carn Menellis, and De Lank quarries on Bodmin Moor are renowned for the building qualities of their granite; that of Dartmoor is a poorer type and not so extensively worked, and Norwegian supplies have the advantage of easier shipment (output of igneous rocks, 1928—Cornwall, 571,446 tons, Devon, 314,772 tons). Cornwall is also famous for its slates which have been worked at Delabole near Camelford from the 16th century. Devon has a large output of

* W. S. Lewis, *The West of England Tin-Mining*, Exeter, 1923.

limestone, a hard variety quarried in the Devonian near Plymouth and Torquay, and a soft Cenomanian stone worked since Roman days in underground quarries at Beer in east Devon (output, 1928—376,123 tons).

Superior in importance to all these is the one really prosperous undertaking of the peninsula, the quarrying of kaolin or china clay. Kaolinisation of the granite masses has taken place in each intrusion but to the greatest extent in the vicinity of St Austell and at Lee Moor, near Plymouth. The process when complete gives china clay, but an intermediate stage, china stone, is common. Although discovered in 1784 a considerable time elapsed before the clay was in great demand outside the Potteries, but, by a fortunate coincidence, circumstances altered at the period when signs of distress first appeared in the tin region, owing to the discovery that the clay was admirably suited for paper filling. A phenomenal increase in the exports to the United States followed; from 130 tons in 1875 they rose to 231,234 tons in 1910. The discovery of further uses quickly followed, notably that of adding the clay to cotton and calico goods to give weight and finish, till the total annual output now approaches a million tons. Most of this is quarried in Cornwall (1928—Cornwall, 725,196 tons; Devon, 62,100 tons), mainly in an area of 30 sq. miles (78 km.^{2}) behind St Austell. Here the countryside is gashed by numerous huge white pits, scores of feet in depth, with enormous conical mounds of waste sand rising at their sides. Mineral lines, aqueducts, settling pits and the dull grey granite drying houses surround one, and with the dreary cottages of dark concrete-like material—an effort to use a fraction of the mountains of sand—are depressing reminders of the price of exploiting Nature, particularly in this county of scenic beauty.

The Oligocene clays of the Bovey basin are probably secondary kaolins, derived from Dartmoor. They are valuable as good potting clays and are extensively quarried near Newton Abbot for shipment from Teignmouth. Most of the clay leaves by sea, that from southwest Dartmoor leaving Plymouth, that from St Austell by Fowey and Par, and that from the western outcrops by Falmouth; Charlestown, Newquay and Padstow are also exporters of small quantities. Fowey handles the bulk of the trans-Atlantic trade and has an important inward movement of coal for drying the clay.

Communications

The traveller who enters the South-West by road is speedily made aware of the restrictions imposed by relief upon communication. Large areas of Exmoor, the haunt of wild deer, are still almost inaccessible by road or rail. Dartmoor and Bodmin Moor, though crossed by a main road, yield the secrets of their wild recesses only to the hardy who are prepared to encounter the pitfalls of dangerous bogs and heather-hidden hollows. Roads are few and often die away into the wild moorland except where a valley bites deep into its edge as in the case of the Dart and better still in the long hollow of the Bovey. Transverse traffic by rail is particularly slow and difficult on the peninsula, for the main lines run the length of the axis, skirting the moors and throwing off branches to the pleasure resorts of the sea; bus services are complementary to, rather than competitive with, the railways. Once the voyager leaves the Taunton Vale or the Blackdown plateau he enters another land, where visions of thatched cob* cottages nestling in smiling hollows are succeeded by those of grim granite or slated structures that tell of wind and storm, where cultivation is absent and cattle graze in the shelter of huge stone or earthen "hedges." He feels that behind the pleasant attitude of the people lurks a sense of reserve, an instinct that is immeasurably strengthened as the Tamar is crossed, for the Tamar is much more than county boundary: it is a racial frontier that was unspanned for many miles from its mouth until the railway came. The reason for this difference between Cornwall and the rest of the region lies in their differing histories.

Population and Historical Changes

The study of human antiquity in the South-West is a matter of long vistas, for recent discoveries point to the existence of human habitation since early Palaeolithic time. The picture is still very inadequate, but it appears fairly certain that towards the end of the Neolithic period the area was introduced to the megalithic civilisation when a link was established with Brittany that for many centuries was an outstanding fact. Both Devon and Cornwall are rich in megalithic monuments especially the extreme west of the latter, and these remains seem to bear some relation to the early mineral prospectors of the Mediterranean. The Beaker-peoples did not reach the South-West, the early affinities of which were with Brittany, and so commenced at this early age that remarkable distinction between the "Downs" country and the "South-West" that has persisted to this day.

The remains of earthworks, hut circles and so on have suffered so much from the attention of tinners and those seeking building material that it is quite unsafe to draw definite conclusions from a study of their characteristics. Traces of Roman occupation

* A mixture of clay and straw used in the construction of walls.

become fewer and more fragmentary west of the River Exe, but during and after that period the Brythonic speech spread to the South-West and from Cornwall to Brittany.

It is possible to recognise traces of the early peoples in the present day population, but until more quantitative results are available little can be said. Around the moors there are to be found survivors of ancient long-headed types, with dark colouring and slender build. They doubtless include representatives of most of the Stone Age peoples of Britain. Scattered here and there particularly in the more remote terminal portions of the peninsula are dark broad-heads, identified on many points of the European coasts as survivors of the Mediterranean prospectors. The harbours of the sunken valleys have a fair proportion of tall, fair men apparently of Norse origin, but where there is such a wealth of old types it seems almost impossible to identify them without a great deal of detailed anthropometric work.

The Saxon colonisation of Devon was practically complete but never extended west of the Tamar. Thus the differences between the peoples east and west of this river that had been gradually growing under the influence of contact with Brittany, Ireland and Wales were accentuated by this infusion of new blood. The succeeding attack by the Normans caused large numbers of the Celtic people to emigrate to Brittany, and further encouraged the peaceful intercourse between the two peninsulas. In the reign of Henry VIII one-sixth of the taxable population in Cornwall was of Breton origin and even to-day Breton place-names and surnames survive, while the old Cornish language is more akin to Breton than to Welsh.

From the Norman period onward the history of the region is closely bound up with that of the country as a whole, and the South-West occupied a foremost place in manufacture and trade for, from the 14th century to the latter half of the 18th century, the woollen trade expanded rapidly. From Taunton to the Tamar the advantages of a relatively dense population based on rich land, good facilities for trade, a suitable climate, abundant water and plentiful supplies of raw material from the native long-woolled sheep created an active manufacture of serges*. The development of overseas trade went hand in hand, and in the 16th century the ports along the Devon coast attained the acme of their greatness, both as outlets of trade and as harbours of refuge from storm and pirates. A vast trade with Europe in tin and with Europe and the

* See R. H. Kinvig, "The Historical Geography of the West Country Woollen Industry," *Geog. Teacher*, vol. VIII, 1916.

East in woollens was soon supplemented by a great export of dried Newfoundland fish. The effects of this period of prosperity and contact with foreign influences may still be traced in the architecture of the older buildings of many of the ports.

The Napoleonic wars had a depressing effect upon the wool trade. As late as 1770 almost every Devonshire cottage had its hand-loom for weaving woollen fabrics; while in 1768, 330,414 pieces of cloth were exported to Europe, about fifty years later the number had fallen to 127,459.

But what historical accident had commenced the character of the people encouraged, for the customary dislike of change was evinced in the restrictions placed upon the use of new inventions. The Industrial Revolution delivered the final blow. The absence of fuel, for the Bovey lignites may be neglected, was a severe handicap, so that to-day, although here and there, as at Wellington, Honiton and Buckfastleigh, traces of the woollen manufacture linger, the greatness of the past is forever gone. With the decay of the woollen industry marched the decline of the overseas trade. During the era of wooden ships of small size the seaports of the west, especially Topsham and Bideford, had all the desirable advantages of timber, safe anchorage, satisfactory location and a population familiar with the intricacies of ocean navigation. The increasing size of ships caused traffic to become coastal in character except in the deeper havens of Falmouth and Plymouth. The former, once an important packet station, now acts as a port of call for orders and a repairing station, though the introduction of wireless has seriously affected the former function. Plymouth retains importance as a port of call for liner traffic, as a naval station and as a centre for the import of many of the commodities necessary to the south-west.

Present Distribution

But if the coming of the era of railways and steamships plunged the peninsula into relative obscurity the appearance of motor transport effected a remarkable rejuvenation. Every year sees the isolation of villages far distant from railways invaded by increasing throngs of tourists. The future of the region appears to lie in the exploitation of its mild climate and varied landscape with a view to acting more and more as a huge sanatorium and recreational district for the great centres of population and as a home for those who seek rest or health in their old age.

The present distribution of population* conforms closely to the

* Total population: Cornwall, 320,705; Devon, 709,614.

physical and economic conditions already outlined. Above the limit of cultivation, approximately 1100 ft. (330 m.), settlement is rare. Along the embayed southern "Riviera" coast the population is fairly dense, and this holds for the inland mining districts. Elsewhere, especially in the agricultural districts, the population is thinly and evenly disseminated over the whole area with occasional concentrations in villages and small towns, consisting essentially of a long and usually wide main street which, on fair or market days, is still crowded with booths and buyers. These small agglomerations are most numerous in the fertile south-east and east and each serves as the market centre for its immediate neighbourhood, but here and there occur larger nuclei, with wider affinities, acting as collecting and distributing centres for a widespread area. Such are Taunton, Exeter and Plymouth and to a lesser extent Barnstaple and Truro*. These larger agglomerations exhibit a development of the smaller manufactures sufficient to meet the local needs. Thus Taunton serves as the shopping centre for west Somerset, the regional control of Exeter extends over east Devon, south Devon as far as Torquay, and north Devon, while Plymouth is the regional capital of south-west Devon and the whole of Cornwall. The general tendency is towards an expansion of the holiday resorts such as Exmouth, Torquay, Ilfracombe and Newquay† and depopulation of the rural areas. The great agglomeration of the "Three Towns," Plymouth, Stonehouse and Devonport, is relatively recent and has grown around the establishment of naval dockyards at Devonport at the end of the 17th century. Changes in the orientation of naval power are slowly sapping the position of Devonport, and the whole future of the Plymouth complex is extremely precarious, for since its industries other than shipbuilding, *e.g.* artificial fertilisers, sweets, soaps, oil refining, cattle-foods, are small and local in market it is tending to lose balance.

Exeter, the old outpost of the early British kingdom of Damnonia, secured a momentum that, by virtue of slightly superior access for the greater part of Devon, has secured for it the administrative functions for the county, supplementing its leadership in matters spiritual. In a similar way the concentration of both temporal and spiritual control of Cornwall in Truro reflects the approximation to industrial conditions and dense population in the west of that county.

The geography of the South-West at the moment affords a

* Populations, respectively: 21,000, 59,582, 210,036, 14,409, 10,843.

† Populations, respectively: 13,606, 39,431, 11,772, 6,637.

remarkable example of the alteration in values of natural advantages with varying circumstances. The old provincial aloofness breaks down grudgingly before the advance of intercourse, and behind the easy-going toleration of the new order there broods in a Celtic twilight that ancient spirit which, from time to time, bursts forth in a resurgence of national consciousness such as that which found expression in the fervour of the Methodist movement in the South-West. Language and literature gone, only the tradition remains, a tradition that inspires a population of whom three-quarters are of families rooted here for centuries, and that acts as a constant stream of energy and spiritual enrichment to this and other countries through those who, like so many of their ancestors, seek wider fields for self-expression.

The interpretation of the existing social complex is simplified when it is realised that no violent disruptive forces have disturbed the normal course of community evolution. The foci of local inspiration and progress have declined in importance with the absorption of the region into the general life of the country, but little change has been effected by the Age of Industry in the actual distribution of population over the South-West. Week by week the small nodal centres gather the inhabitants of their tributary valleys in their market places as they have done through the centuries. From time to time each enlarges its zone of intercourse by another ancient custom—the fair. So the region is knit into one unity by the thread of dependence upon the land. Nevertheless, and often against the will of its people, the peninsula gains a wider outlook. The visitor, now seeking pleasure rather than trade, the drag of the sea felt by the youths of every village, facile road transport and the growth of sport, these offer possibilities for an increase of knowledge from which a wealth of past experience, both good and ill, enables the essence of progress to be distilled.

VI

THE LOWER SEVERN BASIN AND THE PLAIN OF SOMERSET

W. W. Jervis

Limits of the Region

Structure and Drainage Systems

Lying between the old upland region of Wales and the English Plain, this area is separated from the latter by the Jurassic escarpment of the Cotswolds. Though consisting mainly of lowland along the courses of the lower Severn and its tributaries, it rises away from the rivers into imposing hills of varying heights. Throughout the area, the general direction of the lower Severn is from north-east to south-west, but the seaward end of the estuary has an east to west trend, roughly parallel to the strike of the Mendip Hills which divide the low plain of Somerset from land of a similar character in the lower Severn basin.

The basin is a well-defined natural region bounded everywhere by hills, except in the north where the lower Severn enters it and in the south-west where the Bristol Channel cuts across the uplands. On the west the hills form the foothills of the "Welsh Plateau" of Fleure—the foothills of the Brecon Beacons, the Black Mountains and the Radnor and Clue Forests. These hill ranges run from north-east to south-west and are seamed by deep valleys through which the drainage of a large part of the old Welsh Plateau is carried south-east to the lower Severn and its estuary by the Taff, the Usk and the Wye with its numerous tributaries. The Taff and the Usk quickly emerge from the high ground to enter the low coastal plain of Gwent which borders the right bank of the Severn estuary. The Wye and its tributaries, however, after leaving the high land pass through an area of lower relief seamed with low hills—the plain of Hereford—separated from the lower Severn by the hill ranges of Malvern and the Forest of Dean, and finally enter the estuarine flats of the Severn, after having carved out a wonderful gorge through the Carboniferous limestone of which the Forest is mainly composed. The eastern limit—the Jurassic scarp—is part of that great structural line which crosses England diagonally from the Dorsetshire coast to the Cleveland Hills of Yorkshire. That part of the escarpment overlooking the valleys of the lower Severn and

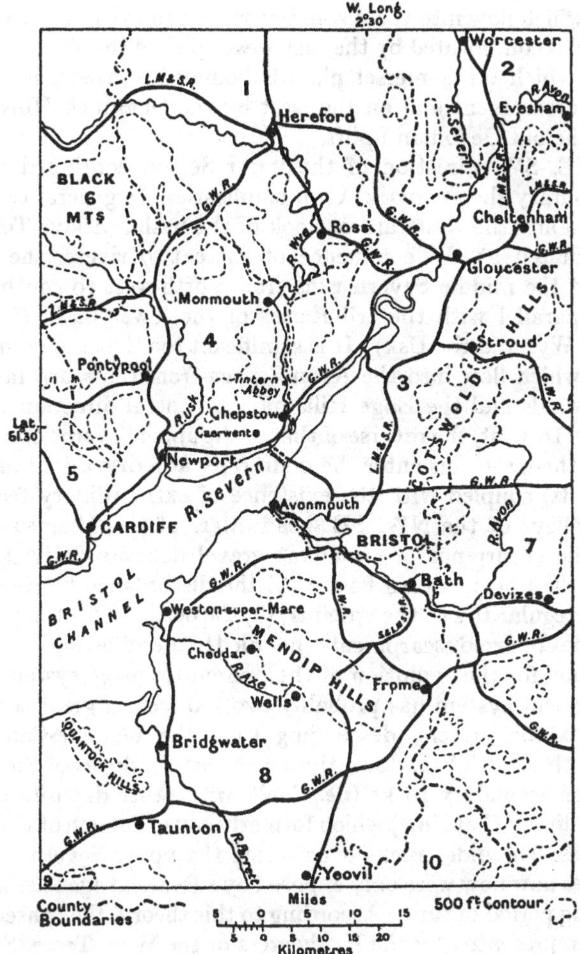

Fig. 21. Plain of Somerset and lower Severn: Railways and Counties. Counties: 1, Hereford; 2, Worcester; 3, Gloucester; 4, Monmouth; 5, Glamorgan; 6, Brecknock; 7, Wiltshire; 8, Somerset; 9, Devon; 10, Dorset.

the Warwick Avon is known as the Cotswolds and trends from north-east to south-west. In places its crest attains an elevation of more than 1000 ft. (305 m.) and its scarp face is notched by numerous streams which flow into the Avon-Severn. In the vicinity of Bristol this trend is complicated by the east-west uplift of the Mendip Hills, south of which the Somerset plain is bounded on the east by the Jurassic escarpment and on the west by the Quantock Hills—the edge of the old Devonian uplift.

The NE.–SW. direction of the lower Severn continued by its chief tributary the Warwick Avon emphasises the general trend of the region and the westward outlook of the basin. Above Tewkesbury, situated at the confluence of these two rivers, the main stream of the middle Severn runs from north-west to south-east, roughly parallel with the tributaries of the lower basin (*i.e.* the Taff, the Wye and the Usk). It is significant too that the numerous streams which flow into the Avon-Severn from the scarp faces of the Cotswolds and the Edge Hills have a general direction south-east to north-west, the reverse of that of the upper Severn. Moreover many of these streams enter the main river at a discordant angle*. These facts, coupled with the existence of extraordinary features in the valleys of the NW.–SE. headwaters of the Thames—such as: (1) the occurrence of abundant gravel deposits derived from rocks in the upper Severn basin; (2) the discordance between the minute irregularities of the streams; (3) the occurrence of dry wind-gaps in the Cotswold escarpment—have led to the following suggested explanation for the evolution of the present drainage system†.

The present system has probably evolved from a great series of roughly parallel rivers, descending from the Pennine-Cambrian divide to the North Sea, down the south-eastern slopes of the great plateau of Secondary rocks (*i.e.* Chalk with other deposits nearer the shore-line of that time) which formerly covered much of England and Wales. The older rocks over which the upper Severn and its tributaries now flow were only exposed by sub-aerial agencies acting over a long period of time. According to this theory, the consequent streams, represented by the headwaters of the Wye, Teme, Severn and Thames, have been intercepted by a cross-channel, the Avon-Severn, which escapes by the gap between the Welsh and Devonian Uplands, while the Severn above Tewkesbury is the former headstream of the Thames‡. Denudation led to the forma-

* *Investigation of Rivers*. Royal Geographical Society. Final Report, 1916.

† *Geology in the Field*, Geol. Assoc. Jubilee Volume, 1858–1908.

‡ W. J. Sollas, "The Severn Estuary and its Tributaries," *Quart. Jour. Geol. Soc.* vol. XXXIX, 1883.

tion of the Chalk and Jurassic escarpments, and exposed the Lias and New Red Sandstone at the foot of the latter escarpment. As a result, the stream draining the country now occupied by the upper end of the Bristol Channel worked its way north-east along the soft exposures at the foot of the Jurassic escarpment. This gave rise to the lower Severn, which in turn captured the upper portions of the original consequents above-mentioned leaving the beheaded rivers to rise farther to the south-east. The direction of some of the original consequent rivers is now followed by other rivers of the lower Severn basin.

The phenomenon of the "tidal bore" is unusually well marked in the case of the tides on the Severn. At spring tides the body of water rushing up the river acquires an increased velocity due to contraction as it passes through the narrow strait at Sharpness. Higher up the river, near Awre, it strikes against the Noose, a large sandbank which occupies the greater part of the bed of the river. The water is therefore ponded back until reaching a height high enough to cover the sandbank it rushes on with increased force. At Fretherne it encounters another sudden contraction of the river bed and the water thus mounts to a height of three feet or more (1 m.) and rushes roaring up the narrow channel. Though smaller bores occur at every spring tide the highest one is at the "Palm tides" in March.

The bore is not considered dangerous to boats if they are afloat in mid-stream, but they are liable to be swamped or stove in if left inshore, as the wave breaks with considerable violence along the banks. Special embankments have been made below Gloucester to protect the land from its attacks.

The coasts of the estuary are typical of a drowned valley region. For the greater part they are flat and marshy and extensive areas of mud flats, in some places extending more than a mile from the shore, are exposed at low tide. A depression of land surface following the Glacial Period seems to have let the waters of the sea into the vast morass into which the Severn formerly flowed, and the valley as far as Newnham assumed a lagoon-like aspect. On the Monmouth side signs of this encroachment of the sea are everywhere apparent and it is clear that the grassy mud flats, now covered at high tide, formerly reached much farther out. In former times the sea at high tides encroached far beyond these mud flats and inundated the slope of the country to the base of the hills known as the levels or moors, rendering the ground useless alike for cultivation or pasture. A short distance from the sea a strong rampart was constructed, about twenty miles (32 km.) in length, and extending from

Sudbrook to the Usk and Rhymney. An inscribed stone, washed out from the soil at a depth of five or six feet on the seaward side of the embankment at Goldcliff, shows that the Romans either built or repaired a mile of it. This embankment was replaced by a stone wall which the sea occasionally washed over at high tide. In the church at Goldcliff a tablet records the disaster of 1606–1607 when "the tide backed by a strong south-west wind swept over the embankment and buried the whole country."

Considerable quantities of silt are brought down by the Severn tributaries. At Tewkesbury there is evidence that as much as 37½ ft. (11·4 m.) has accumulated in the human period. Below Gloucester this deposition leads to the formation of small grass-covered flats which gradually become large enough for enclosure as pastures. This accumulation of sediment around Gloucester has not only rendered the valley more liable to flooding, but the narrowing of the estuary causes the bore to rise to greater heights, which may account for the more frequent flooding.

The Bristol Avon is worthy of note and the last word has not yet been said in explanation of the peculiar nature of its course. The river originates as a consequent stream down the dip-slope of the Cotswold escarpment. On reaching the belt of clays with interbedded sandstones, under which the limestone beds of the scarp disappear, it turns south-westward and flows roughly parallel to the trend of the Cotswolds. Near Bradford it bends again—this time to the west. From thence until it enters the Severn estuary it flows roughly from south-east to north-west, parallel to the trend of the Mendip Hills and the scarp streams of the Cotswolds. The Avon cuts through the Cotswold escarpment and the uplift of Carboniferous limestone rocks at Bristol in two fine gorges (Bath and Clifton), and its valley affords an important route from the west to the English Plain.

Space will not permit of a more detailed discussion on the formation of the Avon gorge, but suffice it to say that the course of the Bristol Avon is an anomalous one from Bradford-on-Avon to the Severn estuary*.

* C. Lloyd Morgan, *Journal of Bristol Naturalists' Society*, vol. IV, 1885; M. Lanchester, *Severn from Source to Mouth*, 1915; A. G. Bradley, *A Book of the Severn*, 1920; *West Coast Pilot*, Admiralty Publication.

Geological Formation

For an area so limited in extent the number of geological formations exposed is unusually large. Rocks of Silurian, Devonian, Carboniferous, Trias and Jurassic ages occur as well as Recent deposits. This is largely due to the important tectonic disturbances to which the area has been subjected at different periods and also to the marked erosion which has been effected from time to time.

The earliest of these earth movements appears to be that of Caledonian age which resulted in the important line of fracture and folding running north and south and brought into position the Malvern complex north of our area and the Ledbury, May Hill and Tortworth Silurian inliers to the south. Two series of folds affected the Devonian and Carboniferous formations; one, running north and south, produced in the north of our area the Forest of Dean and the Bristol Coal basins—the major axis in each case having this direction, and the other series running east and west, has resulted in the marked east and west Mendip periclines in the south of the area*.

The Mesozoic rocks have been much less disturbed by earth movements. They do show, however, gentle flexures, probably the result of a slight rejuvenation of Hercynian movements. There is evidence also of a slight secular movement here in later geological time, a movement following upon, and perhaps related to, the retreat of the ice from the more northern part of Great Britain.

A period of slight elevation followed this, and is evidenced in the raised-beaches at Weston-super-Mare and at Gower, and this in turn was followed by subsidences (post-Neolithic) as indicated by the submerged forests at Barry Docks and the Neolithic land surface at the mouth of the Avon.

Broadly speaking the actual valley of the lower Severn and the Plain of Somerset may be described geologically as the south-western arm of the great Midland area of New Red Sandstone which reaches the sea also at the head of the Mersey estuary and at the mouths of the Tees and Wear. Westward this New Red Sandstone plain laps round the old Palaeozoic rocks of the Welsh Uplands, where they sink into the ancient floor underlying the sands and clays of the plain. To the east the New Red Sandstone rises overlain by Jurassic rocks dipping south and south-east, the lower members of which form the eastern part of the plain, while the upper and more resistant rocks of the series make the escarpment of the Cotswolds. On the west the New Red Sandstone plain

* S. H. Reynolds, *Handbook of the Geology of the Bristol District*, 1912.

is bounded by an important fault which strikes north and south along the east foot of the Malvern range. This fault has let down the newer rocks so that they end with a broken edge against the Archaean, Cambrian and Silurian masses which make up the ridge of Malvern. This fault-line has been traced southwards through May Hill and across the Severn to Tortworth. West of the fault-line, except for the coalfields of South Wales and the Forest of Dean, the lower basin consists for the most part of Old Red Sandstone, arranged synclinally, from which rise isolated outcrops of Archaean rock. The Old Red Sandstone has offered little resistance to denudation and has been worn down to a lowland, the famous plain of Hereford and the Red Land of Gwent. The coalfields of South Wales, the Forest of Dean and the Bristol district lie in basins, defined by a series of folds and in each case the rim of the coalfield is formed of Carboniferous Limestone. In the former the surface outcrop of Coal Measures proper is very extensive, but in the latter case much of the area is covered by New Red Sandstone which offers a good soil for agriculture. Thus the aspects of the coalfields are very different.

The Mendip Hills consist of a periclinal uplift trending north-west to south-east and they exhibit all the characteristic features of a limestone region, such as swallow-holes, underground rivers, caverns and steep-sided gorges, *e.g.* the caves of Cheddar and the gorges of Shipham and Cheddar. Eastward the Mendip anticline disappears beneath a covering of newer rocks. To the north-west the range has been breached by the mouth of the Severn, but islets of Carboniferous Limestone, *e.g.* Flat Holm, prolong the axis and indicate the connection with the coast of Glamorgan. The coastal plain of Glamorgan and Monmouth, the Red Land of Gwent, has a foundation of Old Red Sandstone, but is covered in places by Secondary and Recent deposits. It reproduces many of the features of the Somerset plain*.

Natural Divisions

From the point of view of structure and topography the area may be considered as consisting of four natural divisions:

(1) The lowland plain of Hereford, between the edge of the Welsh Uplands and the Malvern ridge.

(2) The coastal belt of Monmouth with the plain of Morganwg, west of Cardiff, *i.e.* plain of Gwent.

(3) The vale of the lower Severn in Gloucestershire and north Somerset as far as the Mendips.

* S. H. Reynolds, "The Mendips," *Geography*, vol. XIV, no. 79, 1927.

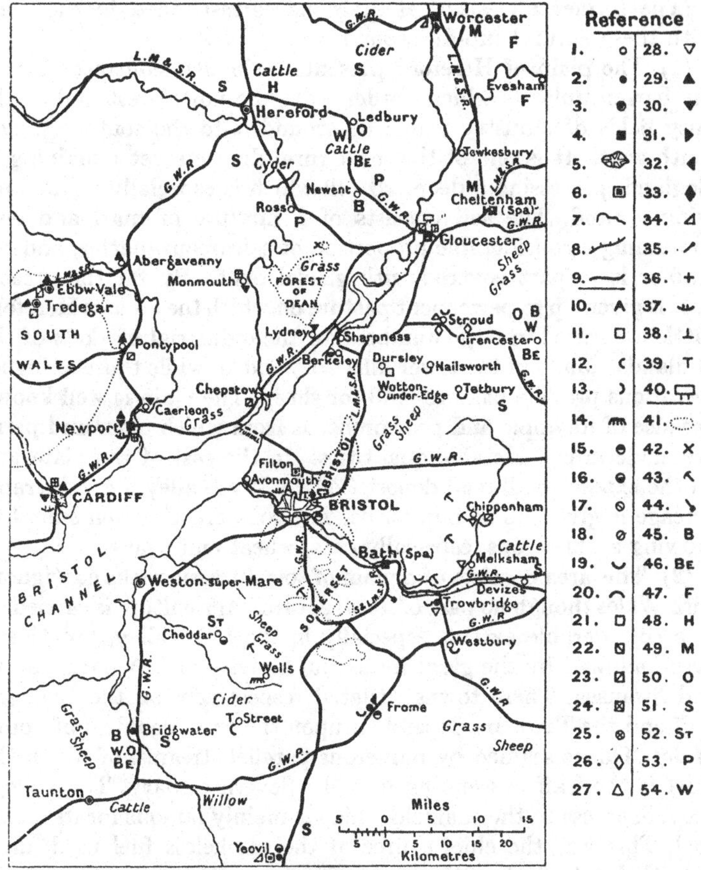

Fig. 22. Plain of Somerset and lower Severn: Towns and Industries. Reference: 1. Towns under 10,000 pop.; 2. Towns, 10,000–20,000; 3. Towns, 20,000–40,000; 4. Towns, 40,000–100,000; 5. Bristol, over 100,000; 6. Aerodromes; 7. Rivers; 8. Canals; 9. Railways. Industries and Crops: 10. Aeroplanes; 11. Alkalis and Chemicals; 12. Boots; 13. Leather and goods; 14. Brushes; 15. Brewing; 16. Cheese; 17. Cocoa and Chocolate; 18. Sugar; 19. Cloth and Woollens; 20. Clothing; 21. Cycles; 22. Motors; 23. Machinery; 24. Paper; 25. Tobacco; 26. Dyeing and Bleaching; 27. Umbrellas and Sticks; 28. Milk; 29. Iron and Steel; 30. Tin plates; 31. China and Earthenware; 32. Hardware; 33. Glass; 34. Gloves; 35. Matches; 36. Furniture; 37. Shipbuilding; 38. Soap; 39. Tanning; 40. Railway waggons; 41. Coalfields; 42. Iron and ochre; 43. Stone-quarries; 44. Fullers-earth; 45. Barley; 46. Beans; 47. Fruit; 48. Hops; 49. Market gardening; 50. Oats; 51. Sugar-beet; 52. Strawberries; 53. Potatoes; 54. Wheat.

(4) The plain of Somerset.

(The former ecclesiastical divisions corresponded in the main with these natural sub-divisions.)

(1) The plain of Hereford presents a singular contrast both to the barren uplands which border it on the north-west and to the busy fields of industry which lie around it to the south-east and south-west. It is one of the most rural districts yet remaining in England, possessing a deep red soil which is especially fertile low-lying ground. The soil consists of a mixture of marl and clay containing a considerable proportion of calcareous matter, and the surface is in part covered with glacial drift. By far the greater part is given up to permanent pasture on which the famous Hereford cattle are raised in large numbers for the industrial markets of the Midlands. Many of the lower hills are wooded, while those of higher elevations provide pasture land for sheep. The plain is well known because of its apple and pear orchards from which cider and perry are made in considerable quantities. In the east of the plain and on the extensive alluvial deposits of the Wye valley a considerable acreage is given up to hop-yards. The soils are also well suited to growing arable crops, especially oats, wheat and beans.

(2) This area consists of a plain of soft Jurassic rocks contiguous with Wales though no part of its structure. Agriculture is carried on to a considerable extent, especially in the Usk valley, but this is overshadowed by the great industrial activity of Newport, Cardiff and Swansea. These towns situated respectively on the Usk, the Taff and the Tawe are dependent upon the great coalfield of South Wales. This is seamed by numerous parallel streams, of which the chief is the Taff descending into the Severn estuary. The valleys provide access to the coalfield which is mainly famous for its steam coal. This was the chief source of the smokeless fuel used until recently by the fleets of the world. The demand is now much reduced and activity is shifting to the west of the Neath valley, where the anthracite is more accessible. A little iron-ore is still mined, and this together with imported ore, largely from Spain, is smelted locally for the making of many types of iron work and for the tin-plate industry. The value of the yearly importation of ores at Swansea alone exceeds £3,000,000*; the chief metal manufactures here are of copper, tin, steel, spelter, tinplate and ships' sheathing. Great quantities of timber are imported for use as pit props and for shipbuilding. Other industries include the manufacture of sails and rope, machinery, chemicals and pottery. Within a thirty-mile radius of Cardiff there is a population of two million people, to

* *The Board of Trade Returns*, H.M. Stationery Office.

support which there is a very large import trade of foodstuffs such as grain, frozen meat, potatoes, vegetables, fruit and other provisions*. Statistics show that Cardiff and Newport are tending more and more to import their own requirements in foodstuffs to the exclusion of supplies from London and Bristol. Newport and Cardiff have extensive docks which are in direct railway communication with the coalfield. These are now under the control of the Great Western Railway, and though it was feared in some quarters that the lack of stimulus resulting from absence of competition might affect industry adversely, yet the Company has already begun further dock improvement work, and there are undoubted advantages in the co-ordination of traffic control in the carrying and shipment of the coal. The fact that the gradient from the coalfield slopes towards the coast has undoubtedly had a stimulating effect on the coal export trade of the South Wales ports, and has also materially helped to make the ports smelting centres. Furthermore, attention must be called to the tendency of the "heavy industries" to migrate from Birmingham to the Newport area.

(3) This district, which embraces in its northern extension the lower part of the Vale of Evesham, consists mainly of a plain floored by New Red Sandstone and Liassic clays and pierced by two uplifts of Carboniferous Limestone—the Forest of Dean and the Bristol coalfields. North of the coalfields the area is devoted to agriculture. Dairy farming on the rich meadows bordering the Severn is the chief industry of the Vales of Gloucester and Berkeley, with orchards on the higher ground. In the Evesham, Cheltenham and Gloucester districts, where patches of warm, sandy loam occur, large areas are given over to market gardening, and small fruit, vegetables and flowers are raised for industrial markets. Special railway facilities ensure the swift delivery of these products to districts both near and far. In the vale country cheese and butter are produced. West of the Severn and north of the Forest of Dean, the soil is less fertile and the relief more pronounced. Mixed farming is the chief activity and on the hills the famous Ryelands breed of sheep is pastured. The valleys of the scarp streams which notch the Cotswolds are wooded, while on the crest plateau the land is either under the plough or used for sheep pasturage. The soil of the Cotswolds for the most part is shallow and strong—locally known as "brash." The Bristol Coalfield lies both north and south of the Avon and is bounded on the south by the Mendips. It is largely overlain by rocks of New Red Sandstone and Liassic age, and mining in this area is complicated by the fact that the

* *The Board of Agriculture Returns*, H.M. Stationery Office.

coal seams have been largely disturbed by faulting. The coal is of varying quality, and the chief mines are found in the neighbourhood of Radstock and Paulton.

The Forest of Dean, which now forms part of the Crown lands, though formerly more extensive, has a present area of about 23,000 acres (*c.* 9300 ha.) of which the Crown is allowed to enclose 11,000 acres. The minerals of the Forest area include coal, iron-ore, building stone and ochre. Coal mining is still a great industry here, but in earlier days, when surface outcrop was essential, the Forest of Dean coal was of great importance. The Upper and Middle Series of house-coal seams are becoming rapidly exhausted. The Lower Series consists of steam-coal seams. Coke-damp is rare and as there is no fire-damp, the miners work with naked lights. The annual output of coal is about 460,000 tons. Since the reign of Charles II any male person of 21 years or upwards born within the Hundred of St Briavels and who, having worked for a year and a day in a local coal or iron mine, has then registered becomes a Free Miner of the Forest of Dean. As such he is entitled to receive from the Crown grants of all "gales" or areas of coal and iron. In 1904 1200 free miners applied for certain gales advertised by the Crown and, the grant having been made, disposed of their interest for a royalty of a half-penny per ton on all coal raised. The principal centres of the industry are Cinderford, Coleford and Lydbrook.

The iron-ore, which is found in the Millstone Grit and in the upper part of the Carboniferous Limestone, has been worked since Roman times. They extracted the metal in such an imperfect way that from the time of Edward I it has been found profitable to rework the old refuse or "cinders," in some cases 50 per cent. of pure iron being obtained.

Up till the reign of Charles I charcoal was used for smelting, but Dud Dudley in his *Metallum Martis* suggested that coal should be used. On the experiment proving successful, the Sussex iron masters, whose supply of timber was becoming exhausted, migrated to South Wales for the sake of the coal.

The supply of Forest iron-ore is now rapidly on the decline. Formerly a considerable part of the timber for the navy was drawn from the Forest, but by 1783 the greater part had been felled and the site laid waste. The waste lands were replanted and at the present day the Forest consists mainly of oaks about 100 years old. In the very heart of the Forest, at Speech House, the Court of Verderers still meets. This Court manages the "Vert and Venison" under the Crown. The last of the deer was killed in 1852 so that the Vert, *i.e.* the timber and pasture, alone remains. The Court has been held since the introduction of the Norman forest law and is probably the most ancient survival of its kind in the kingdom.

Though the coal-mining section of the Forest reminds one of the industrial part of South Wales, the general aspect is of lofty wooded hills intersected by deep dales down which run streams generally reddened by iron-ore, while on the more open commons here and there the miners have built their cottages.

(4) The plain of Somerset is entirely agricultural in character, and consists largely of low-lying alluvial flats which necessitate embankment and careful drainage to protect them from inundation. The coastal levels of different periods are of absorbing interest in that they offer a definite line of approach to the study of the distribution of the different racial types within the area. The western parts are almost entirely given over to dairy farming and the rearing of cattle, but farther east, where the land slopes up to the Jurassic escarpment, a considerable acreage is under the plough and produces oats, barley and wheat. The soil of this rising ground is much less fertile than the low-lying ground.

In the Vale of Taunton Dean, at the south end of the plain, orchards cover much of the district which here reproduces in many ways the scenic features of the plain of Hereford. The famous Cheddar cheese, now made over much of Gloucester and Wiltshire as well as Somerset, is one of the chief products of the dairy farms. The plain, especially in the higher eastern districts, is dotted with numerous market towns, once the seat of the prosperous woollen industry of the West of England. Relics of this industry remain in small-scale manufacture of cloth at Frome, sailcloth at Crewkerne, and lace at Chard. Sugar beet can be grown on all types of land capable of producing root crops. It has proved equally successful on the light soils of the Oolite and on the heavy alluvial soils of the Chedzoy district of Somerset. During the season of 1926, crops averaging 14 tons to the acre and yielding 18 per cent. of sugar were produced.

CLIMATE*

The climate of the region is governed in the main by the fact that the prevailing winds are south-westerly and that it lies in a well-marked track of cyclonic movements. The shape and general direction of the estuary enable the south-west winds to carry oceanic influence far inland. In the winter months this is most

* *The Weather of the British Coasts*, H.M. Stationery Office, 1918; *The Book of Normals* (Air Ministry), Sections I, II and III, 1924; *British Rainfall*, 1921. H.M. Stationery Office, 1922.

strikingly apparent in maintaining an average temperature greatly in excess of what is due simply to latitude and season. There is also a relatively low annual range of temperature of about 20° F. (11° C.). Gales often do extensive damage to shipping. A high spring tide backed by a strong south-west wind is capable of causing considerable damage to shipping as well as extensive inundation of low-lying coastal land*.

Owing to the diversified nature of local relief there is considerable topographical variation within the area, but broadly speaking two distinct zones may be distinguished: (1) the coastal and adjacent vales and lowlands, having the highest mean annual temperature and least daily range throughout the year; (2) the uplands, having the lowest mean annual temperature and the greatest daily range. Harvest as a result is later on the uplands—"As long coming as Cotswold barley"—and in the sequestered valleys round Cheddar, where the strawberry industry flourishes, these plants are in bloom while snow still lies on the adjacent heights.

The heaviest rainfall over most of the region is in autumn and winter, and lack of bright autumn sunshine renders the region less suited to the raising of corn crops than the east coast.

The region lies within the 30–40 in. (760–1000 mm.) rainfall division, but the south of the area is considerably wetter than the north.

Notes on Historical Geography

The lower Severn valley and estuary are thrust as a wedge between the Welsh and Devonian uplands, and as such they functioned to the end of the Danish conquests in this country. The Romans fully realised the strategic importance of the wedge and, once in the West Country, established themselves strongly in the lower Severn basin, thus effectively severing the tribes of Wales from those of the South-West. The Second Legion, working its way down into the West, ultimately settled at Caerleon-on-Usk†. Later a colony was established at Glevum (Gloucester), the first inhabitants of which were probably time-expired men from this Legion‡. Another settlement was formed at Bath, whence they advanced to the lead-mining district of the Mendips. Ferries were established across the Severn at Old and New Passages, the former now marked by the L.M. and S. Railway bridge and the latter by the Severn Tunnel. The New Passage crossing, being situated below the Wye,

* *Publications of Newport and Cardiff Docks Committee.*

† R. G. Collingwood, *Roman Britain.*

‡ F. Haverfield and G. Macdonald, *Roman Occupation of Britain.*

gave direct access to the legionary centre of Caerleon. It is of interest to note that in Roman times the basin was bounded both on the east and west by important roads: by the Fosse Way on the east, and on the west by that from Caerleon, northward along the foot of the Welsh uplands to Wroxeter.

During the period of the Anglo-Saxon conquest, the lower Severn basin was entered by a branch of the West Saxons, who, landing at Southampton, advanced northwards up the rivers of the Hampshire basin and over the chalk downs of Wiltshire to the Cotswolds. Defeating the Britons at Dearham (A.D. 577) near Bath, they descended to the Severn seizing the Roman settlements of Gloucester, Worcester and Hereford. In this way, like the Romans, they divided the Britons of Wales from those of the Devonian uplands.

The contact of the lower Severn valley with the Midlands was early demonstrated. Anglian tribes working west from the east coast rivers, established themselves at Shrewsbury and along the Warwick Avon. They incorporated the settlements of Hereford, Worcester and Gloucester, and the whole of the Midlands from the head of the Severn Estuary to the Humber became the Kingdom of Mercia. On the other hand, at this period the route to the lower Severn via the Thames valley was successfully blocked to the Anglo-Saxon invaders at its eastern end by the tribes there.

During the raids of the Danes London was no longer an obstacle to the invaders. Thus in their attempt at the conquest of Wessex the Danes penetrated the Thames valley and by this route established communication with their fleets in the Severn. The re-conquest of Mercia by King Alfred and his heirs was carried from their base in Wessex north along the Severn valley. The Danes were thus separated from their allies in the Welsh and Devonian uplands.

During the reigns of the Norman and early Angevin kings, and up to the conquest of Wales by Edward I towards the end of the 13th century, the Severn formed a border zone against the turbulent Welsh, controlled by the "Lords Marcher." Newport, Chepstow, Monmouth and Hereford all possessed strong castles commanding the valleys leading into the old upland of Wales. The later historical geography of the region centres chiefly around Bristol.

Significance of Town and Village Sites

Space will not permit of a detailed analysis of the types of town and village settlements within the area, but two towns are of such marked importance as to call for more than a passing reference—*viz.* Gloucester and Bristol.

*Gloucester**

Gloucester is a city of great historical interest and was the rival of Bristol at some periods. Situated on a gentle eminence overlooking the Severn, it marks the site of the lowest road crossing the river and lies on the main route from Bristol to the industrial Midlands and the North. Its importance as a Roman station has already been noted. During mediaeval times its growth was favoured by its position on a river navigable by the shipping of the time, which permitted of the easy export of much of the woollen cloth produced on the slopes of the Cotswolds. In 1580 the city received a Royal Charter extending its jurisdiction from Tewkesbury to Berkeley. Rivalry with Bristol now became very keen since before that time Bristol had held sway over parts of that area. The commercial importance of Gloucester was considerably increased in 1827 by the construction of a ship canal from Gloucester to Sharpness providing a comparatively straight channel in place of the tortuous and sand-impeded Severn. Later improvements enabled vessels of 4000 tons to reach Sharpness and of 800 tons to reach Gloucester, and by smaller canals vessels can now reach Worcester and Birmingham, though in the latter case vessels of 40 tons only. The industries of the city are very varied and of ancient date. The iron trade dates from before the Conquest, tanning was carried on before the reign of Richard II, while pin-making and a bell foundry were introduced in the 16th century, to take the place of the declining cloth manufacture which had been located here since the 12th century. The Severn fisheries furnish the well-known Severn salmon and also those lampreys which are potted and made into pies in Gloucester. In the spring time millions of young eels (elvers) are sold in the city. Gloucester has a very marked nodality, being situated on a navigable river and at the meeting point of numerous routes. But improvements in road and rail services have gradually minimised its nodal importance. The Severn Tunnel took away much of the South Wales traffic, while the Severn Bridge at Sharpness has decreased the coal traffic from the Forest of Dean. As a market centre, however, Gloucester continues to develop.

Bristol

A fact that must have been early recognised is, that of the rivers flowing into the Bristol Channel the Bristol Avon offers the easiest route to the Thames basin and to London, either directly across

* *Victoria County Histories—Gloucester and Somerset.*

the oolitic ridge or indirectly by way of the Vale of Pewsey and the Kennet valley. Here then is the gateway of the West and the apex of the invaders' wedge which has left us with the culture of the Saxon on the one hand and Celtic culture on the other. We have history enshrined in Bristol, and we have that which is not always the concomitant of history—we have romance. It is difficult to imagine anyone, and least of all a geographer, remaining unmoved by meditation on those early voyages whose starting point was Bristol, whose course lay through the Avon gorge to the Severn, and so to the then uncharted seas of the West. Even now in the absolute heart of the city and against the walls of the cathedral rise the tall masts and salt-caked smoke stacks of ocean-going steamers. The port is situated on the land side of the neck of the peninsula formed by the Frome and the Avon, just at the foot of the Clifton Downs where the Avon gorge begins*. Thus in the early times it held a position of strategical importance, and camps of the early British tribes are found on the high points of the district, dominating the passage up the river and affording points from which observation could be kept upon the whole of the lower course of the river. The town was not sited nearer the sea because the flat ground near the mouth of the river, besides being liable to inundation, was exposed to attacks by land or sea. The steep cliffs which line the river on both sides make a landing impossible until one reaches the opening at the foot of Clifton Down. The neck of the peninsula was easily guarded by a castle and the site chosen was fortified landward and seaward. Bristol was walled only about a century before the withdrawal of the Roman legions.

The nature of the trade at this time and in the later Saxon period was for the most part of a purely coastal character. A good deal of trade was done with Ireland from a very early date. This was doubtless due to the fact that the Irish started ahead of the Northern Celts in nautical efficiency, and their comparative proximity to our neighbouring shores probably tempted them to visit and to trade. Bristol stood on the inner margin of a great agricultural belt, commanding its outlet on the west and having easy access along the Avon to the richest corn and cattle lands in the country. The slave trade, which proved so profitable to Bristol, was regarded in the early days as a trade in agricultural implements.

When Henry II obtained Aquitaine as part of his wife's dowry the trade in French wines received a tremendous impetus, Bristol becoming its chief mart, and the street names of Bristol bear their testimony—Wine Street and Corn Street. In the 15th century was

* H. N. Appleby, *The Port of Bristol*, 1925.

formed the famous Society of Merchant Adventurers, the chief members of which seem to have owed their wealth mainly to the cloth trade.

The discovery of America turned the eyes of Europe westwards; in 1630 John Winthrop planted a colony at Charleston and in Maine also a colony was founded by Bristol merchants. In 1620 the "Mayflower" sailed, and by the end of the century Bristol was doing a triangular trade sending goods to Africa, thence shipping slaves to America and returning thence with sugar and tobacco, the former article being refined at the home port. The introduction of Free Trade and the abolition of the slave trade affected the sugar industry very adversely. As long as trans-oceanic shipping depended on wind power no east coast port could compete with those on the west, with their more direct and less foggy route; but with the coming of the Industrial Revolution and the introduction of the steamship the trade of Bristol gradually declined*. The difficulty of entering the port, coupled with the heavy dock dues, diverted trade to the North, and Liverpool outstripped Bristol. In 1848 trade revived owing to the Corporation's buying the harbour, reducing the dock charges, taking over the Portishead and Avonmouth docks from private companies and building a fresh dock out seawards to accommodate the increasing size of ocean-going steamers.

The natural advantages of the port are that, saving one, it is the most inland and centrally situated in the United Kingdom and it serves an enormously extensive industrial hinderland. As the crow flies Birmingham is only 75 miles from the Bristol Docks and within this radius dwell six million people. This together with the wonderful loading and discharging facilities have attracted cargo steamers from about 250 overseas ports. Grain, timber, petroleum, oil nuts, oil seeds, soya beans, tobacco†, ores, phosphates, esparto grass, tinned foods and dried fruits arrive in huge quantities. Moreover the growth of the banana trade over a short period of years is one of the phenomenal features of the development of the trade of the Port of Bristol. With the future possibility of improved water communication with the Midlands and improved rail and road communication with South Wales the utility of the port is likely to be still further enhanced.

The author wishes to thank the Colston Research Society for its support in carrying out regional study in this area.

* *The Port of Bristol Handbook*, 1927.

† *Tobacco and its Associations with Bristol*, Messrs W. D. and H. O. Wills 1922.

VII

THE SOUTH-EAST MIDLANDS

H. O. Beckit

Forty miles from London the traveller by the old North Road (a Roman way long known as Ermine Street) reaches the brink of the wide canted tableland of Chalk which he has been traversing, and finds lying before and beneath him a wide stretch of lowland. Here, just south of Royston, on the confines of Hertfordshire (essentially a county belonging to the London Basin) and of Cambridgeshire, the road plunges down over 400 ft. (120 m.) in three steps, swinging well to the west of north so as to avoid those lowest, flattest and once ill-drained parts of the plain which form the English Fenland on the verge of which, a dozen miles away, lies Cambridge. Along a sinuous line running generally north-west from Cambridge, but nowhere very far from the line of Ermine Street, the margin of the almost featureless Fenland rises, midway across Cambridgeshire and Huntingdonshire, into a weak relief of low tabular hills or rolls of ground separated by the wide flat-floored valleys of meandering streams, whose waters flow north-eastwards across the Fens to the Wash. The North Road is here crossing what is really a great vale, 40 miles (65 km.) wide, in which the shallow, subsidiary valleys of the Great Ouse, of the Nen, and of their many tributaries are cut.

Twenty-five miles farther to the south-west, Watling Street passes north-westward out of London, over or rather through a similar rim of Chalk downs, and across a continuation of the same lowland, here narrowed to less than half its previous width. A wide shelf with sandy soils, two-thirds as high as the Chalk scarp behind it, projects into the lowland along Watling Street as far as Brickhill. This second old Roman road, now forming part of the trunk route running diagonally across England and on to Holyhead in the extreme north-west of Wales, attains but a slightly greater altitude in its passage across the Chalk upland than does Ermine Street, though the summit level of that upland is here half as high again (over 700 ft. or 220 m. above sea-level): it runs, in fact, along the valley of the little River Ver and its streamless extension, *i.e.* through a wind-gap in the scarp, and in this respect is typical of a whole series of relatively low passage ways northwards and north-westwards out of the London Basin, by which run most of the main links of communication, by modern road, canal, or railway, between the Capital and the Midlands and industrial North of England.

The high crest of the Chalk upland near Dunstable in southern Bedfordshire, where Watling Street leaves it, becomes bolder as it is followed south-westward. It soon takes the name of Chiltern Hills, and in 40 miles—the last half of the distance without any significant gap in its skyline—has crossed Buckinghamshire and southern Oxfordshire to the point where it is breached by the water-gap of the Thames below Goring. From many points along this stretch of the escarpment it is possible in clear weather to make out, beyond the great lowland vale at its foot, the gradual rise of another belt of upland far away to the north-west.

The narrowing of the vale, to which attention has already been called, is accompanied by an increase in the average altitude of its floor. It is near this point that the direction of flow of its natural drainage changes. Instead of running north-east to the Wash the rivers flow generally south-west and are caught up in the Thames system. Both here as seen from the crest of the Chilterns and farther west the vale is a conspicuous lowland, only locally diversified by flat-topped hills which form quite minor features of its general relief.

Where it passes across the watershed into the basin of the upper Thames the vale widens somewhat and then narrows again farther west. Its southern margin is still the steep edge of a Chalk upland, the White Horse Hills of northern Berkshire and Wiltshire, oriented almost exactly east and west for 20 miles (32 km.) west of the Thames but then resuming the previous south-westerly trend which was predominant east of that river. The vale is here divided into two by a miniature scarped ridge which the Thames on its way to the much larger Goring gap in the Chalk upland pierces at Oxford. This ridge or line of minor heights may therefore with reason be given the name of Oxford Heights. South of these Oxford Heights lies the Vale of the White Horse, north of them the even wider valley of the upper Thames: from the crest of the White Horse Hills, which in one place rises nearly 860 ft. (about 260 m.) above sea-level, it is possible to see over the whole width of the Oxford Heights and of the two vales north and south of them as far as the Cotswold upland across the Thames in Gloucestershire.

Chalk scarp, neighbouring vale, and Cotswold upland beyond them all swing round more and more to the south and all grow narrower as they pass beyond the area somewhat arbitrarily selected for treatment in this chapter. That area, most of whose southern border has now been outlined, reaches from the edge of the Fenland, not much more than 30 miles inland from the North Sea at the Wash, to within an even shorter distance of an Atlantic inlet, the Bristol Channel. Its length from north-east to south-west thus exceeds 120 miles (190 km.): the varying width of the main lowland element included has been noted above, but in order to define the whole region it becomes necessary to consider the north-western boundary.

Along the north-western edge of the long belt of lowland, about equally small parts of which lie less than 50 and more than 450 ft. (140 m.) above sea-level, there is no such conspicuous line of strong orographical distinction as the face of the Chalk escarpment. The country rises, however, into a series of more or less dissected tablelands sloping up in a direction between north and west to a high Edge (in north-eastern Gloucestershire, northern Oxfordshire, and western Northamptonshire) overlooking the wide lowland of the Severn and Avon valleys. Here regions described in Chapters VI and XII begin.

Corresponding roughly with the three counties named in the preceding paragraph, three sections of tableland may be distinguished under their accepted names of Cotteswold (or Cotswold) Hills, Edge Hill, and Northampton Uplands, with culminating heights above sea-level of about 1000, 700, and 600 ft. respectively. The Cotswold Hills have the highest Edge, and beneath them lies the lowest portion of the Severn plain; Edge Hill, rather less lofty, overlooks the less deep-lying stretches of the Avon valley; while the still lower Northampton Uplands, in that section of them which is here under review, lie also farther inland, opposite the water parting between Avon and Trent, and have west of them no widely excavated lowland, but rather a portion of the general plateau of the inner English Midlands, but little lower than themselves.

All three tablelands, but very little land beyond their north-western Edges, have their natural drainage into the long vale, whose extent has already been indicated, by an elaborate system of valleys which constitute such strong geographical links between them and it that they will here be included along with the great vale under the term "South-Eastern Midlands." The northern portion of the Northampton Uplands, as far as it belongs to the Welland river system, however, will be left out of account. There remain, therefore, to be dealt with in greater or less detail considerable portions of nine out of the ten English counties named successively above, constituting a region 120 miles (nearly 200 km.), in length, with an average breadth of rather over 30 miles, an area of approximately 4000 square miles (10,000 km.[2]), and about a million and a quarter inhabitants.

It has been found convenient to include, in the extreme south, the whole of the White Horse Hills and the valley of the Kennet south of them, so that physically the region may be defined as the middle and upper sections of the basins of the Thames, Great Ouse, and Nen.

The characteristic relief of the area is illustrated by the transverse profiles (Fig. 23), each of which is drawn along or close to a road which crosses the whole area approximately from south-east (on the right hand of the figure) to north-west (on the left). When the road followed penetrates an upland along a gap incised below the general surface level, that upland surface is also shown in broken line. The horizontal scale for all three profiles is approximately 1/300,000 of nature; heights are exaggerated 25-fold.

Explanation of the Land Forms

Alike in the broad physiographical character of the region and in the details of its surface morphology, there is a close and striking correspondence with geological structure. This correspondence depends for the most part upon the way in which the different rocks exposed at the surface have been affected by denudation, and only occasionally shows a direct connection between the tectonic forces of uplift or crustal deformation and the present surface relief.

The region, in short, is a typical scarpland, developed by quite normal sub-aerial processes upon a mass of rocks (clays, sands, and limestones) which are none of them of extreme hardness—as witness the almost exclusive use to-day of imported road metal—but which differ markedly in their power to resist denudation. It is very unusual to find strata lying at other than slight inclinations from the horizontal: gentle south-easterly dips predominate, so that we find a regular succession of geological formations ranging in age from Lower Lias clay in the north-west to Eocene clays and sands in the downfold of the London Basin, of which the Vale of Kennet in the extreme south forms the quite unmetropolitan westerly termination.

The elevated zone of the Cotswold and Edge Hills and the Northampton Uplands coincides with the outcrop of the Middle Lias or Marlstone and of the much thicker overlying massive limestones of the Lower Oolite series, and the Chiltern scarp and White Horse Hills with that of the highest Cretaceous beds, the Upper Greensand and Chalk. The steep ramparts in which both

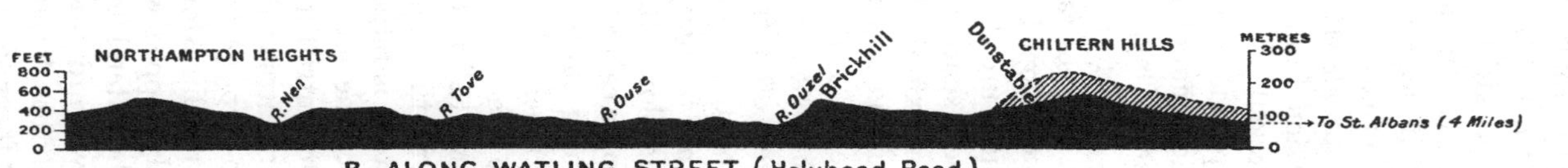

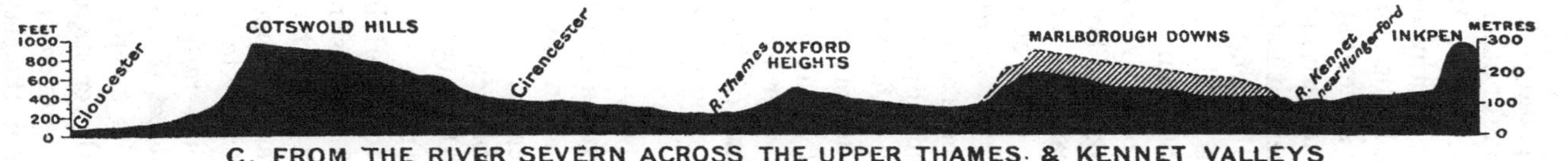

Approximate Horizontal Scale

Miles

5 4 3 2 1 0 5 10 15 20

Kilometres

5 4 3 2 1 0 5 10 15 20 25 30

Fig. 23. Three transverse profiles across the South-East Midlands.

of these zones of upland end towards the north-west are formed of the basset edge of one or other of these two thickest of the relatively resistant rocks of the area. It is to be noted that the long gentle slopes leading down southward and eastward from these scarps or edges are not simple dip-slopes, but that the inclination of the surface is generally somewhat different in degree if not in direction from that of the underlying strata; geological dips are in fact usually a little steeper than average surface slopes, so that even on these uplands the general progression from older to younger rocks as one travels south-eastward prevails.

The great vale between the two uplands is broadly developed where only thinner beds of resistant sandstone or limestone interrupt a series of clays (Oxford Clay, Kimeridge Clay, and Gault). These thin and variant sandy or calcareous beds form the small island hills in the vale, or even, on occasion, a minor range or "cuesta" like the Oxford Heights, which are developed upon a resistant facies of Corallian rocks and show in miniature all the characteristic features of scarpland morphology. A very short southward journey by road across and beyond them from Oxford suffices to impress on the traveller a memory of a series of short steepish rises parted from one another by long gentle descents. The sharp rises are up the faces of the stronger strata, and the long stretches downhill in part along the surface of these beds, here often true dip-slopes; all the time one is passing successively on to younger and younger rocks, *i.e.* climbing stratigraphically if not topographically. Usually at some point midway down the long slopes the inclination diminishes and one comes on to overlying weaker strata, commonly clays, which denudation has left thinning out to a feather edge, so that no distinct break of slope occurs, much less a rise of ground.

Along considerable reaches, on the other hand, of the south-eastern margin of the much larger "cuesta" formed of the lower Jurassic limestones north-west of Oxford, the transition from overlying Oxford Clay to underlying limestone is rather closely accompanied by a slight but distinct rise of the ground—a weak low counter-scarp, as it might be called in contradistinction with the strong high scarp at the opposite (north-western) edge where the Cotswolds drop suddenly to the Severn-Avon plain. This "counter-scarp" usually occurs rather before the limestones are reached, and where it is boldest it is commonly capped with high terrace gravels and owes its steepness to the lateral cutting action of rivers at later and lower stages of downward erosion. This is the result of a recent rejuvenation which has lowered most of the upper Thames vale on the clay by 100 ft. or so, but had time only to allow the rivers to trench their actual valleys in the limestones farther back to about that depth, while between valley and valley the plateau remains upstanding nearly at its previous level.

Another indication that previously to this epoch of rejuvenation and spasmodic valley deepening the limestone plateau graded insensibly into the clay vale is to be found farther west as the divide between the upper Thames and the Bristol Avon is approached. Here well-marked valleys are eroded actually in Oxford Clay, flattish surfaces of which separating the valleys appear to belong to an old smooth surface of denudation which is continued on to other rocks. The long-accepted doctrine of retrogressive erosion seems thus to be extensible beyond its application to the mere lengthening of a river valley headwards.

Throughout the region those rocks which are more resistant are usually so not because of greater hardness but because they are permeable or porous, so that water does not readily collect at the surface and few eroding streams develop, and there is no close-spaced dissection of upland surfaces. Innumerable small elevations occur, capped with one or other of these permeable strata or else with gravels—the latter principally in the clay vales—and possessing reasonably steep sides but remarkably smooth tabular summits of considerable area, from which a panoramic view can only be obtained section by section during a journey round the brink of what is not a peak but a small plateau.

Local tectonic movements, apart from the general south-easterly tilt imposed on the rock structures of the region during repeated Tertiary and post-Tertiary uplifts, although of subordinate importance, do affect certain features of the present relief. A series of north-south upfolds and east-west faults, for instance, has had considerable influence upon the course of valleys in and east of the Cotswolds.

The valley of the River Cherwell or Cher, again, follows the axis of a downward sag in the rocks of the plateau which it traverses before emerging on the Oxford lowland. This sag is continued across the line of the Oxford Heights between sub-parallel anticlinal bulges east and west of it. These seem to have fixed the position of the initial drainage, and are still represented by salients in the scarp of the Oxford Heights on either side of the re-entrant where the Cher, now swelled by the waters of the upper Thames from the west and by smaller streams from the east, pierces those Heights.

The synclinal Vale of Kennet, however, is the largest example of direct dependence of relief upon geological structure. Parallel to it along the extreme southern boundary of the region runs a second line of high downs which at Inkpen reach the greatest height attained by the Chalk in England. Here is the positive expression of a local east-west upfold which by a striking contrast gives rise, farther west, to the inverted relief of the Vale of Pewsey, mentioned in Chapter IV, p. 71.

Apart from flexures and dislocations, there are two kinds of departure from the simple structural type of gently dipping homoclinal strata of varying resistance that complicate the normal scarpland relief. One of these is that some of the rocks were originally deposited in circumstances locally varying, so that changes of lithological facies occur and a resistant scarp-making stratum, say, changes into a weak one, *e.g.* the Corallian beds near Wheatley, five miles east of Oxford. The second complication is even more important, and is the result of geological unconformities, of which the most conspicuous occurs at the junction between Jurassic and Cretaceous rocks. Sometimes two resistant beds not in normal stratigraphic succession follow one another directly, and the resultant hill is higher: sometimes a resistant scarp-making stratum is missing and a wide lowland results on the two weaker rocks not here parted by a stronger. The broad Vale of the White Horse, where Kimeridge Clay and Gault outcrop side by side, is an excellent example, but a still larger one occurs just south-west of the line of Watling Street, between Fenny Stratford and Aylesbury: here Gault, Kimeridge and Oxford Clays all lie in contact, and the widening of the great vale referred to in the introductory paragraphs of this chapter results.

The effects of slow processes of weathering are strikingly predominant over the more localised action of stream erosion. It is the erosive action of streams, however, that has roughed out the pattern of the relief, the vertical measure of which is given by the depth of valleys, almost all of which have completely graded longitudinal profiles, but great differences in transverse section and in plan.

In the eastern half of the region the main drainage is transverse, by way of the Cherwell-Thames, and no explanation of the physiographic history is as adequate or comprehensive as that which regards this as the master consequent stream which has collected the waters of a number of other consequents, such as those which drain the long back slope of the Cotswolds, by a process of successive captures effected by the retrogressive erosion of longitudinal subsequents along the strike of weak rocks and followed by the development of a certain number of short obsequent streams flowing down the faces of the scarps as fast as these were formed at the junction of those rocks with stronger overlying beds.

Steep-sided narrow valleys are confined to the resistant and generally porous sands or limestones, which alone are capable of retaining any but gentle slopes for any length of time, and in which, as has already been summarily pointed out, streams and their erosion valleys are likely to be relatively wide-spaced. Extensive inter-stream spaces here persist in which the degrading action of water must be mainly by underground flow, concentrated no doubt along lines of maximum permeability—joints and the like—but not necessarily productive of continuously graded valleys.

The opinion is held by some observers that even the main valleys of the Cotswold upland are due to solvent action of underground water. There can be little doubt that a number of tributary valleys have so originated, but Clement Reid's hypothesis of a multitude of surface streams existing when percolation was prevented by a frozen subsoil during the Ice Age is also applicable. In the marginal zone beyond the ice sheets, such streams whose surface water has since disappeared with the thaw of a milder climate and the consequent lowering of the water-table, were surely capable of extensive mechanical erosion.

There are, in fact, two quite distinct types of dry valley in the region. One looks exactly like the erosion valley of a stream, with the stream absent: it has a regular grade and quite normal cross profile, dissymmetrical at bends, and these bends are sweeping curves reminiscent of a regularly meandering stream. For such valleys it is reasonable to assume that they were carved by a stream that has since literally sunk into the ground. Other streamless valleys (1) run not in curves but in zigzags, just as if they were following a criss-cross of geological joint planes, have (2) an uneven and in places actually a reversed slope, and (3) a characteristic break of slope at the junction of valley side and floor so as to suggest that the floor is sinking bodily and tearing itself away from the valley walls. These are surely the natural effects of underground solution. Examples may be found, *e.g.*, in the White Horse Hills.

The influence of solution seems to have been exaggerated, partly for want of a realisation that springs, not too far away, to draw out the mineralised water, are essential. The writer suggests that a very common position for such springs is along the line of the deep-sunk stream-tenanted main valleys of the limestone plateaus, but that these main valleys were created by the streams that now run in them.

An interesting contrast frequently occurs in the surface features developed in Cotswold valleys that are cut through the basal horizon of the Inferior Oolite into the passage beds known as the Midford Sands. The land forms produced by mechanical erosion are characteristic of these sandy beds and, when the erosion has been deep enough, of the Upper Lias clay below. The upper slopes, in the limestones, are usually smoother, but manifest a much less complete dependence on surface agents: enclosed hollows and a general non-sequence of slopes, as well as the absence of streams, indicate the predominant influence of underground water, which is no doubt almost entirely solvent and not mechanical in its operation.

In the entrenched meanders of many of the larger Cotswold valleys, the stream to-day frequently fails to sweep consistently round the sloping spurs on the concave side of the curves in its course, or to keep tight up against the foot of the steep amphitheatres of scarp, locally known as "combes," on their convex side. They have consequently been called "misfits," because they appear to be conspicuous cases of a want of proportion between stream and valley. The misfit is more than one of scale, and is essentially one of form. Four possible explanations of the present incompetence of many Cotswold rivers to fill their winding valleys may be given:

(i) Shrinkage of rainfall has been predicated, somewhat arbitrarily, as there is no independent evidence of it having occurred: the explanation is purely speculative.

(ii) Recession of the present water-parting between the Severn and the Thames, to the profit of the Severn, has certainly taken place: according to some writers the drainage system of the Thames originally reached into the Welsh highland before there was any river Severn. Beheading of the present Cotswold streams on an enormous scale like this is hardly suggested by, and is certainly not required to explain the present morphology of their lower valleys. The explanation goes rather beyond the facts to be explained, unless it is allowed that it is not those valleys at something like their present depths that form the problem, but some not impossible but still theoretical ancestors of theirs at considerably higher levels. One at least of the wind-gaps in the Cotswold Edge, that at Farmcote (Lyne's Barn), a couple of miles east of Winchcomb, at the head of a small tributary of the Windrush, strikingly truncates a valley with large entrenched meanders right out to the scarp. It is impossible to avoid the inference that this is the beheaded remnant of the valley of a considerable stream that once rose some miles out over the lowland of the Severn vale, but in the writer's opinion the beheaded portion need not have been as much as 15 miles (say 25 km.) long.

(iii) The present reduced streams flow generally in aggraded valleys. The alluvium in which their channels are cut is more or less constantly water-logged, and the water in it is not stationary, but always slowly creeping down the valley. The cross-section of the alluvium is so enormously larger than that of the stream channel that it is quite possible that a considerable fraction of the flow, as the elaborator of this theory, Dr Lehmann, holds, has become invisible since and by reason of the aggradation of the valley floor. No measurements have yet been made of the rate and volume of this sub-surface flow in the Cotswold valleys. The process certainly takes place to some extent, and is at least a contributory cause of the apparent misfit of stream to valley.

(iv) Although there is no substantial evidence of any change in the total amount of annual precipitation in this part of England, it is certain that sub-Arctic conditions of climate prevailed during portions of the era of spasmodic valley deepening whose stages are represented by the river deposits and buried channels below the level of the Hanborough terrace. During and perhaps for some time after the departure of the ice, and surely beyond the outer limits of regular glaciation, heavy local winter snowfall must have led to floods in spring probably many times greater than those fed by present rains. The total annual run-off in the streams may have been much what it is now, but a quite different seasonal regime may be confidently assumed. It is at high flood that rivers do most of their physiographic work, and it follows that there must have been a time when these Cotswold streams worked at least for part of each year on a much larger scale than they can to-day. The writer believes that this is probably the main cause of the misfits now so conspicuous.

In the clay belts, on the other hand, there are in general no steep slopes to-day and therefore vales rather than valleys. In these weak rocks the rivers first attained a low grade, their valleys were first widened, and aggradation came soonest and was most extensive. It is here, therefore, that we look for and find the principal development both of modern and of ancient alluvial deposits, and both of them, as will be seen later, exercising wide and significant influence upon certain factors of the human geography of the region.

Even in the western part of the region, drained by the tributaries of the transverse Thames, there is a predominance of longitudinal subsequent streams: this is only one of a number of facts which strongly suggest that the present belted scarpland relief is a phenomenon of long standing and has been developed in more than one physiographic cycle. The latest cycle has as a whole reached at least a mature stage, and in the great vale with its generally weaker rocks an excellent example of a local "peneplain"—in the original sense of that term—occurs.

East of the upper Thames basin the main drainage is broadly longitudinal, though even the Ouse and Nen systems show traces of an older transverse drainage in the wind-gaps through the Chalk escarpment and possibly in some stretches of their upper and middle courses. It is, however, more likely that such departures from the characteristically longitudinal drainage as are exemplified in the River Ouzel or the reach of the Great Ouse above Bedford are due rather to irregularities, morainic or otherwise, in the surface of drift through which they have since the Ice Age been re-incising their channels.

In any case it seems permissible to trace, all over the north-eastern half of the region, the influence of the great submergence that gave rise to the depression of the Fenland and the Wash, drawing to itself the drainage of the great vale for quite 50 miles (80 km.) back from the Fenland border of the region. It is, however, reasonably certain that this tapping process has come near its maximum extension, and despite the views of some theorists, no capture of the upper Cherwell by a feeder of the Ouse above Buckingham seems within the range of possibility.

Whatever was its pre-glacial condition, the lowland immediately south-west of the Fens has been, since the largely aggradational episode of the Pleistocene glaciation, rather less completely reduced to the stage of peneplanation than the upper Thames lowland farther west.

It is noteworthy that the present Ouse-Thames divide is morainic and probably replaces an earlier one of tectonic origin, farther west. On either side of the present divide there are, in short, distinct types of lowland, varying in the degree and character of the adjustment of drainage to structure; and it is surely significant that the extension of these two types corresponds broadly with the known degree of glaciation. In the east, ice seems to have overspread the whole country at least once, but in the western part of the area—now drained by the Thames—almost all the evidence suggests that whatever considerable masses of land ice invaded it from outside did so during the earlier stages of glacial advance, and that later phases might more properly be called effects of local nivation rather than regional glaciation.

Soils and Agriculture

There is a corresponding contrast with regard to the characteristic soils of the heavily glaciated east and the little glaciated west. The former, it is true, show variety, from heavy, often chalky boulder clay to light glacial or fluvio-glacial sands or gravels, but the variety of these largely travelled soils has a decidedly irregular and even patchy distribution. This is notably the fact in the vale and the lower-lying portions of the neighbouring uplands, much of the higher portions of which are relatively free from glacial deposits. Here, and farther west on the lowland as well, the soils are essentially sedimentary, and show a close dependence upon the underlying rocks out of which they have been produced by weathering. There is consequently a succession of zones or belts of soil corresponding with the belts of limestone, sandstone and clay beneath. On the clays, *i.e.* in the lowland, deep and particularly fertile soils occur, but are usually heavy to work and hard to drain: the lighter soils of the uplands, usually sandy or calcareous, are lighter, drier, and easier to work, if less productive and shallower. It was on these soils that the earliest efforts towards tillage seem to have been made, and in some degree modern conditions, which have long been tending to throw arable back into grassland, are moving once again in the same direction. Recent increase of areas under permanent grass is most notable on the heavy clay soils.

Superficial deposits other than glacial drift exercise considerable local influence on the character and value of soils throughout the region: it is perhaps only necessary to mention the warm dry soils of old river terraces and other gravelly deposits, and the fine-grained, often water-logged alluvium of much bottom land, as neither occurs continuously over wide areas. Large stretches, however, of the chalk plateaus of northern Berks and Wilts, where not trenched by valleys, bear a residual soil—the so-called "Clay-with-Flints"—of very different composition and texture from the subsoil.

It is virtually certain that the heavy undrained clays of the lowlands were covered primevally with forest or swamp, while the uplands, the home of the earliest inhabitants, were relatively clear. In most of the region the plateaus are still for the most part little wooded, save on the steep slopes of the sides of valleys cut below their general surface, or again in the far north-west, on and beyond the limits of the region, in parts of the old Rockingham Forest of the Northampton Uplands. Here oaks, as well as the beeches and firs characteristic of the Cotswold heights, are common, but oak and elm are the commonest trees in the lowlands, where they grow in the hedgerows or in scattered patches of woodland still preserved in places, often not wholly for economic purposes. Altogether there are enough trees in the lowlands to give them a false appearance of woodland when viewed from any moderate height overlooking them.

Despite this quite possible but false inference from distant general views, only a very small fraction of the area is either forested or otherwise waste land. Nearly all is enclosed farmland, most intensively cultivated in the slightly drier east, but almost everywhere the seat of a more or less highly developed mixed farming. Specialisation of culture, in any broad regional sense, is probably diminishing, except in so far as it is exemplified by the gradual increase of pasture; this is very marked but by no means confined to districts long famous for dairy farming like the Vale of Aylesbury in north Bucks, or the Vale of Kennet, which has more recently become a rival source of milk supply.

Space is not here available for minor agricultural differentiations, even if they possess some local importance: it must suffice to call attention to two among many quotable facts of significance regarding the distribution of live-stock. Cattle are specially numerous in a band of country running transversely across the region so as to cover most of northern Bucks, where there is notably little arable, and western and central Northants: with this is associated in the first of these two counties the dairying industry already referred to, and in the latter the leather industry and especially bootmaking. Sheep raising, though not so markedly as it was once, is still characteristic of the limestone uplands; and local wool has still some industrial importance on the borders of the Cotswolds, *e.g.* at Witney.

Minerals

The region possesses but small resources in the form of accessible minerals to set beside the proceeds of agriculture and stock-raising. Possibly in the east, and even probably in the north-west, but almost certainly not in the centre, deep-seated Coal Measures may one day be found and revolutionise the economic situation; but neither is there any local supply of mechanical power from coal to-day, nor can a substitute ever be furnished by the generally small and always gently flowing rivers of the South-Eastern Midlands.

A quite minor source of wealth lies in thin beds of ironstone in the Middle Lias and Inferior Oolite (Northampton Sands) at a number of localities near to the north-western boundary of the region, but the beds are easily exhausted and their principal development is rather beyond its limits to the north-east*. Good building stone is now hardly to be found except in the massive Oolitic limestones which, with certain thin flagstones or "stone slates" long quarried in the Cotswold area, form the material of the very characteristic and charming grey stone houses and cottages of that district. With modern conveniences for carriage, some of the quarries here send away stone to a distance on a considerable but hardly a large scale; others burn the limestone for cement.

The growing demand for cheap building material in the rather stoneless south-east of England has led to the development at various points in the great clay vale of brickmaking on a much larger scale than merely local needs require. This naturally occurs close to one or other of the main trunk lines of railway, and especially where there is a reasonably short haul for bricks and tiles to London.

Communications

An excellent system of communications has grown up, conspicuously more elaborate than, and different in character from, what the region could need for its own internal service, or even for that of linking it with its immediate neighbours. Such a feature as the four parallel lines of rail on the old North-Western main line, now a part of the L.M.S. system, where it enters the region near Tring is a striking instance of a great trunk line for through traffic right across the district, not even traversing it by way of the largest centres of population, but, like the ancient Watling Street (its ante-type of Roman days), cutting right across the grain of the land on its way from London to the now populous industrial North.

Transverse links like this are in fact the most important of all the chain of artificial or improved routes in the region, and this in spite of the much greater natural ease of movement in a longitudinal direction, at any rate along the lowland belts. Passage in a transverse direction is often markedly assisted by a lay-out along or close to transverse rivers (the Ouzel, the Tove, and a branch of the Nen farther north, in this particular case), which have, so to say, imposed a sort of transverse grain on the broader longitudinal banding of upland and vale. The gaps in the scarped uplands, referred to in the introduction to this chapter, and probably also originated by rivers of the past, conduce to the same end.

* *Vide* Chaps. X and XI.

Railways seem, with the solitary exception of the canals whose construction dates just one stage farther back, to be the first and only lines of route which did not shun the alluvial bottom lands. They have indeed made a fuller use of the valleys in their search after gentle gradients, but they do not commonly follow all the windings of the valley: a good example is the line between Oxford and Worcester, where it threads the windings of the Evenlode valley just east of the Cotswolds proper, along an unbroken succession of alternate embankment and cutting but without any tunnel till it has passed the water-parting between the Thames and Severn drainage.

Ever since the far-off days of the Roman occupation of Britain there have been roads, built like the modern railways for through traffic but with much less ambitious engineering feats, in the relatively difficult transverse direction; but the very oldest tracks of all, and much of the close network of modern roads, use where possible the easy grades found along lines parallel with the major alignments of the surface relief. Such an ancient track was the Icknield Way running, with its westward continuation the Portway, the whole length of the region close under the Chalk scarp, on a natural bench formed usually of the underlying Greensand. An even smoother, lower line of passage roughly parallel with it is followed, but not from end to end, by portions of the Great Western railway system. Other roads avoid inconvenient gradients by following the skyline or the summit of a plateau, a habit almost completely alien to the railroads.

Routes of all kinds are, of course, controlled primarily by the needs of traffic, but in this scarpland region at all events they may be said to show a degree of adjustment to the facts of physical geography quite comparable with the physiographical adjustment of rivers to geological structure.

Distribution of the Population

With only three towns of over 50,000 inhabitants, none with 100,000, and less than twenty in all with as many as 5000, the region has a very much lower proportion of urban population than most of England. A number of older populated centres with even fewer than 5000 inhabitants—*e.g.* Buckingham, long superseded as a county town by Aylesbury—have the character of towns rather than villages, but nearly two-thirds of the population is rural, settled either in small villages or unconcentrated; and most of the urban centres are dependent, indeed parasitic, upon the agricultural pursuits of the adjacent districts.

The average density of population, excluding large and small towns, is about 200, and including them, little over 300, to the square mile (say 75 and 120 per km.^{2}), *i.e.* considerably less than half the mean density for England and about five-eighths of that for the whole of Great Britain. Altogether the region has about the same population but a slightly smaller area than the state of Connecticut or, to make a European comparison, about a quarter of the area, but one-third of the population, of either Denmark or Switzerland.

It is only on its north-western confines, where, as has been already suggested, it has a less conspicuous physiographical boundary, that the region has undergone any extensive industrialisation. Here in Northampton, Kettering, and smaller satellite towns near by, boot manufacture on a large scale has found its principal home in England.

In other parts of the region the distribution of population is directly or indirectly dependent on farming needs and opportunities. Practically all populous centres are either markets serving the local trade of the districts round them or else recent growths based upon traffic requirements—of the railways particularly: the workshops of Swindon on the Great Western and Wolverton on the L.M.S. railway are the largest examples.

An older disposition of smaller but often considerable settlements at nodal points in the road system, and especially in critical positions such as gaps in hill ranges or traditional river crossings, also shows the close connection between population and organised communications.

The foundation of the modern distribution of population, however, is to be found in the comprehensive network of parishes dating, with only minor modifications, at least from Anglo-Saxon times. It is plain that this territorial division represents an adjustment of small agricultural communities to the locally diverse conditions of soil, water supply, and natural vegetation.

Of these three considerations the second, water supply, is no doubt of vital importance but has possibly been over-emphasised, although it is true that both excess and deficiency of water at or near to the surface have exercised significant control of human settlements. Population, for instance, tends to be thin both on alluvial bottom lands and on dry uplands with notably permeable soils, but to be grouped in chains of villages on eminences—notably on river terraces—in the vales, or along the perennial streams and occasional "bournes" of plateau country like the White Horse Hills.

The occurrence of many villages and even some small towns at or close to the line of junction between a stratum of clay and some overlying porous bed, with resulting development of springs, is commonly explained as the outcome simply of the favourable conditions for water supply. The advantage resulting from diversities of soil within a short distance, in the form of a wider range of opportunities for the local farmer, however, has been generally under-estimated. It may even be regarded as the main factor in explaining the existence of another series of aligned settlements where the limestone plateaus in the north-west of the region begin to rise out of the great vale on the Oxford Clay.

The keynote of the South-Eastern Midlands is its essentially rural character. Such internal contrasts as it does afford are in complete harmony with the alternate belts of scarped upland and vale arising naturally out of the geological structure. Wholly inland, it may properly be regarded as a part of the English Midlands, but a juster characterisation would be effected by regarding it as a borderland, a zone of passage and in some ways of transition, between the "metropolitan" London Basin and the largely industrialised inner or true Midlands farther to the north-west. The creeping fringe of outermost London has hardly invaded the region save where the "garden city" of Letchworth lies, significantly, just beyond one of the larger gaps in the upturned rim of the London Basin; in mid-Northamptonshire, on the other hand, industrialisation begins to be a conspicuous feature.

Over much of their area, and from more than one point of view, the South-Eastern Midlands have retained more of the character of the England anterior to the Industrial Revolution than the regions adjoining. The greatest changes of recent date have for the most part arisen out of the growth of lines of through communication forming, as it were, so many bridges across a still rural, non-metropolitan, non-industrial zone which in itself seems like a surviving fragment of an England of the past.

VIII

EAST ANGLIA

Percy M. Roxby

Definition of the Region

East Anglia, as defined in the regional scheme adopted for this volume, coincides closely with the historic kingdom of that name. The coincidence is not fortuitous, for the East Anglia of Anglo-Saxon times was a definite physico-political entity. Its area of characterisation—a relatively open plain or rather low plateau dipping gently eastwards to the North Sea from the blunted escarpment of the Chalk—had a formidable framework of natural frontiers or borderlands. The North Sea defined it on the east and north, the bay of the Wash on the north-west. Its western borderland was the vast swamp of the Fenland basin, "a labyrinth of black wandering streams, broad lagoons, morasses submerged every spring-tide, vast beds of reed and sedge and fern, vast copses of willow and alder and grey poplar, rooted in the floating peat." To the south was the great forest of Essex corresponding closely to the exposure of the heavy London Clay. In two directions its isolation was less pronounced:

1. In the south-east where, however, the narrow belt of alluvial swamp in the lower valley of the Stour formed its mark against the East Saxon (Essex) Kingdom, and
2. In the south-west where the critical neck of open downland, about ten miles in width, between the Fenland and the Forest, connected it with the Chilterns and the central boss of the Chalk.

With the absorption of the East Anglian Kingdom, the depletion of the forest and the drainage of the fens, the connotation of East Anglia naturally tended to be widened, but for the purposes of this essay East Anglia consists of Norfolk, Suffolk, and that portion of south Cambridgeshire which belongs essentially to the upland edge of the region. It excludes, however, the small fenland portion of north-west Norfolk.

Physical Geography

Although East Anglia is a land of low relief, its physical geography presents many features and problems of special interest which are very clearly related to the study of its historical evolution and economic development.

Geologically it is among the youngest regions of England, and its initial form must have been determined by the earth movements of Tertiary times. A prominent feature of the area is the changing direction of the strike of the Cretaceous rocks. In south-west Suffolk and south Cambridgeshire, the strike is SW.–NE. continuing the line of the Chilterns.. In north Suffolk, it is S.–N., and

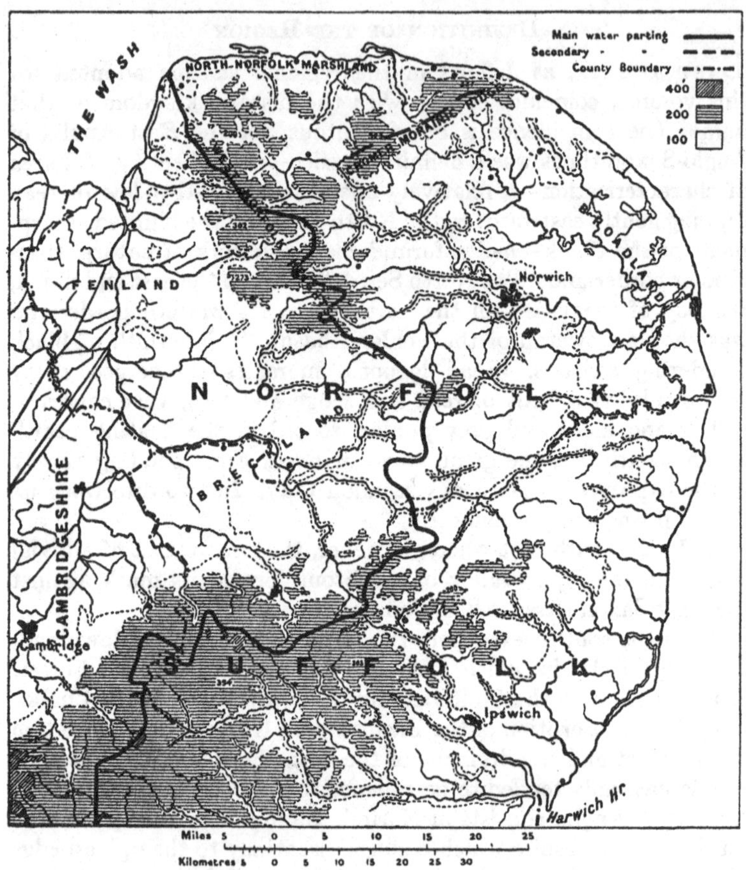

Fig. 24. East Anglia: Relief and Drainage.
(For names of rivers and towns, see Fig. 25.)

finally, north of the depression occupied by the Little Ouse and Waveney, the strike is deflected SSE.–NNW. through Norfolk. The Wash, which significantly marks another change of strike to SE.–NW. in Lincolnshire, is essentially an arm of the sea created by river breaching on a large scale, and accentuated by the late subsidence which affected all East Anglia.

Solid Geology

The whole of western and central East Anglia rests on a foundation of Cretaceous rocks. The oldest formation with which we are concerned is the Lower Greensand which determines the scenery and soils of north-west Norfolk to the east of the Fenland. It contains one clay series (the Snettisham Beds), but its dominant character is that of sands and ferruginous sandstones which form two minor escarpments, and give rise to the typical heather-clad uplands developed round the King's estate at Sandringham. The houses in this rather distinctive sub-region are largely built of the dark-brown massive Carstone or "gingerbread" stone, which, apart from flint, forms almost the only valuable building-stone in East Anglia. It has been particularly used in the construction of the modern watering-place of Hunstanton.

Chalk is the dominant formation of East Anglia, and occupies nearly two-thirds of its entire surface. It is, however, only in the west that it directly determines the character of the scenery and the soil and gives rise to the phenomena of a chalk country so characteristically displayed in the Downlands of southern England. In south-east Cambridgeshire and south-west Suffolk it forms a well-pronounced hilly belt over 300, and in places 400 ft. (90 and 120 m.) in elevation rising fairly sharply from the Fens, although the three distinct scarped ridges, corresponding to the outcrop of the Totternhoe stone, Melbourn rock and Chalk rock, which distinguish the belt south-west of Newmarket towards Hitchin, are no longer clearly defined. Then comes the remarkable depression round the valley of the Little Ouse (*v.s.*), and on either side of this for a considerable distance the Chalk forms no more than a very low sand-covered plateau (Breckland), merging westwards almost imperceptibly into the Fens. From about Swaffham northwards, however, it again assumes the form of a definite ridge (the West Norfolk Heights) touching 300 ft. at its highest points, and terminating in the Hunstanton cliffs. The western edge is much blunted and it is the Lower Greensand rather than the Chalk which presents a low escarpment to the Fens. Relatively free from drift, the West Norfolk Heights form gently rolling uplands with typically rounded contours. Combes and dry valleys like that of Ringstead near Hunstanton occur, and there are some instances of intermittent streams or winter-bournes. Some horizons of the Chalk in west Norfolk are particularly rich in flints and at Grime's Graves, three miles north-east of Brandon, occur the celebrated prehistoric flint-pits (sunk into a particular flint-layer about 40 ft. (12 m.) from the surface) to which reference is made later.

From this upland edge on the west, the Chalk dips gently eastwards, becoming increasingly plastered with boulder clay. Thus the junction of the Chalk with the succeeding Tertiary deposits which occur along a fairly regular line drawn southwards from Weybourne on the north coast is blurred and has little geographical significance. Subterranean borings have proved that east of this line the downward slope of the Chalk surface increases much more rapidly, and that the Chalk disappears below the water-level to the east of a line drawn southwards from the coast near Mundesley.

The Tertiary formations overlying the Chalk in the eastern third of East Anglia are essentially shallow-water deposits. The London Clay of Essex only affects the south-eastern border of our region, though there are extensions of Eocene deposits under the Crag in east Suffolk and east Norfolk. "Crag" is a collective term given to the series of sandy, shelly deposits of Pliocene age which form the principal East Anglian Tertiaries, and were laid down under marine conditions which became estuarine in the later phases*. The geographical effects of these Tertiaries are relatively small. It is only on the slopes of the more deeply-cut of the river valleys such as those of the Bure and the Wensum that they directly influence vegetation, but they have a considerable effect on the water supply, tending in eastern Norfolk and Suffolk to diminish the amount which can be obtained from the underlying Chalk.

Glacial Deposits

The increasingly arctic conditions indicated by the later Pliocene deposits were the prelude to a glacial invasion or series of invasions which have profoundly affected the topography, river-drainage, and soil conditions of modern East Anglia. No part of the country affords a more interesting and important field for the study of both the phenomena and the consequences of glaciation. Only the largest features and results can here be indicated.

1. The low plateau projecting eastwards between the Wash and the Thames estuary was exposed to the full force of the advance of ice-sheets from the area now occupied by the North Sea. These ice-sheets contained material derived from the main focus of glaciation in Scandinavia, but carried also erratics from the contributory ice of Scotland and northern England. At a later stage, there was almost certainly invasion by an "inland" ice-sheet from the north-west via the Fenland.

* The estuary was probably that of the extended Rhine.

2. The sequence of glacial deposits, which are very complete in the districts round Cromer and Ipswich, has not yet been fully interpreted, but probably indicates three distinct glaciations, perhaps corresponding to the Mindel, Riss, and Würm glaciations of Penck and Brückner.

3. The deposits derived directly or indirectly from the ice-sheets and their subsequent melting are of an extremely complex and varied character, and it must suffice to distinguish the chief regional variations (see Fig. 25, p. 170):

(*a*) The Cromer Ridge, a conspicuous belt of picturesque, hummocky hills, trending WSW. from near Weybourne and Mundesley, has an average width of five to six miles, and a length of 15 miles (24 km.). The extreme elevations reach over 300 ft. (90 m.). The ridge is essentially a morainic dump, and its medley of sandy and gravelly deposits probably represents the "terminal moraine of the North Sea Ice Sheet."

(*b*) Round the Cromer Ridge, but especially on the south-east side, lies a wide area of glacial loams and brick-earth, probably representing outwash material from the moraine. This area has some characteristics of the European loess belt, and is one of the most fertile and earliest settled portions of East Anglia. Loams of similar character are spread over the low plateau of east Suffolk south of the Broadland, but are more interspersed with pure sands and gravels which form typical heaths in the neighbourhood of Dunwich and Orford.

(*c*) The deep mantle of boulder clay lying over the centre of Norfolk and Suffolk (the plateau of "High Suffolk" and "High Norfolk") was carried in all probability by the later inland ice-sheet from Lincolnshire, and thence derived its distinctly chalky character. The boulder clay area forms the rather featureless plateau of central East Anglia, but the soils are mixed in texture, and differ from the stiff tenacious clays of the London Basin or the Weald. They are mainly strong loams. Originally this central belt was the woodland region *par excellence* of East Anglia, presenting much greater obstacles to settlement than either the light maritime tracts or the open, if rather sterile, uplands of the western borders.

(*d*) In western East Anglia the glacial drift thins off as it approaches the main watershed, and there is a considerable area of wind-drifted sand associated with gravels. Here, however, we must distinguish two sub-regions:

(i) The low plateau of Breckland between the southern Chalk Escarpment and the West Norfolk Heights where sand and gravel rest directly on the Chalk. The soil is thin, gritty and unstable, and from an economic viewpoint the area is "marginal," *i.e.* it will not be cultivated unless high prices make the tillage of poor land profitable.

(ii) The "Good Sands"* region of north and north-west Norfolk lying between Breckland and the Cromer Ridge. Initially it was less attractive for agricultural settlement than the Loam Region, but, under the scientific treatment introduced by the capitalist farming of the Agrarian Revolution, it attained high celebrity, and is pre-eminently associated with the Norfolk four-course system.

River Drainage

The river drainage of East Anglia presents many problems arising out of the complications due to relatively recent glaciation. Essentially the chief watershed follows the Chalk Escarpment, and an important secondary watershed is determined by the morainic Cromer Ridge. The main rivers of central and south Suffolk, such as the Deben, Gipping, and the Stour as a whole (although this exhibits alternations of transverse and longitudinal stretches), like the rivers of north Essex, are clearly consequent on the dip from the Chalk Escarpment and their drainage lines were probably initiated in Miocene times when the early salient features of the topography were produced. Their lower valleys are markedly drowned as the result of the post-glacial subsidence. On the western side, the Nar and Wissey in Norfolk, and the Thet, Little Ouse, and the Lark in Suffolk, are obsequent streams, flowing into the Great Ouse. The latter as the master subsequent river of the easily eroded clay belt between the Cretaceous and Jurassic scarps is probably responsible for considerable capture of head-waters of eastward-flowing East Anglian streams, and its obsequent tributaries are still working backwards into the retreating Chalk Escarpment. In Norfolk, the Wensum-Yare system seems on the whole to be consequent on the main dip, but its present form has certainly been affected by the earlier glaciation. The Cromer Ridge is one of the best examples of a morainic water-parting in Britain. From it the Glaven, Stiffkey and other smaller streams flow northwards, while the Bure drains southwards to a common outlet with the Wensum-Yare and Waveney in the former bay of Broadland. In general, the drainage of Norfolk is to a marked extent due to post- and inter-glacial features.

The river course most apparently anomalous and difficult of interpretation is that of the Waveney. It forms part, as already indicated, of the remarkable through-valley of which the Little

* The term is Arthur Young's.

Ouse occupies the western portion, and has its counterpart in an ancient depression of the Chalk. Its present form was probably determined in Pleistocene times, and it is possible that the depression was at one stage an important outlet for imprisoned glacial waters from the north.

Coast-line

Of the many factors which have shaped the interesting and picturesque East Anglian coastlands, the following are of outstanding importance:

1. The loose and easily eroded materials composing the cliff sections which readily crumble under the combined action of surface water and wave attack.

2. The southward movement through tidal and wave action (especially marked during strong north-easterly gales) of shingle and sand, derived in part at any rate from the waste of the East Anglian coast and of cliffs much farther north.

3. A very definite post-glacial subsidence of some 70 ft. (21 m.) which continued probably to Middle Bronze Age times. Evidences of this are seen in the drowned lower valleys of the Stour, Orwell, and Deben. The northern coasts show a relatively symmetrical outline, but there is no doubt that Norfolk too was affected by subsidence. There is, for example, the submerged forest of Brancaster, and the fact that as late as the Roman occupation the alluvial flats of Broadland formed a wide bay with great inlets reaching in the drowned lower valley of the Wensum-Yare to the site of Norwich. All evidence points to the gradual progress of sedimentation from north to south under the influence of longshore drift of eroded material. Thus the silting-up of Broadland was primarily due to the growth southwards of a great spit or storm-beach of shingle and sand which gradually formed between the open sea and the bay an embankment sufficiently firm to allow the foundation of the town of Yarmouth on it in the 11th century. The formation of this barrier restricted the influx of the tides and facilitated the deposit of material by the numerous rivers entering the quiet waters of the bay. The greater part of it now forms valuable reclaimed marshland, but the deeper hollows of the original bay are still occupied by the charming reed-bordered sheets of water known as "The Norfolk Broads." The same process is exhibited in the long shingle spit known as Orford Ness which diverts the mouth of the Alde twelve miles south of its original outlet at Aldeburgh. A similar accretion at Landguard Point on the northern (Suffolk) side of Harwich Harbour, formed by the

combined estuaries of the Orwell and the Stour, was checked by the construction in 1866 of a jetty across the spit, which has resulted in the accumulation of an immense amount of shingle on its northern side. Except in so far as man interferes to maintain important sea-gates such as this, the normal process on the south Suffolk coast would seem to be that of the gradual conversion of open estuaries into lagoons and ultimately mud-flats and the progressive regularisation of the shore-line, such as has already been accomplished in Norfolk*.

Probably in no region of Britain save in Holderness (Yorkshire) has coastal recession within historic times been so marked as in East Anglia. In the section which marks the seaward edge of the Cromer Ridge between Weybourne and the northern limit of Broadland, the greater part of six villages has been washed away since the Norman Conquest. The mean annual loss has been calculated at various points at from one to nearly five yards, and the total loss since the Roman occupation is probably from two to three miles. The Cliff sections of Suffolk have suffered almost equally from marine erosion, and the considerable mediaeval seaport town of Dunwich has virtually disappeared, though it is probable that the accretions represented by the North Norfolk Marshlands, the Broadland flats, and the various shingle spits and sand-dunes more than counterbalance in point of area the land which has been lost.

Climate

From the climatic standpoint, East Anglia belongs to the larger "province" of Eastern England which is distinguished, as a result of its low relief and its geographical position in relation to the moisture-bearing winds from the Atlantic on the one hand and to anti-cyclonic influences from the Continent on the other, by its relatively warm summers, cool winters, low rainfall and rather high proportion of sunshine.

Generally speaking, westerly winds greatly predominate. The ratio of winds with a westerly to winds with an easterly component has been calculated at Lowestoft as seven to four, on the

* The north coast of Norfolk is in some ways exceptional. For the probable explanation of the westward-projecting shingle and dune headlands of Blakeney Point and Scolt Head, see the important paper by J. A. Steers on "The East Anglian Coast," *Geog. Jour.* Jan. 1927. It should be noticed that the East Anglian Memoirs of H.M. Geological Survey are now appearing in a new series. Valuable papers on various aspects of the geology and physical geography of East Anglia will also be found in the *Transactions of the Norfolk and Norwich Naturalists' Society*, the *Journal of the Ipswich and District Natural History Society*, the *Quart. Jour. Geol. Soc.* and *Proc. Geologists' Assocn.*

basis of twenty years' observations. There is a marked tendency for East Anglia to be affected by an extension of the continental anti-cyclone, especially in the spring months when cold and searching east winds are a common phenomenon. The North Sea does not greatly modify these continental influences. In the shallow basin south of the Dogger Bank, the temperature of the waters is quickly affected, and the water off the East Anglian coast is said to be colder in midwinter and warmer in midsummer than along any equal stretch of coast in Britain. It is especially cold in spring when the tidal drift from the north is reinforced by cold water escaping from the Baltic after the melting of the winter ice. This circumstance, in conjunction with the prevalence of easterly winds, helps to explain the marked lag in the upward grade of temperature over East Anglia in spring as compared with the slow descent in autumn. It should be noted that the North Sea fisheries, so important to the East Anglian coast towns, are largely controlled by the relative temperatures of the waters. The mackerel, for example, comes north into Norfolk waters in May when the North Sea temperature begins to exceed the temperature in the Channel and returns south in October–November when the North Sea temperature is falling rapidly. The herring arrives in East Anglian waters from the north in September.

The duration of sunshine is an important factor in agriculture, and the eastern counties have a distinct advantage in this respect, the average daily total of 6·05 hours for the critical months June–September being exceeded only in the south-east with 6·5 hours. The high proportion of sunshine in September, 5·1 hours, is particularly important in late seasons.

The seasonal distribution of rainfall in East Anglia is of considerable significance. The wettest month is October, and this rainfall is associated with the passage of cyclonic storms. But the chief feature of the rainfall here, as compared with the country as a whole, is the high proportion which the summer fall bears to the total for the year. The late winter and spring months are markedly dry, and this contributes greatly to the successful cultivation of wheat and barley. It is an important point that the variation from the mean rainfall is less in Eastern England than in any other part of the British Isles.

Natural Vegetation

The generally light nature of the soils, the distinctly low rainfall, and, in the coastal regions, the prevalence of strong winds were factors operating to prevent East Anglia from becoming that "sea of woodland" which the clay vales of the Midlands, the London Basin, and the Weald presented to the men of the Neolithic and Early Metal Ages. Both the ecological and the historical evidence point to the fact that only on the boulder clays of central Norfolk and Suffolk was there oakwood forest of a comparatively dense type; and it is certain that this was more easily and earlier cleared than the historic forests of the south-east, centre and west of England. Outside High Norfolk and High Suffolk, we may think of essentially light forests in the Loam Region of north-east Norfolk and the Greensand Belt of the north-west. Apart from the alluvial and marshy lands on the north coast and in Broadland, the greater part of the remaining area was either open heath or grassland, forming a very considerable percentage of the whole. Three sub-regions of this "open" character are of special importance:

(*a*) The belt of chalk-downs along the Suffolk-Cambridgeshire border around Newmarket—the East Anglian Gateway, as it may be called—where grassland of the Wiltshire Downs type predominated, although rather on the slopes than on the summit-levels.

This grades into (*b*) Breckland, already defined, where drifting sand, thinly covering the Chalk, produced what can best be described as "heath," although dry grassland occurs in patches. The flora of the country round Brandon, Thetford, and Mildenhall is in many respects peculiar, containing more than ten plants practically confined to this area. This may be a survival from a post-glacial steppe flora, and probably it "exhibits the nearest approach to steppe-conditions to be found within the British Isles."

(*c*) The maritime belt of Suffolk widening in the south-east. Here the glacial sands and gravels and the underlying Crag impart to the soil a very light character, more favourable to heath than forest. Much heath and common still remain, although a much larger proportion has been utilised for agriculture than in Breckland. For this sub-region the old name of the "Sandlings" may conveniently be retained, and the boundary between it and the "Woodlands" is approximately along the line of the main railway from Ipswich to Beccles.

Natural Regions

This brief sketch of the physical geography of East Anglia thus reveals distinct sub-regions such as Breckland, the Loam District, and the Sandlings. They may be considered as comparable in content and character to the well-known *pays* of the Basin of

Paris, although lacking an equal heritage of historical associations. They do, however, indicate geographical realities and may be treated as entities for the study of settlements and of agricultural development. Reference will frequently be made to them in the remaining sections dealing with the historical development and modern economic geography of East Anglia, and their main characteristics are indicated in a summary facing Fig. 25 (p. 170), where their approximate boundaries are shown.

Historical and Human Geography

In the study of the evolution of man's relation to his physical environment, few English regions surpass East Anglia in interest. Its present human geography is so essentially the outcome of long centuries of development that, even within the small compass of this essay, it seems best to approach its salient features historically.

In the important prehistory of East Anglia, Breckland stands out as the chief region of primary settlement. The absence of forest, the dry heathland interspersed with numerous little meres which favoured fowling, the abundance of good flint, and, after the introduction of agriculture, the suitable soil of the valleys of the Little Ouse, Lark, and smaller streams were conditions which made it specially attractive to primitive man. The country around Brandon, Thetford, and Lakenheath contains a very considerable proportion of the authenticated Palaeolithic sites of Britain; and the immense number of implements (chipped axes, knives, and borers) found in and around the valleys of the Little Ouse and Lark show that no part of Britain was more favourable to Neolithic culture. That even in the Neolithic Age this region was in trade contact with other centres of population by way of the Chalk Escarpment is indicated by the relatively immense size of the famous "flint factory" at Grime's Graves, north-east of Brandon (20 acres—8 ha.—in extent and containing 346 pits).

The new cultures and racial types characteristic of the Early and Middle Bronze Age (approximately 2000–1000 B.C.) have a characteristic distribution in East Anglia, closely repeated by the later Teutonic settlements. The well-known Beaker burials are found (*a*) in the valleys of the Little Ouse, Lark, and Wissey; (*b*) in the East Norfolk Loam Region and round the Suffolk estuaries, indicating approach by the Fenland rivers and the coastal inlets respectively. Innumerable Round Barrows on the Chalk belt and the occurrence of Breton axes and of gold ornaments show that East Anglia participated fully in the relatively advanced culture which developed on the open uplands of southern England.

Along the slope of the chalk beneath the forest limit runs the track or series of tracks known as the Icknield Way, one of the most ancient of the prehistoric roads of England, linking from Early Bronze, if not from Neolithic, times, the Chalk Downs of Wilts-Dorset with the heathlands around Thetford, approximately its East Anglian terminus. It must, however, have been closely linked with an ancient track, the precursor of the Roman Peddar's Way, which continues its natural line along the West Norfolk Heights to Holme-next-the-Sea, close to the entrance of the Wash (see Fig. 25 and also Map of Roman Britain (Ordnance Survey—Second Edition)).

In the Early Iron Age tribal territories began to take definite shape in accordance with the larger physical features of the country, and settlement areas began to be linked politically in harmony with their natural space-relations. The East Anglian Gateway had increasingly the character of both a cultural and political frontier. The remarkable series of earthworks which cross the chalk neck transversely between Forest and Fen (Fig. 25) is now definitely assigned by archaeologists to Anglo-Saxon times, *i.e.* to the period of the Mercian-East Anglian wars. Their structure, with the fosse on the outer side, shows that they were intended to defend the East Anglian plain on its one vulnerable aspect. The longest (7½ miles, or 12 km.) and most formidable of these "Cambridgeshire Dykes" is the famous Devil's Dyke across Newmarket Heath. Both this and the outer Fleam Dyke have been shown to contain Roman remains under their banks and all are clearly later than the Icknield Way. But it has also been demonstrated from the study of coin distributions that the approximate line of the Devil's Dyke was the frontier of the Iceni, whose tribal territory probably covered much the same area (Norfolk, Suffolk and south-east Cambridgeshire) as the later Kingdom of the East Angles. The natural boundary persisted into Anglo-Saxon times and was then fortified. So too the defensive line of the River Stour, which formed the frontier between the East Anglian and East Saxon Kingdoms, is proved to have been important as a cultural boundary in the Early Iron Age*.

The Roman occupation of East Anglia was in the nature of an episode. The road system shows the continued military importance of the Cambridgeshire entry at the southern apex of the Fens,

* I am indebted for this revised statement about the Cambridgeshire Dykes and the Stour boundary to correspondence with Dr Cyril Fox whose *Archaeology of the Cambridge Region* is a masterly study of prehistoric distributions in relation to primitive physical conditions. The later date of the Devil's Dyke has been demonstrated since the appearance of his book. See the *Cambridge Antiquarian Society's Communications*, vol. XXVI.

where the so-called Via Devana intersected the Icknield Way and the parallel Street Way, usually esteemed Roman, but quite possibly a prehistoric "summer" trackway alternative to the Icknield. The relative paucity of Roman remains, especially of villas, in East Anglia probably reflects to a large extent the ruthless devastation of the Icenian country after the famous revolt under Queen Boudicca. In the final stage of the Roman occupation, when the Saxon pressure on the east coast was already becoming severe, the importance of the Wash entry is indicated by the selection of Branodunum (Brancaster) as the site of a fort under the control of the Count of the Saxon Shore, although there does not seem authority for the statement often made that it was his headquarters. Other forts controlled the entries into the bay of Broadland and the Stour-Orwell estuaries.

The Anglo-Saxon period both repeats the large features of the prehistoric invasions and lays the foundations of all subsequent development. On the assumption that the distribution of Anglian cemeteries, as shown in Mr Thurlow Leeds' maps, reflects the extent and character of colonisation during the Pagan period (approximately A.D. 450–650), we have a picture strikingly similar to that of the Early Bronze Age. Important groups occur, on the one hand, in the Lark and Little Ouse valleys, and, on the other, round the Suffolk estuaries and the borders of Broadland. The two latter, in close proximity to Ipswich and Norwich respectively, certainly indicate the nuclei of Suffolk and Norfolk. In the central boulder-clay region the cemeteries are significantly few; they are rarer in High Suffolk than in High Norfolk where the forest was less dense and where glacial gravels and brick-earth frequently occur in the interior valleys. But the essential unity of our region led to an early amalgamation of the different groups of settlements into the East Anglian Kingdom with a definite military frontier across the neck of Chalk. The inner and shorter line of defence known as the Fleam Dyke was apparently constructed at this period. The Fens had relapsed, after Roman times, into their original condition. In the 10th century Abbo, a monk of Ramsey, could speak of East Anglia as almost water-girdled. For a brief period in the 7th century, under Redwald, who established suzerainty over all the English Kingdoms south of the Humber, the frontier was advanced to hold the critical bridge-head of Cambridge and the southern Fens were incorporated, but East Anglia lacked the great geographical advantages which enabled Wessex to become the real maker of England. Nevertheless, in the Danish period, East Anglia was able to retain its identity, and, in the subsequent administrative reconstruction of England by Wessex, the original entities were preserved in the essentially "natural" counties of Norfolk and Suffolk. These had significantly a common ecclesiastical centre, first established at Dunwich, then later moved to the ancient town of Thetford, at the terminus of the Icknield Way, and finally placed at Norwich, whose position at the head of the Broadland estuaries and controlling both land and sea-routes fitted it to be a great regional capital.

Within an area of characterisation so rigidly defined, the successive and relatively numerous groups of colonists, Anglian and Danish, were compelled by force of circumstances to settle more closely than was the case in Mercia or Wessex whose western limits were capable of great extension. By the time of the Great Survey (1086)* East Anglia was the most populous region in England. On the basis of Mr Ellis' estimates, the density of population in Norfolk and Suffolk was nearly twice that of the counties of the southern Midlands and Thames basin and many times that of the counties of the Welsh Border and the North. Confirmatory evidence is afforded by the closeness of the nucleated villages to each other and the relatively immense number of parish churches.

Yet within East Anglia there were significant regional differences of development. In Breckland the vills of Domesday are still confined to the favoured valley sites of prehistoric times, and the dry heathlands long remained open sheepwalks, a character which they have by no means lost even to-day. In greatest contrast are the rich loam districts of east Norfolk and parts of south-east Suffolk and certain island-like tracts in Broadland. Perhaps no part of England, unless it be the plain of south Somerset, shows a more intensive rural settlement than the triangular tract north-east of Norwich with Aylsham and North Walsham as its centres. In the marsh-encircled district of Flegg, only eight miles by six in extent, there were in 1087 no less than 27 distinctly marked parishes, mostly with Danish names.

The initial impetus to economic activity in East Anglia provided by relative density of population within a limited area was stimulated in the mediaeval period by its relation to the Lower Rhineland. The transference of the national capital from Winchester to London was symbolic of an eastward movement of the centre of economic and political power and of an increasing orientation of trade to the countries reached across the narrowest part of the North Sea, which became much more marked in the centuries

* The Survey of Norfolk, Suffolk and Essex, much fuller than for the rest of the country and known as Little Domesday Book, was perhaps not completed until 1100.

following the Norman Conquest. Flanders, closely linked with the great fairs of Champagne and eastern France, with the Hanseatic organisation of the Baltic and the Venetian organisation of Mediterranean and Eastern trade, greatly affected the development of East Anglia. As late as the 17th century, Reyce, the author of a quaint "Breviary of Suffolk," habitually refers to the North Sea as the "Flemish Sea." The breweries of Ghent and Bruges became a great market for barley grown on the light soils of eastern Norfolk and Suffolk, which are pre-eminently suited for this crop, and East Anglian wheat was similarly exported to feed the industrial population of the Low Countries. Above all, its ports became centres for the export of the greatest of English mediaeval products, raw wool, which was also the commodity most demanded by Flanders. The Breckland region was long famous as a sheep district. Much of the English wool exported, however, came from more pre-eminent sheep regions in the West and North, brought thither by pack-horse routes converging on the East Anglian Gateway. It is significant that the greatest of English fairs was held at Stourbridge close to Cambridge, at the southern apex of the Fenland basin, at the point where East Anglia made contact with the Midlands and North.

The transition from the active export of raw wool to the establishment of the celebrated textile industry of East Anglia in the later Middle Ages was essentially a natural process. The prime factors were—first, accessibility to raw material, much of it derived from outside the region; secondly, intimate commercial contact with Flemish merchants and industrialists; and thirdly, the presence of a relatively dense population and the accumulation of capital as the result of active continental trade.

In one important respect indeed, East Anglia was badly handicapped. There was an almost complete absence of water-power. In the long run this disadvantage, in conjunction with an essential change in the value of her space-relations, proved fatal, but even in the heyday of her industrial greatness it prevented the development of woollen manufactures requiring milling, for which water-power was essential. The East Anglian industry was essentially one of worsted goods, made from long wool which was "combed," as distinct from true woollen goods such as broadcloths made from short wool, "carded" and "milled." This, as Mr Kinvig* has shown,

* R. H. Kinvig, "The Historical Geography of the West Country Woollen Industry," *Geog. Teacher*, vol. VIII, 1916. For the economic organisation of the Stour valley textile industry, see the late Prof. G. Unwin's article on "The Economic History of Suffolk" in the *Victoria County History*, an invaluable work of reference for both Norfolk and Suffolk.

was the essential difference between the East Anglian industry and that of the "West Country" which, developing rather later, co-existed with it for many centuries. The two regions produced fabrics of quite a different character and the underlying cause was the presence of water-power in the one case (Cotswolds-Mendips-Downs) and its absence in the other.

Within East Anglia itself, two distinct foci must be distinguished. The first is east Norfolk, with Norwich as its metropolis, which developed a great variety of pure worsted manufactures such as damasks and camlets, and, later, mixed fabrics of worsted* with silk such as crapes and bombazines.

The second district is the Stour valley on the borders of Suffolk and Essex. The Stour, flowing from the western uplands, affords a little water-power, and probably for this reason we find the region in Tudor times producing not only worsted fabrics such as serges and shalloons, but kerseys (the name taken from a small Suffolk village) and baizes which were in the nature of rough woollen fabrics. These required a little water-power for scouring, but not the full milling process, and they were not true broadcloths in the West Country sense.

For four centuries (14th to 18th), East Anglia was the premier industrial region of England, and some of its present characteristics can only be interpreted in the light of that fact. The evidence of its material wealth in this period is written large on the face of the land. It is a region of noble churches, built of local flint with free-stone imported from the Oolitic belt for corner work, while the close contact with Flanders led to the early adoption of brickwork, as seen in the numerous manor houses and guildhalls.

The wide ramifications of the textile manufacture, not only creating a capitalist class but, as an essentially domestic industry, affecting the life of the villages for miles around the chief centres, developed a distinctive East Anglian public opinion on national affairs. The financial and foreign policy of the Tudors was closely scrutinised. The creative part played by East Anglia and the Eastern Midlands in the movement which produced the voyage of the "Mayflower," the organisation of the Eastern Counties Association and of Cromwell's Ironsides during the Civil War are prominent illustrations among many of the leadership of this region during many critical centuries of English history. It produced too a very high proportion of statesmen and men of affairs, and Mr Havelock Ellis, in his study of the distribution of British genius, gives the East Anglian focus first place. The varied agri-

* From the little town near Norwich.

cultural, industrial, and commercial interests of East Anglia and its close contact with one of the most active and important regions of the Continent gave it a relatively intense economic and political life and put it naturally in the vanguard of forward movements.

One of the most far reaching of these contributions was made at a time when its industrial supremacy was already beginning to pass to the newer textile regions of the North with whose resources in power, both of water and coal, it could not compete.

Agricultural Geography

The modern phase of East Anglian economic life begins in the 18th century with the Agrarian Revolution. Centuries earlier, the growth of industry and pressure of population had impelled extensive enclosures in the fertile "Norfolk Loam" districts. But the new needs and the new resources of the 18th and early 19th centuries pointed to the more intensive use of the sandy soils of the east and south-east of the region, and also north-west and north Norfolk. It was the latter district described as the "Good Sands" region which was more particularly the scene of the New Agriculture and of the celebrated Norfolk Husbandry.

Until late in the 18th century it was largely open, and a good deal of it was characterised metaphorically by the dictum that two rabbits fought for a single blade of grass. It was here that the famous Coke of Holkham and other great landowners effected a revolution of agricultural method which was soon adopted in other regions of a similar light-soiled character, such as the Yorkshire Wolds, and helped to make possible the feeding of England during the Industrial Revolution.

The essence of the new system was to make the new root-crops, which East Anglia again owed to the countries across the North Sea, the basis of the whole rural economy. The cultivation of roots allowed sheep to be run over the light, sandy soils, and so prepared them for a cereal crop, particularly in the case of north Norfolk for barley. Clover and exotic grasses were also introduced as an essential part of the rotation proper to light soils. At the same time extensive reclamations of the adjacent coastal marshland were carried out and the rich pastures thus afforded were treated as an adjunct of the arable farms of the uplands. With the new supplies of winter fodder the breed of sheep was enormously improved, and sheep, turnips, and barley have ever since been the outstanding characteristics of north Norfolk farming (see summary on p. 171).

The new rotation and methods naturally affected the much older system of east Norfolk, but there are some interesting differences of regional development. In this Loam Region, with its better and stronger, although still relatively light soils, wheat is less subordinated to barley, which, however, remains the dominant cereal. In Marshall's time the east Norfolk rotation is given as: wheat, barley, turnips, barley, clover, rye, grass, broken up about midsummer and fallowed for wheat in rotation. But the most striking difference is the reliance on cattle instead of sheep. The grand object of husbandry was the fattening of bullocks (a name used to cover all kinds of cattle) for Smithfield (London) and other markets. These came by drover routes from long distances, especially from Scotland (Galloways and Highland cattle) and were bought by the east Norfolk farmers at celebrated fairs, particularly that of St Faith's near Norwich, at the end of the summer, for stall feeding on local turnips during the winter. "The whole system of the Norfolk management," writes Marshall*, "hinges on the turnip crop," and again, "The loss of the turnip crop deranges the whole farm." This, with certain modifications, still describes the situation. But the Loam Region like that of the Good Sands has its summer pastures. A study of the regional statistics for 1925 brings out very strikingly the different emphasis on sheep and cattle between the two arable regions and their adjacent marshland annexes. In the table on p. 168, the statistics of C and F^1 (the "Good Sands" and the North Norfolk Marshland) and of D and F^2 (the Loam Region and Broadland) have been respectively grouped together. The marked contrast between the use of the two great alluvial lowlands has an interesting analogy in that between Romney Marsh (sheep) and Pevensey Marsh (cattle).

The northern region is the classical district of large scale capitalist agriculture, distinguished by large farms, often running to 800 acres (324 ha.), let out on leases rigidly defining the rotation to be followed, and with substantial farm-buildings. Partly because these farms were essentially arable, involving much labour at harvest time, and partly because much of the land now brought into cultivation by the new methods lay at a considerable distance from the original village settlements, this region and also the chalk uplands, similarly developed, were the principal scene during the early 19th century of labour gangs specially imported for the harvest. In the long settled eastern region on the other hand,

* Marshall's *Rural Economy of Norfolk* (2 vols.) is the best contemporary account of the Norfolk Husbandry. See also Kent's *Norfolk* in the *County Reports* of the Board of Agriculture.

although the yeoman class greatly dwindled, the farms remained smaller and the system defined by leases has been less rigid.

The response of the other regions of East Anglia to the changes summed up in the phase "The Norfolk Husbandry" varied according to their soil conditions and situation. The East Suffolk "Sandlings" or "Sandlands" (to use ancient regional names) developed on lines in many respects similar to those of north Norfolk. An old established dairy industry gave place to a system of close folding of roots by ewes in preparation for barley. There, too, root-culture was recognised as the foundation of the barley crop, and barley and sheep became the main objectives of farming. In the estuarine area of the south-east, in close touch by sea with the London market, vegetables and particularly carrots (one of the cultures originally introduced from Flanders) adapted to light "warm" soils, formed an instance of regional specialisation. In the west the Chalk Downland of the Cambridgeshire border was mainly enclosed between 1790 and 1815, almost entirely for large scale arable farming, and this region (G) is still to-day one of the most exclusively arable districts in Britain. In the period of high prices during the Napoleonic Wars, extensive enclosures took place also in Breckland; but its sterile, shifting sands, unrelieved by any mixture of clay, have not proved susceptible of any remunerative crop rotation. It has remained chiefly a region of heath, rabbit warrens, and "catch crops."

The great increase of regional specialisation made possible by the Agrarian Revolution is strikingly shown both by the contrast between East Anglia as a whole and the Midlands (enclosures mainly for arable and permanent pasture respectively) and by the varied response of the different sub-regions within East Anglia.

The chief features of East Anglian farming established by the Agrarian Revolution remain unchanged, and are a striking tribute to the scientific perception which then laid down a form of agricultural practice essentially suited to the soil and climatic conditions.

From about 1875, owing to the import of relatively cheap foreign corn and meat, economic conditions have certainly been unfavourable to the East Anglian farmer except for the interlude of the Great War, but it is not easy to see how the essentials of the system can be altered. East Anglia is not a region of naturally "convertible husbandry," able to make rapid adjustments to changing conditions.

Yet some changes are occurring, and more are indicated. Even in the "Good Sands" region there has been some modification in favour of milk production and some conversion of arable upland

Agricultural Statistics of Sub-Regions of East Anglia, 1925

Per 100 *acres of cultivated land*

	Arable	Per-manent pasture	Wheat	Barley	Oats	Total corn crops	Total root crops	Sugar beet	Cattle	Sheep
A^1. "High Norfolk"	74·4	25·6	10·5	21·2	5·1	39·1	14·9	1·1	14 +	24 +
A^2. "High Suffolk"	77	23	14·9	18·9	5·6	48·4	6·9	·9	11 −	26 +
B. Breckland	71	29	8·7	9·3	6·6	31·4	13·4	6	10 −	19 +
C. "Good Sands" Region / F^1. North Norfolk Marshland	78·4	21·6	6	26·2	6·8	40·3	17·6	1·2	11 +	63 −
D. East Norfolk Loam Region / F^2. Broadland	70	30	10·2	17·2	7·6	36·7	15·2	1·7	18 −	8 +
E. East Suffolk "Sandlands"	70·4	29·6	9·1	13·9	9·2	38	11·6	2·8	13	33 −
G. The Chalk Downland ...	87·7	12·3	15·7	19·5	8·5	47·3	10·5	·9	5 −	45 −
H. The Greensand Belt ...	59	41	7·1	17·6	4·5	31	12·1	1·4	18 +	50
East Anglia	73·5	26·5	10·3	18	6·7	39	12·8	2	12 +	33 +

NOTE. "Total corn crops" include wheat, barley, oats, mixed corn, rye, beans and peas.

"Total root crops" include turnips, swedes, mangolds, and sugar beet.

The boundaries of these regions as indicated on the map are necessarily rather arbitrary. In the case of the North Norfolk Marshland the parishes generally extend into the uplands of the Good Sands Region, but the dividing line has been drawn along their southern limits. The Broadland Marshes, although here grouped with the Loam Region, are also to some extent used by upland farmers in A^1 (High Norfolk).

to permanent pasture. The chief hope at the present time is centred in sugar beet. The climatic and soil conditions of East Anglia as a whole closely approximate to those of some of the famous sugar-beet regions of the Continent. A good turnip soil is likely to be a good beet soil, while climatically the combination of relatively sunny summers with abundant showers and a comparatively dry early autumn is said to be a favourable factor. Sugar beet in East Anglia is at present in an experimental stage. The high proportion in Breckland (B, see Table) no doubt reflects the elasticity of a marginal "catch-crop" region, ready to try any promising venture. The figures for east Suffolk are significant, but the question is whether the cultivation of sugar beet can be "consolidated" before the withdrawal of the present subsidy. Beet culture certainly harmonises with the general structure of East Anglian agriculture, and, as in Denmark and Germany, the refuse beet from the factory could be used in support of the cattle industry. The chief hope of East Anglian agriculture probably lies in a closer approximation to the Danish system of arable farming in relation to dairy and meat production, since this would not involve any violent change of method and is appropriate to the physical conditions.

Modern Industries

The period during which the new agriculture was making steady progress witnessed the rapid decline of East Anglia's historic textile industry. The application of power machinery to the worsted trade gave Yorkshire an advantage against which Norwich struggled in vain, and by about 1840 almost all the fabrics for which the region had long been famous were successfully transplanted to the North. The displacement of the major industry, however, left in parts of East Anglia a large amount of technical skill which could be utilised in other forms of manufacture where power was not as yet so essential.

The silk factories established by Spitalfields firms in Sudbury about 1780 and, later, the introduction of mixed fabrics such as fustians and drabbet and of the weaving of horsehair and coconut fibre, too coarse to be manipulated by machinery, represented attempts to utilise the inherited skill of the Stour valley villagers under changed conditions. Apart, however, from the manufacture of nets and sails for the fishing fleet at Lowestoft and other coast towns, industries of this character, continuing the old textile tradition, are now of comparatively small account in East Anglia. Far more important are those directly related to her premier activity of agriculture. They comprise two main groups: (1) The manufacture of agricultural implements; (2) Milling, malting and the preparation of food-stuffs and fertilizers.

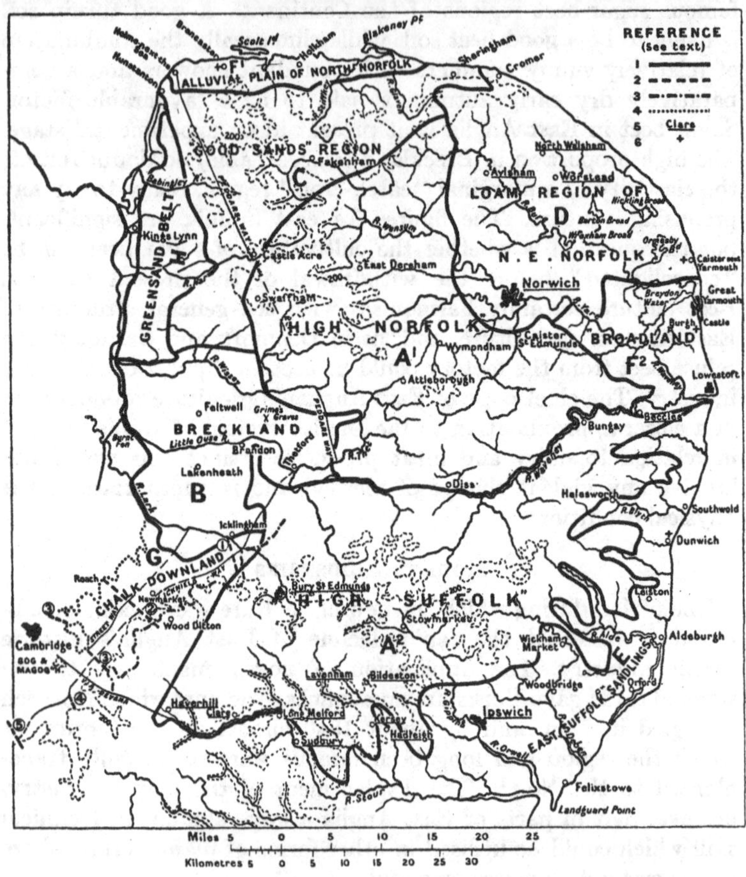

Fig. 25. East Anglia: the sub-regions and certain historical features. Reference: 1, Approximate boundaries of sub-regions; 2, Cambridgeshire Dykes across "Gateway" of E. Anglia (*viz.* (1) Black Ditches; (2) Devil's Dyke; (3) Fleam Dyke; (4) Brent Ditch; (5) Heydon Ditch); 3, approximate NW. limit of primitive forest; 4, approximate SE. limit of fen; 5, places prominently associated with former East Anglian worsted industries; 6, Roman sites.

The making of agricultural machinery in East Anglia has been an important industry since the late 18th century when works were established at Leiston and Ipswich (the Orwell works) to supply seed-drills, ploughs, etc., adapted to the needs of the new farming.

Later the output of agricultural machinery came to exceed the local demand and many East Anglian firms acquired a world-wide fame when the exploitation of the virgin soils of the Americas and other "new" lands began on a commercial scale. In recent years the competition of American machinery, specially adapted for the needs of Prairie and Pampas farmers, and the loss of the Russian market have been adverse factors, but it remains a great industry. Its location in East Anglia, at a considerable distance

---

Summary of Sub-regions of East Anglia

A[1]. High Norfolk.
Boulder clay, inclining to be heavy and tenacious, but with much local variation and interspersed with gravels and brick-earth. Originally "Woodland"; settlement later and more sporadic than in D. Mixed arable farming; many commons.

A[2]. High Suffolk.
Same characteristics as A[1], but soils on the whole heavier especially in south. The Suffolk "Woodlands" as distinct from "Sandlands" (east) and "Fieldings" (west).

B. Breckland.
"Blowing sand" on chalk. Much uncultivated heathland. Early prehistoric settlements; numerous meres (fowling) and small fertile valleys; abundance of flint. Agriculturally "marginal." Catch-crops. Region of experiments (sugar beet, tobacco). Rye often replaces wheat.

C. "Good Sands."
Not initially attractive to agricultural colonists. Productive under high farming: roots, barley, sheep. Large farms.

D. "Loam Region."
Light but generally very fertile and easily worked soils of glacial loams. Early occupied and developed. Villages thickly clustered. Numerous small ports. Much early enclosure. Roots, barley, wheat, and especially bullocks.

E. East Suffolk "Sandlands" or "Sandlings."
Light soils of glacial loams, sands and gravels underlaid by Crag. Some fertile and early settled tracts, penetrated by long estuaries favouring maritime trade. Much heathland.

F[1]. North Norfolk Marshland. F[2]. Broadland.
Valuable alluvial land. Grazing season in marshes April–October, providing rich summer pastures for arable farms of C and D respectively.

G. Chalk Downland: "The East Anglian Gateway."
Originally dry upland pasture. A frontier zone. Late enclosure in large fields. Essentially arable farming: wheat, barley, sheep.

H. The Greensand Belt.
Broken hilly country, with wooded ridges and heathland and some fertile valleys with relatively good pasture. Villages largely built of local carstone.

from the main centres of iron and steel manufacture, is explained in part by the character of the local demand and in part by the weight and bulk of agricultural machinery, making it cheaper to assemble and put together the raw materials in the immediate neighbourhood of the chief English market. Some of the firms have the additional advantage of water-frontage on the Suffolk estuaries.

Milling and malting are older industries but they have tended to follow similar lines. They have gravitated increasingly to the ports where foreign supplies of wheat and barley supplement the local crops. The great malting establishments on the Deben (Woodbridge) and the Orwell (Ipswich) are especially well-placed. They are close to the best and long famous barley districts of East Anglia, can readily import from the Mediterranean and the Near East and are within easy access of London, the greatest malt market. Ipswich also specialises in the manufacture of field fertilisers. Although the industry is now based on imported supplies, it was originally developed out of the phosphatic nodules (coprolites) found between the Red Crag and London Clay in south-east Suffolk. Thus while Norfolk pioneered the new farming Suffolk initiated the industrial processes ministering to its development. Norwich, however, has its fair share of this class of manufactures, including its far-famed mustard and malt vinegar from barley. In the aggregate East Anglian agricultural industries are very considerable and should be much increased if sugar-beet cultivation is thoroughly established*.

Modern East Anglia has many compensations for the loss of her former industrial supremacy. The well-tilled fields painted by Constable, Crome, and many other artists of a notable school, have escaped the devastating phases of modern industrialism. Few countrysides are richer in historic associations or in the evidence of human achievement over long centuries. There is no sharp separation between rural and urban life. Industry is linked with agriculture and many a beautiful country town like Bury St Edmunds has developed modern industries without losing the evidences of its antiquity.

The population is fairly well distributed, and, in spite of a considerable rural exodus during the acute phases of agricultural depression, still reaches a density of 337 (East Suffolk), 243 (Norfolk), and 181 (West Suffolk) per sq. mile†. The towns are of moderate

* For the economic geography of modern East Anglia and adjacent counties, see J. Bygott's *Eastern England*.

† 130, 94, 70 per km.^{2} respectively.

size, and their growth has not been too rapid for social well-being, *e.g.* Norwich (120,000), Ipswich (79,000), and Yarmouth (60,000).

What the future holds in store for our region must depend primarily on the factors affecting English agriculture. The immediate horizon may be clouded, but the increasing emphasis on the national need for a revival of agriculture and of greater attention to rural problems is perhaps ultimately more significant. No part of England stands to gain more from such a movement than this which has played so great a part in the oldest and most fundamental of English activities.

IX

THE FENLANDS

F. Debenham

THE lower portions of the basins of those rivers which flow into the Wash form a region sufficiently well defined to be given a name derived from the chief feature of the area, the Fenlands.

These rivers are the Witham, the Welland, the Nen and the Ouse, together with their tributaries. If the 50 foot (15 m.) contour be followed round from near Wainfleet on the west side of the Wash to the neighbourhood of King's Lynn on the south-east side it will delimit fairly well the area under consideration, leaving out the upper courses of the rivers but enclosing many small areas or "islands" above the 50 foot level. The Fenlands will thus be seen to be about 70 miles long by 30 miles broad (110 × 50 km.) and it includes nearly one million acres (404,000 ha.) of dry land, all of which is so similar in conformation, climate, history and productions that it may be taken as a natural region although politically it is divided up amongst five counties.

Its structure is very simple, consisting as it does of a flat plain of recent deposits, both alluvial and glacial, which occupies a shallow basin carved out of the Cretaceous and Jurassic rocks, bounded on three sides by low hills of those rocks. These underlying strata dip gently towards the south-east in this region and have some effect on the surface features. The variable resistance of the basement rocks to the agents of erosion has resulted in the Cretaceous Chalk remaining as uplands of comparatively steep slope to the south and east of the region, while the older but softer Jurassic clays to the west appear as gentle dip-slopes for the most part.

This is not a complete explanation of the relief, however, since to the north rise the Lincolnshire Wolds of Chalk, the continuation of those on the Norfolk side of the Wash, showing that the ancient drainage system had completely cut through this barrier to form the Wash itself.

It is therefore clear that our region owes its general topography to an old river which flowed to the north-east through the Wash and probably joined the ancestor of the Rhine and the Thames before reaching the sea. Once this river had broken through the Chalk its history was a perfectly normal one, that is to say it cut down the soft Jurassic clays to a broad flat basin which, however, narrowed to the north-east where it passed through the Chalk, and such is the present form of the basin if the 100 foot contour be traced over the region.

The residual gravels of this river indicate that its gradient was considerably sharper than is that of its chief descendant the Ouse,

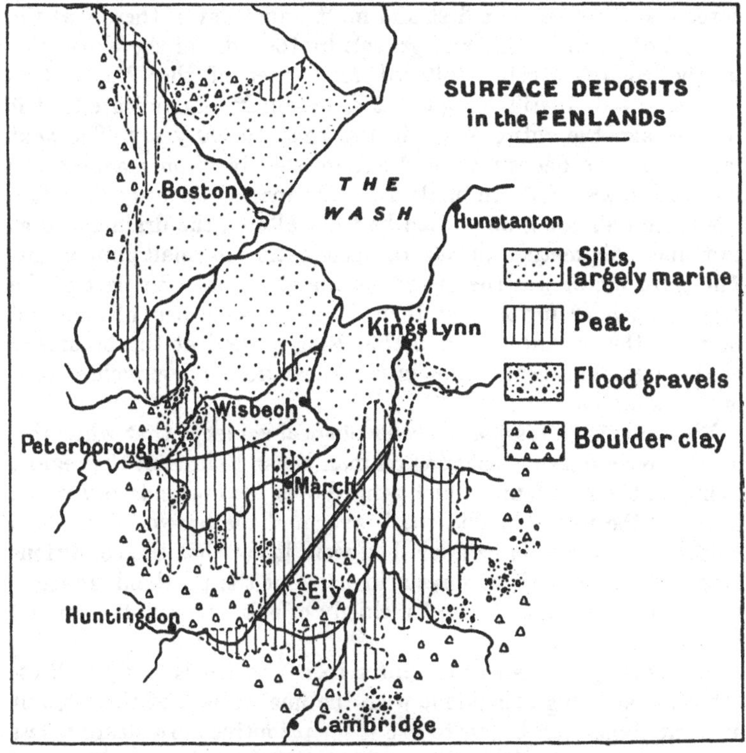

Fig. 26. Surface deposits in the Fenlands.
The scale is approximately 1 : 1,126,000 (1 inch = *ca.* 18 miles).

but subsidence of the land had reduced this gradient and had let the sea well into the gap between the Chalk uplands before the next great event in the geological history of the area.

During the Ice Age the broad basin of the river was more or less completely filled with an ice-sheet which at first was moving from the north and east bringing with it rocks from the Scandinavian peninsula, but later on was more local in origin and laid down boulder clays largely derived from the Chalk of Lincolnshire.

Nevertheless the retreat of the ice left the basin very much the same in general topography as before, and the renewed but more sluggish erosion has done but little to alter the main outlines, though it has re-distributed much of the glacial debris.

At this stage we have to imagine a much deeper embayment of the sea than is now the case and we may regard the building up of the present-day Fenlands as beginning under the influence of three main factors. The first and most important of these was the deposit of marine muds and gravels by the tides of the North Sea in the comparatively sheltered forerunner of the Wash. This process is indeed still going on, and measurements indicate that it will reclaim the entire Wash in less than 10,000 years. The next factor was the deposit of land silt by the rivers on reaching the stagnant marshlands embanked by the marine deposits, and this would be still going on if man had not altered the drainage so as to reduce this method of natural reclamation to small dimensions. The third factor was the growth of marsh plants over part of the area in such profusion that there is now a wide belt of peatland between the marine silts and the rising ground at the boundary of the area. The peat varies in thickness from a few inches to as much as fifteen feet.

We know from fossil and archaeological evidence that while this natural reclamation was going on man began to inhabit the region, living on the rich fauna which ranged over the lowlands and which included the urus, the elk, the beaver and the brown bear. Such animals would not have flourished were it not that the conditions favoured the growth of forests in parts of the peatland, remains of which in the form of trunks, chiefly of oak, are still dug up and used as gateposts.

The coming of man as a hunter to the region is fairly well established as being coincident with the final retreat of the ice, but at what stage man began to use it for primitive agriculture is less easy to determine. But if we skip over the misty centuries of prehistory from the time of the hunter to some time before the advent of the Romans we find that an advanced state of agriculture had been reached.

Some by no means rough embanking and draining had been done, comparable possibly with the vast defensive embankments and ditches, such as the Fleam Dyke and the Devil's Dyke, which run down to the impassable fens on the south-east side. There were established settlements of tribes in and around the fens of whom we know a good deal. The Romans improved upon the embankments and there seems little doubt that before they left the country many parts of the Fenlands were a rich source of agricultural products, more especially on the marine silts, nearer the sea than the peat. The settlements in fact must have formed a fringe to the Wash itself, isolated to some extent from the higher lands by the belt of peatlands, which were still more or less a waste of marsh and mere, teeming with fish and fowl but not fit for agriculture except on the "isles" of clay and gravel which here and there rose above the flood level.

For some reason, probably connected with the subsidence of the land, the forests of the Neolithic Age had disappeared by historical times and the vegetation then, and until quite recently, consisted of sedge plants together with water-loving trees and brushwood. The fen indeed was covered with dense sedge and intersected by sluggish streams or shallow lakes, while here and there on firmer ground were thickets of willow and alder, a paradise for waterfowl, and all of it liable to flooding in the winter.

Such a region, favourable to its inhabitants who knew its vagaries but hostile to the newcomer who did not, was naturally at once a refuge for the pursued and yet a prey to small marauding parties. So we find it figuring in these rôles very often up to and even after the Norman Conquest. The dreary wastes attracted the ascetic souls of monks like St Guthlac to found a Crowland, the comparative isolation of the island of Ely offered peace and security to Aetheldreda for her monastery, but at the same time gave harbourage though not sanctuary to the lawless and the outlawed as long as they were in small parties. No invasions on a large scale could approach via the fens, but, as the monks and simple fisherfolk found to their cost, it was ideal country for the small marauding bands of Danes who for scores of years worked their longships up the sluggish streams to loot and destroy and later even to settle, and leave a slender mark upon the culture and language.

The Fenlands can even claim to be the last of the Saxon strongholds to be held against the invading Norman, and the tales of Hereward the Wake and his resistance to William the Conqueror, however legendary they may be, show that military operations in the region could wreck the reputation of even the most experienced of leaders.

The climate of the Fenlands is, in general terms, that of much of East Anglia away from the sea coast. That is to say it partakes of the "continental type" with a tendency to extremes of temperature in winter and summer. Although south-westerly weather is the rule here as in most of England, a marked feature of the eastern side of England is the prevalence at certain seasons of cold, dry east winds, the product of high pressures over the Continent. They are therefore associated with clear skies, and the Fenlands as a rule enjoy a high percentage of sunshine. The rainfall on the other hand is the lowest in England, being rarely over 25 in. (625 mm.) in the year, but it is well distributed, the least rain occurring in spring and the most in autumn. These conditions favour the growth of cereals, especially of wheat, but there are few crops which could not be grown on the fens, and more than one great change in the crops grown has taken place, such for instance as the invasion of the Fenlands by the potato and more recently by the sugar beet.

This fertility was recognised at an early date and it was equally well understood that without artificial drainage the area suitable for agriculture would be very small. The Fenlands in fact from being a waste have become the most artificial region of the British Isles and the most fertile, and no account of the region can be complete without reference to the drainage. In the beginning the schemes, which at that time consisted merely of drainage ditches and protective banks, were quite local in extent. The area between the Witham and the Welland was undertaken first, and parts of it were certainly drained before the time of the Romans who probably added the sea banks which are still traceable near Long Sutton. But it was not till the 17th century that the matter was viewed in a broad way and attempts made to drain really large areas, when the family of the Earls of Bedford took up the schemes and finally succeeded in reclaiming about 100,000 acres (over 40,000 ha.) of what is now known as the Bedford Level. The system adopted, under the guidance of Cornelius Vermuyden, a Dutch engineer, was to make straight cuts to shorten the rivers with minor drains feeding them in a herring-bone pattern, and the permanence of much of their work is obvious from a study of any map showing the drainage now. The most prominent of their works was what is known as the Old and New Bedford rivers which now divert the Ouse at Earith and run straight to Denver, where there is a large sluice which keeps the sea water out of the old Ouse and allows it to run up the new cuts and spend its flood in them and between them.

But the promoters found many difficulties, some of which are still present and have modified the economic history of the region. One was that any scheme of drainage to benefit one area was liable to flood the neighbouring areas, and at that time the Fenlands could not be dealt with as a whole. Another trouble was that after the draining had begun to take effect the loss of water from the peat made it shrink, even to as much as six feet, and the level being thus lowered the minor drains no longer flowed by gravity but the water had to be pumped up to the main cuts. Yet these, which were merely engineering difficulties, were comparatively small matters at first compared to the active opposition from the local inhabitants; indeed, it may be said that until very recently every scheme has been fiercely opposed by the occupiers of the land to be drained. There is nothing remarkable in this fact for the inhabitants of an undrained area made their living by fowling and fishing and peat cutting, not by agriculture, and the change meant either ruin or a complete change of occupation for them. Acts of sabotage on the ditches and embankments were frequent, and at times the work of the engineers has been entirely held up by an active minority of interested persons. The first pioneers with these extensive works, or "adventurers" as they were called, made but little out of the business, but the value of the land increased enormously and attracted population of a good type. In the 18th century further big works were undertaken partly because of the failure of some of the former work due to silting up of the channels. These later schemes were on the "catchwater" principle, that is to say the water from the uplands is caught before it gets to the lowlands and led away by canals which approximately follow a contour to the nearest outfall, and they are especially associated with the name of the engineer Rennie.

The combination of the two systems is not satisfactory since the first system requires plenty of water in order to keep the cuts from silting up and the second system robs it of that water. As usual the compromise is a very expensive one and though it is successful in reasonable seasons much of the land is dry only by a narrow margin and the pumping expenses are high. The pumping was originally done by hand or horse power, but this gave place to windmills, following the example of Holland, and these again to steam pumps in recent times. The general outlines of the drainage system can be seen on a map, but much of it is too complicated to be represented by any simple method of cartography.

The Fenlands began to change their character in Stuart times, and though it was always a slow and sometimes a painful process the original fishermen and fowlers were ultimately replaced by tenant farmers and in some more favoured places by small-holders. Roads began to take the place of the shallow waterways and the area of sedge to decrease, retreating to the meres, such as Whittlesea Mere, which defied the efforts of engineers to drain it until late in the 19th century. At the present time there are only one or two small areas of the original fen, the most famous being Wicken Fen, about 12 miles from Cambridge. The type of agriculture which evolved in the Fenlands was intensive, partly because of the high value of the land when money for drainage had been spent on it, and partly because of the character of the soil. But the crops grown are very varied and have altered from time to time with changes in demand and foreign competition. Wheat has always been a staple crop with oats and barley as secondary cereals; but with the increase of railway facilities some of the agriculture took on the character of market gardening, the chief crop being potatoes, which now occupies the premier position as a product in parts of the Fenland. The land and climate is equally suitable for sugar beet, and since the Great War this crop has invaded the Fens at a great rate. It remains to be seen whether it will endure as an economic crop when the assistance granted by the Government is diminished or ceases. At all times there has been a tendency to grow a wide variety of crops, especially those which are too exhausting for less rich soils. At one time a great deal of flax was grown; the curious plant, woad, was grown for its dye, and nowadays a great deal of buckwheat is raised. Root and leguminous crops are grown for rotation purposes and to supply the stock of which there is a larger quantity than one would expect on land which is pre-eminently arable. Of still more specialised crops we may instance the fruit industry of the loose higher soil of the marine silts, especially centred round Wisbech, the bulb industry of the Spalding area and the pure market gardening of the lighter soils near main lines of railway.

The intensive type of cultivation means a greater density of population than is usually associated with farming industries and the Fenlands rarely have less than 200 persons per 1000 acres (404 ha.), and in the fruit-growing districts much more. The distribution of the people is controlled by two factors, the accessibility of transport and the nature of the soil. The towns and villages have to be on the higher and if possible more gravelly soil and none are on the pure peat lands, but if they are to be centres of population they also have to be centres of transport facilities. Moderate sized villages are the rule, and it is only in the market towns which also have rail facilities that the size is likely to grow beyond 5000.

Of manufactures and industries in the large sense there are none in the Fenlands, even the farm implements which are required in such large quantities coming chiefly from outside the area, as also the artificial manures. Building material is naturally scarce and the brickmaking factories of Peterborough district and other clay centres can perhaps rank as a minor industry, but it is not one which will ever cause great concentrations of population.

The fact that the lower lands are farmed from the higher ground is responsible for the curiously uninhabited appearance of much of the region, the prospect usually consisting of field after field of dark earth or dense crop, separated by ditches and unrelieved by signs of habitation other than an occasional chimney stack of a pumping station or a skyline of thin elms or willows marking slightly higher land or more probably the bank of a larger drainage ditch.

On the other hand, where there are ridges of land above the main fen level, the villages may be almost continuous as they are on parts of the Isle of Ely and from Outwell to Wisbech. Stranger still is it to see a long line of buildings perched on the artificial embankment of one of the major streams and sharing what appears to be a precarious foothold thereon with the road or even the railway.

Such a region is bound to have an effect on the character of its inhabitants, but it is not easy to generalise. From comparatively short contact with them one would be apt to say that their first and chief characteristic is their capacity for hard manual work in the open. The long distances from home to work, the bleak weather in which much of it is done and the comparative lack of amenities does not seem to render them discontented. Only the greatest of injustices have stirred them in the past to riot or disturbance. Their isolation makes them less ready at conversation and *camaraderie* perhaps, but in place of the highly developed social instincts of the truly industrial areas they have a faculty for village *esprit de corps* which is in many ways superior in its qualities. On the whole one may say that the simpler life and the absence of large communities has produced a contentment which might be the envy of the inhabitants of the great industrial cities.

It may seem curious that one of the chief disabilities of the area for carrying a dense population is its lack of a suitable water-supply. The fen water itself was used for a long time and was doubtless responsible for much of the bad reputation for health which the Fenlands once bore. The matter was taken in hand by the health authorities about the middle of last century and the larger towns now have a better but not a very high standard of water supply. The villages, which are usually on gravel and silt, obtain their small requirements from shallow wells, but when a larger supply is needed it can only be obtained by using the river water well filtered or else by transporting water from springs beyond the boundaries of the Fenlands proper. Underneath the Fenlands there is nothing but impervious clay for many hundreds of feet and therefore no chance of a sub-artesian supply.

The general appearance of the region can be imagined fairly well from the description of its industries, but it is necessary to distinguish between the peatlands and the silt lands. In the latter the stronger growth of trees is responsible for quite a different landscape, and the farmhouses are dotted about much more promiscuously than is possible in the true peat fens. As one approaches the coast of the Wash the trees and farming land end suddenly at an embankment beyond which is the salt marsh in various stages of reclamation. Here can be seen the process which has in the past formed nearly all of our region. The tides of the Wash sweeping down from the north and east bring finely divided silt in suspension and gravels along the bottom. The flood tide flows over the sandy flats and drops some of its load before the less powerful ebb tide takes back the remainder. The process is greatly assisted by the growth of the samphire (*Salicornia*), which checks the flow of the water, and helps to bind the loose debris together, and there is a regular succession of plants coming into play as the banks rise. Besides the natural reclamation there is at present some artificial reclamation being done in the south-east corner of the Wash, and various other schemes have been considered in the past, even to the plan of reclaiming the whole of the Wash in much the same way as the Zuyder Zee is being dealt with on the other side of the North Sea. A certain amount of fishing is carried on in the Wash itself, but the navigation is difficult, and with the growth in size of shipping the two chief ports, Boston and King's Lynn, lost their prominence as ports for East Anglia and Lincoln; but they still, together with Wisbech, have a local importance as ports of export and import. The inland towns have various reasons for their position, either as bordering the fen, like Peterborough and Cambridge, as rail centres such as March, or as the nucleus of a district separated from others by the fen, such as Ely.

The religious life of the Fenlands has had much to do with its history, and besides the cathedrals of Ely and Peterborough there are many churches which bear evidence of the religious feeling and the energy of the fenlanders.

The Fenlands are therefore unique in the British Isles in topography and origin and to the true geographer present a study which is of very great interest, nor will he journey over the region for long without learning something of the charm which its wide spaces have. Should he begin his study by viewing the prospect from the tower of the Cathedral of Ely, whence he can see the peat levels all round him, he will easily add to his gifts that most estimable quality of imagination, for it needs but little effort to see the Isle of Ely surrounded not by rich lands but by waste of marsh and mere and stagnant streams, a region of difficulty, now transformed to one of plenty.

X

LINCOLNSHIRE

John Bygott

LINCOLNSHIRE, the second largest English county, is admirably adapted for treatment as a geographical entity with well-marked sub-regions, though in a work like this better co-ordination is obtained by including the Lincolnshire Fens with the Fenland of adjacent counties*. Considered as a whole, Lincolnshire is still primarily an agricultural community, the commercial interests of Grimsby or Immingham and the iron-smelting of Scunthorpe being relatively recent developments. No part of Lincolnshire is too high for exploitation, rarely is there soil so poor as to repel absolutely. There is no great climatic drawback. Lincolnshire is part of the drier east of England where there is no heavy rainfall to restrict cereal growth, but it is not so dry that roots or any normal British crop will not flourish. Her farms and farmers have long been renowned, her best stock has been bought at high prices to improve the breeds of distant continents, and if as producer of beef, bread, and beer, she does not preponderate as she did when Arthur Young wrote the reports of his famous tours†, the causes are economic rather than geographical. Heavy taxes and big labour bills, competition from the more cheaply-produced grain and meat of newer lands oversea, in this county, as elsewhere in the British Isles, tend to blunt and to negative geographical factors which have been favourable to agriculture in the past.

It will be convenient to examine the main physical regions, which can easily be followed on Fig. 27. Two, the Marsh and the Wolds, will be examined in more detail than the others. This will give types of a lowland and an upland community and will serve as some basis for comparison when noting other divisions.

THE MARSH

Skirting the shore-line of the Humber and the North Sea is a stretch of marshland, which superficially is a mass of silty alluvium, varied in places by glacial clays and gravels. A narrow coastal

* The south-eastern portion of the County is described in Chapter IX (The Fenlands).

† His *General View of the Agriculture of the County of Lincoln* (1799), may be cited here.

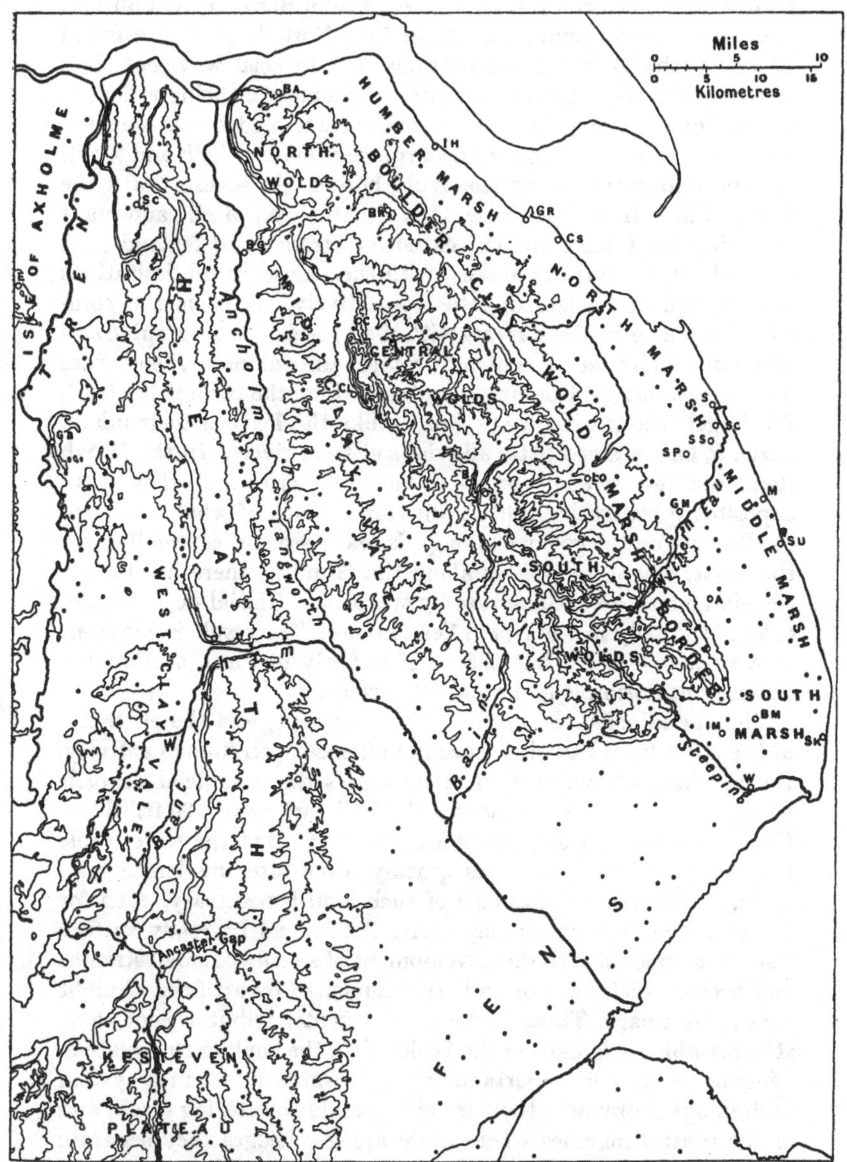

Fig. 27. Lincolnshire: relief and chief rivers, towns and villages. Contours are drawn at 50, 100, 200, 300, 400 and 500 feet, O.D. Places mentioned in text are shown by circles and initials. The dots are villages not mentioned.

plain from 6 to 10 miles wide, it extends from Barton to Wainfleet*, and is continued round the Wash into Norfolk as the seaward border of the Fens. It is essentially a pre-glacial wave-cut shelf covered with more recent deposits, its foundation being a flexure of the Chalk forming the Wolds, whose eastern edge was once the shore-line, as is proved by the occurrence of marine shells and fossils in sand and gravel along the Wold border. In several ways the Marsh differs from the Fens. It is a flat region of silt and clays overlying the Chalk, but the extensive stretches of fenland peat and old river gravel are absent, with their agricultural limitations and control of settlement sites respectively. It lacks the comparatively long Fenland rivers with their special physiographic and economic significance centring round embankments, elaborate drainage canals, and pumping systems. Like the coastal Fenland, it is largely the gift of the sea, but, unlike the Fens in their inland parts, it has received little alluvial soil from rivers. In the Marsh there are few streams of even moderate size, and they have deposited comparatively little alluvium. Much of what is classed as alluvium on the geological maps is really sea-silt, especially near the coast, where, as well as along the Humber, there has been a certain amount of reclamation by warping†. Glacial deposits also occur, especially in the North Sea Marshes. The result is considerable variety of soils, but the relief is fairly uniform, so that the general impression is monotony of surface.

The varied soils, the nature of the drainage, and the character of the coast have had their special influence upon human activity. Borings have shown that in many places the superficial deposits which rest on the substratum of Chalk are over 100 ft. thick. Except the boulder clay, they are porous, so that the water-table is near the surface and rises quickly, even after moderate rain. Thus, careful subsoil drainage of such land is necessary, even for pasture. The impervious clay easily floods, and its ready surface flow-off is conducive to the development of a series of small streams and becks, which have often been augmented by artificial straight cuts for drainage. These becks are frequently tributaries of short streams which originate in the Wolds or in the boulder clay plateau fringing them. The importance of such watercourses in a system of drainage is obvious; they are also useful for watering stock, and in the coastal marshes where there are few hedges they separate fields. Settlement has generally been along these streams, either

* Ba and W (Fig. 27).

† The technical term for fertilisation by silt left by turbid water admitted to and then drained off poor lowland by warping drains or channels.

on "islets" of glacial deposits or on the boulder clay margin west of the Marsh. The coast is low and the spring tides are often very high, so that where there are no dunes artificial protection by banking or groynes has been necessary; near the outfall of the larger streams rough sluices are some protection against abnormal tides. There are no natural harbours along the whole coast of Lincolnshire: Grimsby and Immingham*, though for centuries their havens had been the rendezvous of small ships, have grown into modern ports around artificially created docks. The sandy beach between the Humber and the Wash, though unfavourable for the growth of large ports, has helped fishing villages to develop into watering places and holiday resorts† for the industrial folk of the Midlands, the ozone-laden breezes and the frequent anticyclonic weather being largely due to the continental outlook of this coast.

The coastal plain is here referred to as the "Marshland," but locally the term "marsh" is generally applied merely to the grasslands bordering the coast. From this point of view, the economic significance of the Marsh is limited and to some extent negative. These coastal lands, even when drained, are fit for little except pasture. The best are excellent pastures, their value as such being frequently emphasised by Arthur Young. Along the North Sea there is a peculiar saltiness in the grass; nowadays this saltiness is mainly due to the onshore breezes and is as beneficial to cattle as the rock salt which is put into their feeding troughs for them to lick; but previous to drainage and protection, when the tidal waters flooded and remained on the land for some considerable time, it was not beneficial, and Young speaks of the injurious effect of pasturing stock on the salt-impregnated grass. On the maps names like Saltmarsh Farm and Salt "Ings" are common and significant. In Domesday Book there is mention of salt pans in these marshes, such salt being made by evaporating the water which was conducted from natural lagoons or from the sea into shallow clay-lined pans or pools. This was an important industry when the salting of meat for winter use was necessary owing to lack of root crops and other winter fodder. Mention‡ is made of a Lincolnshire salt industry as late as Elizabethan times, but it subsequently died out before the competition of the Cheshire and foreign salt deposits.

The pastoral part of the Marsh has never invited settlement.

* Gr and Ih (Fig. 27).

† *E.g.* Cleethorpes, Sutton, Mablethorpe, Skegness (Cs, Su, M, Sk on Fig. 27).

‡ Bullein, *Bulwarke of Defence against all Sicknes* (1562).

The Humber Marshes, which are narrower than those of the North Sea coastal plain, do not contain villages, the only dwellings being isolated cottages, often in connection with the brickyards. The Humber warp makes good brick clay, and as small craft are conveniently loaded at the yards, many brick and tile kilns developed round Barton and elsewhere along the estuary. Despite this industry, which is less important than formerly, the Marsh is here of only limited importance. Between it and the Wolds there is a low plateau capped with boulder clay and glacial gravel and referred to by Arthur Young as the "Clays." This higher land is relatively fertile, though the clay is heavy to work. The plateau varies from 50 to 100 ft. (15–30 m.) above sea-level and is dissected by streams which, with patches of glacial gravel, have determined village sites. The parish boundaries of these villages extend to the coast and include part of the Marsh, which is a pastoral offset of the more agricultural land farther inland. The isolated dwellings on the Marsh are of what Mr Peake* would term the squatter type and are occupied by brick workers, small-holders or cottagers. Along the North Sea, where the coastal plain is wider, there are villages on the Marsh. Some are on a raised terrace of alluvium and glacial drift which is about twice as high as the ordinary marsh level of 10 ft. above O.D., and where the patches of glacial gravel furnish good sites reasonably free from flood. In the South Marsh, where the boulder clay is nearer the coast, certain villages on the clay have the suffix "le Marsh," and some have the same name as others a considerable distance away in the Wolds†. In these typical Marsh villages, as along the Humber, the agricultural land is mainly on the clay and the pastures are on the silt nearer the coast. The typical Marsh villages are long and straggling and might be classed with Mr Peake's squatter or heath villages. Their frequent "by" suffixes tell of Scandinavian influence, but there would be considerable expansion after enclosure in the 16th and 18th centuries and after reclamation and drainage. Such expansion and squatting is shown by the common name of adjacent villages (*e.g.* Saltfleetby) distinguished by suffixes like "All Saints," "St Peter's" and "St Clement's,"‡ this distinction occurring when squatting or expansion from the older parish made the newer sufficiently important to have a church and a distinctive name of its own. The complete absence of marsh names from the Domesday

* *The English Village*, 1922.

† *E.g.* Gayton, Gayton le Marsh, Irby, Irby le Marsh (G, GM, I, IM, on Fig. 27).

‡ SS, SP, SC (Fig. 27).

list of churches is significant when compared with the number in the Wold valleys and on the Wold and Limestone upland borders. Domesday mentions villages in the Marsh, especially along the streams, and Saltfleet* is noted as a port with some coastal trade, but settlement in the Marshland was not sufficiently close or permanent to warrant the building of a church, even if the difficulty of transporting suitable building materials could have been overcome. The coasts of eastern England for long were subject to attacks from the Norsemen, and this was unfavourable to the growth of villages, but when such raids became a thing of the past, small villages grew up wherever a tiny cove or stream mouth offered shelter to the little boats then in use. Longshore fishing and a primitive commerce provided a somewhat precarious livelihood. Some of these coastal settlements of course became more important than others, and places like Barton, Grimsby, and Immingham are mentioned as assisting Edward III with men and ships in his French Wars. Nowadays few of the Lincolnshire coast villages are engaged in even longshore fishing beyond a little shrimping or cockle-gathering, the fishing on this coast being mainly from Grimsby by means of deep-sea trawlers financed by individual fish merchants or companies, a condition different from that in Cornwall and Devon, where many of the fishermen still own their boats or combine in shares of them. Thus, the dwellers of this coastal plain are now more dependent upon the land than upon the sea. During the "season" there is scope for boatmen and others who cater for visitors at the watering places, but this is only a temporary activity confined to relatively few. In the Marshland villages, where small-holders and little farmers predominate, dairy produce is raised for Grimsby and other towns, and where the soil is suitable, as in the rich loam of the Middle Marsh, market gardening is practised. However, the Marsh is essentially a grazing area, and for cattle rather than sheep, though in the drier and better drained parts numbers of the latter find pasture in the summer, being folded on turnips on the bordering clays in winter, an example of transhumance.

The economic significance of the boulder clay plateau between the Marsh and the Wolds has been considerable. On its gently sloping eastern border, where the erosion of numerous small streams has been very active, there is a line of villages, and for some distance along the Wold border there is a second line where patches of glacial gravel remain. The glacial deposits on this plateau generally form a light, easily-worked loam, though in

* S (Fig. 27).

places there is a rather stiff clay soil. Favourable soil, convenient water supply, and reasonably dry sites for settlement, explain the presence of villages with churches here at Domesday, when the neighbouring Marsh was relatively bare of population. The villages on the "Clays," as Arthur Young called this region, are fairly large, and as a rule larger than those of the Wolds, though their farms and fields are smaller. Even when their boundaries do not extend into the Marsh, many of their farmers rent marsh pasture. Young refers to this, and emphasises the fact that graziers exploiting the Marsh as a rule lived out of it. The Clays are less under rotation than the Wolds and support many small farmers, who grow crops according to their needs rather than according to fixed plan. Beans, wheat, and turnips are the chief crops, though potatoes are largely grown, and in some of the larger villages considerable attention is paid to market gardening. Several farmers engage in dairying, especially for milk which is sent to Grimsby, but many of the cattle from these villages pastured in the marshes are fattened for market. The practice has been to winter them upon roots and cake in the foldyards and then to turn them out to grass in the spring, the east winds helping to evolve a hardy race of stock, mainly Lincoln Reds of the kind so highly praised by Young as quick to fatten, hardy, and good milkers. Many small market towns, such as Barton, Louth, and Alford*, have grown up around this region, which carries the railway line connecting Grimsby with Peterborough and London. Before the railway era, ports like Barton shipped produce from the Clays and the Wolds coastwise to London, and along the Humber-Ouse system to the growing industrial centres of west Yorkshire.

The Wolds

The Lincolnshire Wolds are a very interesting region and afford much material for tracing the influence of geographical factors upon human endeavour. Their physical features are varied, and much more so than would appear from a small-scale generalised map of England, which gives little idea of the escarpment or of the number and nature of the valleys. Study of both the escarpment and the valleys is necessary in order to interpret the distribution of villages.

The Lincolnshire Wolds are a belt of upland about 45 miles long and 5 to 8 miles wide (70 × 8 to 13 km.), flanked on the east by the low boulder clay plateau previously noted and on the west by what we shall subsequently examine as the Mid Clay Vale. The Wolds are part of the general system of scarped Chalk uplands

* Ba, Lo, A (Fig. 27).

which stretch almost continuously from Wiltshire to Flamborough Head. The mass is Chalk, but where streams and springs have been especially active, there are exposures of the Lower Cretaceous and other rocks (Carstone, Tealby Limestone, Tealby Clay, Spilsby Sandstone, in descending order). The alternation of limestone and clay is important in connection with spring formation, especially along the escarpment, where the springs have had such a significant economic influence. Because the Wold is essentially a tabular mass of Chalk dissected by springs, streams, and the other factors of erosion, there is apparent alternation of ridges and valleys. But imagine the valleys filled in by the material eroded from them and the conception of a tableland is obvious. Were it now a tableland, many factors favourable to settlement would be absent. Much of the Wold is mantled by glacial deposits, so that to some extent the original features of erosion have been modified, and the soils have also been influenced (see Fig. 27).

The North Wolds have few streams, and along their edge the escarpment is lower and more regular than farther south. They are now almost destitute of glacial deposits, so the rain has had every opportunity to sink into the permeable Chalk instead of running off to feed streams as it does where there is boulder clay on the Central and South Wolds. Hence villages in the northern section lie along the scarp base, and elsewhere on the upland there is practically a blank on the population map compared with those well-marked lines of villages in the valleys as well as along the scarp of the other divisions. The distribution of Wold villages is essentially little different from what it was at Domesday, and then there was a fair proportion with churches, an index of permanent settlement of some duration. The Wold villages are smaller than those of the Marsh or the Fens and are relatively more compact. They are smaller because the Wold farms are larger and on the whole less intensively cultivated than those on the lower lands, so that not only are small-holders and little farmers rare but fewer workers in proportion to area will be required for cultivation. The villages are not straggling route villages stretching along a main road, but are generally grouped around cross-roads or situated in the blind end of one of the numerous tributary stream valleys of the Central or South Wolds. They are mainly typical valley villages where the original controlling factors of settlement were fertile alluvial soil, adequate water supply, and a sufficiently dry site for the dwellings. The scarp-side villages are more longitudinal in their plan, but the grouping tendency of the valley village is also noticeable in them. Villages along the scarp are generally between the 50 and 100 ft. contours in the North Wolds, and between the 100 and 200 ft. contours in the Central Wolds, so that they are well away from the lower and damper lands of the bordering Clay Vale. Springs, and often the headwaters of a Clay Vale stream, furnish water; soil from alluvium or from talus washed down the scarp slope provides soil, and the position on the slope would afford a sheltered situation compared with the bleaker ridge and upland plateau. Probably water supply was the dominating factor in the sites of the earlier Wold villages, though it happened that soil and other advantages were present where water was found. It is significant that in Domesday records there is mention of many mills in the Wold parishes, and they would be water-mills. These early settlements were selective and much of the Wold long remained Down pasture or sandy heath. The Wold villages have always had a predominantly agricultural population. In the days when village craftsmanship had not been stifled by cheap mass-made goods and when there was no delivery by motor van from the market town, village craftsmanship held its own in the larger Wold villages.

While the crests of the ridges and plateaus generally lack habitations, yet in the North and Central Wolds they sometimes carry isolated farmsteads, locally known as "tops." These are part of the parishes of which the valley villages are centres, and are similar to the squatter outposts of the Marsh. They generally date from the enclosure and amelioration of the inferior Wold land at the end of the 18th century. Such enclosure is geographically and economically instructive. When the increasing urban and industrial populations showed the need for greater food production, considerable areas of inferior land were brought under cultivation. Much of such land on the Wolds had been of use for little except sheep walks or rabbit warrens. During this adaptation to more intensive use, the grass was burnt and the ashes allowed to work into the soil, which was marled or mixed with chalk, as the numerous disused chalk pits testify. It was drained, enclosed, and let on favourable leasehold terms to tenants, who were thus encouraged to cultivate and to fertilise it with that care which soon made the Wold agriculture famous. Good farm buildings were erected, as on the Brocklesby* estate of the Earl of Yarborough, which ranked with the Norfolk Holkham estate of the Earl of Leicester among pioneer efforts of the Agricultural Revolution. The Brocklesby estate was also a pioneer in scientific afforestation, more than 27 million trees having been planted at regular intervals between

* Br (Fig. 27).

1787 and 1927. The timber has been periodically renewed as exploited, an example of the manner in which sandy soils little suitable for agriculture can be utilised.

Among the boulder clay deposits of the Wolds are frequent patches of glacial sand. Such sandy tracts were largely given over to rabbit warrening and the surrounding land was devoted to the upkeep of the rabbits, several villages being noted by Arthur Young as having over 1000 acres (405 ha.) under warrens. The rabbits were sent to the towns as food and their skins utilised as furs. At one time, market towns like Louth, Brigg, and Caistor* were seats of a lucrative industry connected with dressing the skins, those of the silver-grey rabbits being particularly esteemed.

It is interesting to note that there are iron deposits in the Lower Cretaceous series (Claxby Ironstone) of the Central Wold scarp. In the middle of the 19th century attempts were made to exploit this ironstone, which yielded about 30 per cent. of metallic iron, but galleries had to be driven into the hillside and the enterprise was abandoned when the more accessible deposits at Scunthorpe† were worked. The open quarrying of the latter place was much cheaper, so that what at one time seemed an opportunity for industrialising part of the Wolds passed. At Claxby‡ a terrace of red-brick houses, uncommon in Wold villages, is a reminder of accommodation begun for the iron-workers.

Physical features are more varied and complex in the Central Wolds than in the other divisions. There is greater stream development, marked by more clearly defined valleys on the gentler eastern slope, and thus by greater number of villages away from the scarp. The escarpment is best developed in the Central Wolds, where it is more fretted by streams, so the more regular line of the lower North Wold scarp is lacking. Nowhere is the scarp too steep for the transverse roads, though some of these have rather sharp gradients. In the South Wolds the scarp gradually ceases to be a prominent feature, and most of the villages here are of the valley type, those in the Bain valley often being built on alluvial cones. The Central Wolds are noted for several fairly steep post-glacial ravines, which have little economic significance except that their abrupt slopes are wooded, mainly with conifers. Streams, when their former courses were dammed by glacial drift, cut these ravines through the chalk spurs which had resulted from previous erosion. The streams of the South Wolds are less mature than those of the Central division. There are instances of capture in

* Lo, Bg, C (Fig. 27). † Sc (Fig. 27). ‡ Cl (Fig. 27).

the Steeping and Withern Eau basins, though the streams are nowhere so large as to influence lines of communication or gap settlements as in the Downs. However, erosion has been active and has exposed some of the lower strata where the Chalk is less high. In the south and west of the Wolds boulder clay covers the Kimeridge as a rule and combination of these clays has influenced the soil. The brown soil of the Carstone, and the bluish grey of the Kimeridge contrast with the lighter soil on the Chalk. The soil of the bare Wold is generally a sandy loam, composed mainly of quartz with some flint where the chalky boulder clay does not occur. It is naturally poor and its crops are the result of much marling and manuring. When soil like that of the Wold is brought under cultivation, phosphates and nitrates must be constantly renewed by means of artificial manures, so that the cost of farming is relatively so great that only fairly extensive cultivation paid even when British agriculture was at its zenith. After the Industrial Revolution, to use capital for the betterment of Wold land was remunerative to landlord and tenant alike. Now, however, the competition of cheaply produced corn from oversea, apart from high labour bills, handicaps the farmer to such an extent that scientific experiment and progress is generally impossible. Many Wold farmers have dropped out of the race, and the character and size of the Wold farms, often with their 30–50 acre (12–20 ha.) fields, are unfavourable to development on intensive lines and by small-holders, the cost of adapting the farms and the soil, hitherto under extensive cultivation, to the requirements of the smaller man being prohibitive, whilst to turn back again into sheep walks would be equally undesirable. Though most Wold farms are relatively large, those of the South Wolds are smaller where villages are closer together. The essential features of Wold farming are observance of crop rotation, especially the combination of roots with corn, so that sheep may be folded on turnips during winter and gradually distributed over most of the fields of the farm. Barley is generally grown in preference to the longer-rooted wheat, though when prices were good a fair amount of Wold land was devoted to wheat. Many Wold farmers still rent pasture in the Marsh or the Ancholme Carrs for summer feeding, and bring cattle to the home foldyards for winter fattening. Such was a common practice when Arthur Young wrote, when the more substantial farmers often owned their own Marsh land. The large foldyards of the Wold farmsteads were specially adapted to this method, and with their spacious barns and granaries, testified to the activity of the mixed farming of the Wolds, important bases of whose wealth were the famous Lincoln Red shorthorn cattle and the long-woolled Lincoln sheep, strains of which have been largely used to develop the flocks and herds of New Zealand and Argentina.

The Mid Clay Vale

West of the Wolds is a lowland which may be termed the Mid Clay Vale. It has been mainly carved out of the soft Kimeridge and Oxford Clays by subsequent streams working backward from the Humber and Witham. Not only have such streams eroded much of these clays, but development of their westward-flowing tributaries has widened the valley by causing some recession of the Wold scarp eastward. That this is a pre-glacial valley is evident from the deposits of boulder clay and glacial gravel which mantle its slopes. In the south-east of the Vale these deposits form a low plateau, considerably cut up by streams upon which lie strings of villages parallel to the main direction of the valley. Much of the Mid Clay Vale consists of the basins of the Ancholme and Langworth, the latter a tributary of the Witham. The Langworth and its tributaries are factors in village sites, whereas the Ancholme is practically negative in this respect. In the south of the Vale there are considerable deposits of boulder clay which have been dissected by the streams, but there is not absolute flatness, and thus there has not been necessity for villages to avoid the land near the streams. On the other hand, there are practically no villages along the Ancholme. In the north of the Vale villages are either on the boulder clay skirting the Wolds or along the eastern edge of the limestone Heath uplands. The Ancholme has been canalised with a straight cut, and there are many small lateral drainage cuts, but formerly this was little better than undrained Fen. The drained land here is known as "carrs" or "ings," tracts being so named on the Ordnance maps with the adjacent village name as prefix. Near the new Ancholme or straight cut there are isolated dwellings of the squatter type as on the Marsh. The carrs are noted for pasture and are often held along with Wold farms of the scarp-side villages. Farms in the Vale are not large and there is little specialisation, the farming being mostly mixed, with emphasis on milk-producing when there is access to the railway.

The Heath, Kesteven Plateau and West Clay Vale

West of the Mid Clay Vale are limestone heights, comprising the Heath or Cliff Range, which extends southward to the Ancaster Gap, and then merges into the limestone Kesteven Plateau. In many respects these limestone heights are similar to the Wolds in their economic bearing. There is development of an escarpment with its springs and consequent village control. In the Kesteven Plateau there are stream valleys which at Domesday as regards water-mills and population were as important as those of the Wolds. There were the same stretches of poor soil and heath pasture which experienced the amelioration and enclosure noted on the Wolds, so that now they are good sheep and barley lands. However, there are many differences between the two regions. The chief points wherein the Heath contrasts with the Wold are its lack of valleys, lack of woodland, lack of glacial deposits, regularity of scarpline and ridge. A feature the Wold lacks is the transverse gap where the Witham breaches the upland at Lincoln*, a feature which is generally held to mark the former course of the Trent. Lack of valleys is due to lack of streams, a result of the extensive outcrop of porous limestone, which has no impervious boulder clay mantle as is found in the Central Wolds. There is some undulation on the east, where the upper valleys of the Mid Clay streams extend westward towards the Heath in shallow depressions, their head-waters often marking village sites. Thus, on the Heath there is no settlement except on the borders, a contrast to the villages in the heart of the Wolds. The Heath escarpment is a factor governing village sites. The lower slope of the scarp is Upper Lias Clay, above which are Northampton Sands and Lincolnshire Limestone. This combination of clay below limestone leads to springs, which, as on the Wolds, have influenced the distribution of villages along the scarp. The character of such villages is similar to that of the Wold villages.

West of the Heath escarpment the Lias forms a long belt, essentially a vale, and in places a low plateau; but erosion by certain tributaries of the Trent, by the middle Witham and Brant, has carved a series of minor valleys, so that the more resistant strata stand out, giving the country an undulating appearance. In general the Lias is a thick mass of bluish clay, but stratified in it are some beds of limestone, often containing iron, which resist erosion and give diversity to the landscape. The Lower Lias forms a vale of heavy clay land, with much pasture, especially around the streams, where the red-brick houses are a contrast to the stone houses and stone fences of the limestone area in the upper Witham valley of the Kesteven Plateau. North of this West Clay Vale is a low plateau, bordered by a low cliff which skirts the Trent. This plateau is naturally a region of warrens or commons, capped with blown sand, where there has been little settlement. Since the local

* L (Fig. 27).

ironstone has been worked, a considerable population has concentrated around Scunthorpe.

There is evidence that the iron deposits in the Lincolnshire limestone were known in the past. In Domesday there is mention of forges at Stowe, and in 1911 remains of a primitive forge near a human skeleton were found at Scunthorpe. The ore was never systematically worked until the middle of the 19th century, the first blast furnace being opened in 1861. The deposit is a ferruginous limestone, and contains about 30 per cent. of iron. The ore is self-fluxing when the lime content is about 12 per cent. The fact that this ore can be worked in open quarries has contributed to the development of the district, which is now one of the most important sources of supply in England, despite considerable distance from the coalfields. The Scunthorpe iron is very suitable for bars and wires, the presence of manganese rendering it particularly adaptable for the latter. Before the war it was mainly used for foundry purposes, but much pig iron for basic steel has since been turned out.

The most continuous of the deposits cover an area of about 3 sq. miles, of which more than a third has been worked out. The ferruginous limestone is sometimes 30 ft. thick, but the working is only about half that thickness. It has been estimated that between 1880 and 1916, ten million tons of the Scunthorpe ore were worked and that there is a reserve of rather less than twice this amount*.

The Northampton ironstone, which extends from Northamptonshire into Rutland and Lincolnshire, is worked in the south of the last-named county, but on a very small scale compared with Northamptonshire. It is estimated that from 1855 to 1916, rather more than seven million tons of this ore were worked in Lincolnshire. There is no industrial region here as at Scunthorpe. The ore is utilised elsewhere, Kettering and Wellingborough being smelting centres of what are broadly called the Northamptonshire ironstone deposits. (See also Ch. XI, pp. 212 and 213.)

Isle of Axholme

In the north-west of the county is the Isle of Axholme. Like the Fens, it is essentially a low plain, and much of it is the result of reclamation. As in the Fens, human settlements are generally on raised islands, but whereas in the Fens these islands are of gravel or clay, in the Isle of Axholme they are rock *in situ*, a hard Keuper marl which has resisted erosion more than the surrounding softer

* R. L. Sherlock, *Man as a Geological Agent* (1922).

beds removed by streams and by normal weathering. In places they are covered with blown sand and gravel, which have acted as small natural reservoirs, and before drainage gave immunity from flood. This is not the place to discuss the history of the difficulties and triumphs of the drainage of these former swampy lands. It is sufficient to say that drainage and reclamation by warping have rendered them exceedingly productive. Their agriculture is similar to that of the Fens. They are producers of wheat, potatoes and other vegetables. Arthur Young refers to the growth and manufacture of flax, but this is a thing of the past. Not so, however, the prosperity of the small proprietors to whom he referred in such laudatory terms. Small-holdings are common in the Isle, and around Epworth* there still survives the curious custom of cultivating fields in strips of different crops. Except for pigs, the Isle is not a stock-rearing region, land being too valuable for tillage. The Trent, with which many of the drainage canals are connected, is a natural water outlet for Axholme†. Gainsborough‡ was once a thriving river port with shipbuilding, but the produce of the Isle is now mainly distributed by railway.

Towns

Little can here be said about the urban side of Lincolnshire. Grimsby and Immingham are modern ports whose prosperity has been made possible by the activity of the former Great Central Railway§. Though to some extent they are dependent upon the coalfields, much of their trade, apart from the fishing industry of Grimsby, is based on Baltic timber and Danish dairy produce. Lincoln is an old Roman station and ecclesiastical centre whose importance has been increased by the railway and by the development of agricultural and other engineering works. For an analysis of the ports and for some estimate of the geographical bases of Lincoln's growth, reference can be made to a volume by the writer of this essay, in which also the agricultural geography of the county is analysed somewhat fully||.

Future Development

Future development in Lincolnshire would seem to be concerned with the development of its agriculture and stock-raising along lines which would co-operate with manufacture of the raw products as in Denmark. There is no geographical reason why a county like

* E (Fig. 27). † *Vide infra*, pp. 309–311 with Fig. 40. ‡ GB (Fig. 27).
§ *Vide infra*, pp. 319–321. || *Eastern England* (1923).

this should not become as great a producer of bacon, butter, and eggs as Denmark, and as active a producer of beet sugar as Germany and other continental countries. There are equal geographical advantages of soil, climate, and access to markets as in the continental countries named. Some progress has been made with beet cultivation in Lincolnshire to feed sugar factories opened at Brigg, Bardney and at Spalding. The crop is not sufficient, and to keep the factories working there has been import of cane sugar for refining. This has led to temporary increase in barge traffic on the Ancholme. Not very successful experiments have been made with bacon factories in the Fens and elsewhere. The main difficulty is that breeders have not been able to supply sufficient suitable pigs, so that at times the factory must remain idle. A great handicap is the conservatism of English farmers, their suspicion of Government interference, of technical education, and of such devices as co-operative factories and co-operative marketings, which have been successful in British Dominions like New Zealand and Canada as well as in Denmark. Also, most farmers have little capital available for experiment and a great number can barely struggle on in the conventional way which, "good enough for the father," seems equally well adapted to the son.

Greater industrial developments are possible around Scunthorpe, and geologists life Prof. Kendall have shown that there are undoubtedly extensive coal deposits under the west and centre of the county even if at considerable depth. However, Lincolnshire is primarily a community of workers on the land, and probably will remain such for a long time to come.

XI

THE NORTH-EAST MIDLANDS

P. W. Bryan

THE English Midlands, like ancient Gaul, may be readily divided into three parts. Of these, two form very clear-cut natural entities but the third is of a much more indefinite character. The first of these three parts is the Central and North-West, consisting mainly of the Birmingham plateau and the North and South Staffordshire Coalfields, a region largely industrialised. The second is the South-East, a region lying in the main between the Cotswolds and the Chilterns, the central feature of which is the Oxford Clay plain. This area is for the most part an agricultural entity, and therefore contrasts markedly with the first region. When we come to examine the third section of the Midlands, that of the North-East, we find an area which partakes of the outstanding characteristics of both of the foregoing regions. While it is in some places highly industrialised and in others wholly rural in character, neither manufacturing industry nor agriculture can be regarded as dominant in any broad view. It is this very diversity of occupation which makes the region of so much interest to the geographer—a diversity which is as clearly due to geological factors as its unity is due to its subsequent morphological development.

From the geographical point of view any attempt to define the exact boundaries of a region is usually a failure, but none the less all areas have certain very clearly defined central or focal points about which one can group the various parts. The North-East Midlands have two such major points. They are Leicester in the Soar valley, about which the region for the most part groups itself, and Nottingham in the Trent valley. Because of the decline of its staple industry and the transport developments which have made available the lines of the Derwent and Erewash valleys, Nottingham, which historically was the more important of these two centres, is to-day being steadily overtaken* by Leicester, whose staple industries have not been so seriously affected by industrial depression and whose position is more favourable in regard to transport facilities.

* A study of the bank clearings for Nottingham and Leicester show that the Leicester clearing figures, at one time smaller than those of Nottingham, are now greater.

Let us then examine the distinctive subdivisions of the region which are more or less grouped about and related to these two centres. As it is impossible in a short essay to treat of the whole area in anything approaching detail the writer proposes, after dealing in a general way with the subdivisions of the region and the relationship of the region to the rest of the country, to discuss in rather more detail that more central portion of the region related to Leicester with which he is most familiar.

Administratively the North-East Midlands consist of the whole of the county of Leicestershire with Leicester roughly centrally placed therein and of parts of a number of adjoining counties. To the north of Leicestershire the area is occupied by the south-eastern part of Derbyshire and the southern part of Nottinghamshire, to the east by the major part of Rutland, and to the south by the northern part of Northamptonshire. To the south-west that portion of ancient Watling Street which now forms such a splendid motor highway between Atherstone, High Cross and the upper Avon is the county boundary and serves to separate the North-East from the highly industrialised Central Midlands.

Topographically the central feature of the East Midlands is the valley of the Soar. This valley has a roughly north-south course bending slightly east and then north-west in avoiding the ancient rock mass of the Charnwood. Running approximately at right angles to this general direction is the flat alluvial stretch of the Trent between Burton and a point below Nottingham (Fig. 28). These two stretches of alluvial lowland formed by the Soar and Trent are in the rough form of a capital letter T tilted slightly towards the west. Round the stem of the T the main subdivisions of the area lie. To the west of the T there is a mass of high ground known as Charnwood Forest, very little of the forest character of which remains. From its highest point, the summit of Bardon Hill (912 ft. or 278 m. above sea-level), much of the region is in view.

Here one is, as it were, on a peak of a rocky timbered island lying mainly east and south-east of the viewpoint. This apparent island is the Charnwood Forest, and the illusion of a surrounding sea is striking when the low ground is veiled on a hazy day. Much of this apparent flatness is of course due to the height of the observer's position, as we shall see in studying the region in detail, but relatively to the Charnwood the surrounding areas are low-lying.

Looking eastward from Bardon summit we have the Charnwood as it were at our feet—ridge beyond ridge separated by deeply cut valleys. In the middle distance we see the wide stretching clay lands centrally placed, in which is the alluvial valley of the Soar.

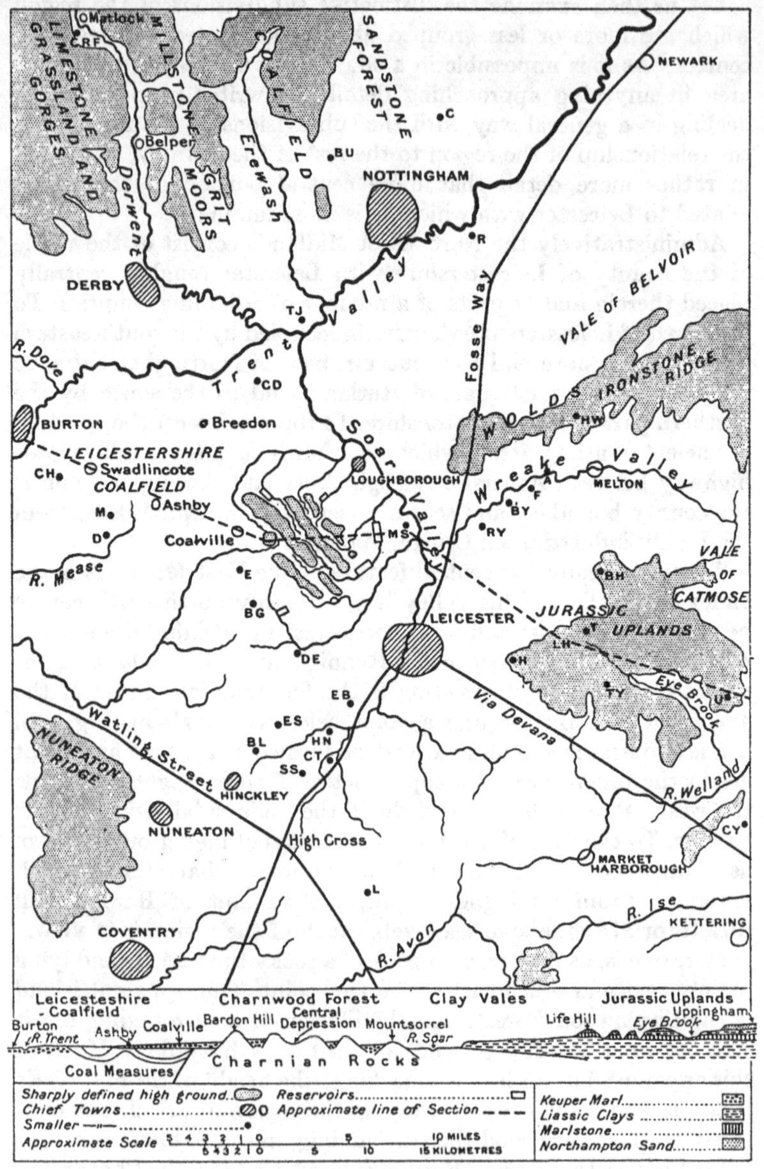

Fig. 28. Sketch-map of the North-East Midlands with an approximate geological section. (Key to the towns on p. 203.)

This section of the picture is more or less continuous from the southern edge of our region away to the Trent valley in the north. Beyond the lowlands we may see the ground rising north-eastward to the Wolds and the Melton ironstone ridge, and south-eastward and eastward to the marlstone, sandstone and limestone uplands.

Southwards the hill country disappears, but the lower ground becomes flatter, corresponding to a change in the structure. Towards the south-west the Nuneaton ridge, which is beyond the limits of the area, appears, and here and there in the middle distance, as at Bagworth and Ellistown, the overhead gear of a colliery indicates the extension to southward of a concealed coalfield. Westward we see the main portion of the coalfield around Coalville, Ashby and Whitwick, one of those very few coalfields in Great Britain which have little manufacturing industry. Beyond the coalfield lies Burton-upon-Trent, just above the junction of that river with the Dove. To the north the lowland seems continuous, broken only by an isolated hill. The lowland is the great stretch of the middle Trent valley between Burton and Nottingham with the lower portion of the Derwent valley between Derby and the Trent. The isolated hill is that of Breedon, formed of Carboniferous rocks, an outpost of the Pennine system farther north; the great rolling outlines of the Pennines can be glimpsed in exceptionally fine weather, forming a clearly defined northern framework to our mental picture.

Geologically regarded, the region is mostly an arrangement of younger rocks (chiefly Triassic and Jurassic) around the south-east corner of the more ancient Carboniferous system constituting the Pennine moorlands. This view would, however, if unmodified, neglect two points of the highest importance, the one economical and the other physiographical. The ridges and peaks of the Charnwood are made of ancient volcanic materials related to the old floor upon which much of modern Britain has been built, while the Leicestershire Coalfield, west of the Charnwood, is a southward projection of the Pennine rocks into the younger formations of the Trias. If we still keep our mental picture of the Charnian island then the Pennine region can be thought of as mainland,

Key to the Towns: A, Asfordby; BG, Bagworth; BH, Burrough on the Hill; BL, Barwell; BU, Bulwell; BY, Brooksby; C, Calverton; CD, Castle Donington; CH, Church Gresley; CRF, Cromford; CT, Croft; CY, Corby; D, Donisthorpe; DE, Desford; E, Ellistown; EB, Enderby; ES, Earl Shilton, F. Frisby; H, Houghton on the Hill; HN, Huncote; HW, Holwell; L, Lutterworth; LH, Life Hill; M, Moira; MS, Mountsorrel; R, Radcliffe on Trent; RY, Rearsby; SS, Stoney Stanton; T, Tilton; TJ, Trent Junction; TY, Tugby; U, Uppingham.

and a useful concept of the whole area can be formed by thinking of all the rest as sea. This sea would consist of Triassic marls and Recent alluvium around the Charnwood and in the Trent and Soar valleys, undulating here and there. To the south-east the region could be likened to a wave, the rising front of which is formed by the limestones, sands and clays of the Lower Lias, while the crest and back is made of the marlstones, sands, clays and limestones of the Middle and Upper Lias and the Inferior Oolite. An escarpment, that of the marlstone, is the actual crest line where it is best defined and can be well seen near Tilton, north of the Uppingham road, and on the north-west slope of the Melton ridge. This Melton ridge, projecting directly westward from the main mass of the uplands, is the only feature to break the symmetry of the wave. Within this broad concept of Carboniferous mainland, volcanic island, Triassic backwash, and rising Jurassic wave one can readily fit the structural details of each sub-region in due proportion.

Starting in the north-west of the area we have the middle Trent valley between Burton and Nottingham, with the lower parts of the Derwent and Erewash valleys entering the main valley approximately at right angles from the north. These two valleys offer striking contrasts and bring out clearly the outstanding geological features of the country north of the Trent. The Derwent rises and flows for much of its length in the region of the Dales, steep-sided valleys formed in the rolling grass-covered uplands of white limestone. Before reaching Derby the river leaves this belt and crosses eastward into the flanking zone of Millstone Grit with its rugged outlines and cover of heather and gorse. Farther east lies the coalfield whose southern limit is in the vicinity of Derby and Nottingham; and beyond this the zone of the Bunter sandstones that reaches southward from Sherwood Forest. To the south of this region of dale and heather-clad moor, coalfield and sandstone forest lies the alluvium of the Trent valley flanked by Keuper marls, in which the beds of gypsum give the waters of the Trent that peculiar hardness so prized for brewing purposes. We thus find this industry extensively developed along the Trent valley at Burton, Nottingham and Newark, the gypsum itself outcropping at such points as Radcliffe and East Bridgeford in Nottinghamshire and Chellaston in Derbyshire*. The finest gypsum or alabaster in the Middle Ages was extensively used for carved work, but to-day it is mostly mined for plaster of paris.

* The process of putting gypsum into water used for brewing is for this reason called "Burtonisation."

Nottingham, the chief centre of the middle Trent, occupies a very striking geographical position. The Trent, a meandering river of the plain, is liable to floods and difficult to cross. Standing partly on a hill, good for defence, Nottingham marked a convenient point at which the river could be crossed, and one at which parties could rest and equip before entering the fastnesses of Sherwood to the north. One of the salt ways used for the distribution of salt from such centres as Cheshire and Worcestershire seems to have crossed the Trent at this point in pre-Roman days. Probably because of the swampy meandering character of the river the Romans entirely neglected Nottingham, and their great route to the north, Ermine Street, passed along the drier uplands to the east by Grantham and Lincoln, where it was joined by the south-west and north-east line of the Fosseway from Leicester and High Cross. With the coming of the Anglo-Saxons and the Danes by the water-ways from the east, Nottingham's riverine position again brought her to the fore, a position enhanced subsequently by the building of the bridge.

In the Middle Ages Nottingham laid the foundation of her cycle, motor, and machinery industries of to-day. Her smiths, working with the charcoal of Sherwood to smelt their iron-ore, became famous. To-day the charcoal is replaced by the great coalfield which perhaps more than any other in Great Britain is capable of future extension. The coal mining of the Erewash valley, in the Coal Measures to the north-west of Nottingham, is spreading steadily eastward under the sands of Sherwood Forest beyond Mansfield and Ollerton, below which ancient haunt of robber and outlaw, bear and wolf, the Top Hard Seam or famous Barnsley Bed of Yorkshire is mined at an economical depth and with a thickness of some six feet. In connection with the Coal Measures beds of fire-clay and ganister, important in the potteries and brickworks of the Midlands, are found. Much of the coal burned in London originates in this field and passes southward through Trent Junction which lies a little to the south of and about midway between Nottingham and Derby at a point where Derwent, Trent, Erewash and Soar concentrate, a natural nodal point on which railways from all these valleys centre.

Although the iron industry is again coming to the fore, when we think of Nottingham to-day we think not so much of iron as of hosiery and lace. The invention of the stocking frame by the curate of Calverton, near Nottingham, was followed by that of Strutt for the making of ribbed hose, and with the coming to Nottingham of Arkwright and Hargreaves, the inventors of spinning machines, the hosiery and cotton industries became firmly established. They further developed at Derby, Cromford and Belper in the Derwent valley and at many other centres in the vicinity, now helped by the abundance of power made available on the coalfield. To-day Nottingham is making a determined effort to develop her hosiery trade.

Derby has more particularly developed as a route centre. Possessing the chief works of the Midland Railway system it also manufactures the finest of all road vehicles, the Rolls Royce car. A glance at the map will show that if we take a straight line from the London Docks to that great agglomeration of population around Manchester this line will pass but a mile or two west of Derby and Leicester. Topographically this straight line very nearly corresponds with the line of least resistance. From London, Bedford is reached by the breach in the chalk at Luton. Thence by way of Wellingborough, Kettering, and Market Harborough, Leicester is reached after passing through the oolitic uplands at a point where the headstreams flowing north-east to the Wash have produced through erosion a partial breakdown. From Leicester the Soar valley continues the north-westerly line to Trent Junction and Derby in the Derwent valley. This valley forms a relatively easy line of communication across the south-western end of the Pennines by Matlock and Buxton and the Goyt valley to the Manchester region. This line is followed by both rail and road between the Capital and the second largest population centre in the country; and since rail traffic for the north on reaching Trent Junction does not pass through Nottingham but follows the more direct line of the Erewash valley to Chesterfield and Sheffield, Nottingham is to some extent sidetracked under modern conditions. We have seen that this was also the case under the Romans. But this sidetracking can easily be over-emphasised. Nottingham stands on a river which with the completion of the canalisation of the Trent can be reached by large barges from the sea; it is at the junction of two sharply contrasted geological areas, the coalfields to the north and the widespreading agricultural areas to the south and east. The cattle to the south plus the oaks of Sherwood to the north gave rise to her important tanneries. In consequence of these facts Nottingham is a great natural market, and with the development of her coalfield eastward her importance is bound to grow.

Let us turn now to a wholly different region, the Charnwood. This region consists, as we have seen, of a mass of ancient rocks half buried in the much younger marls of Triassic times. Topographically and geologically the area presents a certain symmetry which is not at first evident. Study of a detailed relief map shows that the general trend of the relief is from north-west to south-east and consists of three longitudinal depressions separated by four regions of upland accentuated by a series of well-marked summits. These uplands and ridges are rugged, steep, and barren, and contrast sharply with the gentle outlines of the valleys between, which are often fertile and well cultivated, producing either a rich grass or ordinary crops, and have all the appearance of being part of the adjoining lowlands. In this sharp contrast is to be found one of the great charms of the Forest area and one of the reasons why it performs the useful economic function of being a playground for the people of Leicestershire, the other being the fine air available in some of the more elevated villages, such as Woodhouse Eaves (*vide* Fig. 29).

The symmetry of the area becomes more marked when we study its drainage. The central depression rises from either end to a low water-parting which separates the waters of the Black Brook flowing north-west from the Ulverscroft stream flowing to the south-east. In the flanking depressions we find similar streams but with a double water-parting. In that to the east the northerly part of the depression is occupied by a headstream of the Burleigh Brook while the southern part provides a channel for the Brand stream. The stream arrangement here, however, differs from that of the central depression in as much as, wedged in between the headwaters of these streams, there is to be found the headwaters of a third stream, the Wood Brook, which cuts its way out directly through the uplands to the Soar at Loughborough, filling the Nanpantan Reservoir on its way out. On a very small scale this stream bears the same relationship to its neighbours as that borne by the Reuss to the headwaters of the Rhine and Rhone. If we examine the westerly Charnian depression by Groby Pool, Markfield and Whitwick we find a similar stream arrangement. It is further of great physiographical interest to note that the Charnian streams, with the exception of the Black Brook, flow for some distance along the smooth, wide, marl-covered valleys and then turn at a sharp angle to cut their way out of the region through a bar of ancient, resistant rock.

It is clear, then, that the Charnwood possesses symmetry and great contrast. Both are structural in origin. Geologically the area is best pictured as a series of rude horse-shoes placed one within the other, the open ends towards the north-west. The depressions correspond to the less resistant arms of the shoes, the ridges to the more resistant. The development of the present relief may be briefly pictured. In Archaean times the district was probably occupied by a series of volcanic islands, from the vents of

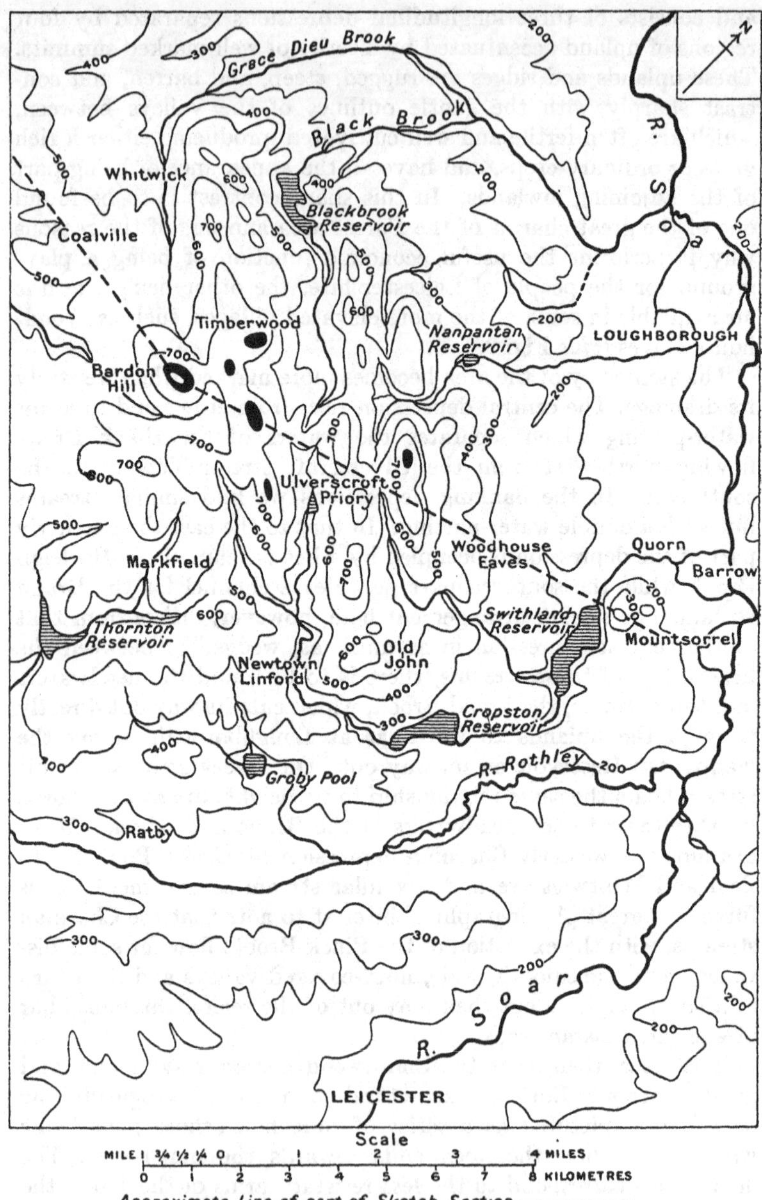

Fig. 29. Sketch-map showing the relief of the Charnwood Forest district (note the orientation). Land between 800 and 900 ft. is shown in black.

which the bulk of the material in the form of lavas and ashes which form the present rocks was ejected. Originally more or less horizontal this material was ridged up subsequent to its deposition into a crude dome. Normal erosional processes would break open the dome and form a series of ridges and valleys. These were later depressed and buried under deposits of Keuper Marl in Triassic times. Further erosion is still exposing slowly to our eyes this curious fossilised pre-Triassic landscape*.

Here and there in the Charnwood at various dates in its chequered history have occurred upwellings from the molten magma within the crust. It is this material that has given rise to the many quarries of the region which result in the high position occupied by Leicestershire as a producer of igneous rock. At Mountsorrel in the east of the region, Bardon in the west, and Groby in the south, granites and similar rocks are quarried for road metal and paving sets. At isolated points away to the south of the Charnian district, in the pastoral country at Enderby, Croft, Huncote, and Stoney Stanton, masses of syenite come to the surface and are probably closely related to the movements of the liquid magma below the Charnwood. These frequently form striking hills and are extensively quarried. A further aspect of economic activity worthy of mention is to be found in the reservoirs. These, conveniently placed on the edges of the uplands, furnish the neighbouring towns with water, though this resource has long since ceased to prove adequate for the needs of the growing population, and Leicester, to mention but one, has had to embark, in cooperation with Sheffield, Nottingham and Derby, in an elaborate scheme for bringing water from the headstreams of the Derwent far to northward on the Pennine uplands.

As we have seen, the fertility of the Charnian valleys is dependent on the presence of the red marl. With the exception of the infertile Black Brook valley it is found in all of them and at a higher elevation than in the surrounding districts†. Its presence in the other valleys is explained by the occurrence of the rock bars which all the Charnian streams save the Black Brook have to cross to reach the low ground. These belts of resistant rocks have forced the streams to expend their energies in cutting gorges instead of grading their valleys and removing the marl.

The Charnwood is thus a region in which both scenic contrasts and economic activities are strikingly related to physical history.

* W. W. Watts, "Charnwood Forest: a Buried Triassic Landscape," *Geog. Jour.* vol. XXI, 1903, pp. 623–36.

† See Section, Fig. 28.

Turning now to the districts west and south of the Charnwood we find a widespreading lowland which can be conveniently divided into two sections without clearly defined boundaries, though sharply contrasted in their main activities. Looking westward from Bardon summit we have the Leicestershire Coalfield with its main pithead gears and miners' dwellings clearly visible around Coalville and stretching away westward round Ashby, Swadlincote, Church Gresley, to distant Burton. Southward the pithead gears become fewer and finally disappear and we see nothing but long dreary stretches of pastoral country developed on the Keuper Marls. In the far distance the dim outlines of the Nuneaton ridge limit our area. The coalfield is a geological basin, the upturned edges of which rest on the ancient Charnian mass at our feet and represent a period when that mass was probably an island in the Carboniferous sea. How far this coal was also laid down to east of the Charnwood is not known, but if laid down it is thought to have been subsequently removed, so that the chance of ever developing a coalfield in the Soar valley seems very remote. In the coalfield there are three zones running north-west and south-east or parallel to the Charnian axis. Two of these zones are productive. The central one, which forms an anticline running approximately through Ashby-de-la-Zouch, contains the lowest measures, largely unproductive, and separates the western zone around Coalville, in which the coal measures are concealed under the Triassic marl, from the eastern zone around Donisthorpe, Moira and Swadlincote. In this latter district the clays associated with the Coal Measures are suitable for the manufacture of drainpipes, firebricks and saggars, and thus give rise to one of the distinctive industries of the area. The actual junction of the coal with the Charnian rocks to the west is somewhat complicated by a great fault which runs from the vicinity of Whitwick via Bagworth to Desford, as a result of which the Coal Measures are thrown down against the ancient rocks to the east. A similar line of fault separates the unproductive central measures from the prolific western zone, which is itself rather extensively cross-faulted, thus adding to the difficulty and expense of working.

Throughout this field the coal is of a quality admirably suited for household purposes, and for that reason the bulk of the coal mined moves out of the region for domestic consumption. Doubtless it is due to this fact and the reasons giving rise to it that we do not here find a region of great manufacturing activity; but the area is covered with large villages or small towns which are in the main the homes of the miners and of others depending on or providing subsidiary services in connection with the major industry.

In the southern part of the area are the quarries already referred to, and in many of the little towns are extensions of the boot and hosiery factories of Leicester. Such centres are Hinckley, Barwell and Earl Shilton. In all probability the ruling factors in this expansion are the lower county rates and the surplus labour available.

Apart from these developments, cattle rearing on the extensive grasslands to be found on the marly clays is the chief occupation, with tillage to a limited extent where sands and gravels of glacial origin give a certain diversity. Scenically the area is rather depressing owing to the long uninteresting stretches of flat country resulting from the typical Keuper Marl conditions.

Crossing now south-eastwards over the Soar valley and leaving it for detailed consideration later, we enter a district which in many ways contrasts sharply with that which we have left. We have crossed a geological boundary, the line separating the Triassic clays from the Lower Lias. This latter, although mostly clays, contains bands of limestone and ironstone which give greater diversity to the landscape without removing its dominant feature—that it forms part of the great Midland grass vales and is famous fox-hunting country. The vale of Catmose far to the east is succeeded by that of Belvoir to the north of the Melton ridge and the Wreake valley to the south of this ridge, while the country around the important hunting centre of Market Harborough has few equals from this point of view. In all these areas the Lower, Middle or Upper Lias clay is the dominant geological formation and it is here that we find the Cottesmore Hunt, the Quorn, and the Pytchley, to mention but a few of the names outstanding in hunting annals. Throughout this region the main industry is cattle-tending, both for dairy purposes and beef. In the latter case the cattle are bought in the spring, fattened on the rich grassland during the summer and sold off as winter approaches, leaving the farmer with little but hunting to occupy his time till the succeeding spring*.

Eastward from the clay vales the ground rises to the much more diversified country flanking the slopes of the oolitic ridge. This area is mostly composed of the sands, clays, limestones and marls of the Middle and Upper Lias. To use the symbolism already given, it is the front and crest of the breaking wave and exhibits in consequence a variety of relief which is in striking contrast to that of

* See *Agricultural Atlas of England and Wales*, J. P. Howell's map of 2-year-old cattle. Leicestershire has the greatest density, 209 per 1000 acres, of such cattle.

the last two districts considered. The numerous gradients of one-in-seven or less on the by-roads help to emphasise the fact that we are here in a region of steep slopes sufficiently marked to refute Ruskin's dictum that Leicestershire is flat and uninteresting. A good general impression of the area can be obtained by a journey on the ridge road from Leicester to Uppingham and thence onward to Peterborough.

From Leicester to near Uppingham the country consists of clay-covered flat-topped stretches with steep slopes of marl, as in the Tilton ridge, or sand leading down near Uppingham to clay-floored valleys of which the Eye Brook is a striking example. The region is typical of the weathering resulting from stream action in nearly horizontal strata of varying resistances. Here and there boulder clay and some glacial sands give a certain variety to the flat surfaces of the plateaus, but for the most part the predominance of clay makes the region a grassland and this is accentuated by its elevation above sea-level. Much milk is produced, and along the highroad a familiar feature is the little platform for loading the milk cans into lorries. Cattle and sheep herding with a very little tillage complete the occupations of the people and the rather poor-looking houses and villages present a sharp contrast to the large farms and prosperous-looking villages on the richer and more varied soils developed from the sandstones, clays, marls and limestones of Rutlandshire to the east.

If we return for a moment to Tilton we can examine the development of a resource which is of growing importance to the area and to the country. Wherever the marlstone rock-bed and the Northampton Sand are sufficiently rich in iron-ore they are now worked. With the advent of the Thomas and Gilchrist process of smelting low-grade ores, first applied to those of Cleveland, the Jurassic ores found both in the marlstone of the Lias and the sands of the Inferior Oolite became of great importance and to-day furnish about one-third of the total ore consumed in this country*. The ore is low grade, somewhere around 20 per cent., but it is easily and cheaply worked where it outcrops or is under slight cover, or has been exposed by stream action. To the east of Tilton the ironstone is covered by the Middle Lias and by boulder clay. Here the ironstone is worked in a long cutting across a field. The surface cover is removed from one side of the cutting to the other, the iron-ore extracted and in this way the fertility of the area is maintained though the general level is reduced by the thickness of the ore bed, *i.e.* some six feet.

* See *Report of Inspector of Mines*.

Similar workings, though on a much larger scale as a rule, are to be found in the Rockingham Forest country, south of Uppingham at Corby, in the Desborough, Kettering, and Wellingborough districts farther south, where there are also important iron furnaces, and south of the Melton ridge at Holwell. At Asfordby in the Wreake valley are large works specialising in the manufacture of piping, and using a blend of the local ores smelted with coal from the northern fields. The moulds are made from sand obtained at Bulwell, on the Bunter sands north of Nottingham.

Before dealing with the Soar valley region one should draw attention to the great stretches of Wold country lying along the Melton ridge and extending eastward almost to the Soar by Loughborough. This consists partly of the oolitic limestone and partly of great stretches of calcareous boulder clay. In its western half it is a flat and uninteresting upland covered with a network of fields with little or no variety. It is crossed by the ancient Fosseway from Leicester to Newark and thence to Lincoln. In all this Wold country both to north and south of the Melton ridge the manufacture of Stilton cheese is of prime importance and is said to be localised through some property which the soil gives to the milk.

We can usefully conclude our brief account of the North-East Midlands with a short description of the Soar valley. This valley derives many of its distinctive features from the fact that it has cut out a flood plain in its meandering course below the general level of the surrounding country and has exposed in its banks the edges of the limestones of the Lower Lias which form a clearly marked minor escarpment in parts of its right bank. At Barrow-on-Soar these beds are of great interest to geologists since here have been found many early Jurassic reptilian remains. To the economist the special interest of the region lies in the great lime works which utilise this bed to manufacture an excellent hydraulic cement from which drainpipes and artificial pavement are made, and which has the special property of hardening under water and is therefore of great value in underwater work.

The Soar, whose headwaters rise close to the ancient Roman centre of Venonae (High Cross) where the Fosseway crossed Watling Street on the water-parting between the streams flowing north to the Trent and those flowing south to the Avon, sweeps as we have seen in a rough flat bow round the eastern side of the Charnwood. Situated about the middle of the valley is Leicester, on a site on the east bank probably selected by the ancient Britons as a camp because it was a dry gravel terrace (of glacial age) lying well above the Soar flood water. Through this camp the Romans carried the

Fosseway on its way from High Cross to Lincoln, and to it also they carried the only other road which roughly paralleled Watling Street but did not start from London. This road was the Via Devana from Colchester to Ratae Coritanorum, the Roman name for Leicester, by way of Godmanchester and Medbourne. As a terminal point of this route on the Fosseway Leicester must have been of great importance to the Romans, an importance indicated further by the number of Roman remains found in and about the city. With the coming of the Angles and later the Norsemen along the northern waterways the approach to Leicester changed from the high grounds of the south-east to the Trent and Soar valleys. In the 6th century it was captured by Angles from the Trent (possibly from Nottingham) and in the 9th century by the Danes. Both Angles and Danes have left many traces of their occupation here as elsewhere in the place-name terminations "ham," "ton," and "by." The termination "by," of Danish origin, is very common along what were then navigable streams, as at Sileby, Rearsby, Brooksby, and Frisby in the Soar and Wreake valleys.

It is difficult to trace the beginnings of industry in Leicester, but there is little doubt that in the 13th century, because of the excellent nature of the pastures in the vicinity, Leicester wool was rated more highly than most other English wools and the woollen industry was here well established. The reputation of the district for high-grade wool was still further enhanced in the 18th century through the introduction by Robert Bakewell, who was born near Loughborough, of the new Leicester Breed of sheep. To-day Leicester sheep are to be found in most parts of the world where wool is extensively produced. This early industrial growth is now represented by the important wool-spinning industry of the city. We have here an interesting example of how a slight superiority, due in the first place to geographical factors, is subsequently intensified by the hard work and patience of those connected with the industry, an illustration of "To him who has shall be given...."

It was not until the 17th century that hosiery was introduced, now one of the two staple industries of the city. Hand framework knitting developed in about one hundred villages in the county but more particularly in Leicester, Loughborough*, Hinckley and Castle Donington. With the coming of steam the concentration in the larger centres became more marked, and to-day Leicester is by far the biggest centre in the country for the manufacture of

* Loughborough has since developed as a hosiery, engineering, and a bell foundry centre.

woollen hosiery. Here are to be found the headquarters of firms with a world reputation in this line.

With the rich pastures of the alluvial Soar and Wreake valleys and the adjoining clay lands is to be connected the development of the boot and shoe industry, which constitutes the other principal activity of the people of Leicester and the neighbouring centres. The boot and shoe industry, which at one time was entirely a hand industry, has to-day become a highly organised example of machine production through the application of both steam power and inventive genius to the solution of its problems. The many complicated pieces of mechanism used in the industry have given rise to the boot and shoe machinery industry which provides much employment in Leicester and supplies machinery to the boot trade through the country. Leicester may perhaps be regarded as the headquarters of both the boot and shoe manufacturing industry and also of many of the biggest firms engaged in the distributing trade. These latter firms probably control on a conservative estimate at least twelve hundred retail shops in different parts of the country.

Another well-established subsidiary industry is that of hosiery machinery.

In addition to these major industries the city possesses many other manufactures, some of which are independent and others are subsidiary. Of these one may perhaps mention among others the printing, malting, elastic webbing, rubber and iron industries.

Surrounded by great pastoral and agricultural areas Leicester, in addition to her manufacturing activities, is a great market and distributing centre. With the development of new transport facilities in the shape of motor buses this aspect of her economic life is becoming more marked each year. To-day Leicester is the market centre for the whole of the area we have discussed with the exception of the Trent valley in the north. This side of her development is typified in the great market-place in which markets are held three times a week and to which and the adjoining shopping centres the country folk flock on Wednesday to do their shopping. Her periodical cattle and sheep fairs are among the most important in the Midlands and emphasise the pastoral occupations of the surrounding districts.

In our necessarily rapid survey of the chief subdivisions of the North-East Midlands we have been forced to sketch in somewhat broad and sweeping strokes the main outlines of each region. It is but just to add that within these broad outlines there is an infinity of detail on which it is not possible to touch in an essay of this character, but which would amply repay detailed study.

XII

THE NORTH-WEST MIDLANDS

R. H. Kinvig

FOR the purpose of the present essay the area of the North-West Midlands is taken to include two main units, which, although rather dissimilar in their geographical characteristics, are now linked together for various purposes because of proximity and the economic strength of the dominating unit. These chief sections are: (1) the Birmingham or Midland Plateau, with its surrounding river basins of the mid-Severn, the Avon and the Trent, and (2) the Plain of north Shropshire.

The latter unit, a lowland for the most part and occupying a significant position on the Welsh Borderland, has had a different evolution from that of the Birmingham district. It has, however, always had a natural focus at or near Shrewsbury, whereas Birmingham has only attained its existing supremacy in relatively modern times. Life was formerly much more flourishing around the lowland margins of the Midland Plateau, along the river valleys—the old highways—and here are to be found the older historical centres, Worcester, Warwick, Coventry, Tamworth, Lichfield and Stafford. The Plateau was rather an obstacle to be avoided until the Industrial Revolution revealed to the full the riches of this watershed region, and thus gave Birmingham, which had been developing as the chief centre within the area, an immense superiority over all its marginal rivals.

This Midland Plateau or peneplane, which comprises most of south Staffordshire together with the adjoining parts of Warwickshire, Worcestershire and east Shropshire, has been best delimited by a line (Lapworth's Line) generalised from the 300 ft. (91 m.) contour. Thus defined, the area has an elliptical form with the longer axis, about 46 miles (74 km.), running roughly NNW.–SSE. from a point near Stafford to one near Stratford-on-Avon, while the shorter axis, about 34 miles (55 km.), trending WSW.–ENE., stretches from Nuneaton to Kinver Edge.

While the main emphasis will be laid on the core of the region as thus defined, attention must also be given to the surrounding valleys of the mid-Severn (including eastern Shropshire), Avon, and upper Trent, which are indissolubly bound up with the physical and human evolution of the Plateau; and indeed the boundaries of the area in the widest sense must be regarded as extending westwards to the line of high land between the Malverns and the Wrekin, and south-eastwards to the oolitic escarpment.

On its north-western side a neck of high ground, which may be referred to as the Watling Street isthmus, rising to over 600 ft. (180 m.) in parts, connects the Plateau on the one hand to the Wrekin mass and to the mountain country of south Shropshire and Wales on the other. This barrier, which tends to separate the north Shropshire lowland from the other section of the West Midlands, is, however, breached by the anomalous valley of the Severn.

Structure and Drainage

Viewed as a whole, the Plateau is highest round its margin, especially in its south-western section where the greatest height is attained in the bold ridge of the Clents with Walton Hill reaching 1036 ft. (316 m.). The difficulty thus presented to communication with the Severn is illustrated by the existence of the steep incline (1 in 37·5) at Blackwell on the Midland main line, while at Old Hill, on the Great Western, there is a gradient of 1 in 50. On the contrary, the most open side of the Plateau is around Tamworth on the north-east where the broad valley of the lower Tame opens towards the Trent.

The main system of drainage is formed by the slow inward-flowing rivers of the Tame and its chief tributaries, namely the Rea (on which Birmingham was originally founded), the Cole, Blythe and Bourne, which, apart from the Rea, converge on the main river near Coleshill, a fact which accounts for the earlier significance of this centre*. From here the Tame continues the same northerly direction defined by the Blythe, and the combined stream with sluggish features reflecting the "sulky Trent" reaches the latter ten miles beyond Tamworth (*vide* Fig. 30).

The north-western section of the Plateau is drained separately to the Trent, via the Sow, by the northward flowing River Penk; but elsewhere the western margin belongs to the Severn drainage system. Thus, in the central portion are the Stour and its tributary the Smestow, while farther south is the Salwarpe. These flow directly to the Severn, but in the extreme south the Arrow, with its feeder the Alne, enters the Avon. There is a pronounced contrast

* The selection of Hams Hall, adjoining Coleshill, as the chief generating station in the regional electricity scheme is an important illustration of the significance of this focus under modern conditions.

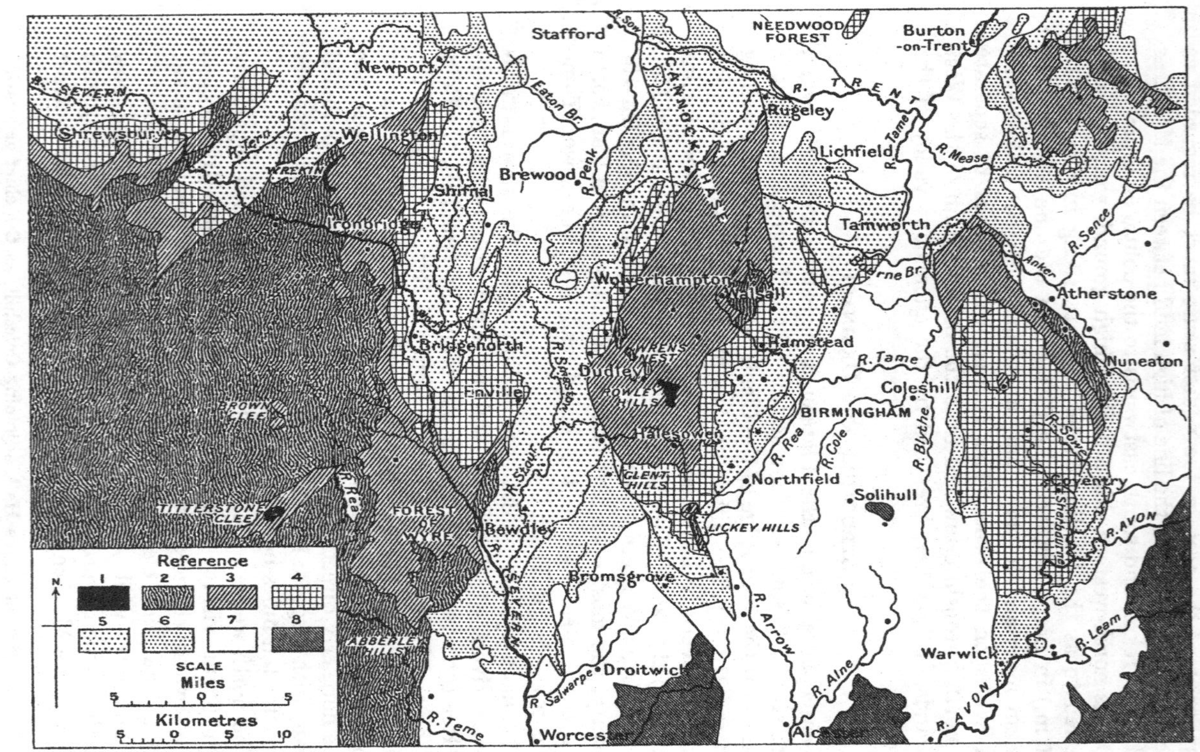

Fig. 30. Geological map of the Birmingham District (after Lapworth). Reference: 1, Igneous rocks; 2, Pre-Carboniferous; 3, Coal Measures and Lower Carboniferous; 4, "Permian" and Carboniferous Red Measures; 5, Bunter; 6, Keuper Sandstone; 7, Keuper Marl; 8, Lias. (*Note:* The black dots are towns.)

physically between the streams flowing to the Trent and those to the Severn; for while the former have the appearance of maturity, occupying broad valleys with gentle gradients, the latter, especially the Stour, Arrow and Salwarpe, are more youthful, being confined in their upper portions to ravine-like valleys. Economically the steeper gradients together with the relatively greater and more constant supply of water in the Severn rivers have given them a marked advantage over the others in the provision of good water-power. These differences are to be explained, first, by the windward situation of the western streams, and secondly by the fact that the Trent tributaries have to travel at least three times the distance of those of the Severn before reaching levels approximating to that of the sea; while finally it must be recognised that since glacial times the middle and lower sections of the Severn have undergone rejuvenation.

Within the Plateau two major upland units may be distinguished, those of south Staffordshire and east Warwickshire, while a third is formed by the smaller mass of Wrottesley or Tettenhall to the north-west, with a general elevation of over 400 ft. (say 120 m.). These represent, with respect to the Mesozoic rocks, broad anticlinal areas with intervening basins; and all the dominant feature lines have a north-south (Malvernoid) trend. In many subsidiary features, however, the trend is rather Charnoid (NW.–SE.) in character, or in some cases Caledonoid (NE.–SW.)*.

The south Staffordshire Plateau, which is by far the largest and most important unit, is subdivided by the broad basin of the upper Tame which extends diagonally from the neighbourhood of Wolverhampton south-eastwards towards Birmingham. The northern (or Oldbury-Cannock) section is relatively low-lying, much of it being less than 500 ft. (150 m.), but there are still stretches of higher ground, particularly in the dome-like region of Cannock Chase, with several parts exceeding 700 ft. (210 m.).

The surface of the southern (or Dudley-Northfield) section is higher and far more varied. Its most conspicuous feature is the ridge of high ground, often over 700 ft. above sea-level, skirting its western margin from Upper Penn through Sedgley, the Wren's Nest, Dudley Castle Hill and Rowley to Frankley, whence the line is continued by the high ridges of the Clent and Lickey Hills. Breaching this mass of high land on the west are the two well-marked amphitheatre-like basins of the upper Stour and Arrow which have been noted previously.

* See L. J. Wills, *Physiographical Evolution of Britain* (1929), p. 76. The nomenclature is that suggested by Prof. C. Lapworth, and the termination "-oid" indicates that direction only is implied.

The east Warwickshire Plateau is separated from south Staffordshire by a north-south Trias syncline now occupied by the joint Blythe-lower Tame system. The upland extends for about 14 miles (22 km.) from Atherstone southwards to the neighbourhood of Berkswell, and averages about 6 miles (10 km.) in width. While nowhere very high (averaging 400–500 ft.) it presents a steep edge on its north-eastern side between Atherstone and Nuneaton, where its ancient hard rocks are faulted against the Trias. The smaller upland mass of Wrottesley or Tettenhall is separated from the central plateau by a syncline drained in the north by the Penk, and in the south by the Smestow, a tributary of the Stour. Although lying outside the Midland Plateau proper the upfolded mass of Enville, rising in parts to over 650 ft., may be noted as forming a structural link between south Staffordshire and the plateau of south Shropshire.

It will be evident that the drainage of the area is primarily a consequent one owing its origin to the domes and basins formed in post-Triassic and probably Tertiary times. The radial systems thus initiated from the uplands, and draining into the north-south synclines, had become well established before the Glacial Period. The joint Blythe-lower Tame occupies the most clearly marked and most continuous syncline; and the combined stream enters the Trent in the stretch where this river is changing its character from that of an old consequent, in the upper portion, to that of subsequent. Within the south Staffordshire dome the radial drainage is in general well marked; but here modifications in the structure have aided the special development of both the upper Tame and the Stour. The relatively extensive depredations on the part of the latter are to be attributed to the local absence, owing to faulting, of the more resistant beds of the Upper Coal Measures and Permian formations. The course of the Tame, on the other hand, appears to have been influenced by a sag in the original Trias cover, probably connected with the wide concave bend in the eastern boundary fault of the coalfield which is well marked in this area.

The River Severn presents anomalous features particularly in the striking contrast between the upper stretch, above Buildwas, where the river exhibits a fairly open, serpentine valley and a gentle flow, and the middle section to Holt Bridge where the stream is confined to a narrow valley with a straight course and a swift current—especially marked in the classical gorge section between Ironbridge and Bridgnorth. It has been amply demonstrated by Dr L. J. Wills that these anomalies, as originally suggested by Prof. Lapworth, are to be explained by glacial intervention. Thus the upper Severn originally flowed northwards into the Irish Sea while the pre-glacial Severn, east of the Shropshire-Hereford plateau, consisted of a north-south synclinal stream having as its headwaters the present Worfe system draining from Shifnal and Tong. The gorge at Ironbridge originated as an overflow channel from ice-dammed lakes formed during successive phases in the retreat of the ice-sheet towards the north and north-west, and ultimately became the permanent channel for the waters of the upper Severn. As a result largely of this increased volume of water the mid-Severn, together with the lower section below Worcester, has undergone rejuvenation*.

Outline of the Geology

Geologically the West Midlands form pre-eminently a Triassic country, this formation occupying the larger proportion of its area. Apart from the region of north Shropshire, however, it is the Carboniferous formation which has played the deciding rôle in its economic evolution.

The Midland Trias was accumulated in an irregular basin, shallowest towards the south-east, where the beds are banked up against remnants of the old Mercian highlands (exposed stumps of which still remain in the Lickey and Nuneaton ridges) but deepening towards the north-west. Towards the end of the epoch, however, that is in the Keuper Marl period, these highlands were almost entirely covered by deposits. As already shown, this Triassic sheet was subsequently warped to form several low and wide anticlines; the New Red rocks have been removed from the crests of many of these so as to expose horsts of the underlying Palaeozoic formations. Thus were laid bare the "Visible" Coalfields of south Staffordshire, east Warwickshire and mid-Severn (including Coalbrookdale and the Forest of Wyre). In parts denudation has proceeded far enough to reveal still earlier formations which formed parts of the pre-Carboniferous floor, as in the case of the Archaean and Cambrian of the Lickey Hills and Nuneaton and the Silurian inliers of Walsall, Dudley Castle Hill and the Wren's Nest within the south Staffordshire dome. It may also be noted here that flanking each of the coal-fields is a margin of rocks of doubtful Permian age. These consist in the main of red sandstones and marls together with sheets of breccia —the latter being perhaps the most characteristic deposit of this Midland group. Geological opinion is still divided as to whether

* For details see L. J. Wills, *Quart. Jour. Geol. Soc.* (1924), vol. LXXX, pt. 2, pp. 274–314; and also, *Physiographical Evolution of Britain* (1929), pp. 226–8.

these rocks may or may not be grouped entirely with the Upper Coal Measures*. In a few areas there are igneous intrusions, possibly of Tertiary date, notably at Rowley Regis in the heart of the Black Country, where the laccolitic mass of basalt and dolerite rises conspicuously to a height of 876 ft. (267 m.).

Of the Triassic rocks the Bunter is generally the more prominent in the north-western sections of the region (excepting however the synclinal area of Brewood), while the Keuper is supreme in the south-east. The soft Mottled Sandstones of the Bunter usually form vales, and one of the most famous of these (formed of Upper Mottled) runs through the north-west side of Birmingham. Here the even-grained sands with an adequate coating of iron-oxide and clay are eminently suitable for moulding; and they have therefore an important connection with the metal trades of the area. The Pebble Beds form characteristic upland heaths in Cannock Chase and Sutton Park to the north and east of south Staffordshire, while they produce the striking escarpments between Wolverhampton and Bridgnorth. The Keuper Marls outcropping in broad belts give rise to gently rolling, monotonous and heavy country, as for example in the Arden Forest of central Warwickshire. The Droitwich area illustrates best the nature of the salt deposits within these rocks, and brine has been pumped from this district at least since Roman times. The outcrops of the more consolidated Lower Keuper Sandstone usually form pronounced ridges as, for example, the narrow band which runs through the heart of Birmingham, continuing thence in a general north-easterly direction to Sutton Coldfield. This rock type has undoubtedly had a marked influence in determining the original sites of settlements within the region, its advantages consisting in the provision of a well-drained and relatively open position, good building stone, a medium soil and easily accessible supplies of water for domestic use.

A notable feature of the coalfields in the region is that the Coal Measures lie directly on an irregular floor of Silurian or earlier formations. In parts, especially at Dudley Castle Hill and the Wren's Nest, sharp anticlines of Silurian rocks, with bands of limestone, project as pleasantly wooded hills in strange contrast to the sombre features which mar the surrounding Coal Measures. In the South Staffordshire Coalfield, which extends for about 25 miles (40 km.) from Rugeley to the Lickeys, the Productive Measures are thickest in the north (*c.* 2000 ft.), but diminish to

* *Vide Memoirs of Geological Survey—South Staffs. Coalfield* (1927); *Birmingham* (1925); *Coventry* (1923); also C. Lapworth, *Geology of the Birmingham Country* (1913), for details on all these matters.

about 500 ft. in the south. Crossing the field from east to west, just north of Walsall, are the Bentley Faults which divide the area into two sections—(*a*) the Cannock Chase Coalfield to the north, and (*b*) that of Dudley to the south. The distinction is an important one economically and historically since it is only the latter which contains the Thick or Ten-Yard Coal, on which the industrial triumph of south Staffordshire was based. The best supplies of this have now been exhausted; and as a producing area the Dudley section has been displaced by Cannock Chase. This Thick Coal, while having the appearance of a single seam, is in reality composed of thirteen or fourteen distinct seams which, in the Cannock Chase area, are separated by thick masses of barren strata*.

Extensions of the "Visible" Coalfield have been worked during the last fifty years beneath the Triassic cover of Cannock Chase, and also beneath the barren Upper Coal Measures and Permian outside the eastern and western boundary faults. Thus, to the east the Thick Coal is being worked at Sandwell Park and Hamstead within five miles of the heart of Birmingham, and to the west the same seam is worked at Baggeridge, while the Hilton Main Sinking, four miles north of Wolverhampton, has opened up a new area to the north-west. In these marginal extensions the greatest depths of the workings exceed 2000 ft. (610 m.), whereas within the "Visible" Coalfield practically all the seams lie within a depth of 1500 ft. (460 m.) from the surface.

Formerly, clay-ironstones—usually in the form of nodules—were extensively worked, particularly in the Gubbin Measures below the Thick Coal, but the best of these have now been exhausted and the present production is negligible, so that at the present day the foundry iron used in the district comes largely from Derbyshire and Northamptonshire, while steel comes from South Wales, Sheffield, or abroad.

In the East Warwickshire Coalfield, which is mainly synclinal in structure, the seams in part resemble those of the south Staffordshire field, and they include in the south the compound Hawkesbury seam. Economically, however, the northern part of the district is probably the best. Extensions have been developed chiefly on the south-western side where the coal is concealed beneath the barren Upper Measures and Permian. Surrounding

* Cf. coal outputs, 1924: (*a*) Cannock Chase, 6·06 million tons; (*b*) S. Staffs. and Worcester, 1·73 million tons. In 1864 the output in the latter was estimated at 10·3 million tons; while in the same year about 1 million tons of ironstone were raised in this area. For various sections of the S. Staffs. Thick Coal, see *Mem. Geol. Surv. Birmingham* (1925), Plate III.

this basin, but, closely adjoining it, are the towns of Tamworth, Atherstone, Nuneaton and Coventry.

The Shropshire coalfields, chief of which are the Coalbrookdale and the Forest of Wyre, or Bewdley, have now little economic value although historically they have had a memorable share in the development of the region. The origin of coke-smelting at Coalbrookdale, it may be noted, is connected with the fact that here the "sweet clod" coal of the Lower Measures suitable for coke-making was found near the surface.

Glacial deposits, derived almost entirely from the north-west, are widespread throughout the region. While it is difficult to generalise regarding the respective distributions of boulder clays and of sands and gravels, it may be said that the latter generally prevail in the "glacial fringe" area of the south and east. In the Arden country, with its heavy Keuper Marls, pads of gravel drift which fill up hollows in the old surface appear to have played a significant part in the location of settlements, as, for example, in the cases of Coleshill and Solihull, the size of the villages being regulated by the thickness of the pad which determined the available water supply*.

The Human Evolution of the Region

The North-West Midlands with large areas of "woods and chilly clayland" had little attraction for early man. True, certain parts of the Plateau, where lighter soils exist, were suitable for settlement and were actually occupied, as for example the northern Clents, but the encircling river valleys and wide forest belts proved effective barriers against movements towards the area. Of the larger woodlands Arden occupied the south-east, probably extending from the Avon north-westwards as far as the Rea. To the south-west in the Severn valley lay an extensive area of which Wyre forest is but a fragment. The Brewood forest on the north-west was linked to that around the Wrekin and the Coalbrookdale Coalfield; while the Black Country itself as well as the East Warwickshire Plateau was moderately well wooded. Extensive prehistoric finds are therefore limited, in the main, to the higher lands outside the region—the limestone moorlands of north-east Staffordshire, the uplands west of the Severn, and the oolitic escarpment south of the Avon.

Conditions had not changed greatly within the Plateau during Roman times despite increased communications. The famous Watling Street, forming the main trunk road from the south-east

* Cf. P. E. Martineau, in Introd. to *Solihull and its Church*, by R. Pemberton, 1905.

to the north-west, avoids the main mass of the Plateau and crosses it in the Cannock Chase district of the north, running thence to Viroconium (Wroxeter) in Shropshire. There was however another road, the Rycknield Street, which traversed the area from south to north, running from the Fosseway, through Alcester at the junction of the Alne and Arrow, and thence past Birmingham and through Sutton Park to join the Watling Street at Letocetum (Wall) near Lichfield. While it seems certain that the road passed through the western part of Birmingham, its actual course is quite in doubt; but in any case there is no evidence to suggest any settlement here at the period. Joining this road a little to the south-west of the city was another from Worcester and Droitwich, and most probably connected with the salt trade of the latter. Thus was defined the main trunk line across the country from south-west to north-east which has since become such an important feature. These roads did not, however, materially alter the nature of the area, and it has to be regarded as one of the most thinly peopled parts of Roman Britain and one in which the Roman civilisation has left little impress.

An exception has to be made in the case of northern Shropshire, for here was to be found Viroconium, the third or fourth largest Roman city in Britain. This great centre lay at the end of Watling Street on the Severn, west of the Wrekin, and was admirably chosen for guarding the line of the river and holding back fierce British hill tribes. Although strategic in its origin, nevertheless the military life of Viroconium was apparently short lived, and for the greater part of the period it flourished as a Romano-British country town. Apart from this centre, northern Shropshire is not rich in Roman remains; but the divergence of the area from the rest of the North-West Midlands is apparent even at this early date.

The Mediaeval Period

The effective settlement of the area as a whole began in the Anglo-Saxon period. Of the two groups of invaders, the Angles advancing via the Trent had the easier route into the Plateau through the open Tame valley, and they therefore dominated the greater part of the region. On the south-west, however, the Avon valley was occupied during the 6th century* by the Hwicca group of West Saxons who had entered via the Thames valley, and thus

* The evidence of the cemetery at Bidford-on-Avon apparently indicates the occupation of this area quite early in the 6th century. *Vide* J. Humphreys, in *Archaeologia*, vol. LXXIII (1922–23). For the general spread of early Anglo-Saxon colonists, see E. Thurlow Leeds, *Archaeology of the Anglo-Saxon Settlements*.

the south-western margin of the Plateau became a boundary zone between the two waves of colonists. It was successfully crossed by the Saxons in parts, especially in the south where the valley of the Arrow enabled them to penetrate to the outskirts of Birmingham; and to the present day the areas of Kings Norton and Northfield remain a projecting strip of Worcestershire.

The original Anglian kingdom of Mercia included most of the territory watered by the upper Trent and the Tame and had Tamworth as its capital, while Lichfield, to the north-west, was the seat of the first bishopric in the district. Ultimately the boundary was extended to include the Hwicca settlements to the south-west, and finally the whole of the Midland area was welded into a single kingdom of Mercia. When this was finally carved into shires the division was based on the waterways with fortified towns as focal points. Thus the area was shared between the three counties of Stafford, Warwick and Worcester; and it is interesting to note that Birmingham, occupying a central position on the Plateau, lies at the convergence of the three counties, from all of which parts have been taken in the building up of this great Midland metropolis.

But this development has only come about gradually. The original settlement of the city on the left bank of the small river Rea apparently dates from the Anglo-Saxon period. The precise reason for the selection of the site remains rather obscure; but it was outside the main forested area and, as previously noted, its nucleus is based on the Keuper Sandstone at a point where the hard rock comes close to the river, thus giving a relatively good crossing point in an area which was very liable to floods. In the Domesday record the manor of Birmingham held quite an insignificant position, but the following centuries saw it rise to local prominence as a trading centre, market rights being secured from the Crown during the 12th and 13th centuries. This commercial aspect of the city's early growth was an important one and tended to become increasingly so. It certainly had a rival from this point of view in Coleshill, placed farther east on a wedge of glacial gravels at the drainage centre of the Plateau, and lying nearer the main highway from London towards the north-west. The immediate neighbourhood was however forested or swampy; and while Birmingham's site was drier, its centrality with regard to the whole region told increasingly in its favour, so that it became a centre for roads from many directions. As a manufacturing town progress was very slow and it is not until Tudor times that definite evidence is given of its connection with the iron trade. Its earliest industry was leather, based on its cattle market, oak-bark from the surrounding forests, and its river on which tanneries were established*.

On the western frontier of the Plateau Wolverhampton also grew up as an Anglo-Saxon settlement, placed on the Permian Sandstone where a prominent rise gave a commanding position for its famous collegiate church whose existing foundation dates from the 10th century. Within the coalfield area Dudley and Halesowen were among the few places of any note. The former grew up around its castle-crowned hill and under the paternal care of its baron, while the life of Halesowen was based on its monastery which had a typical site in the wide amphitheatre-like valley of the upper Stour. What was to become the Black Country was still a picturesque woodland region whose mineral resources were barely known.

The really important centres in mediaeval times were outside the Plateau. Amongst these Coventry held a very high rank. Occupying a key position on the upper Avon and near the main route to the north-west, it had been important since at least the early 14th century, when the Benedictine Priory had been established, while later it became a centre of the cloth trade and the commercial metropolis of the Midlands. Lower down the Avon, Warwick's fame, based on its castle built on a spur of Keuper Sandstone, is well known. In the south-west Worcester was dominant. Standing on a long strip of level ground raised above the Severn, exempt from floods, and commanding the route up the Teme valley and so to Wales, it was a significant road centre. Ecclesiastically its diocese, which included the southern section of Warwickshire, was based on part of the old Hwicca domain, the remainder of which fell to Hereford. Lichfield on the north continued the ecclesiastical pre-eminence it had acquired in the old Mercian days. Placed near to the Roman road centre at Wall, and surrounded by the rich lowlands of the Trent basin, the chief parts of the city, including the cathedral, occupy the rising ground formed by the outcrop of Keuper Sandstone. It had a conveniently central position in relation to its huge diocese, which comprised Staffordshire and Derbyshire with the northern parts of Warwickshire and Shropshire, but while it remained the centre of the see its position was challenged by Coventry, so much so that from the 12th to the 19th centuries the see had the joint title of "Lichfield and Coventry†."

* See W. Barrow in *Birmingham Institutions* (1911), pp. 3–58; also *British Association Handbook, Birmingham* (1913): sections by B. Walker and others. For map of old roads see B. Walker in *Trans. Birmingham Archaeological Soc.* vol. LII (1927).

† For a considerable period prior to the Reformation the diocese was known as "Coventry and Lichfield."

The Growth of Industrialism

Transferring our view-point to the early 18th century, a very different picture from that just sketched is obtained, since even before the momentous changes in industrial processes which marked that century had been made, the centripetal forces which were to give the core of the region its dominating position had been exerting their influence. Thus, Birmingham and surrounding districts, including Dudley, Wolverhampton and Walsall, had become important centres of iron-working. Nails were a staple product, but in addition there were many goods of both iron and brass associated with this hardware district.

It is important in connection with this development to recall briefly the essential factors on which the iron trade depended at that period. Charcoal was fundamental in the production of pig- and bar-iron, but coal was used by the smiths in the production of finished articles, *e.g.* nails, locks and bolts, chains and various agricultural implements. These were made from the bar-iron which, before coming into the hands of the smith, needed cutting or rolling—processes carried out in mills worked by water-power. Although there was still some wood within this area for smelting purposes, supplies of pig- and bar-iron chiefly came from the mid-Severn district, having been made there or imported from South Wales or Sweden. Actually two-thirds of the bar-iron used in the British metal industries came from abroad, and water-transport up the Severn was a vital factor in this trade. The main interest of south Staffordshire was therefore in the finishing processes, and of these, slitting and rolling were largely carried on in the Stour valley at Lye, Cradley and Halesowen, with good water-power on the west of the watershed, while Birmingham to the east of the divide was rather concerned with smith's work (as witness Leland's classical reference, *c.* 1540) for which coal was used.

The Coalbrookdale area in which were found all stages of iron manufacture, particularly the earlier ones, probably exceeded the Black Country in importance at this period, for it combined the advantages of all the raw materials with those of good water-power and water-transport. Bewdley, however, derived most profit from the Severn traffic and it was the chief iron-marketing centre for the whole of the West Midlands*.

Birmingham's manufactures included various products of brass and copper in addition to those of iron, the former having been introduced during the 17th century. It is possible that the city's

* T. S. Ashton, *Iron and Steel in the Industrial Revolution* (1924), and W. T. Jackman, *Transportation in Modern England* (1916), vol. I.

relative disadvantages as compared with the truer Black Country centres in obtaining sufficient supplies of iron, were a factor in turning the attention of its artisans already skilled in metal working to the alternative medium for producing the artistic novelties so fashionable after the Restoration.

In accounting for the growth of Birmingham since mediaeval times the significance of the site as a market, previously examined, must be constantly borne in mind. Its earliest manufacture, as already indicated, was dependent on agriculture, and its original iron trades also appear to have been connected with the needs of an agricultural community—bits, nails, horse-shoes and various farming implements. In the further development of both the iron and the brass the possession of good local supplies of sand from the Upper Bunter, for casting purposes, was certainly of importance. Apart from these factors due significance must be given to the notable increase in population—and that of a very virile type—from 5000 to 15,000 between 1650 and 1700. Anomalous as it may appear, it was as a result of the early insignificance of Birmingham that, following on the Restoration with its accompanying religious persecution, it received a large influx of Dissenters who had been driven from the corporate boroughs. The city was then governed, as it continued to be until the 19th century, as a small country town, and it possessed neither charter nor restrictive craft gilds. Those who came into it at that period were men and women of strong character, and it is certain that the life of the district received a permanent benefit from their coming.

Statistics show that in 1720 the West Midlands (including Shropshire) possessed about 20 per cent. of the iron furnaces and 40 per cent. of the forges out of the totals for England and Wales. Thus before coal entirely replaced charcoal and before steam-power removed the tyranny of water-power, the area had obtained a good lead and was ready to take full advantage of the discoveries begun by the Darbys at Coalbrookdale early in the 18th century and culminating in Watt's development of the steam-engine during the seventies from the experimental to the commercial stage. Thenceforth intensive concentration of all phases of the iron industry within the coalfields became the rule, and amazing progress was therefore made in the development of the Black Country on account of its great advantages—the possession of rich and cheap supplies of Thick Coal, ironstone worked with the coal, and limestone as a flux from the Silurian hills at Dudley and elsewhere.

Thus grew up the true Black Country of the 19th century as described by Mackinder—"a great workshop both above ground and below; at night it is lurid with the flames of iron-furnaces;

by day it appears one vast loosely-knit town of humble homes amid cinder heaps and fields stripped of vegetation by smoke and fumes." Figures bear out vividly enough the extent and fundamental characteristics of this material progress, for by 1806 there were 42 furnaces in the Black Country alone, and by 1858 a maximum of 182 had been built, of which 147 were in blast.

Accompanying these developments and fundamentally related to them were the essential improvements in transport, both within the industrial area and as connecting links with the outside world. Canals dominated the situation until about 1840, when railways assumed the leading rôle. The Birmingham canal system is a network of waterways chiefly within the mining area, but forming a complicated chain between Birmingham and Wolverhampton through Smethwick and Tipton, with various extensions north-eastward through Walsall and other centres towards the Trent and Mersey Canal, and southwards to Halesowen and the Stour valley. The whole system (159 miles or 256 km. in all) occupies three main levels, all over 400 ft. (120 m.) in height, while it includes 216 locks and several tunnels of which the greatest, Lapal Tunnel, is 3795 yards long. These details reflect the essentially adverse physical conditions underlying the system which, under modern conditions, have become increasingly evident. In their heyday the canals were wonderfully prosperous, although this prosperity rather depended on the great mineral resources which the area enjoyed; and since the seventies of last century the greater part of the cost of maintaining this expensive system appears to have been borne by the L.N.W. Railway which acquired the control of the Canal Company in the momentous year 1846. The traffic on the system is relatively large compared with other inland navigations, but it is essentially local, travelling an average distance of about eight miles only, and it consists of bulky raw materials, of which coal forms one-half, going to various works on the canal banks. The Birmingham canals are connected with the four estuaries by various links, and in recent years many plans have been suggested for the improvement of these waterways. Particularly has the route south-westwards to the Severn been advocated on account of its shortness. Physically, however, such a route suffers more severely from steep gradients than routes in other directions; and since 1927 the 100-ton barge scheme to the Severn appears to have been definitely dropped by the Birmingham authorities*.

* For the latest available statistics and also for other recent proposals for improving the waterways (most of which have come from the National Council for Inland Waterways, founded in 1921 with headquarters in Birmingham), see G. Cadbury and S. P. Dobbs, *Canals and Inland Waterways* (1929), pp. 109 and 140 ff.

As a railway territory the Plateau is now divided between two great groups—the L.M.S.R. and the G.W.R.—the former generally controlling the northern half and the latter the southern section, while the systems interlock in the Black Country and especially along the critical line between Birmingham and Wolverhampton. It must be noted that the main trunk line from south-west to north-east is in the hands of the L.M.S.R. This route, leading via the open Tame valley, gives the area a very significant connection with the industrial region of the North-East Midlands.

To the west, Shrewsbury has a railway network almost rivalling that of Birmingham itself, being the focus of seven different lines, but chiefly those north–south along the Welsh border and east–west via the Severn. The G.W.R. main line through Shifnal and Wolverhampton forms the only direct connecting link between these respective railway foci.

Survey of Existing Conditions—Industrial and Agricultural

We may now make a broad survey of the different sections comprised within the whole region at the present day from the point of view of their relative population and the outstanding features in their economic life (*vide* Fig. 31).

The huge agglomeration of people in the central mass of Birmingham together with the Black Country to the north-west is very evident. As already seen, there has always been a distinction between these two sections of the hardware area, and during the 19th century, while the Black Country proper was essentially the heavy iron region, Birmingham and a number of adjoining towns and villages were chiefly engaged in the manufacture of an amazing variety of small metal wares. On the human side the existence of a large number of small manufacturers was an essential feature. At the present day, however, such a description does not serve as a true picture of the district, for vital changes have been in progress during the past fifty years and are still very much in evidence. The period of the seventies of last century marked the beginning of the "new industrial revolution" which, in the Midlands especially, gathered increased force as a result of the Great War. Among the local factors responsible for this radical change in economic structure, the exhaustion of the best seams of thick coal and of the ironstone has been fundamental; and, as previously indicated, the chief centre of coal production has now shifted northwards to Cannock Chase. Of the wider factors the coming of steel and the increasing competition of foreign industrial centres have affected

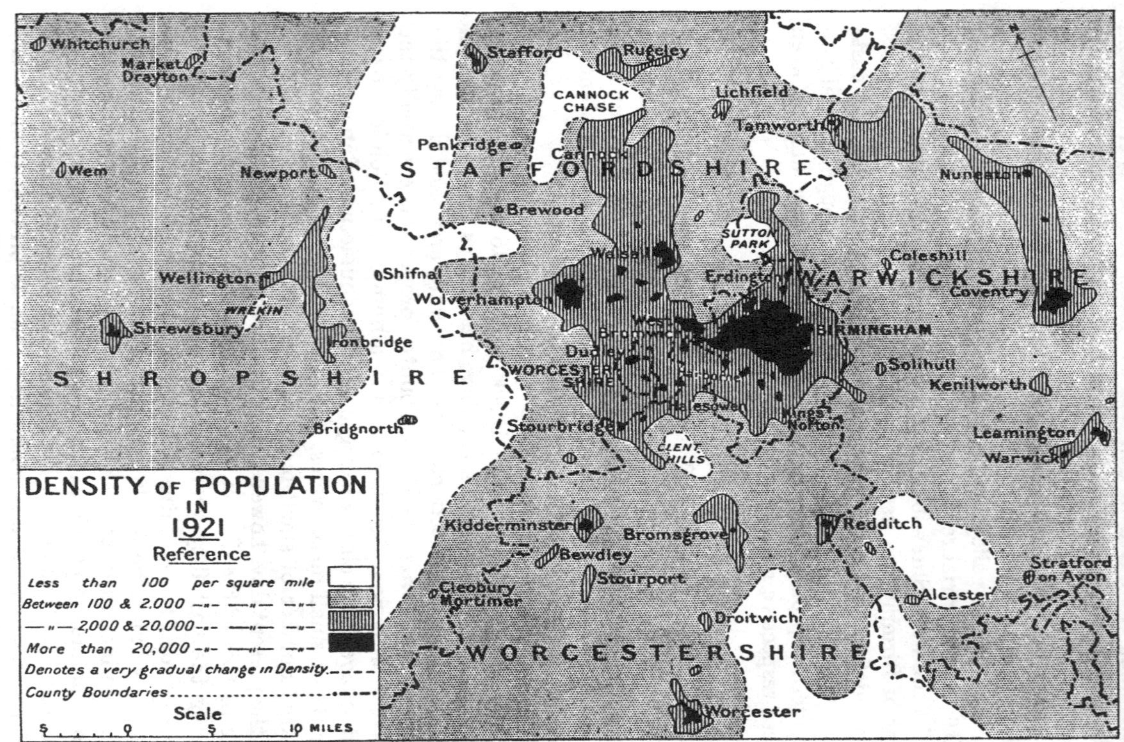

Fig. 31. The North-West Midlands: density of population.

this Midland district more acutely than other centres. The handicap of transport, indeed, which formerly tended to differentiate Birmingham from the Black Country has more and more spread to the whole industrial area and separated it from those nearer the seaboard. Thus no longer is the Black Country a region "lurid at night with the flames of iron-furnaces," for at the present time only three or four furnaces are in operation. Nor, indeed, is Birmingham exclusively or even chiefly a hardware town.

Yet the region has not declined; rather has it enjoyed a great and continuous development since about 1886. In place of the old staples, many new industries have been attracted, and the district has thus become concerned with the production of highly composite articles and of the semi-finished materials necessary for their manufacture. Typical industries at present therefore include those producing motors, cycles, rubber, artificial silk, ready-made clothing, electrical equipment, machine tools, munitions, wireless apparatus, food and drink. Moreover, as these industries require large plants for their manufacture, the small unit once the characteristic human feature of the district has become less and less so. Thus, the Birmingham district which played so conspicuous a part in the industrial changes initiated during the 18th century has also been peculiarly sensitive, by reason of its geographical characteristics, to the forces of the "new industrial revolution" in which it has led the way*.

On the eastern margin of the Plateau lies the second industrial area based on the East Warwickshire Coalfield. In this belt Atherstone and Nuneaton have various small industries including textiles; but Coventry outside the "visible coal" area is the chief manufacturing centre. This ancient city, although dwarfed now by Birmingham, has continued to develop by adapting itself like its great neighbour to changing circumstances. It was indeed the pioneer in the development of the cycle and motor-car industries which, with artificial silk, have displaced its earlier trades. The city has now a population of 128,000 and appears to be moving forward rapidly, having recently had a notable extension of its boundaries; but economically, although on the margin of the Birmingham country, it gradually tends to come more under the influence of the West Midland metropolis.

In the Coalbrookdale area, including Ironbridge, Madeley and Wellington, the population is only relatively dense. While

* See G. C. Allen, *Industrial Development of Birmingham and the Black Country* (1929), particularly pp. 278–314 and statistical tables, pp. 457 ff.

historically the area will have a lasting significance, economically its iron industry has now sunk to small proportions, for its natural resources are small and the Severn has lost its former rôle. Apart from the cast- and wrought-iron trade which still persists, there is an interesting group of industries, including bricks and pottery, depending on the Coal Measure clays.

Elsewhere within the region, and to its great benefit, the population map indicates low densities characteristic of agricultural areas, except in the various urban units, most of which are arranged as outposts around the margins of the Plateau (*e.g.* Stafford, Tamworth, Warwick, Worcester) to recall a regime which is past. In the west Shrewsbury is supreme.

Agricultural practice is necessarily influenced by the existence of the dense masses of population with their demand for dairying and market-garden produce; but interesting differences may be detected between various rural districts arising from variations in the physical conditions of soil and climate.

In the case of Shropshire, whose lowlands of the north and east are included in this essay, three different agricultural regions may be distinguished. In the extreme north (including Ellesmere, Wem, and Whitchurch) are the strong lands floored chiefly by Keuper Marls or rocks of Lias age, while wide areas are covered by thick boulder clay. Heavy soils are thus the rule and dairying is all important, the cattle density being high—33 per 100 acres; indeed, the area is but a continuation of the Cheshire plain. In marked contrast to this region is the lowland on the east of the county—with Shifnal and Bridgnorth as typical centres. Here light, easily-worked loams and sands prevail, derived from Triassic or Permian sandstones. Climatically the belt is distinguished by its relatively light rainfall—26 in.–30 in. (650–750 mm.) annually, particularly during the harvesting period of August and September. Arable cultivation thus predominates, the area under crops forming 48 per cent. of the total, while permanent grass accounts for only 37 per cent. Wheat and barley and root crops are raised extensively, and along with the production of these goes the fattening of sheep which forms an essential feature of the agricultural system. Thus, while there are about 38 sheep per 100 acres (40 ha.) the cattle density is only about 14. Here then is an arable region which, although it has suffered a decrease in population since 1881, has been able to resist the worst evils of acute agricultural depression. The remaining lowland, that of central Shropshire, including Shrewsbury and extending westwards to Oswestry and eastwards to Newport, is essentially an area of mixed soils—heavier loams and clays alternating with lighter sandy soils. This mixture is largely due to the existence of varied glacial deposits derived partly from the Welsh, and partly from the Irish Sea ice-sheets. As regards rainfall there is a range from about 37 in. in the west to about 26 in. (925–650 mm.) annually in the east. Agriculturally, therefore, the region is a highly diversified one, both permanent grassland and arable cultivation being important, while the density of both cattle and sheep is relatively high (32 and 21 per 100 acres respectively). A study of this rich and varied region has been recommended by Sir D. Hall to those who declared English farming to be a lost and unprofitable art.

These rural regions have not always been lands of peace; their position on the Welsh border and athwart the Severn highway formerly denied them that; and Shropshire's plains have been battlefields, "its houses are moated and loop-holed, and its very churches have been used as keeps and watch-towers*." Shrewsbury in its famous river loop has combined many functions, being fortress, route-centre, market, and to some extent at least an industrial town based on its small coalfield. Since Anglo-Saxon times it has played a leading rôle in the life of the area; and to-day its railway network gives some measure both of its pre-eminence and its independence.

To the south-west of the Midland Plateau the Kidderminster and Bromsgrove areas of Worcestershire are intimately bound up with Birmingham economically. There are found soils of a light sandy nature—typical turnip soil—derived in the main from Keuper and Bunter sandstones and recalling in character those of eastern Shropshire. Arable cultivation therefore predominates; but on account of the closer proximity to the Black Country, market gardening tends to be the characteristic feature. In the Bromsgrove section, with Birmingham but 10 or 12 miles away, small holdings producing fruits as well as vegetables have become the distinctive feature since 1890. Farther to the south-west and south the heavy clays of the Red Marls form the pastures for large herds of milking cattle.

The eastern and south-eastern sections of the Plateau form pre-eminently heavy clay-lands, developed either on soft Permian rocks or on Keuper Marls. Thus in the Arden country permanent pasture is the rule, forming in some cases up to 80 per cent. of the total area; and the milk finds a ready market close by. But, as previously noted, glacial sands and gravels are also scattered throughout the district, and where these exist in large patches—as

* W. W. Watts, *Shropshire* (1919).

around Solihull and Coleshill—market gardening and especially the production of potatoes form a noticeable feature.

On the northern and north-western sides of the Plateau the general distinction between the Keuper Marl region of mid-Staffordshire, with its dairying, and the light land areas as around Lichfield and Penkridge is again evident. While the former is predominantly pastoral the sandier soils of the latter areas rather favour the production of cereals and root crops, although here conditions are much affected by the great demand for milk and meat amongst the huge industrial population lying immediately to the south.

Finally it may be noted that of the total population of nearly 1¾ million people composing the great conurbation of Birmingham and the Black Country, close on one million are found in the city itself, which now, since 1911, covers 68 sq. miles. As already indicated, many factors have contributed to the growth of this great city, so unique in its position—at a height averaging about 450 ft. (137 m.), near the main watershed of England and away from any navigable river. Now its commercial supremacy is confirmed by its central position in a radial system of both roads and railways apart from canals. On the industrial side this centrality combined with the wealth accumulated in the past has offset the disadvantages arising from its diminished local resources and its distance away from the sea; and above all its leaders have shown marked resourcefulness in adapting themselves to changing industrial conditions. For the future, Birmingham's responsibilities as the religious and intellectual capital of the area are many and varied; and not the least significant of its many functions or of its opportunities is that of guiding the development of the Midland Region, covering some 1700 sq. miles (or 4400 km.^{2}) and coinciding broadly with the Plateau and its margins, the largest region for which a Joint Town Planning Committee has so far been formed*.

* The author would like to express his thanks to Mr C. Redmill, M.Sc., for assistance in compiling the population map of the area, and to various students for suggestions.

Note on Authorities: In addition to the works already quoted, the following are indispensable for any detailed study of the region: *Trans. Birmingham Archaeological Soc.*; *Proc. Birmingham Nat. Hist. and Philos. Soc.*; *Victoria County Histories.*

XIII

WALES

H. J. Fleure

The hill countries of Europe keep alive inheritances from the remote past, though in most cases the languages of the plains have spread up the valleys and have ousted ancient forms of speech save for a few words used in the farmyard or the kitchen. There are, however, a few old tongues surviving here and there, Basque on the Franco-Spanish frontier in the west, Romansch, Ladin and Frioul in the Eastern Alps. In none of the cases just mentioned does the ancient language gather around it powerful emotional associations affecting large numbers of people, but this is the case in Wales. We may say that in many ways Wales is the refuge and repository of ancient heritages England once possessed, but we must also say that associations gathering in comparatively recent times around the Celtic language, still widely spoken and possessing a growing literature, have helped to keep up and even in some ways to accentuate distinctions that are rooted in a long and involved history.

Structure and Land Forms

Wales is thus essentially the hill country on the west of the English Midlands. It is recognised as a separate administrative unit for some purposes and in certain of these it is made to include Monmouthshire, which, for purposes of the law courts, is in England. Without Monmouthshire (534 sq. miles or 1377 km.^{2}) Wales has an area of 7467 sq. miles (19,264 km.^{2}), with the sea on the south, west and north, and the English Marches on the east. It is almost entirely built up of Palaeozoic rocks, and a large proportion of its surface stands more than 600 ft. (say 180 m.) above sea-level, while its coast-line is nearly 400 miles (640 km.) in length. The contrast with England, which has such a large proportion of its surface below the 600-foot contour line and so much of that surface formed by post-Palaeozoic rocks, is thus very marked. Much of England has been built up, in geological history, against the nucleus formed by Wales, and, in the same way, Wales north and west of the chief Carboniferous outcrops and formed of pre-Carboniferous rocks has been built against the still more ancient nucleus that is preserved as the Isle of Anglesey. The NE.–SW. trend lines and the absence of post-Carboniferous rocks suggest that west and north Wales is an old block and to that old block, of which parts of eastern Ireland are also held to be remnants, the name of St George's Land has been given. That this ancient land was reduced to a peneplane in Mesozoic times is reasonably clear, though opinion has wavered as to whether it did or did not remain above sea-level during the great submergence that led to the deposition of the English Chalk. There is no doubt as to subsequent block uplifts converting the peneplane into a plateau, or rather a series of plateaus, for, in several parts, around a higher plateau, which may have a cliff-like edge, there lies a lower or coastal plateau forming most of the land of Wales that lies below the 600-foot contour. Whether these uplifts were continuous or interrupted by intervening depressions is more than can be worked out here; it is however important to notice a late sinking movement, that seems to be post-glacial and that may have been preceded and interrupted by periods of uprise. It has been a widespread view that this sinking movement came to an end in the Bronze Age, for whereas the older and more deeply submerged post-glacial forest has dubious associations with some "Epipalaeolithic" flints, the younger and still coastal submerged forest gives indications of polished stone implements. When the cooling of north-west Europe's climate about 1000–600 B.C. is recalled, as well as the probability, according to Gams and Nordhagen, that a late phase of slight regrowth of glaciers is to be dated at about that time, it seems not unlikely that that was a period of uplift of the land, and the last sinking movement seems to need reinvestigation as to the date of its closing phases. The last sinking has brought the sea up the estuaries of the south-west, and its influence in reducing the area of lowland has been supplemented by the process of consolidation or settling down of the loose accumulations of glacial boulder clay which seems to have been very extensive in Cardigan Bay.

Whatever may have been the crust movements which have contributed to the present form of Wales, it is to be noted that the regions above the 2000-foot (or say 600 m.) contour line have sharply worn edges, deep-worn cirques (cwms) cut right into the heart of the mountain, so that these parts stand out above the general plateau level, and their cirques often have little lakes, while these hardly ever occur in the many cirques found at lower altitudes. Just as these ancient remnant-heights stand out above the general plateau, so parts of the higher plateau stand out here and there above the coastal plateau, notably in the neighbourhood of St David's and in Cardiganshire.

Anglesey (Môn) is low-lying, with Holy Island, though only 720 ft. above sea-level, standing out strikingly above the worn NE.–SW. ribs of ancient (Caledonian) highland that alternate with still lower lines of marsh or water, in some of which, including the Menai Straits, Carboniferous rocks occur. Across these straits is Arfon, the land over against Môn, with its long mountain range culminating at Snowdon (3560 ft. or 1085 m.). This range, lens-shaped, extends NE.–SW. and is built of Lower Palaeozoic rocks with thick layers of lava of Ordovician age. It is distinguished by the deep cuts through the range forming a series of passes, low watersheds between turbulent rivers, which receive the waters of streams from the black cwms that give an effect of wildness to the peaks and ridges quite unusual save in mountains of far greater height. The Carnarvonshire ranges taper south-westward into the peninsula of Lleyn, which has some hills formed of masses of crystalline rocks, and it terminates in Bardsey, the isle of the saints. The main mountain-range is marked off by a low line, running NE.–SW. from Capel Curig to Beddgelert, from the subordinate high line with Moel Siabod (2860 ft. or 872 m.) as its culminating point. The deep clefts of the range have several long lakes, some of which at least owe their existence in part to morainic dams. These contrast strikingly with the little lakes of the cirques.

West Merionethshire is part of a dome of Cambrian grits around which curves a great belt of volcanic masses. The broken dome is to some extent marked off on the east from the grits and lavas of the volcanic belt by what has become the valley line of the Eden and Mawddach rivers. The southern part of the volcanic belt is marked off from the rest by a large fault system in which have been denuded out the low lines from Corwen and Bala to the western sea. This fault system and the igneous rock together have given Cader Idris specially sharp contours, though it does not rise above 2927 ft. (893 m.). Its cirque-lakes on both flanks are very fine examples. Aran Mawddwy (2970 ft. or 905 m.) and Arenig (2800 ft. or 853 m.) are other outstanding elements of the volcanic belt.

It may be stated, generally, that in Carnarvonshire, Merionethshire and their borders the volcanic areas are those which remain outstanding above the general plateau level floored by grits and shales of Ordovician and Silurian age, overtopping the 2000-foot contour only in the Plynlymon area and at a few other spots, for the region of the heights of Berwyn has igneous rock and is in a sense an outlying part of the igneous belt.

The older maps of this area of Ordovician and Silurian rocks are unsatisfactory; they show a very large stretch of Ordovician whereas the greater part of that area is Silurian and the Ordovician appears in the Berwyn mountains and on their southern slopes, in denuded anticlines in north Cardiganshire and in a large area of south Cardiganshire, north Pembrokeshire and the Towy valley. The plateau generally has a good deal of boulder clay, and around its higher areas, *e.g.* Plynlymon, are cirque-like formations which hardly ever contain lakes, though just below the summit of the moorland on Plynlymon itself (2469 ft. or 752 m.) is a cirque-lake, Llyn Llygad Rheidol. After the peneplanation and, possibly, submergence of the plateau, it was uplifted, probably in stages, and then dissected by streams, the valleys of which have been deepened through the increase of cutting power when the lower or coastal plateau was uplifted, then modified by ice action and, subsequently, by rivers again. Practically the whole surface of this plateau is impervious to water, for this characteristic is common to the glacial boulder clays and the Ordovician and Silurian grits and shales. There are on it many patches of high moorland which are ill-drained and often form sphagnum bogs, and the result is that parts of the plateau form much more effective barriers to human intercourse than do the more mountainous regions of igneous rock.

Around this plateau as a nucleus were gathered folds of post-Carboniferous date running, broadly, north to south along the English border and east to west in south Wales. In south Wales we have the long oval basin of the coalfield narrowing and thinning westwards in Pembrokeshire. On the north side of the syncline emerge the Millstone Grit, the Carboniferous Limestone and the Old Red Sandstone, the last with the northward-facing scarp of the Brecon Beacons (2862 ft. or 872 m.). The syncline itself is largely high land, for among its rocks are the very hard Pennant Grits (in the Coal Measure series). The southern border of the great syncline is generally lowland, and the sea has broken in to form Swansea Bay and Carmarthen Bay. The drainage of the coalfield follows a remarkable scheme which will be sketched briefly below. Glamorganshire south of the coalfield (the so-called Vale of Glamorgan) is largely floored by Triassic and Liassic sandstones and limestones which appear in Wales, elsewhere, only in one of the low lines of Anglesey and on the east and west flanks of the Clwydian Range, which itself forms an uplift having a Silurian axis with Carboniferous rocks on its eastern side.

The folding of post-Carboniferous date around the older rocks has already been said to give N.–S. folds in England, illustrated by the Pennines, as well as E.–W. lines (the Armorican trend lines) in the south. These, with the NE.–SW. lines mentioned earlier, help to give us the clue to the relief pattern of Wales. In the north-west of Wales, on the whole, the NE.–SW. lines dominate the pattern with Menai Straits, the Snowdon Range, the Bala fault line and Cader Idris as leading features. On the Ordovician and Silurian area of the plateau, the worn and often broken anticlines have axes N.–S. and pitch southwards, but curve round to the E.–W. direction in the south. Some of the synclines with thick grits have remained outstanding. In this part of Wales the relief pattern is thus a mixture of a NE.–SW. scheme of lines, illustrated specially well by the Wenlock Edge on the English Border, and a scheme of lines curving round nearly parallel to the coast of the southern half of Cardigan Bay. One of these lines, north of the Pembrokeshire parts of the coalfield, forms the Preseli Hills with extensions to St David's Head and Ramsey Island; these hills and their extensions are rich in igneous rocks. Igneous rocks, with sulphur springs, also occur in Radnorshire.

Here and there in the plateau region east and west lines are visible, but they are clearly subordinate to those running N.–S. and NE.–SW. Study, however, soon shows that there is also a series running NW.–SE. and these have been looked upon as vestiges of the original, or consequent, drainage lines from the Snowdon range which is presumed to be the remnant of a high and much larger mountain land. The fact that some of these NW.–SE. lines can be followed from Wales right across England to the English Channel is interesting in this connection*.

The South Wales Coalfield possesses NE.–SW. valleys but also, farther east, valleys which run NNW.–SSE. parallel with one another. They are separated by moorland lines of Pennant Grit which become higher towards the north-east, attaining a summit over 1900 ft. above sea-level. They are connected with fault systems which have been shown to be of post-Liassic date, *i.e.* they are a good deal later than the foldings of the coalfield area. Needless to say, other facts, such as the arrangement of minor anticlines and synclines within the coalfield, and the southward spread of glaciers from the great glacier-parting formed by the Brecon Beacons, have contributed both to the arrangements of the relief pattern and to the character of the valleys which are typically deep and narrow, often with a U section.

It has also been suggested that those rivers of the South Wales Coalfield which run NNW.–SSE. may have had their courses

* *Cf.* Chapters VI and VII (pp. 116 and 140).

determined on a cover of deposits of Cretaceous age since removed by weathering, *i.e.* this may be a superimposed drainage modified by utilisation of fault lines.

The heights of the Brecon Beacons (2907 ft. or 886 m.) on the north side of the South Wales Coalfield, and forming a sharp southern boundary to the old plateau of central Wales, have scarp edges of Old Red Sandstone; behind them southward the Carboniferous Limestone does not stand out much. On the eastern side of the coalfield however the Carboniferous Limestone lip of the basin is decidedly the outstanding feature and the Old Red and Silurian rocks are in the lowland.

Most of the rivers of Wales utilise weak lines of structure and some at least have complex histories of captures and recaptures, related to the changes involved in the uplift of the coastal plateau in many instances, but post-glacial in a few such as the Dylife in west Montgomery.

The River Usk is in rather a different category as the collector of streams from the northern and eastern side of the coalfield. Special mention must also be made of plateau-edge streams such as the Rheidol running into the middle of Cardigan Bay enriched by the waters of the northern part of an old drainage line stretching along the worn and broken anticline. In this way water that previously reached the sea from Devil's Bridge in about 60 miles was taken thither in 12, with a consequent great increase of its cutting power, so that a remarkably sharp gorge has been formed. This capture seems associated with increase of cutting power through uprise of the coastal plateau. The shore in north Wales is low for a considerable distance from the Dee, but Great Orme's Head, of Carboniferous Limestone, stands out, likewise Penmaenmawr to the west, and there is usually a cliff edge. In Cardigan Bay the shore-line is often the cliff edge of the coastal plateau, but the eating away of the boulder clay land which seems to have existed here in late glacial times has left certain results. In the first place the lines of former continuations of the present westward-flowing rivers have been lines of specially successful maritime advance, whereas the former low water-partings between the tributaries of successive streams remain as low ridges with their ends submerged. One of these, Sarn Badrig, is about 19 miles long. In the next place the sea has sifted out the fine sandy clay from the coarser material of the boulder clay which is heaped up in storm-beaches that may be near the foot of the cliff or may lie, as at Borth, a few miles away. In this fashion the drainage from the cliff base at Borth has been obstructed and as a consequence there is an extensive moor or bog, some parts of which are influenced by salt water. The margins of the bog land towards the estuary of the Dyfi are large armeria-pastures; to the west lies the storm-beach accumulated chiefly under the influence of waves driven by the prevalent south-west winds; and at the north end of the storm-beach, near the mouth of the river, we find an expanse of sandhills. Prof. O. T. Jones has demonstrated important results of the draining of great lakes of water dammed in by an Irish Sea ice-sheet and finding successive outlets as that sheet diminished.

The pebbles of the storm-beaches betray their origin from boulder clay once laid down in the bay, for they include stones from Anglesey, north-east Ireland, and even Ailsa Craig and the Ross of Mull, material obviously brought southward by an ice-sheet through the Irish Sea. The Irish Sea ice-sheet worked its way inland to some extent in south-west Wales, and here also the coast is often cliffed, with the cliffs broken fairly frequently by bays. In Pembrokeshire, especially, we find sunken valleys, the chief of which is Milford Haven. Cliffs are the general rule along the south Wales coast, but, as elsewhere, they are varied with storm-beaches, sand dunes and more or less estuarine inbreaks.

The bog or moor described for Borth is perhaps the chief of its kind, but a number of smaller ones occur near the shore behind storm-beaches. High moors occur on many parts of the plateau with indeterminate drainage, the Plynlymon region being notorious in this connection. There is however another very large bog at Tregaron (Cards), behind a great boulder clay dam that crosses the Teifi valley. Thus a great deal of land in Wales is moor or bog, while steep rock and scree slopes also take up large areas and limit severely the habitability of the country. This limitation is increased by the fact that such large areas have impervious soil which is heavy and poor, whether it be derived from boulder clay or from the Ordovician and Silurian slates and grits.

The Old Red Sandstone area in Wales with its great scarps and its glacial clays is fertile only in patches, and Brecknockshire contrasts strikingly with Herefordshire in fertility and habitability. The newer rocks flooring the vale of Glamorgan, with less influence of boulder clay, give a soil of greater fertility and the climate supplements these favourable factors; the cliffs often show Liassic rocks unconformably placed over Carboniferous Limestone.

Climate

With its sharp land relief near the western sea Wales is naturally a land of heavy rainfall and a belt from north to south from Snowdon to the hills of the South Wales Coalfield gets over 60 in. (1500 mm.) of rain nearly everywhere, while 80, 100, and even 150 in. (2000, 2500, 3750 mm.) of rain are features of various hill areas and high valleys. South Pembrokeshire and south Glamorgan enjoy more sunshine and receive less rain, but these are almost the only patches save the Severn, Wye and Usk valleys near to the English border that get less than 40 in. On the north coast, however, Anglesey averages a little under 40 in., helped in this respect by its low-lying surface, and the vale of Clwyd with the Denbigh moorlands to rainward as well as Deeside with the Clwydian range to rainward gets less than 30 in. The general climatic problem in Wales as regards agriculture, health, etc. is thus excess of rain rather than drought. The temperatures are mild near the sea, especially along the west and south coasts, which often lie well sheltered beneath the hills and hardly ever feel a real north-east wind. The highland gets much more cold, but snow rarely lies for long even on the heights save in shadowed spots. The rainfall maximum is in autumn, which is a season of very mild temperature, but the proximity of the sea gives cool conditions in spring and the spring phenomena are often a fortnight later in west Wales than in the same latitude in England.

The Early Human Occupation

When we try to picture Wales before its surface coverings were altered and adapted by man, we realise that, once the modern type of climate had definitely superseded the previous glacial phase, the valleys would cover themselves with trees where swamp did not inhibit their growth; these trees would clothe most valley slopes but could hardly establish themselves on the wind-swept heights. Thus the tops of the plateaus would be fairly clear of trees though these latter would have spread higher at a distance from the sea than they could near it. Add to this that the plateau tops in the interior are for the most part considerably more than 1000 ft. above sea-level, and that the tree limit seems to have lain in many places as high as the 1500-foot (460 m.) line, and we understand the sparseness of any early populations in Wales save perhaps around the coast. Indeed, never from the earliest human beginnings does it seem that Wales can have been a human unit in any great degree. It must be looked upon as originally a more or less barren highland, with its fringes towards the sea and towards England touched by streams of culture from time to time, and with the older cultures taking refuge in its valleys that are lost among the moorland hills. Long ago it would seem that the streams of culture along the coasts were as important or more important than those from the English side, but since the Normans came and probably from still earlier days it has been the influences from the English side that have been the stronger; and so it comes about that it is west (north-west, west and parts of south-west) Wales that preserves most of the distinctively Celtic and pre-Celtic heritages of culture.

Another reason for the paucity of early population in Wales is the fact that no flint occurs in the deposits of the country save on the storm-beaches which include material from the Irish Sea ice-sheet. Even here the flint is for the most part in small pebbles. A few finds of grit stones which seem to have been shaped for use as implements have suggested that there were alternatives to flints, but it is by no means certain that they are human artifacts at all. A few flint-chipping floors have been uncovered near favoured beaches, chiefly in Cardiganshire and Pembrokeshire, and the little flint points have been studied; but it would be rash to say more than that they might be "Epipalaeolithic," though that stage of culture may have continued very late in Wales, and it seems that chipped flints may have remained in use till perhaps Norman times. The flint-chipping floors at any rate indicate early utilisation of the coastal flints. When the art of stone grinding and polishing spread westward, Wales was no longer at the same disadvantage, for, at Penmaenmawr and in various places in the areas of volcanic rocks in NW. and SW. Wales at least, there was material that lent itself to the making of implements. Some very fine polished stone axes have been found in Wales. Another, probably more or less contemporary, evidence of man's interest in Wales is provided by the great stone monuments. Dolmens occur in the plain of south Wales, in Gower and Pembrokeshire, as well as in Carnarvonshire and Anglesey, and some seem to be related perhaps rather to those of Ireland and of the Baltic than to those of Brittany and the Iberian Peninsula. This suggestion is supported by the fact that a certain number of holed axe-hammers have been found in Wales and these are a special feature of the Baltic culture of the earliest days of metal. We may add to this the fact that one or two bronze weapons and some flint and chalcedony arrowheads of Irish workmanship have been found in west Wales, and an Irish gold lunula at Llanllyfni in Carnarvonshire. A few stray beaker pots have been discovered in Wales, and these no doubt show penetration from England chiefly along the coastal plains of north and south, but some other later types of Bronze Age pottery seem more akin to those of Ireland and Scotland. Tumuli are numerous near the ridgeway roads of Wales, and some of the mounds have been shown to belong to the Bronze Age. They indicate a period in which men still lived on the open highland above the forested valleys.

Generally, then, the country was fringed by Irish and English culture in the Bronze Age and we do not feel able to say that it had much of a culture of its own. Finds from excavations in Wales are notoriously few and this suggests that bronze was highly valued and usually remelted rather than left buried, and that leather was probably used in place of pottery for most purposes, while we may be fairly sure that the population was small.

A feature of certain districts in Wales is the occurrence of circles of standing stones; they are usually included with dolmens as "great stone monuments," but there is no indication in Wales that all are of the same age. In this connection reference must be made to the recently established fact that the stones of foreign origin included in the great and composite monument at Stonehenge were carried there from the Preseli area in Pembrokeshire, where it is thought they already formed some monument, *i.e.* were consecrated. It may be noted in passing that some of the Welsh stone circles occur in lead-mining areas, a point which does not encourage surmises as to great antiquity for these particular circles.

Wales in Early Historic Times

Earthworks which have been compared with those of the Côte d'Or in east France are numerous in patches here and there in Wales, but are by no means generally distributed. Dr Willoughby Gardner has been able to show that the fortress at Dinorben in north Wales dates in the main from the period when the Romans were near-by but is the work of native or at least non-Roman peoples. The same view has recently been reached for the earthwork at Cissbury, Sussex, by Mr H. W. Toms, of Brighton, and it seems likely that the earthworks of Wales represent work, under orders, of an early Iron Age population in contact with Roman culture, but maintaining a considerable amount of the native heritage. As it seems that some of the invaders of the early Iron Age who came from north Gaul to England spoke a Celtic tongue that was related to Welsh, it is tempting to hold, lightly of course, the hypothesis that the Iron Age earthwork-builders introduced the Celtic speech of Welsh type into the country round about the period when Roman influence was spreading over Britain. Dr Wheeler has been able to show that, soon after the Roman conquest of Wales, the conquerors were able to reduce their garrisons in the country and so they would seem to have reached some form of *modus vivendi* with the earthwork builders who, in one case at least, were specially strong in an area where Roman remains are very few and doubtful. But readers must be referred to Dr Wheeler's books and papers for details on this subject. Mr David Thomas has found numerous traces, in systems of counting and in farmyard and other words, of a variety of Celtic allied rather to Erse, and he is inclined to think that a Celtic tongue akin to Erse (*i.e.* Gaelic) was spoken in Wales before Welsh. This view is not improbable in view of the many prehistoric links there must have been between Ireland and Wales, but some students think these Gaelic traces are due to immigrants of late Roman times from Ireland.

The linguistic story of Wales is no simple one. There may be survivals of Gaelic-Celtic as Mr David Thomas thinks, and even of pre-Celtic as he seems to think too. There are Latin borrowings and apparently many Anglo-Saxon borrowings in Welsh, and the Welsh tongue, like every other one, has by no means remained unchanged through the centuries. It is akin to the languages used in parts of south England in immediately pre-Roman times; it has disappeared thence, but has established itself in Wales and now forms the centre of a mass of emotional associations for large numbers of people in west Wales.

Much has been imagined concerning pre-Christian religion in Wales, but it is not profitable to say more than that Môn (Anglesey) was evidently of special religious importance and that in Wales, as elsewhere in Celtic lands, the early Christian teachers seem to have taken over a good deal from older cults. A church and churchyard within a stone circle at Ysbyty Cynfyn, Cardiganshire, illustrate this taking-over rather aptly, and there are several cases of churches similarly placed. Further, it seems likely that St David's was a place of some importance in prehistoric times with possibly a number of alternative landings to suit different states of wind and tide. The headland at St David's gives every indication of its ancient importance, and as the early Christian preachers apparently moved between Ireland, Wales, Cornwall and Brittany, using St David's as a Welsh station, it is likely that the Irish-Welsh intercourse of early Christian times was a continuation of something much older, and that it in turn led to the rise of St David's as the metropolitan centre of the Church in Wales and as a pilgrim centre with chapels at various landings. The development of the religious supremacy of Rome in England from the 7th century onwards was followed by the growth of Roman supremacy, expressed through Canterbury, in the Church in Wales. The attempts of Giraldus Cambrensis to stem the tide are a part of history.

Dr Fox is publishing year by year in *Archaeologia Cambrensis* important studies of successive portions of Offa's Dyke, a boundary of Mercia over against Wales. He has shown that the Mercian influence penetrated deeply into what later became Radnorshire and this may be considered in its possible relation to the fact that the Welsh language has almost disappeared from large parts of that country in spite of its mountainous character which might have been expected to make it a refuge for the ancient tongue.

It would seem that many Scandinavian elements affected the coasts of the Irish Sea in the time between Roman and Norman rule in Wales and there are many Scandinavian place-names in south Pembrokeshire, Gower, south Glamorgan and on patches of coast even elsewhere. From Norman times onwards the influences working on Wales came overwhelmingly from the English Border and it is to be noted that such influences have had several alternative lines of entry along the north or the south coastal plain, along the Dee, the Severn, the Bishop's Castle gap, the Wye, or the Usk. The entries along the north coastal plain via Chester, and the south coastal plain via Gloucester and Chepstow, and the central entry via Ludlow seem to have been of special importance in mediaeval times, and Shrewsbury was also of great importance in its relation to Wales. In modern times the railway has made Shrewsbury a most important junction and the most effective focus for the whole of Wales, though it is on the English side of the border.

The multiplicity of entries of Anglo-Norman influence helped to fractionate Welsh Wales, which of itself had been hopelessly divided into principalities and lordships, largely determined by the relief of the land. Comparing Wales with Scotland, we note in Wales the Berwyn, Plynlymon, and Brecknock ranges, separating the Severn and Wye valleys from the numerous valleys that fan out to the sea, north, west and south. In these latter valleys there is no place with communications sufficiently good to have allowed it to develop as a focus of Welsh social and political expression, and the hill ranges named above have prevented such a centre from arising in the Severn, or in the Wye, valley; and these valleys with their openings out to England have rather tended to dilute the Welsh characteristics with imported features. In Scotland, on the other hand, the Midland Valley and Strathmore, with Perth,

Stirling, Edinburgh, and Glasgow, and St Andrew's on the east coast, have given an admirable focal region for Scottish life. Scotland was thus enabled to survive as a political entity, and has preserved its Law to the present day. The little Welsh valleys, however, through their remoteness, have become centres of preservation of the Welsh language in numerous dialect-forms, whereas the openness of the Midland Valley of Scotland, and the manifold contacts of its great centres with the world outside have led to the disappearance of Gaelic speech from that region and from most of the Highlands, with the exception of a few remote glens and some western isles. The linguistic map of Wales from the 1921 census illustrates the survival of the old language, and the contrast with Scotland is emphasised when we reflect that only some 50,000 people at most still use Gaelic to any extent (Fig. 32, p. 252).

Before proceeding to sketch the development of the characteristics of modern Wales it may be well to refer briefly to its old characteristics and divisions.

The People of Wales

Among its people one finds on the Plynlymon moorland, among the Black Mts. of Carmarthen and elsewhere in remote spots, persons who still carry features characteristic of the men of late Palaeolithic times. They have dark hair and eyes and often swarthy skins, their big heads are very long, narrow and high, their brow-ridges often stand out strongly, their cheek-bones and noses are broad, and their mouths usually prominent. On the moorlands of Denbighshire there are very tall, dark, long-headed people, with narrow features, who seem to represent some ancient stock; the type can be matched in the Western Highlands of Scotland, in parts of Ireland, on Exmoor, and, apparently, in Spanish Castile. The general basis of the Welsh population consists of what are commonly known as "the little dark people," moderately long-headed, smooth-featured, and of slender build. Along the Bala cleft, in the Towy valley, and elsewhere, one finds tall, broad-headed people, often with rather light colouring, and strong brow-ridges; they are apparently related to the people who made beaker pottery about 2000 B.C., and whose burials of that period are found in certain parts of England, as can best be followed by studying the map by Dr Cyril Fox in *Archaeologia Cambrensis*. Dark, broad-headed people, often of tall stature, are characteristic of a number of coastal patches, south Glamorgan, north Pembrokeshire, north-west Merionethshire, etc.; they may be interpreted as a residue of some coastal drift, probably of the Bronze Age.

Anglesey, Lleyn, mid-Cardiganshire, south Pembrokeshire, Gower, and south Glamorgan also have fair, strongly built, moderate-headed people, no doubt the descendants of Scandinavian settlers, or of previous migrants along the Norsemen's tracks. Strong fair types are also notable in the Severn and Wye valleys, but we have no means at present of deciding how far they may be interpreted as Anglo-Saxon immigrants or to what extent they may represent the descendants of Iron Age, Celtic-speaking invaders (Fig. 33, p. 258).

Welsh folk-tales abound in descriptions of the moorlands, and of the clash between their ancient traditions and those of iron-using invaders in the valleys below. This aspect of Welsh life is worked out in a paper by the present writer, "Ancient Wales, Anthropological evidences" (*Cymmrodorion Society*, 1917). The Mabinogion gives us echoes of contacts between Wales and Ireland, and the *History of the Kings of Britain*, by Geoffrey of Monmouth, also has embedded in it many an old tale, but in both cases the work of mediaeval artists has doubtless deeply altered the original stories.

Before the establishment of Norman power, Wales was held to consist of three principal divisions, Gwynedd in the north-west, reaching south to the mouth of the Dyfi, and east to the mouth of the Dee, Dinefawr in the south-west, from the mouth of the Dyfi to the mouth of the Severn, and Powys, including the basin of the Dee, and that of the upper Severn and upper Dyfi, and stretching south to the Wye. These three main divisions were further split up into cantrefs, to some extent foreshadowing some modern counties, and the cantrefs were divided into commotes. In a number of cases it is possible to interpret the commote as an area with a nucleus of moorland and a fringe of valleys. The names of some of the ancient divisions, small as well as large, remain in common use, and are still often of value in distinguishing what may be called cultural regions.

The Welsh people in pre-Norman days were organised on a basis of kinship, and Mr Seebohm in *The English Village Community*, and in *The Tribal System in Wales*, sketched an interesting picture of their social organisation. The homestead or tyddyn seems to have stood out as a unit in contrast with the agricultural village of the plains of England and parts of western Europe. This is apparently to be correlated with the preponderating importance of stock-raising in wet hill-country. The result is that only here and there, chiefly along ways from England, do we find the typical agricultural village, with its memories of common arable land held in strips. The rural agglomerations of people in Wales are rather of the nature of groups around a bridge, a ford, a haven, or a sacred spot, *i.e.* they are rather germs of towns which have not grown than agricultural villages in the strict sense.

There are many indications that, of old, a part of the population used to move up to the moorlands for the summer, and a high pasture in Wales bears the name "hafod" (lit. summer-place). With the fall of the leaves, migration set towards the lowlands, with their winter residences which were often distinguished by names like "Hendre," "Pentre," etc. The migration of flocks of sheep between moorland and lowland is still a marked feature of Welsh farming, but it is not now accompanied to any extent by movements of people other than a few shepherds. Moorland and valley farmers take over appropriate numbers of sheep at the beginning of the appropriate season. It will thus be seen that, even in modern Wales, there are many rural features of life deep-rooted in a long past. They, along with the Welsh language, have thus far withstood the ceaseless and inevitable pressure from the east, which has operated since Norman times, and which must now be considered in some of its effects on modern Wales.

Historical and Social Considerations

The history of Wales is outside the scope of this article and should be studied in Prof. J. E. Lloyd's well-known work. The effective conquest of Wales was carried through by Edward I, and, despite the efforts of the brilliant and mysterious personality of Owain Glyndwr, was never undone. During the disorders of England known as the Wars of the Roses, Wales suffered severely, and her leaders rallied to the Lancastrian side at the head of which stood Henry Richmond of the Welsh house of Tudor. It was thus, in a sense, a Welsh house that came to rule England and the Tudors effected the legal union of England and Wales in 1535–42, by which the semi-independence of the lords of the Welsh marches was made to cease and the government of parts of Wales was assimilated to that of English counties. The intention of this Tudor effort was to assimilate Wales to England, and the native language, in particular, was looked upon as an obstacle to be abolished if possible. Some Welsh customs of inheritance were specifically retained. Glamorgan, Carmarthen, Pembroke (*i.e.* the districts along the south coastal plain), Carnarvon, Anglesey and Merioneth (*i.e.* the refuge of the Welsh princes which had been most completely conquered), Flint, at the northern entry into Wales, and Cardigan were already shires; and now Brecknock, Radnor, Montgomery and Denbigh, erstwhile specially under the power of Lords Marchers,

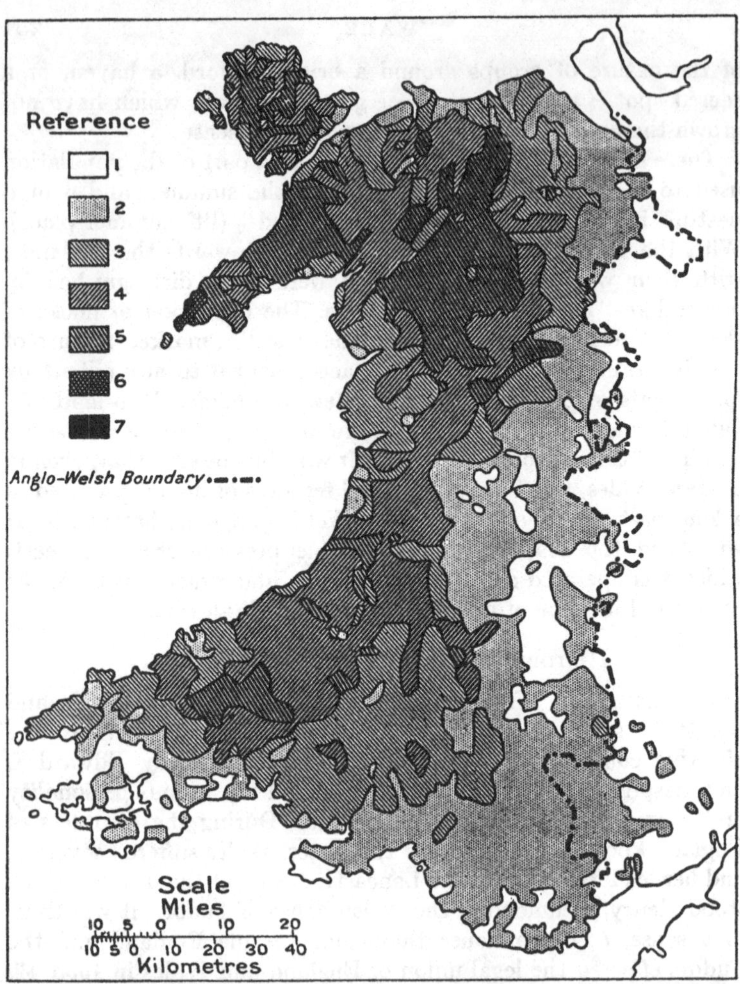

Fig. 32. Wales: Distribution of the Welsh Language; Percentage of the Population able to speak Welsh: 1, nil; 2, 0 to 10 per cent.; 3, 10 to 30 per cent.; 4, 30 to 50 per cent.; 5, 50 to 80 per cent.; 6, 80 to 100 per cent.; 7, 100 per cent. Map by Trevor Lewis, first published in *Annales de Géographie*, vol. xxxv, 1926, p. 415.

Explanation of Fig. 33

The "clock" symbols indicate characteristics of native groups in various localities, more than one symbol often being inserted to show what appear to be different types in a locality. *Dark shaded circles but for lack of space would be spread all over the map, indicating "the short, dark long-headed" person.*

The cephalic index is shown by shaded segments of circles. Pigmentation: 1, Light hair and blue eyes; 2, brown hair and blue or grey eyes; 3, brown hair and eyes; 4, usually very dark hair and eyes; 5 and 6, red hair—5, when cephalic index is below 79; 6, when above 79.

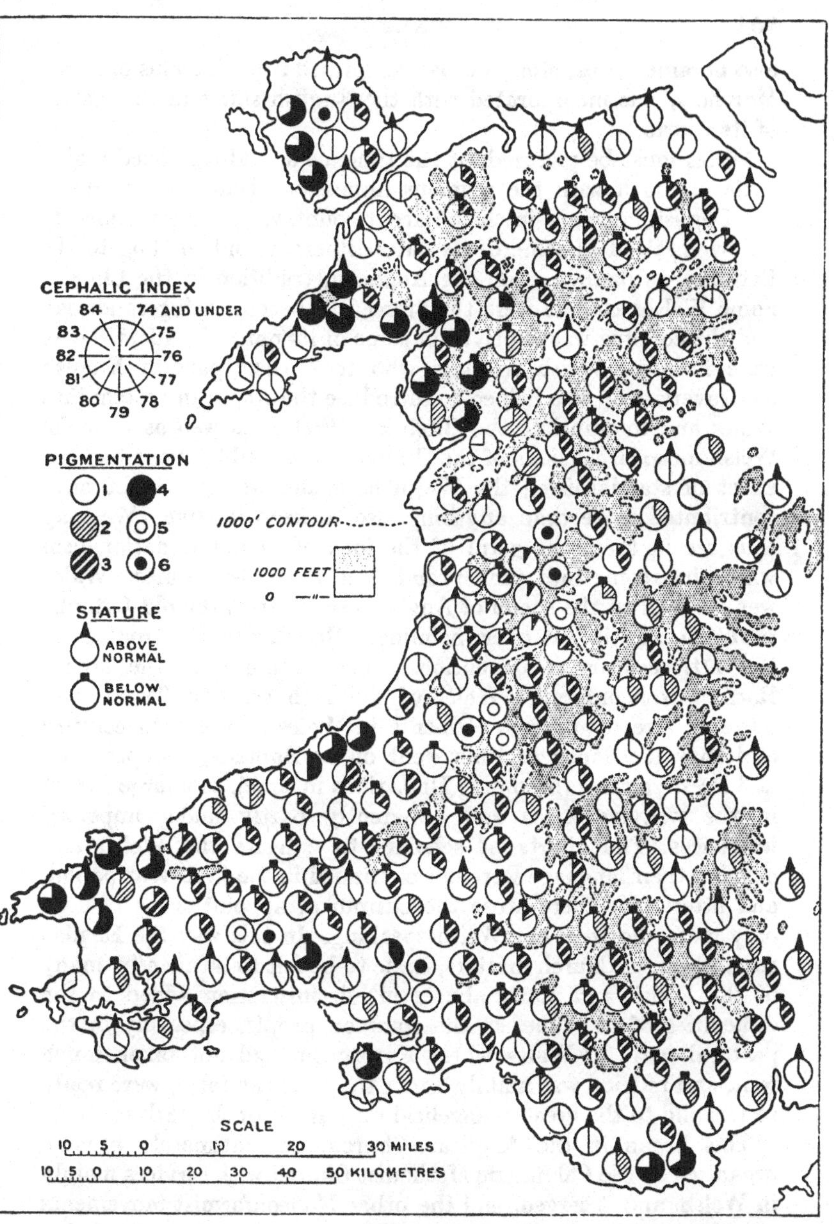

Fig. 33. A Tentative Map of the Distribution of Racial Types in Wales, prepared by E. G. Bowen, M.A. (For explanation see foot of opposite page.)

also became shires; Monmouthshire, formed from domains of Lords Marchers, was incorporated with the English shires in the matter of its courts.

Wales must be pictured, at the time, as an old-fashioned region little in touch with the religious discussions being carried on in the English language in the adjacent country, and thus inclined to follow the lead of some of its old aristocracy and to cling to the Catholic, as distinct from the Anglican, tradition in the Church. Some Welsh scholars were thus forced to become refugees abroad and the country was deprived of some of its natural leaders. There were however Welsh scholars who took their part in English controversies or who helped to introduce the Anglican system into Wales and to translate the Bible into Welsh, as well as to make Welsh metrical versions of the Psalms. These publications had the effect of standardising the Welsh language, and no doubt have contributed more than anything else to keep it alive. We may thus see in them the germ of the idea of linguistic nationalism which has come into such prominence in modern times. Wales was inclined to the shade of Anglicanism nearest the old Catholic tradition, and it was overwhelmingly Royalist in the troubles of the 17th century, thus curiously contrasting with the ardent Radical Nonconformist Wales of the 19th century. There were however already puritan elements in Wales in the 16th century and they developed as time went on as Baptists, Independents and Presbyterians (becoming Unitarians in time), to a large extent in the weaving areas and apparently partly under imported influences. The efforts of scholars to educate the Welsh and develop their language were not continued in the Civil War period, and afterwards there arose the custom of appointing as bishops only men who were not Welsh-speaking. In this way the English connection, in Church matters, came to be felt to be alien by many of the Welsh and many abuses in administration added to the difficulties of the time, so the common people, especially in the pastoral parts of Wales where Puritanism had not taken much root, and religion was mainly under the Anglican form, were ready to respond to the great evangelical movement of the 18th century.

This began in the Anglican Church, but ultimately became organised as the Calvinistic Methodist Church with services mainly in Welsh, and it served, and the other Nonconformist movements reinvigorated by the Evangelical Revival served too, as rallying grounds for religious, poetic and to some extent political expression in the 19th century, the period of linguistic nationalism in most parts of Europe. The land-owning class had become Anglicised, and so a cleft deepened between aristocracy and democracy in Wales, and was marked in language and religion as well as in class-consciousness on both sides. During the last 50 years the Welsh linguistic tradition has grown once more in the Anglican Church, which in the 20th century has ceased to be the State Church in Wales and has been given an organisation of its own. The gathering of Welsh feeling around Nonconformity and in opposition to the Anglicised landowners made Wales Radical in politics in the 19th and early 20th centuries, and it may be mentioned that the appeal of Mr Gladstone's religious and moral enthusiasm to Welsh feeling was a most powerful one. Periodicals of many kinds and literature in the Welsh language developed apace in the later 19th century and voiced the demands of Wales for special legislative treatment in various educational and social matters, leading up to the creation of a national scheme of secondary education and a national University just before the century's end. The orographical difficulties of Wales led to the establishment of a large number of secondary schools, often small and struggling, and to the creation of four University centres; thus illustrating that fractionation which has apparently been a permanent feature of Welsh life. Another concession to Welsh feeling was the establishment in the early 20th century of a National Library and Museum, the former in Aberystwyth, the eldest University centre and a little focus in the narrow westerly neck that unites the two main areas (NW. and SW. areas) of Welsh speech, the latter in Cardiff, Wales' largest city.

The rise of modern industry based upon coal-power in the 19th century transformed South Wales and gave it an enormously increased population with a considerable immigrant element partly derived from the poorer rural areas and partly from outside Wales. For several decades already there has preponderated in the country this mining and industrial population, with problems and interests in many respects very different from, and even in some points conflicting with those of rural Wales. As a consequence of this, if any separatist tendencies had developed, as they did on the bases of linguistic and religious nationalisms elsewhere, Wales would have been dominated completely by the coalfield; it is interesting in view of this that political separatism has not developed strength in Wales and the nationalist movements have been mainly social, educational and literary in their scope. This gives the Welsh people certain marked characteristics and interests that distinguish them very markedly from the masses of the English democracy. The nationalist revival has been responsible for the rise of the "Eisteddfod," a gathering to hear and judge competitions primarily in poetry and singing. This movement looks back to bardic contests of long ago for its prime inspiration, but it has spread on new lines and in a measure has become a university of the people.

It has thus happened that large numbers of Welsh people in the humbler ranks of society have contrived to keep a certain cultural tradition and enthusiasm of their own which is rarely found among similarly situated people in England, and this very real but rather intangible characteristic may be said to be the main distinguishing feature of modern, or at least of recent, Wales. In the last generation the social and political problems accompanying the large agglomeration of population in the coalfield, and the change from personal to impersonal capitalism in the same area, have occupied men's thoughts and have set new ideas in motion. The result is that while the immigrants from rural Wales into the coalfield, and, to some extent their descendants, still carry on the Welsh cultural tradition, a stream of acute political controversy and of new and often revolutionary views on social and industrial matters sets strongly in South Wales. On the other hand, in rural Wales the weakening of the old landlordism and the increased freedom of expression in religion, politics and education of the farmers, many of whom are possessed of some capital, makes for a swing to the right in matters of opinion, and there is now a deep cleft between rural Wales and a large element in the coalfield. There is thus less than ever an opportunity for any separatist movement, and the Welsh tradition tends to continue its effort rather for literary and religious expression.

Economic Aspects

At the dawn of modern industrialism Wales was mainly concerned in farming, growing some corn, but being more concerned with the rearing of cattle, and especially of sheep. There were market towns, smaller and spaced farther apart than in the wheat lands of the English plain, with small industries such as tanning and weaving. There were also numerous little ports, some like Aberayron being practically new creations of the early 19th century in response to increased opportunities of both coastwise and foreign trade. In some of these little ports furniture-making was a characteristic accompaniment of shipbuilding. Agricultural development was fostered by the transport of lime by sea to many areas. Mining, for lead containing a little silver, had been important at certain periods in north Cardiganshire and in Flintshire, and copper mines had been worked in Anglesey, while Carnarvonshire quarries yielded unrivalled slates. The growth of railways, however much it has been impeded in Wales by the difficulties of gradients, has led to the decay of maritime trade in the little ports, which however often survive in relative prosperity as small market centres. The industries of small towns have been almost destroyed by the competition of large centres in England. The lead and copper ores of Wales have felt the competition of foreign products, and metalliferous mining has almost disappeared. On several counts therefore there is every reason for a drift of population away from the little towns of rural Wales; this drift is supplemented by one from the poorer parts of the country, the people of which found themselves left behind in resources and opportunities in modern life. This trend of population is exemplified in the following figures which show that outward drift from the counties named exceeds natural increase.

	1891	1901	1911	1921	
Merionethshire	49,212	48,852	45,565	45,087	(census includes some tourists in 1921)
Montgomeryshire	58,003	54,901	53,146	51,263	
Cardiganshire	63,467	61,078	59,879	60,881	(census includes some tourists in 1921)

In agriculture less land than ever is devoted to corn crops, the chief of which is oats. The breeding of horses was formerly of importance but plays a far smaller part now that mechanical traction has become general. The large tracts of moorland, often unenclosed, help to maintain the sheep-rearing industry; whereas in the whole of England mountain and heath land forms only about 9·75 per cent. of the total area connected with farming, in Wales the figure is 33·4 per cent. In England, again, of the land under crops and grass in 1918, 52·7 per cent. was under permanent grass, while in Wales the figure was 66 per cent. Both figures would be higher now. On the other hand, in 1918, England had 483 sheep per 1000 acres (405 ha.) of land connected with farming, while Wales had 852, and in Brecknock, Radnor, Montgomery, Merioneth and Denbigh the number rose well above 1000. The difference in numbers of cattle was not marked either way, but there has long been a tendency to send cattle to England for fattening purposes. The Hereford breed of cattle has established itself on the Welsh side of the border, but the old black breed has maintained itself, especially in the hilly regions of the south-west. Centuries ago the goat was a great feature of Welsh rural life, but it has almost entirely disappeared. The rural situation is again changing thanks to the spread of motor buses and mechanical transport and the consequent increase of work on the roads and revival of rural inns, repairing shops, etc., and thanks also to the diminution of isolation resulting from the spread of wireless; there is also side by side with this an effort to revive social life in villages through women's institutes and varied activities gathering around churches and schools, through the spread of adult extra-mural education from the universities, and the deliberate fostering of small rural handicrafts*.

Before the beginning of the 19th century the iron trade had begun to develop in South Wales, utilising at first wood from the forested slopes of what has become the coalfield, and iron from nodules found in some beds of the Coal Measures. Among the early centres of the iron industry we may note Pontypool situated where the Afon (river) Lwyd breaks through the eastern rim of the coalfield and Merthyr Tydfil in the northern part of the coalfield. It would appear that supplies of wood for this iron industry had seriously diminished when, at the end of the 18th century, the utilisation and export of coal began, and ensured the continuation of the iron works. One of the great difficulties was that of transport which was at first done along hill tracks, largely by mules and ponies. In the last years of the 18th century, however, canals were opened from Pontypool to Newport and from Merthyr Tydfil to Cardiff, with the result that these coastal towns linked their fates with the coalfield and grew rapidly as coal exporters as well as in relation to the iron industry. The eastern valleys of the coalfield were developed industrially earlier than the western ones, and at first Newport grew faster than Cardiff, but the latter city has long since gained the premier position for reasons we must try to estimate later on. The coal of the region east of the Taff valley is bituminous, but north-westwards it becomes richer in carbon, and west of the Neath it has the form of anthracite. Development of the mines of steam coal and anthracite has been a main feature of the life of South Wales in the latter half of the 19th century, during which the Rhondda valleys were exploited and acquired a large population poorly housed, along the narrow valley bottoms for the most part, in what has become almost a continuous urban ribbon. The Cynon and Taff valleys are in some degree through-ways owing to the relations of their heads with the Neath valley on the one hand and with the depression between Mynydd Llangynidr and the hills of the coalfield proper on the other. The Rhondda valleys, on the other hand, are culs-de-sac or were such until a long tunnel was constructed; this increased their isolation and the

* See recently published reports on rural industries by Miss A. M. Jones and by W. F. Crankshaw (textiles only).

environmental conditions with poor housing and large "tips" of mine-waste prevented the growth of a sufficiently varied population in these valleys, in which few people reside unless they are connected with the mines or are doctors, ministers, shopkeepers or teachers. Directors of coal companies and even some managers tend to live in the more open country if they can. Moreover, the population naturally includes large numbers of young miners, and the 1911 census showed that the proportion of females to males was only 84 per cent., a very atypical state of affairs. These features help to explain the cleft between miners and directors of mines and the psychological isolation of the former, and they also help to interpret the keen enthusiasm of some elements among the miners for singing, education, and political controversy as outlets for mental energy. It is found that the elements which drift into the coalfield and remain there include specially large proportions of the "little dark Welshman," as this type seems better able than others to withstand the cramped conditions and to find modes of expression in these difficult surroundings. The rise of the Rhondda as a producer of steam coal led to the predominance of Cardiff as a coal port, and to the growth of Pontypridd at the junction of the ways from the Rhondda valleys with the ways down the Taff and Cynon valleys. This illustrates the manner in which the coalfield valleys have developed, each more or less apart from the others and all looking to Cardiff and Newport in the main, but there are "cross-ways" in the coalfield, relatively low lines east to west, as for example from Abergavenny via Merthyr Tydfil to the Vale of Neath and from Pontypool to Hengoed. A public need is however keenly felt for co-operation between the valleys in roads, water-supply problems and general administration, and the new motor-bus services are both a response to this need and a stimulus to fresh developments in this direction.

The increase in size of steamships led in the last half of the 19th century to a great change in the coalfield from dependence on local iron-ores to dependence on foreign and especially Spanish ores. Meanwhile, also, the smelting of foreign ores of copper, tin, etc. was developing rapidly. The metallurgical expansion in the South Wales Coalfield has led to many changes*. With importation of ores it became less practicable to concentrate smelting in the northern part of the coalfield and both the Newport and Port Talbot areas, on the coast, took up this work. Some, but not a great deal, has been done at Cardiff, which with its associated ports of

* On all these questions see also A. E. Trueman, "Population Changes in the Eastern part of the South Wales Coalfield," *Geog. Jour.* 1919.

Penarth and Barry was until recently very fully employed in exporting coal and in the commercial organisation of the coalfield. The rise of Port Talbot as a metallurgical centre led to the tunnelling of a way from the north-west of the Rhondda to the Afon valley so that Rhondda coal became available at Port Talbot. At Swansea the smelting of imported copper ores with local anthracite had begun before the Industrial Revolution, but it developed enormously for a while in the 19th century and there grew around it the treatment of tin, lead and zinc. In 1911 the copper industry occupied 2841 persons in the Swansea district, while about 1000 were engaged in zinc industries, and 1270 in nickel industries. The making of tinplate on a basis of Siemens steel had become a large industry by 1911 in the Swansea and Llanelly districts and the industry is highly organised with what is called vertical integration, *i.e.* with a unitary control over most of the successive processes in the preparation of the steel and the tin and the plating of the tin on the steel. Swansea's population was 143,997 in 1911; the boundary was extended in 1918 and the 1921 total was 157,554, while the estimate for 1925 was 163,200. Alongside of the metallurgical industries, Swansea concerns itself with export of anthracite and in recent years with importation, storage and distribution of mineral oil for power purposes. Llanelly, on a broad estuary hampered by sandbanks farther west, has developed copper smelting and tin-plating and has grown from insignificance at the beginning of the 19th century to become by far the largest Welsh town outside Glamorganshire, with 36,504 people in 1921.

Industrial Wales after the War of 1914–1918

The population of Wales excluding Monmouthshire in 1901 was 1,714,800, in 1911 it was 2,025,202, in 1921 it was 2,205,680. The corresponding figures for Monmouthshire were 298,076; 395,716; and 450,794. Of the population of Wales Glamorganshire with its county boroughs included 859,931; 1,120,910; 1,252,481 at these successive censuses. As part of the population of Carmarthenshire also is engaged in coal mining and associated industries, it is evident that the mining and industrial element of South Wales is enormously preponderant in the Welsh population and becomes still more so if Monmouthshire is added to Wales. In 1911 no less than 214,348 persons in South Wales were engaged in coal mining, 42,317 in steel and iron work and 23,042 in general engineering, illustrating the concentration of the population on this type of work. The increases of the population represent the results of industrial development of pre-War days. The post-War industrial position of Britain is nowhere more difficult than it is in South Wales. Trade depression in Britain and the shrinkage of world credit and markets has made British users seek to economise in amounts of coal burnt, and oil has superseded coal in British industry here and there, while shipbuilding has been at a low ebb. The home demand for coal has thus remained, at best, stationary; and the new coalfield in South Yorks and Nottingham has cut down the sales from the older coalfields to some extent. But it is especially the decline in the coal export business that has troubled South Wales. Many ships, including the Royal Navy, burn oil fuel and the steam coal of the Rhondda no longer finds maritime transport ready to absorb all it can produce. The demand for coal in Russia no longer helps South Wales, Italy does all it can with hydro-electric power, Germany has increased its utilisation of its home lignite and its demand for British coal has slumped since the Ruhr complications ended. France gets coal deliveries from Germany and is making use of hydro-electric power; other countries have tried to develop any coalfields they may possess. All these difficulties have made their full force felt in the last two years. Part of the depression is to be ascribed to the extreme individualism of past generations which led to small colliery companies with poor equipment and no resources for research work, part to wasteful methods of using coal which increase industrial costs enormously, part to lack of selling organisations and costly ill-arranged schemes of transport. But in addition to all these factors there is the changed world situation, and it is undoubtedly threatening the future of South Wales and maintaining a situation already difficult and explosive because of bad housing, high accident rates and lack of personal contacts between employers and miners, which would probably have brought a serious crisis had it not been for the scheme of "Unemployment Pay" inaugurated some years ago. It is thought that the population will soon cease to increase, but it will be long before the turn of this tide brings any relief to Wales' problems with its agriculture in a precarious state, its rural population inclined to dwindle, its industrial life half paralysed and its industrial population still growing.

Wales possesses another coalfield, but a small one, on the north-east, with Wrexham as its chief town; but mining and metallurgical and chemical industries extend northwards into Flintshire, and the Flintshire side of the Dee estuary has tended of late years to become more and more an appanage of the Liverpool-Birkenhead complex, with artificial silk and alkali works at Flint, lead smelting at Bagillt, and iron works at Mostyn. Near Wrexham is the mining town of Ruabon and round about it is quite a considerable population of miners. In all, in 1911, the coalfield of north-east Wales employed 14,732 miners, 8335 iron and steel, etc. workers, and 2322 engineering workers. The post-War crisis in this small coalfield has been even more severe if possible than in South Wales, though coal here was not mined for export, but industry in Flintshire is expanding remarkably.

As against these difficulties it is well to mention developments now in progress in the western part of the South Wales Coalfield. The Neath valley is the expression of denudation along a fault zone and this fault separates the anthracite coal on the west from less highly carbonised material on the east. Whereas the bituminous and steam coal trades on the east are faced with depression of trade and restriction of markets, the anthracite area on the west has expanding opportunities, and is able to go forward without too severe a handicap from the past. In place of numerous separate colliery schemes, whether now amalgamated with larger ones or not, the west is in the main under the control of a large organisation that also has interests in the U.S.A. and elsewhere and is interested in research, in propaganda for increasing the use of anthracite and so on. Policy can be so framed as to take into account the situation at any particular time in America or in continental Europe; and metallurgical industries are worked into the scheme so that there may be means of absorbing surplus coal, while the treatment of waste products is also organised. The social aspects of the industry are receiving attention, and it is greatly to be hoped that the new anthracite development may avoid the tragic experiences of the eastern part of the coalfield by attention to housing and to health. At any rate, this part of the coalfield is reasonably prosperous even under present depressed conditions and it is attracting a number of the unemployed from the east of Glyn Neath. We may be at the beginning of a change of the centre of mining and industrial activities from the Rhondda and other valleys east of Glyn Neath to Swansea and Llanelly and the valleys behind them, with schemes of electrical power, highly organised metal industries, and a widespread propaganda for the use of anthracite for heating.

The situation farther east may however be to some extent redeemed by schemes for conversion of coal into more economical forms, with careful utilisation of waste products, by better marketing and other auxiliary arrangements. It is also being met in another way by the appointment of a governmental staff to help miners to find alternative employment. Voluntary agencies are also at work trying to develop new trades. It is hoped that the well-known skill and intelligence of the population of the coal valleys will help it to surmount the present serious crisis. The establishment of an Adult School on residential lines, not unlike the Danish Folk High Schools, was one of the noteworthy events in Welsh life in 1927; its attention is specially focussed on the problems of the miners, but it is characteristically situated at Harlech, N. Wales, in the beautiful country between the mountains and the sea.

XIV

THE PENNINE HIGHLAND

C. B. Fawcett

THE Pennine Highland is the dominant feature in the relief of the northern half of England. It extends through a total distance of nearly 150 miles (*c.* 240 km.) from north to south, from the Tyne Gap to the Vale of Trent, with a mean width of about 30 miles (*c.* 50 km.); the width varies from 50 miles (*c.* 80 km.) west of York to 20 miles (*c.* 30 km.) at Stainmore and north-east of Manchester. Its main structural lines are related to the Hercynian folds, though there are evidences of the influence of other movements. The Highland is composed almost wholly of rocks of Carboniferous age; only towards the north, in the Silurian of Teesdale and the intrusive basalts of the Whinsill and many dykes, are there outcrops of other rock systems; and it is believed that a pre-Hercynian massif of older rocks underlies the North and Central Pennines and has exerted some influence on their development.

In both structure and relief this Highland falls into three main divisions, North, Central, and South, separated by Stainmore Pass and the Craven Gaps respectively; the last of these may conveniently be divided into two parts for geographical purposes since from the High Peak northward its surface rocks are of the Millstone Grit series while the southern part is mainly composed of limestones (Figs. 42 and 43, pp. 324–325).

The North Pennine massif is cut off on three sides, north, west and south, by fault-line depressions, and dips eastward under the Permian rocks of County Durham. To the north it is bounded by the Tyne Gap, which severs it from the Cheviots whose main trend-lines belong to the Caledonian system and not to the Hercynian folds of the Pennines. On the west it overlooks the Vale of Eden in the great fault-line scarp of the Pennine Scar on which are the highest summits of the Pennines (Cross Fell, 2799 ft. or 853 m.). This scarp rises abruptly to a height of from 1500 to 2000 ft. (*c.* 450–600 m.) above the Vale for a distance of thirty miles from Stainmore Pass to the Tyne Gap and is probably the most impressive physical feature in England. Near Stainmore the great Pennine fault breaks up. One branch trends eastward through the Stainmore Pass which separates the north and central sections of the Highland. A second branch stretches south-south-westwards by the western edge of the Central Pennines for over twenty miles and then curves round towards the south-east in the Craven faults.

Thus the central portion of the Pennines, like the northern massif, is bounded to north, west and south by fault-line depressions; and it also dips eastward under Permian rocks. But here the north to south axis of the Pennine uplift is intersected by a transverse axis of crustal movement in the line of the Lake District and the North York Moors, and this central block is broken by intricate series of faults spreading fanwise from its north-west corner and

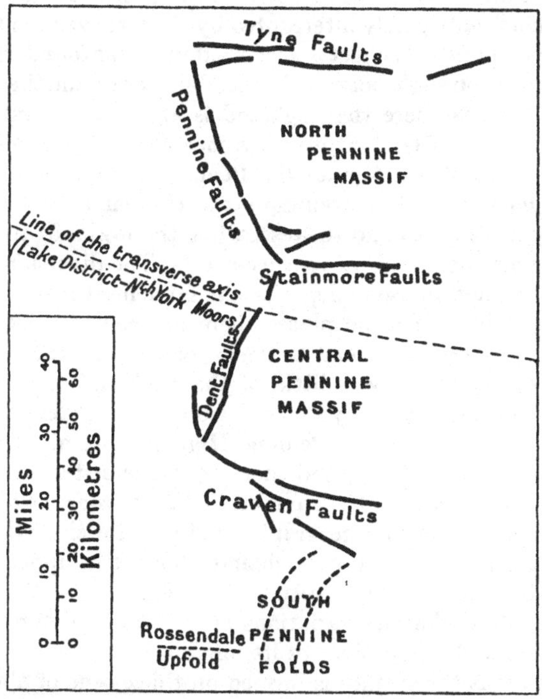

Fig. 34. The Pennine Highland: structural diagram.

crossed by other series of transverse faults, and is much less compact than the northern massif. The whole of the drainage of the Northern Pennines goes east to the North Sea, except for the short torrential becks which gully the Pennine Scar; but that of the Central Pennines is almost equally divided between west- and east-flowing streams, whose valleys are frequently connected by low cols. In the Northern Pennines is the largest continuous area above the 2000-ft. (*c.* 600 m.) level south of the Scottish Highlands, in marked contrast to the scattered residual hills which alone rise to that altitude in the central area, where the general altitude is some hundreds of feet lower. The decrease in the general altitude continues to the Southern Pennines. South of the Craven district the mean altitude of the moorlands is little more than 1000 ft. and few summits rise above 1500 ft. But in the southern half of our South Pennines the Highland rises a little in the High Peak district, where the Peak itself reaches 2088 ft. (635 m.), before dying away to the lowland. The depression between the Craven and High Peak districts causes a marked narrowing of the Highland, only partly interrupted by the transverse ridge of the Rossendale upfold. Its effect is accentuated by the foundering of the older rocks in the angle between Rossendale Forest and the Southern Pennines; so that here the Highland is at its narrowest and the tongues of inhabited land in the valleys meet through it in some places.

During the last Ice Age all the Pennine valleys were occupied by local glaciers; and at Stainmore the Highland itself was overridden by ice from the north-west, while the lowlands to east and west were also occupied by vast ice-sheets. The diversion of many rivers by ice and moraine deposits has influenced many details of the topography. In some places there are considerable areas of glacial drift in the valleys and on the lower slopes; but except about Stainmore this drift covers little of the Highland surface and its thin soils are more directly derived from their underlying rocks.

The surface rocks of the Pennine Highland are all of Lower or Middle Carboniferous age; and there are three distinctive types of moorland depending on the character of these rocks, which are of primary importance in a geographical study; for since the general altitude places most of the Highland above the upper limits of arable cultivation in these latitudes, the character of the surface is more important than its variations in altitude in determining the distribution and occupations of its inhabitants.

In two areas the surface is formed on thick beds of almost pure limestone. These are the northern part of Craven and the district just south of the High Peak. Here we get the characteristic features of a karst topography, with its dry valleys and bare rock surfaces ("clints"), swallow holes and caves, gorges and cliffs ("scars") well developed; though much of the surface is covered by a thin turf which supports good pasture grasses and makes these limestone areas useful for sheep-rearing.

Between these two karst districts, in the lowest part of the Highland, rocks of the Millstone Grit series still cover the whole of the surface. Here, in strong contrast to the dry grassy moors just referred to, the flat tops are mainly occupied by peat bogs and the slopes by heather. These moors have a sour soil and poor vegetation which makes them of little direct value to man. But they send down to the bordering lowlands an abundance of soft water which has been an important factor in the development of the textile industries on their flanks. Here almost the whole of the Highland is now reserved as gathering grounds for the water supply of the valley towns in west Yorkshire and east Lancashire, and it is practically uninhabited. In spite of its heavy and constant rainfall, of more than 50 in. (1250 mm.) per annum on the higher parts, the area does not gather sufficient water for the densely peopled industrial regions, and some of the larger cities obtain additional supplies from farther north.

In the three areas just referred to the surface rocks are fairly uniform; but over the rest of the Highland, north of the Craven district, there is much more variety. The rocks over the higher areas are all of the Lower Carboniferous Limestone series; but the limestone beds become generally thinner towards the north and are also more and more mingled with intercalated shales and sandstones; so that the surface is usually formed on impure rocks. Local variations in the proportions of limestone and other rocks give variety in detail, but broad stretches of one type of rock are rare. On the lower slopes, especially towards the eastern edge, there are extensive areas of the Millstone Grit series which form moorlands similar to those described above. The lower altitudes however allow of some spread of cultivation on to them though their sour soils and bad drainage make them infertile; only a small part is used as gathering grounds for water supply, especially in the district immediately north of the Yorkshire Coalfield where Leeds and Bradford have gathering grounds.

The general surface of the Pennine Highland is remarkably smooth. Though many of the valleys are deep they occupy but a small part of the total area and the moorland between is in some areas almost featureless. Such hills as Ingleborough and Penyghent, and many others, are residual fragments of the grit rocks resting on the limestone plateau in north Craven; and elsewhere there are many scattered hills of similar structure. They are best summed up as "monadnocks" (remnant-hills) rising above the peneplane which is now the surface of this Highland. These remnant-hills, and the narrow valleys ("dales") incised in the raised peneplane, are the features which give variety to the landscape.

The Cheviot Hills differ from the Pennines in their main trend; but they are composed of similar rocks and are in most respects similar to the Northern Pennines in their surface features. Cheviot itself is the weathered remnant of a Tertiary volcanic mountain and is thus structurally distinct; but since almost the whole outcrop of igneous rocks is at altitudes above the limits of cultivation this difference has little effect on the population.

It follows from the facts we have stated that the population of the Pennine and Cheviot Highlands is almost entirely limited to the dales. There is some agricultural activity though there is little cultivation. The county which is most nearly limited to these highlands and the Carboniferous rocks is Northumberland, and its agriculture offers the best measure of that of the dales and moorlands as a whole. At the time of the agricultural census of 1925 91 per cent. of the area of the county was under grass, including in this the moorland and hill pasture; and personal observation justifies the statement that practically all the 9 per cent. put to other uses is in the coastal lowland and valleys at less than 300 ft. above sea-level; so that the dales and moorlands produce little but grass. Practically the whole of the farming is devoted to livestock; for about two-thirds of the small amount of arable land is given to fodder crops and the smallness of the area devoted to wheat, under 5 per cent. of the cultivated land, supports the observation that it is planted only to fill a gap in a rotation. The main dependence is on sheep and cattle, but while the density of sheep is higher in Northumberland than in any other county of England that of cattle is well below the average for the country; and in spite of the proximity of industrial towns the county ranks very low in its proportion of dairy cattle. In fact the dales and the margins of the highlands are more suitable for cattle-rearing than for dairy-farming which is generally limited to more fertile lowland areas; and the dale farmer usually rears cattle and sells them to be fattened on the lowland before their final journey to the great markets of the industrial areas. Sheep are nearly ten times as numerous as cattle; but they also are for the most part in the dales and on the lower areas. Even in late summer when the hill pastures are in full use the number of sheep on the moorlands is small except in the limestone districts, though the whole area is used for summer pasture. The sheep are valued chiefly for the production of mutton and lamb; the wool is generally of poor quality and the fleeces are not large. It has been estimated that wool represents not more than a tenth in value of the produce of these flocks.

The forest resources of the Highland are at present almost negligible. There are some patches of scrubby birch and thorn and other small trees, and small woodlands in sheltered valleys. But in the present century many large plantations of conifers have been established, chiefly in the gathering grounds of the great waterworks, and it is clear that large areas are suitable for afforestation. The chief direct obstacles to tree growth are the stagnant acidic waters of the peat bogs and the strength of the wind on the exposed surface.

These poor agricultural resources are supplemented in a few localities by mineral wealth. Where fairly thick beds of pure limestone outcrop in lower Teesdale there are large quarries from which stone is sent down to the iron furnaces near the river's mouth in quantities which have exceeded 3,000,000 tons in a year. It is noteworthy that the large quarries are in those outcrops of suitable stone which are nearest to the furnaces, though other outcrops show thicker and perhaps more easily worked stone. In a few places lead is still mined, chiefly in the Northern Pennines about Alston; elsewhere the output is now almost negligible, though the mines are not exhausted and a prospect of higher prices for the metal might cause a revival. Other materials obtained are barytes and fluorspar in Weardale, for which the total demand is small, and ganister in a number of scattered quarries. On the whole this mineral wealth is not at the present day sufficient to add much to the resources and population of the dales.

One very important difference between the three main divisions of the Pennine Highland which we have indicated results from the fact that there are important coalfields on the flanks of the northern and southern sections while the Central Pennines are not bordered by any coalfield. The higher parts of the Durham-Northumberland and the Yorkshire and Lancashire Coalfields lie among the Pennine foothills; and the presence of these fields and their associated mining and industrial populations modify the human geography of these parts of the Highland. But any study of these fields belongs to other chapters.

Except on or near to the coalfields there are no considerable towns within the Pennine and Cheviot regions. Most of their markets are in foothill towns on the edge of the lowland. In the wider parts of the Highland there are small market towns in it such as Rothbury on the Coquet, Pateley Bridge in Nidderdale and Settle and Skipton in Craven. The last-named is the largest; it is the nodal town of the Craven Gaps, an important route centre, the principal market town of Craven, and so near to the industrial area that it has some woollen mills; yet its population in 1921 was only 12,012. Except in a few small tourist centres the population of the Highland is decreasing as the attraction of richer areas makes itself felt more and more; and outward migration is still, as it has been for some generations, the only considerable movement of its population.

XV

LANCASTRIA

W. Fitzgerald, H. King and J. Kershaw *

THE term Lancastria is coming into use to denote the related economic regions of two counties—Lancashire and Cheshire. The district historically known as "Between Ribble and Mersey" (*Inter Ripam et Mersham*) was the nucleus of the county of Lancaster which grew with the addition of Amounderness† and other districts. As late as the time of the Domesday Survey there was no county of Lancaster: its present territories were then included within Yorkshire and Cheshire, the River Ribble providing part of the boundary line between these two counties. To-day the district between the Ribble and Mersey is the territorial basis of one of the most highly industrialised regions of the world; while on the southern side of the Mersey industrial Lancastria extends considerably beyond these limits on to the plain of Cheshire.

Considered from the physical aspect the predominant feature of Lancastria is the lowland which intervenes between the Pennine Range and the Irish Sea and retains its essential character over the greater part of Cheshire and Lancashire. This lowland has its counterpart on the eastern side of the Pennines in the rich Vale of York. Each provides a highway from southern Britain to Scotland, and it was this function which gave the region of Lancastria an importance during the earliest days of government in Britain. The Cheshire Plain is delimited on the south by the approach from either side (in Shropshire and Staffordshire respectively) of spurs detached from the Welsh and Pennine Highlands, though a series of gaps permits easy communication with the English Midlands or with the middle Severn valley.

The counterpart to the plain of Lancastria is provided by highland tracts consisting of (*a*) the western flanks of the Pennines southwards from the main trans-Pennine gap (which is indicated by the upper course of the River Aire and is known as the Craven or Aire Gap) to the Cheshire-Staffordshire border, (*b*) the two great western spurs of the Pennines, the Fells of Bowland and the upland of greater industrial importance and of lower average altitude—the Rossendale Fells.

* Mr King has written the sections on "Merseyside and Deeside," "Soils and Agriculture," and Miss Kershaw the section on "The Potteries". The remainder of the essay is by Mr Fitzgerald.

† A district to the north of the lower Ribble.

Thus Lancastria comprises two contrasting physical divisions. In industrial and commercial exploitation, however, the two divisions of lowland and upland have been marked by so close a relationship that each has become complementary to the other. Moreover, when comparing Lancastria with other regions of Britain we find that the human activities of the former, as expressed in commerce and industry, possess a quite distinctive character.

It is convenient to define the eastern limits of Lancastria by the Lancashire-Yorkshire boundary which keeps to the summit moorlands of the Pennines before trending westwards across the Ribble basin to the Bowland Fells. Yet it should not be forgotten that the textile industries on either side of that boundary, in east Lancashire and the West Riding of Yorkshire, are related, especially in the processes of bleaching. Modern transport has minimised the barrier character of the Pennines and permitted the flow of commerce and, to a lesser extent, of industry from one flank of the range to the other.

Territory in Lancashire north of the River Lune is more suitably included in the region of Cumbria. The town of Lancaster, at the tidal limit of the Lune estuary, and at a point where the western projection of the Bowland Forest Fells reduces the plain to the merest coastal strip, marks the northern limit of our region. Though it has granted its name to the Shire, Lancaster has lost importance in affairs of county administration to Preston, more conveniently, because more centrally, placed in relation to the plain of west Lancashire as a whole.

Of excellent fertility though wide areas of the plain of Lancastria undoubtedly are, it is not agriculture which gives the region an outstanding significance in modern industrial England. Within its narrow limits is concentrated the greatest single manufacturing industry of the world, a greatness which is shown even more by complex organisation, including subdivision into highly specialised processes for which particular districts or towns are renowned, than by its enormous output of cotton textiles.

As a result of long and extensive exploitation of the mineral wealth of the region and of the great industrial establishment which that exploitation has encouraged, there is present, particularly between the Rossendale Fells and the Mersey, a density of population which is not exceeded or even equalled by any other industrial region of Britain. The evils of over-crowding, of lack of care in the selection of sites for particular utilities, are features which an awakened public conscience is attempting to remedy with the aid of regional planning organisations.

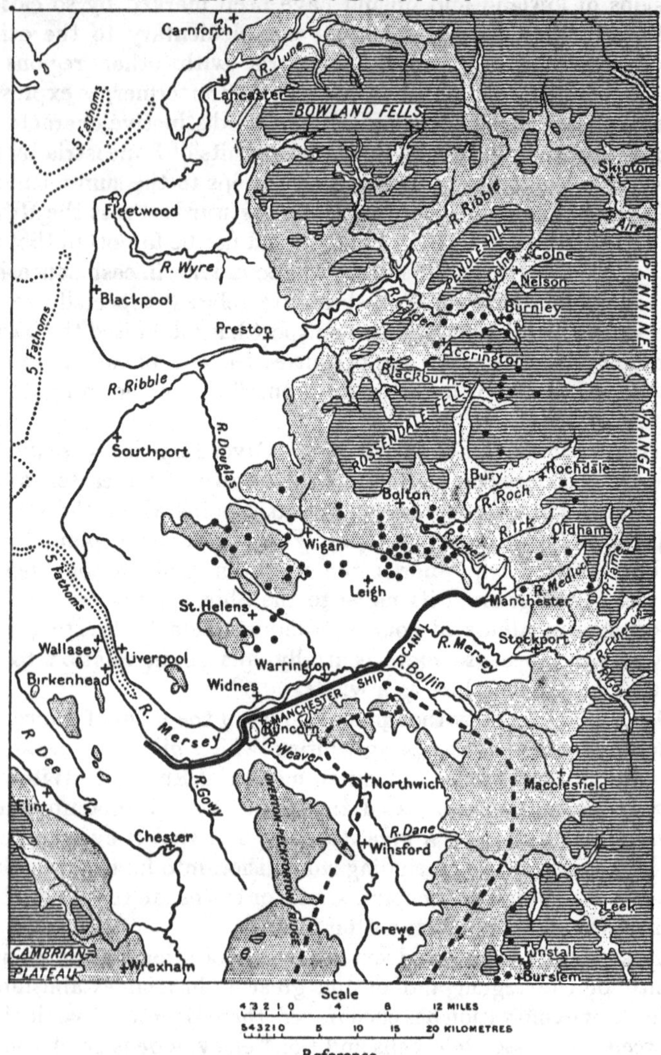

Fig. 85. Sketch-map showing relief, towns, coalfields, etc.

Lancastria, through its choice of industrial specialisation in cotton fabrics, is completely dependent for its raw material on foreign supplies which must be conveyed several thousand miles by sea. This dependence is indicated by the growth of Liverpool, not merely the sole port of first consequence along the extensive coast-line of Lancastria but the principal ocean gateway of western Britain and the leading port of the world where export trade alone is considered.

With the exception of the River Dee, whose headwaters flow from the interior of the Welsh Plateau, all the major streams of Lancastria rise in the Pennines or on the Fells of Bowland and Rossendale. The Mersey, Ribble and Lune river systems lead in a general south-westerly direction to the sea across plains, of which the Mersey lowland is easily the most extensive. Only at its seaward extremity is the Lune valley continuous with the narrow northern extension of the plain of west Lancashire, and where the coastal plain and the lower Lune valley merge Lancaster serves as the focus for north-south lines of communication. On the north and east the upper Lune valley is narrowly confined between the Cumbrian and Pennine highlands, while to the south the broad upland of the Bowland Fells, rising to well-nigh 2000 feet (600 m.), obstructs the principal routes linking the middle sections of the Lune and Ribble basins. On the western or seaward side of the Rossendale Fells the descent is gradual to the plain of south-west Lancashire, a wide lowland, and there is no serious obstacle to easy transport between the lower Mersey and lower Ribble valleys.

Physical Conditions

Considered geologically, the plain of Lancastria is formed mainly of the Trias series, but over the greater part glacial and post-glacial deposits have modified both the scenery and the soil character. This plain is markedly level in places and particularly so behind the coastal belt of dunes in south-west Lancashire. Even where undulating, as in Cheshire, it rarely rises more than 300 ft. (90 m.) from sea-level. In mid-Cheshire occurs a minor but still prominent interruption of the plain level: separating the Dee and Mersey-Weaver basins is a long broken escarpment, sometimes known as the Overton-Peckforton ridge rising near its southern limit to 700 ft. (213 m.). In the Upper Trias rocks, especially those of Cheshire east of this ridge, the Keuper Marl beds beneath the glacial sands and clays contain abundant salt deposits which have provided the basis of an extensive chemical industry in the neighbourhood.

The Lancastrian highlands of the western Pennines, Bowland and Rossendale, represent foldings of an epoch earlier than that in which the Triassic rocks of the plain were deposited. Their rocks are of Carboniferous age, and the most recent deposits of that age, the Millstone Grits and Coal Measures, are widely exposed. From an industrial standpoint the latter feature is very significant, for the Grits and Coal Measures are the most useful of the Carboniferous series. Lower Carboniferous deposits, known as Mountain Limestone, are well represented to the north of the Ribble and form some of the Fell country of Bowland Forest. All highlands south of the Ribble, with the exception of the minor uplands of the Lancastrian Plain, are developed from the Millstone Grits and Coal Measures, and the latter are not discovered to any extent in Lancastria north of the Ribble. A typical representative of the Upper Carboniferous rocks is a coarse-grained sandstone which is exposed over great stretches of high moorland.

Apart from their value as building stone, paving stone and road metal the Grits form a gathering ground for innumerable streams of abundant soft water which is one of the determining factors in the localisation of the textile industry. It is evident that the Coal Measures once had a continuous extension between the coalfields of south Lancashire and the West Riding of Yorkshire and that erosion has completely removed the coal from the summits of the Pennine arch where the older Grits are now exposed. From the arch of the Rossendale upland, the axis of whose folding is north-east to south-west, the Coal Measures have been only partially denuded. Over the lower south-western extremity of the upland certain of the richer Coal Measure series are continuous, and there are situated the more important coal-mining centres of the Lancastrian field. East of Wigan the Middle Coal Measures, the most valuable deposits of the series, extend along the southern flanks of the Rossendale Fells and on the edge of the plain are overlapped by the Triassic rocks. In conformity with the embayed configuration of the uplands to the north and east of Manchester the coalfield extends on the eastern side of that city and, southwards, gradually diminishes in eastern Cheshire. On the northern flanks of the Rossendale Fells coal has a less extensive development and is mainly confined to the Burnley Basin, a coalfield of importance though secondary to that of south Lancashire.

The Lancashire coalfield serves local industry and its overseas export trade is almost negligible compared with that of South Wales, a field of very much greater output. There is, however, considerable coastwise trade which includes Ireland. Production of coal has fluctuated in recent years, but a declining tendency is observed: the average annual output is rarely much greater than 17 million tons, a considerably lower figure than that for pre-War years.

Climate

Both in agriculture and the chief manufacturing industry climate is a critical factor. Lancastria receives full benefit from its western position in Britain and from the cyclonic regime of the North Atlantic Ocean. Characteristics which its climate shares with that of other western districts are a small seasonal range of temperature, a high degree of relative humidity and an abundant, well-distributed rainfall*. Temperature shows an average (reduced to sea-level) of 40° F. (4°·4 C.) for January and of 60°–61° F. (15°·6–16°·1 C.) for July. Mild temperature conditions and high humidity of the atmosphere may not be of such consequence for the local textile industry as was at one time supposed, but certainly they are of advantage to the processes of cotton spinning†. Moreover, the abundance of the rainfall and especially the equalisation of precipitation throughout the year are of outstanding benefit to a region which requires an almost unlimited supply of pure water for its major industry.

As a result of the diversified relief in Lancastria there are considerable variations in the amount of rainfall between one district and another. On the lowland of south-western Cheshire the comparatively light amount of 26 in. (650 mm.) is due to the rain-shadow caused by the Cambrian Plateau. Excepting certain local reductions, such as that provided immediately to the east of the Overton-Peckforton ridge by the rain-shadow of the upland, there is a progressive increase of rainfall from the south-west of the Cheshire Plain to the southern slopes of the Rossendale Fells and the western flanks of the Pennines, so that north of Manchester in the upper valley of the River Irwell the annual figure is over 50 in. (1250 mm.). In northern Lancashire there is a similar feature of rainfall increase from the coastal plain to the highland on the east and north-east (in this instance the Bowland Fells).

Abundant rainfall on the western slopes of the southern Pennines and on the Rossendale Fells maintains numerous streams of considerable volume which form the headwaters of the Mersey and Ribble. These streams carry to the plain much water that has

* On the plain of Cheshire a feature of rainfall distribution which is unusual in western Britain is that the highest seasonal precipitation occurs in the summer half-year.

† An important discussion of the influence of physical conditions upon the localisation of the cotton industry is contained in the *Jour. Manchester Geog. Soc.* vol. XLIII, 1927, pp. 8–29. The other papers in this Volume, which is devoted to "Lancastria," should also be consulted.

flowed from the Grits and Coal Measures and has the quality of softness which makes it invaluable for industrial uses. It is a fortunate circumstance that the heaviest rainfall of Lancastria occurs on the Grit and Coal Measure uplands rather than on the Triassic plain, where the supplies of stream water are smaller and liable to possess the unfavourable quality of hardness.

Sub-Regions of Lancastria

In passing to discuss in outline the industrial and commercial significance of Lancastria we are able, for convenience of treatment and without obscuring the essential unity of the region, to distinguish certain prominent sub-regions in each of which the relationships of industrial life to physical and other geographical conditions possess a distinctive character.

South-East Lancashire—the Manchester District

This region of congested urban communities has well-defined limits and generally coincides with the upper division of the Mersey basin. On the east the Pennine moorlands represent a sparsely-settled zone separating the textile provinces of east Lancashire and the West Riding of Yorkshire; on the west, if the industrialised zone around Warrington and Widnes be excepted, a comparatively thinly-populated rural belt intervenes between the lower Mersey-side nucleus of population and that which has Manchester for its centre, and much of this land has hitherto been difficult of settlement owing to wide moss tracts which extend between Manchester and Warrington; on the north the crests of the Rossendale Fells, reaching 1500 ft. (450 m.), serve in a general way to separate the predominantly cotton-spinning centres concentrated round Manchester from the chief weaving towns, all situated in the Ribble basin.

The points of convergence of the Mersey head-streams with each other and with the parent river granted early importance to certain of the modern industrial centres, for the tributary valleys were utilised both by lines of communication and by industrial establishments. Manchester and Stockport illustrate this feature. As a result of the rapid expansion of the textile industry during the 19th century population which was at first confined to the banks of the streams proceeded to utilise the land between the tributary rivers. This merging of the population of neighbouring towns is a prominent feature in south-east Lancashire. There is scarcely a perceptible break in the spread of dense urban population between Manchester and the large neighbouring towns of Oldham and Stockport, while Salford, a city of 250,000 inhabitants, is inseparably joined to its dominating partner in the metropolitan concentration of Greater Manchester (Fig. 36).

At the beginning of the Industrial Age the water of the Mersey head-streams was employed both as a source of power and for various processes of textile manufacture. The second of these functions is still of outstanding importance, the lime-free water being fully utilised in the bleaching, dyeing and calico-printing

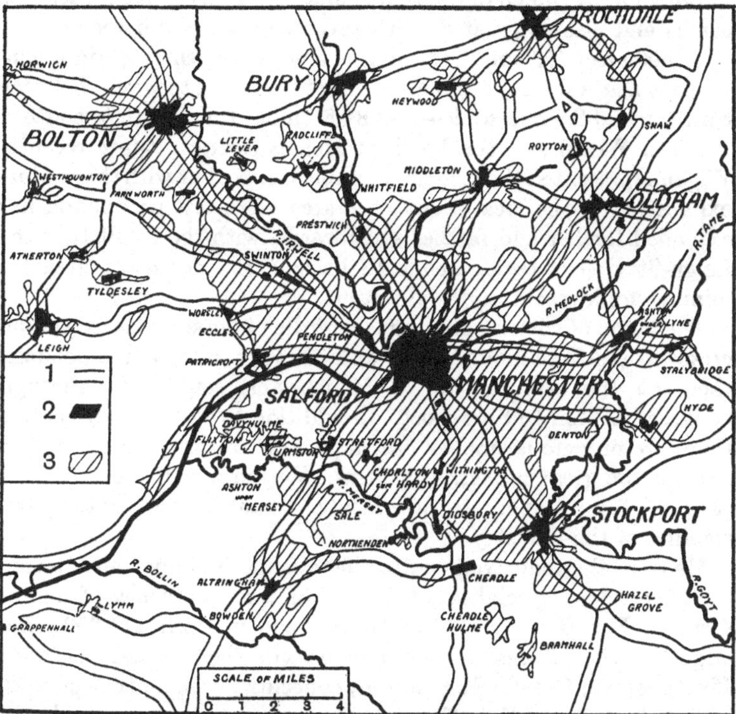

Fig. 36. The Manchester-Salford nucleus. 1, Roads in 1930; 2, Areas of close urban settlement in 1830; 3, Areas of close urban settlement in 1929.

processes. In the immediate neighbourhood of the head-streams and main course of the upper Irwell the clustering of works engaged in these processes is strongly emphasised, and indeed for cotton manufacturing as a whole the Irwell basin is to be considered the very heart of industrial Lancashire. The concentration of industrial undertakings in the valleys both of south-east Lancashire and of the weaving area immediately to the north of the crest-line of the Rossendale Fells is partly explained by the presence of abundant supplies of soft water.

The vast coal resources in the immediate neighbourhood are a further advantage which helps to explain the location of the cotton industry. Many of the more important spinning towns such as Bolton, Bury, Oldham and Manchester stand actually on developed portions of the South Lancashire Field. It has been estimated that for the 200 square miles (520 km.^{2}) of accessible Coal Measures some four thousand million tons of coal represent the available resources. Unfortunately, much of the coal is mined in thin or much disturbed seams and also at great depths, several of the pits descending to over 3000 feet (920 m.)*. Mining is chiefly concentrated in a continuous belt a little more than 20 miles from east to west—between Swinton and Skelmersdale—and 8 or 9 miles from north to south—between Wigan and the Leigh-St Helens district. Quite recently Wigan represented the principal mining centre, but with the gradual exhaustion of the local seams, the area of richest production has extended farther and farther southwards, with the result that the Leigh-St Helens region tends to become the dominant area for mining employment and coal output.

The home of cotton spinning is a girdle of large towns extending crescent-like from Bolton eastwards through Bury and Rochdale, then, on the eastern side of Manchester, southwards through Oldham, Ashton to Stockport. In addition to these larger centres there are numerous smaller but important towns grouped around each of the great industrial centres. Within a radius of 18 miles (29 km.) from Manchester are located about 80 per cent. of the spindles of the industry.

Many of the industrial centres have a special importance owing to independent specialisation in one or more branches of manufacture. For example, Bolton (population in 1921, 177,000), the largest of the Lancashire boroughs outside the "capital" cities of Liverpool and Greater Manchester, concentrates upon the spinning of fine, long-staple Egyptian cotton and has a prominence also in the bleaching industry; while Oldham (population 147,000), greater than Bolton in its output of yarn, is occupied more with coarse and fine "counts," though the tendency now apparent towards fine spinning will probably be very marked in the near future†. The northern and eastern districts of Manchester still gain for the metropolis a high place among the spinning towns, but the chief function of the city is concerned with the commercial control of

* More than 50 per cent. of the output of coal is produced from seams less than 4 ft. thick.

† Of the 60 million spindles of Lancashire in 1928, Oldham claimed rather less than one-third.

the industry. Manchester is both the repository of the yarns and fabrics of the industry as a whole and the one great market for these finished products. Liverpool maintains its place as the principal market for raw cotton, especially for the American varieties which provide the staple supply of the industry. Within Manchester, in addition to the warehousing and marketing of raw and manufactured cotton, there is the big industry of packing and "making up" of the finished products, and this requires a delicate sense of the varying needs of Lancashire's foreign customers in all parts of the world.

The central position of Manchester within a girdle of industrial towns goes far to explain its rank as the commercial capital of south-east Lancastria. After the Industrial Revolution it became the distributing centre for the raw cotton and manufactured products, and this function was facilitated by the early established railway connections with the port of Liverpool. Towards the close of the 19th century there were clear signs that the city was about to sacrifice its own industrial character as a cotton-spinning centre through serving increasingly the commercial needs of its region; but, largely as a result of new advantages brought by the construction of a ship canal linking Manchester to the ocean, a revival of industry began and this time in the realm of engineering, with the result that by 1921 there were twice as many persons engaged in metallurgical and engineering industries as in textile processes.

Before the construction of the Ship Canal along the line of the Mersey-Irwell at the close of last century the cotton industry laboured under the disadvantage of big additions to the cost of imported cotton incurred both at the port of Liverpool and on the railway to Manchester (by the three alternative routes). The Ship Canal is to be considered as an extension of the dock system of lower Merseyside giving to Manchester a new function—that of a sea-port. Over its length of 35 miles (56 km.) a water depth of 28 ft. (8·5 m.) is now provided and vessels with a dead-weight capacity of 12,000 tons are using the waterway daily. More than one-sixth of the cotton entering Britain moves directly to the Manchester Docks, and in this amount is included more than 50 per cent. of the Egyptian cotton imports, used largely in the fine spinning of the Bolton district. But for other products also—petroleum is a notable instance—Manchester has become the gateway to a densely-settled hinterland. Large tracts of practically unused land close to the Ship Canal suddenly became attractive to manufacturers in quest of accessible sites for their establishments. The park-land estate of Trafford Park in 1894 gained a frontage on the newly-constructed Canal and Manchester Docks with the result that to-day about 130 industrial undertakings are established there and its day-time population is over 30,000 work-people. A similar development occurred at many other points along the line of the Ship Canal and the growth of industry in this neighbourhood is only at an early stage as yet.

Accompanying this new industrial expansion there proceeded a great extension of the city boundaries in order to absorb rapidly developing residential districts. The southern half of the area of the present city of Manchester represents the additions made in the late 19th and early 20th centuries*, and by this extension to the south Manchester spread from its nucleus around the Irwell, Irk and Medlock valleys right up to the banks of the Mersey and across an almost featureless plain falling almost imperceptibly to the west. The Mersey is only a temporary check to the southward expansion of the city, for the need to accommodate an ever-increasing population by the provision of new housing estates will lead inevitably to the incorporation of neighbouring rural districts in Cheshire.

The Middle Mersey Industrial Region

The cotton-spinning industry of south-east Lancashire overlaps on to the Cheshire Plain between Stockport and Macclesfield, both of these towns being situated on the edge of the foot-hill country of the western Pennines. Westwards from the Stockport-Macclesfield line we pass rapidly into a pastoral district. The rural character is maintained for a few miles, but gives place in the district of the River Weaver to a vigorous industrial development based on the exploitation of the Cheshire Salt Field (*v.* Fig. 35).

The position of the Lancastrian section of the salt field has already been indicated. It occupies essentially the Weaver valley and is linked to the Mersey waterway by the Weaver Navigation. The River Weaver, which was canalised in the early 18th century mainly for the purpose of transporting salt to the Mersey estuary, shares with the Aire and Calder system of canals and canalised rivers the distinction of being one of the few inland waterways carrying heavy traffic in England to-day. From the north of Cheshire, at Lymm close to the River Mersey, the field stretches southwards with increasing width to the Northwich district. Farther south, both in southern Cheshire and northern Shropshire, the salt deposits have an extensive, if not continuous, develop-

* Rusholme was added in 1885, Moss Side, Withington, Burnage and Didsbury in 1904 while, on the east, densely populated Gorton and Levenshulme were not incorporated until 1909.

ment. Production reaches a maximum between Nantwich and Northwich, the latter town being the industrial focus of the field, while Winsford farther south is another important centre of salt extraction. The deposits are found in two beds each some 100 feet (30 m.) in thickness and varying in depth between 200 and 500 feet below the land surface.

From Roman times down to the present day salt has been a most valuable product of the Cheshire Plain. The shipments from Liverpool between 1870 and 1880, when the overseas salt trade was at its height, represented 10 million tons, one-quarter of this quantity going to British India and a similar fraction to the ports of eastern North America. These destinations are still important in the export trade: much salt is sent to Nova Scotia and Newfoundland for use in agriculture and the fishing industry. In recent times the method of extracting the salt has altered, and now it is by the regulated inflow of water to the salt beds and by the collection of the brine so formed that the mineral is obtained.

Apart from domestic use the utilisation of salt in industry takes many forms and is responsible for the establishment of large-scale heavy chemical manufacture both on the salt field and nearby, on either bank of the Mersey, at Widnes and Runcorn. Furthermore, salt is the foundation of the manufacture of bleaching products, an industry which has been greatly stimulated by demands for such products from the neighbouring cotton region.

The mid-Mersey heavy chemical region is now the home of alkali manufacture in Great Britain. When the industry was founded local salt was utilised in a now largely abandoned process for the production of chlorine, an important bleaching agent required by the cotton industry. Other and later methods have been fully exploited at such towns as Northwich and the twin towns of Widnes and Runcorn on opposite banks of the Mersey close to where the Overton ridge approaches the river. Runcorn and Widnes, each devoted to a variety of chemical enterprises, provided the nucleus for the establishment of a great chemical corporation whose works are now well distributed over the salt field and in its near neighbourhood.

The chemical industry of the middle Mersey supplies materials not only for the cotton industry but also for a great tanning industry located in the Runcorn district, the principal centre in Britain for leather output; for glass manufacture, mainly at St Helens, a northern outpost of the middle Mersey chemical region and a principal centre of the South Lancashire Coalfield; and for soap manufacture at Widnes, Port Sunlight—on the Mersey near Liverpool—and

Warrington. The latter ancient town has a wide and varied basis of industrial interests, though some of its 19th-century occupations have disappeared. Shipbuilding and the manufacture of sail-cloth until recently served to remind of former port activities. Wire-drawing is the largest single industry, while the cotton-spinning zone of Lancashire has here a south-western outpost though the production of yarns and fabrics is very small compared to that of any other town of similar size concentrating upon cotton processes.

In its relations to communications crossing the plain of Lancastria Warrington has always retained a critical importance. Centrally situated on the plain it is the principal focus of north-south routes which require to avoid both the wide estuarine reaches of the Mersey and the moss lands (*e.g.* Chat Moss), which occupy a very large proportion of the stretch of country between Warrington and the outskirts of Greater Manchester. The site of Warrington gained its initial importance as the first point up-stream where bridging was easy, and indeed in ancient times a ford (Latchford) provided a crossing about a mile above the earliest bridge. Bordering the right bank of the Mersey Warrington is separated from the Ship Canal by a flat-floored valley, approximately one mile wide, across which the great loops of the Mersey meanders are coiled; and this avoidance of the town by the most recently constructed highway between Manchester and the sea is rather to the detriment of Warrington. On the other hand, Warrington retains the advantage of providing the passage-way for the only direct highroad so far constructed between Manchester and Liverpool.

The very recent development of Widnes and Runcorn as industrial centres is due in no small measure to the construction of the Manchester Ship Canal which passes between them. Now that Runcorn is a port it has become a convenient store-house for chemicals awaiting export from the mid-Mersey region. The Weaver Navigation which leads to the Mersey at Runcorn from the centre of the salt field carries much of the raw and manufactured products of the heavy chemical industry.

Merseyside and Deeside

H. King

Viewed in its larger aspect the southern coastal region of Lancastria has always possessed great focal importance. The obstacle of the Mersey marshes restricted intercourse between the North and South, and Chester was the natural seaward terminus of the route through the Cheshire Gate, especially as Cheshire's relations were with the Midlands and South rather than with the North. Chester commanded the routes across forested North Wales and formed the base for coastwise movement to the North and West and, later, to Ireland. But the progressive silting of the Dee, in spite of several preventive measures and the creation of new outports under the control of Chester, had led to the rise of Liverpool as the recognised port of the North-West before the great expansion of maritime commerce which followed the discovery of the New World.

Liverpool started with the severe handicap of natural disadvantages. The estuary was choked with sandbanks and the only safe approach was by a channel along the north Wirral shore. The river was also difficult of negotiation by small sailing craft. In the second place, there was a dearth of good natural arteries of communication with the interior. The Mersey was navigable only as far as Warrington and the Weaver to a point a few miles above Frodsham. By land it was about 18 and 36 miles (29 and 58 km.) to the North Road at Warrington and Preston respectively, and the 40 miles which separated it from the increasingly prosperous industrial valleys of the Pennines had to be negotiated over very bad roads. On the other hand, it shared with Bristol the advantages of the new orientation of trade which followed the opening up of the New World; and later, during Britain's troubles with Holland and France, the greater security of its approaches deflected traffic from London and the Channel ports. It produced men capable of visualising and seizing new opportunities, and the working of the notorious "Great Trade Triangle"* brought commercial and industrial prosperity; whilst at the same time the coastwise and Irish trades were not neglected.

The construction of canals and turnpike roads in the 18th and 19th centuries eliminated the difficulties of access to the hinterland and temporarily adjusted the problem of transport between the port and its markets. Eastern Lancastria and to a large extent the Midlands and the West Riding of Yorkshire came within its sphere. The dock system, expanded as occasion demanded on the site of former marshes, now extends for several miles along the right bank of the estuary. In the middle of last century port developments began on the Wirral shores of the river. Birkenhead rapidly developed along the banks of Wallasey Pool and the closest

* Liverpool ships sailed to West Africa with cargoes of trinkets, muskets and spirits, and there took on board slaves for the West Indies. They returned to Liverpool with cargoes of molasses, tobacco and cotton. (See P. M. Roxby in *Geography*, 1927, p. 94.)

connection between the opposite shores is maintained by ferries. More recently Bromborough Pool, another creek in the estuary, has been developed in the interests of the extensive and varied activities of Port Sunlight. Still farther up-stream the rise of Ellesmere Port is witness of the advantages to industry of a site which marks the junction of ocean and inland waterways. Industrial activity has progressed with the development of shipping, since Merseyside possesses unique advantages both for assembling raw materials and marketing the finished products either at home or abroad. The industries all exhibit in some form their dependence on the port. They include flour-milling, sugar-refining, shipbuilding and repairing, general engineering and oil-extracting; the manufacture of margarine, oil-cake, soap, chemicals, tobacco, artificial silk, ships' tackle and foodstuffs of all kinds. In addition the inhabitants engage in a multitude of trades, all necessary to the upkeep of a mercantile marine and of the fabric of an immense conurbation.

The monopoly of the water-front by docks and warehouses has caused an inland localisation of Liverpool's industries. On the Wirral side of the estuary the same tendency, though still evident, is perhaps not quite so marked. Because of this peculiarity lighters play an insignificant part in trans-shipment and there is little doubt that the relatively high cost of alternative means of transport is detrimental both to commerce and industry. The problem of transport is always present and its most recent features are: first, the building of the Manchester Ship Canal; secondly, the linking up of the opposite shores of the estuary by the Mersey Tunnel in order to avoid the congestion and delay occasioned in the existing system of luggage ferries; thirdly, the concentration of municipal attention on road construction, perhaps not unconnected with Manchester's undoubted success in capturing, through the agency of the Ship Canal, the traffic of the inland towns from Wigan eastwards; fourthly, the dredging of the Mersey channel during the last thirty years in view of the increasing size of modern liners. Liverpool, owing to a combination of causes, has recently lost to Southampton much of the trans-Atlantic passenger traffic. It is claimed, however, that the new Gladstone Dock, at the lowest point of the dock system, is said to be capable of meeting all the requirements of the world's largest shipping, and will to some extent mitigate existing congestion.

The 19th century thus saw a phenomenal increase in Liverpool's commerce and industry, and the city now covers an immense area. In the early stages of its growth a site of boundless possibilities was irretrievably sacrificed to the need for easy access to work. The unfortunate result and the social problems, of which some are directly attributable to the city's haphazard expansion, have aroused public concern; and increasing attention has recently been given to town planning in all its aspects. On the outer margins of the city there has been noble provision of open spaces, and the expansion of the suburbs is now under considerable public control. Many of the suburbs of the port, such as Wirral and the narrow belt of sand-dunes along the west Lancashire coast, come under the control of outside authorities.

Conditions along the Dee estuary have been in comparison fairly stationary. The region is the natural focus of Flintshire and Denbighshire and there has always been some industrial activity based on the coal and other minerals which gravitate towards its shores. In addition to its administrative functions Chester has varied light industries. The smelting of lead and copper, based on the mineral resources of the hills behind, are old-established industries along the western bank, and recent ventures in iron-smelting and artificial silk manufacturing are indicative of faith in an industrial future. It would appear that the fullest realisation of such hopes of development depends on several factors, the most vital of which is perhaps the solution of the transport problem, either by improvement of the Dee navigation or by some alternative scheme.

The Ribble Basin

W. Fitzgerald

Within the basin of the Ribble is located the northern division of the cotton province of Lancastria, including all the principal weaving centres. From the industrial standpoint the most important district of the Ribble basin lies on the northern slopes of the Fells of Rossendale Forest. The western limit of industrial concentration is at Preston, and from there to Colne on the western flanks of the Pennines a chain of towns, comparable to the girdle of spinning towns in south Lancashire, extends eastwards. It should be noted that the majority of the weaving centres are situated towards the southern margin of the Ribble basin.

It is convenient to distinguish between (*a*) the upper and middle districts of the Ribble basin, where the tributary valleys are numerous but narrow and where relief is rugged and definitely that of a highland country, and (*b*) the lower Ribble valley, wide and open to the sea, with Preston the port and dominating industrial centre.

The main stream of the Ribble system flows southwards between the western Pennines and the Bowland Fells, but from near Hellifield it follows a south-westward course to the sea. The notable absence of towns of importance in the immediate neighbourhood of the main stream of the Ribble until Preston is reached forms a very interesting feature of urban localisation within the Ribble basin*. Even the apparent nodality of the place of junction of the Ribble, Hodder and Calder valleys has not stimulated the development of a town or smaller nucleus. The Colne-Calder valleys with drainage tributary to the Ribble are separated from the main stream above the point of convergence by the high mass of Pendle Hill (1800 ft.). In addition to its industrial prominence, indicated by the clustering of great weaving centres which include Burnley, Nelson and Colne, the Colne-Calder depression, leading to the Craven or Aire Gap in the Pennines, provides comparatively easy access for routes between the West Lancashire Plain and the West Riding industrial region. Farther west than Burnley but still on the northern slopes of Rossendale is the group of big industrial towns which includes Blackburn, Darwen and Accrington, all served by tributary streams of the Ribble.

Though now devoted to the manufacture of cotton goods the towns of the Colne valley—Nelson and Colne—were in earlier times engaged in the worsted industry, which became localised in the West Riding but provided an offshoot in Lancashire through the Aire Gap. The concentration on weaving and lack of important subsidiary industries is a marked feature of Nelson and Colne. Census returns for these towns indicate that the numbers of men and women engaged in textile occupations are very nearly equal. This position is in strong contrast with that of Accrington, in the Blackburn group of towns, where more than twice as many women as men are employed in textile occupations, owing to the attractions for male labour of coal-mining and metal-working.

Burnley (population in 1921, 103,000), on the Calder above its junction with the Colne, is the rival of Blackburn for predominance in the weaving industry. Its employed population is mainly concerned with cotton fabrics, yet though weaving is the mainstay of industrial life coal-mining and engineering absorb one-third of the available male labour supply. The textiles produced are mainly the cheaper grades of cotton cloths, including the famous "Burnley Lumps" and these goods are intended primarily for the Indian and Chinese markets. There is a tendency, however, for the cheaper

* Clitheroe, a cotton town standing back from the Ribble in the neighbourhood of Pendle Hill is the only centre of any—and even then a declining—significance.

qualities of weave to be replaced in importance by finer and more expensive fabrics. As the result of a whole series of difficulties arising in the Far Eastern markets the Lancashire textile industry has been obliged to accept a much diminished demand from these parts of the world. The Burnley weavers, like the Oldham spinners, are beginning to attend to the more discriminating buyers of cotton goods, and in the finer fabrics, without doubt, Lancashire will be able for long to maintain its pre-eminence.

In the Blackburn-Accrington group of towns the basis of industrial life is wider and more varied than in the Colne-Calder zone. Cotton processes again are paramount, but weaving, more so even than in Burnley, is a woman's occupation. A very considerable importance attaches to calico-printing, an industry which, so far as the Ribble basin is concerned, has been localised in and around Accrington since its introduction to Lancashire about the end of the 18th century. Blackburn—the town of the Blake Burn and rather larger than Burnley in population—possesses commercial significance in relation to the organisation of the weaving industry, and its intermediate position between the extremities of the weaving zone is of definite advantage in this connection. The wide area served by its Chamber of Commerce is evidence of this commercial function and extends to Accrington, Darwen, Chorley and Clitheroe. Weaving in Blackburn is characterised by greater variety of fabrics than in the corresponding industry of Burnley: fine cloths as well as coarse qualities ("dhooties") for Far Eastern markets are produced. Spinning is another notable feature of the town and here we may point to an interesting feature of the distribution of spinning in the Ribble basin as a whole. This occupation increases in importance from east to west: it is almost excluded from the Colne valley towns, is an appreciable feature of Burnley (500,000 spindles), receives greater attention in Accrington (700,000 spindles) and reaches its maximum importance in the Blackburn (over 1 million spindles) and Preston districts (1,900,000 spindles). In Blackburn and neighbouring towns, more particularly Darwen, the manufacture of paper has become a prominent industry, largely as a result of the very suitable water supply from the Rossendale hills, and much of the wood-pulp and esparto grass used is imported by the neighbouring port of Preston.

In the lower Ribble valley Preston, standing well apart from the Blackburn and Burnley groups of industrial towns, ranks very high in the production of textiles and especially in the quality of its fabrics. This town of 123,000 inhabitants produces the finest and most expensive specimens of the weaver's craft; and in very recent years, with the sudden popularity of artificial silk Preston has quickly developed a new side to its textile activities. The proportion of spinning to weaving is larger here than in the other weaving towns and still there is room for a very important and many-sided engineering industry. The manufacture of the means of modern transport—motor vehicles, steam tractors, tram-cars—is appropriately associated with a town which has functioned as one of the principal nodal points in the communications of northern England.

Long before Manchester and Liverpool had gained prominence Preston was the recognised focus of social and commercial activities in Lancashire. Its position on road and railway routes is comparable with that of Warrington on the Mersey. The main railway line to Scotland west of the Pennines leads from Crewe and Warrington to Preston, on the right bank of the Ribble, and then continues almost due north to Lancaster. At Preston this north-south trunk line meets the east-west system (the Lancashire and Yorkshire section of the London, Midland and Scottish Railway) which, by traversing the Pennine Highland, serves the needs of the two great textile regions of northern England.

Preston retains some importance as a port, though its maritime activities are very minor in comparison with those of Merseyside. The estuary of the Ribble is much encumbered with sandbanks, but one channel, the New Cut, has been straightened and dredged so that a large dock of 40 acres (16 ha.) situated 16 miles (28 km.) from the sea can accommodate the smaller types of cargo vessels, not exceeding 5000 tons carrying capacity. New dredging plans are in preparation to render Preston accessible to large cargo vessels. Even if bereft of its maritime and industrial activities Preston would still gain prominence in northern England as one of the principal market towns. It serves as the collecting and distributing centre for the stock and agricultural produce of a wide area including the very fertile district of Fylde (*i.e.* the Field) situated between the Ribble estuary and the River Wyre. The eminence of Preston in commerce and industry is recognised by its rank as the administrative centre of Lancashire outside the County Boroughs.

On the Lancashire coast north of the Ribble two urban centres merit some mention. Fleetwood, where the Wyre enters the sea, has become in a short space of years the foremost fishing port of western Britain. Its very modern fleet of trawlers engages in local and distant fishing and the herring catch reaches vast dimensions. Farther south, Blackpool, on the straight low coast of the Fylde, has a population which, fluctuating seasonally, is easily greatest in the summer half-year, when its resort attractions are sought by the textile workers of both Lancashire and Yorkshire.

Lancaster and the Lower Lune

Space does not permit of more than the briefest reference to the lower Lune valley of which Lancaster is the natural centre. This small though ancient town (population 40,000) at the northern limit of the Plain of Lancastria is of quite minor industrial importance when compared to Preston. It retains a very small activity in cotton spinning, but its main industrial occupation is the manufacture of linoleum and oil-cloth. A small importance also attaches to its port activities. The Lune is tidal as far up-stream as Lancaster, but the estuary is silted except for a shallow channel which permits coasting vessels of not more than 800 tons to use the port of Lancaster, 10 miles (16 km.) from the sea. Glasson, much farther down the estuary, has accommodation for larger vessels.

Lancaster, gathered round an eminence on the southern bank of the Lune, was chosen as the site for a military station during the Roman occupation of western Britain. Throughout history the critical position of the town in regard to north-south communications on the western side of the Pennines has always been recognised. To-day the main line of railway from London to western Scotland crosses the Lune on the western side of Lancaster and then skirts the coast as far as Carnforth. Beyond there is a climb—one of the most difficult in the railway system of England—to the saddle between the Cumbrian and Pennine ranges. An alternative route by road leads from Lancaster up the Lune valley and after a steep ascent enters the head of the Eden valley, and so to Carlisle.

Soils and Agriculture

H. King

Thus far attention has been concentrated on vast conurbations in the east, north and south-west. Agriculture especially in Cheshire and west Lancashire equally forms part of the economic entity of Lancastria. Proximity to large markets has stimulated production from arable and pasture lands alike. The soils exhibit all shades of variation between the extremes of the relatively sterile rock-waste of the Lower Keuper, Millstone Grit and the Pendleside series on the one hand and the prolific arable light loams of west Lancashire and north Cheshire and the stronger grassland loams of the Keuper Marls on the other. "*In situ*" soils occur only on the summits and higher slopes of Bowland, the Pennines and Rossendale and on the Audlem anticlinal of east Cheshire and north Staffordshire. The latter, a high anticline trending NE.–SW., effectively blocked the progress of the ice-sheet from the Irish Sea area*. Consequently its dip-slope towards the south-east is drift-free, and in the Potteries presents a fault-broken sequence of Carboniferous rocks, followed near Longton and Stone by the Keuper Marls. The rest of Lancastria is thickly covered with glacial drift, excepting only the more elevated Triassic outcrops, of Wirral and Delamere particularly, and occasional outcrops of Lower Carboniferous which indicate the continuation, after a slight change in direction, of the axis of Rossendale towards Liverpool. In the centre and west of Cheshire, in south Lancashire and the Fylde the general effect of the drift has been to soften the contours; but in eastern Cheshire, at elevations above 200 ft. (60 m.), to the east of Delamere and around the uplands generally glacial sands and gravels in one of several forms—moraines, eskers, drumlins, deltaic fans—have occasioned topographic irregularities and determined the sites of settlements†.

The sands and gravels in their natural state yield mechanically weak soils which formerly supported only "heath" associations. The boulder-clays vary greatly both mechanically and chemically according to derivation. They may have a high lime content when derived from the Carboniferous Limestone as in the upper Ribble valley, or may be siliceous if they originate in grit country. Generally they form rich loams of mixed origin and varying colour. In west Lancashire and north Cheshire they yield warm dry loams and owe their origin mainly to the Trias. Farther north and east the tendency is towards a stronger loam and in the low-lying Dee valley the drift almost approaches the nature of a true clay. Everywhere they probably supported a woodland vegetation. Between the lower Ribble and a line drawn from Liverpool to St Helens, and sporadically to the south of it, there occurs above the boulder clay a deposit of very fine sand which is usually regarded as representing re-sorted glacial sands and the wind-scattered stuff of pristine sand dunes. The formation does not transgress the ridge which forms the left bank of the Douglas. Lying on the surface of the ill-drained plain, these sands are in turn often covered by still more recent accumulations of peat. Peat recurs extensively behind the coast north of the Alt and

* The effects of ice are seen at heights of about 1800 ft. (400 m.).

† See H. W. Ogden, "Cheshire Villages," *Jour. Manchester Geog. Soc.* vol. XXXIX, 1925, p. 125 *et seq.*

continues to the shore below the existing line of sand dunes. It is also found in various parts of Cheshire, extensively in Chat Moss and along the Mersey banks, in Leyland Moss to the east of the Douglas and in Marton Moss in the Fylde. Both formations are of cardinal importance in agriculture.

At the beginning of the Industrial Revolution the need for a high degree of local self-sufficiency everywhere compelled some arable cultivation prior to a protracted ley*. Large upland areas remained unreclaimed in spite of centuries of secular enclosure. The mosses and meres were largely undrained or at best formed the roughest of pasture, and the Mersey, Dee and Ribble estuaries were flanked by flats of salt-marsh. Industrial prosperity has completely revolutionised these conditions. It is unfortunately only too true that wanton waste due to reckless building and inundation of thousands of subsided acres in mining districts are amongst perhaps the minor general penalties of this prosperity. It is true also that incalculable harm is done by fumes, and that many farmers' accounts are balanced by the compensation received from the offending firms. But the amount reclaimed from river, morass and heath is at least as great as the loss, and certainly forms superior agricultural land. Judged solely from the standpoint of agriculture, industry is a boon. Intensive farming is rendered possible. The abundance of relatively cheap fertilisers goes far to provide the means; assured markets are the incentive.

Two areas of Lancastria are devoted almost exclusively to arable cultivation.

1. The arable farms of Cheshire† are found chiefly in the area defined by the Mersey on the north, the Delamere ridge on the west and a line drawn through Northwich to a point south of Altrincham on the south. An extension runs off towards the foothill region of the south (south-east Cheshire and north-west Staffordshire) where there is a considerable sandy element in the drift. The rich sandy loam is generally managed under a four course rotation; one root crop, usually potatoes, two cereal crops, chiefly oats but a little wheat, and grass for either one or two years. Cheshire has a dairying tradition and the dairy is there one of the chief interests of both arable and grass farming.

2. The arable belt of south-west Lancashire‡ lies west of the line of the Douglas-Glazebrook, and sweeps up the Mersey valley

* There was of course even then a pronounced concentration on animal husbandry in the north and on hilly districts generally.

† See figures for Daresbury in Table, p. 294.

‡ See figures for Ormskirk and Prescot.

towards Manchester. The soils are similar to those of north Cheshire, with a tendency to become lighter towards the north, where in addition there are considerable tracts of peaty land. The management also is broadly similar to that found in Cheshire, but two peculiarities should be noted. First, on the light "hungry" soils on the north a *three* course rotation is becoming popular. This consists of (1) potatoes, (2) oats or more rarely wheat, and then (3) grass for one year only*. Italian rye grass instead of perennial rye grass is sown and two heavy crops are usually obtained. Secondly, the dairy tends more and more to be excluded, and the crops to become the only end in view.

In both these arable areas the tendency during the 19th century was towards the almost complete elimination of permanent grassland. The farmer is now allowed perfect freedom in the disposal of his produce; his crops are almost entirely sold off the farm and manure purchased in the towns. Oats are the chief cereal crop, perhaps less on grounds of climate than of economics. Oats find a ready sale in the neighbouring towns and in the surrounding dairy districts, and oat straw is in great demand for litter. In the aggregate rye becomes a crop of some significance in south-east Cheshire and north-west Staffordshire. It is threshed by hand to avoid crushing the straw, which commands a high price for packing in the Potteries.

Potatoes form the chief root crop almost to the exclusion of turnips and mangolds. The peaty soils of south Lancashire are the potato soils *par excellence*, but potatoes figure in the rotation of all arable farms. Early varieties are produced on the sandy soils of Cheshire, which appears to be less liable to late frost than Lancashire.

In favourable locations the farmer cultivates intensively a small portion of his holding with a view to producing vegetables for the industrial markets. Vegetables are produced in this way in the Mersey valley, round Altrincham, Ormskirk and the environs of Liverpool. All kinds of soil are utilised for the purpose, but the mosses are particularly valuable and a small portion of Chat Moss was divided into 5 acre (2 ha.) farms which specialised in the production of celery, greens and vegetables for the Manchester market. The true market-gardening of the horticultural type is found at Wallasey and around Chester.

To the exigencies of space alone is due the discussion of the rest of Lancastria under the one heading of permanent grasslands. Strictly the region demands careful subdivision. In some regions

* See figures for Ormskirk.

such as Wirral and on the mixed soils near the Potteries, arable land is quite as important as grassland*. In other districts, especially in the moss districts of the Fylde, and in Leyland Moss in south Lancashire, considerable areas are devoted purely to arable cultivation. The other extreme is found in the Dee and Ribble valleys, the fell slopes of Bowland, south Cheshire and parts of north Staffordshire, where over 95 per cent. of the land is under permanent grass.

Lancashire and Cheshire are each more heavily stocked with cattle than any other county in Britain, but within the grassland areas there is some variation in the objectives of management. Everywhere the dairy is important and has developed enormously with the increasing demands of surrounding industrial markets. In the industrial valleys of the Pennines and of Rossendale and around industrial towns everywhere farmers of small-holdings literally wring a livelihood out of the begrimed pastures which often almost fail to separate the towns. These are the pathetic but sturdy rear-guards of agriculture, against advancing and victorious industry.

The breeding of cattle is necessarily associated with dairying, but in the vicinity of towns the calves pass to the butcher when still very young. In Cheshire and the lower Ribble valley, a greater number are reared to maturity than in east Lancashire but the true breeding centre is found north of the Ribble†. In the valleys and on the lower fell slopes of Bowland and so through the Craven district into Yorkshire the rainfall is rarely less than 40 in. (1000 mm.) and generally over 60 in. (1500 mm.), whilst increasing elevation further renders arable cultivation precarious. Consequently cattle, general-purpose Shorthorns, form the keystone of north Lancashire agriculture, and no less an authority than Sir A. D. Hall has paid the industry a great tribute in describing the quality of "the noble groups of heifers and milch cows on nearly every farm." The area is very heavily stocked and is one of the sources whence town dairymen draw their supplies of deep-milking cows, and the butchers their bullocks, largely via the markets of Skipton, Settle, Hellifield and Preston. The maintenance of quality in face of the incalculable waste of high-grade breeding cows entailed by the system of stall-feeding in town dairies is a further tribute to the work of these Lancashire breeders.

Bowland also shows greater devotion to the sheep industry than

* See figures for Wirral and Sandbach.

† See figures for Clitheroe. This point is not indicated clearly in the figures, because Petty Sessional Divisions include lands of widely different types.

*Agricultural Statistics of Selected Petty Sessional Divisions**

Crop acreage and animals per 100 acres of cultivated land.

	Wheat	Oats	Rye	Potatoes	Other roots	Temporary grass	Permanent grass	Dairy cattle	Other cattle	Sheep	Pigs
Ormskirk (SW. Lancashire, northern belt)	10·9	29·1	—	24·4	1·0	17·5	10·8	3·3	3·5	0·6	6·8
Prescot (SW. Lancashire, southern belt)	12·6	27·3	—	12·1	1·7	23·7	19·2	5·4	4·4	10·2	6·6
Daresbury (S. of Warrington)	6·4	25·9	—	11·0	3·6	25·1	26·3	16·1	8·3	4·9	5·5
Sandbach (S.E. Cheshire)	3·7	17·2	2·3	7·8	3·2	15·5	48·5	21·4	11·8	5·9	9·2
Wirral	2·1	12·4	—	4·4	4·0	16·7	58·3	17·8	10·6	33·3	7·6
Kirkham (includes the Fylde)	1·7	8·2	—	3·7	1·9	8·3	73·4	30·3	15·5	27·3	17·1
Clitheroe	—	0·3	—	0·1	0·1	0·6	98·7	21·2	15·4	93·2	6·1
Bolton-by-Bowland	—	—	—	—	—	0·4	99·5	15	10·6	106·2	1·9

* The divisions unfortunately do not coincide with "natural" divisions. They have only administrative significance. The table has been prepared from statistics for 1925 supplied by the Ministry of Agriculture and Fisheries.

any other portion of Lancastria*. From Derbyshire northwards a type of hardy cross-bred general-purpose sheep is found, with local variations, on the limestone summits and on the rougher pastures of the steep slopes of the Grits and Coal Measures. They are capable of foraging under all conditions of climate except snow. The general practice is to sell off the surplus sheep in October for fattening on lowland pastures. Sheep-raising is most often carried on in conjunction with cattle-rearing in Bowland, owing partly to the survival in some districts of an archaic system of land distribution which ensures to (or forces upon) each occupier a share of every type of land from the valleys to the fell summits.

The Fylde†, perhaps more famous than any other part of Lancastria for its subsidiary industries of pig and poultry raising, was formerly the granary of Lancashire, but is now almost entirely under grass. It illustrates, perhaps in an extreme form, the reverse process to that observed in the arable districts. On the stiffer loams the tendency during the 19th century was to abandon arable cultivation in favour of grass.

A study of the agriculture of Lancastria thus illustrates the contention that the "norm of regional specialisation is use adapted to intrinsic conditions." Within the limits of such use constant adjustments are necessary. The management of arable farms has long been directed towards the maintenance of large numbers of transport animals in towns. Mechanical transport is inexorably ousting the horse although the process may exhibit periodic fluctuations. The cattle industry may conceivably absorb the resultant surplus of fodder, but the extra cost of transport will further reduce the farmers' margin of profit. More land may be put down to grass, but a great deal of south Lancashire at least is inconvertible arable land. The three shift system, traceable in the figures for the Ormskirk division, with its greater return of roots and cereals over a given period, is only a partial solution. The final solution may lie only in a drastic change in the crops themselves.

Regional Planning

W. Fitzgerald

It is not possible to close the study of Lancastria without reference to the social and civic problems raised by the industrial development within the region. The close grouping of expanding towns round Manchester-Salford, round Liverpool and in the

* See figures for Bolton-by-Bowland.

† See figures for Kirkham.

Ribble basin has made necessary some form of co-ordinated regional planning within which town planning is subsidiary. What is lacking at the moment is the realisation that within a geographical unit such as Lancastria co-operation among the regional committees is not only desirable but necessary.

The Manchester Regional Committee embraces a great number of authorities in east Lancashire, Cheshire and the West Riding, and extends as far west as Warrington. Liverpool and almost all the remaining authorities of south-west Lancashire have also recently formed a regional committee. Wigan, one of the outposts of industrial Lancashire, is frequently considered along with the south-western communities of the County, though its interests with highly industrialised south-east Lancashire are at least equally strong. Its flooded areas, a consequence of subsidence in the coal-field, present a problem which is common to other mining districts farther east. To the west the higher ridges above the Douglas gorge are becoming residential and their vantage points, commanding a coastal view across a foreground of open plain, are increasingly visited during leisure hours by the townspeople of Wigan. This town could render great service by urging some form of co-ordination of the Manchester and SW. Lancashire regional schemes. It cannot remain indifferent to the fate of the open belt to the west any more than of the industrial region to the east. This open belt retained its identity as an agricultural zone throughout the developments of the 19th century, but it is now threatened by impending encroachment from all sides. People must be housed and it may be impossible to retain this plain as a "lung" of entirely open land between the communities of the coast on the one hand and industrial Lancashire on the other. It should at least be possible to ensure its planning on scientific lines before it is submerged by industrial and urban expansion.

It is also clearly desirable that the south-west Lancashire scheme should be related to developments in the Wirral Peninsula and on Deeside. Central Wirral is at present a valuable break between the lower Merseyside urban group and the expanding industrial district on the western flanks of the Dee estuary. The development of the latter is covered by the programme of the Deeside Regional Planning Committee, but there is at present no effective provision for the maintenance of the rural belt of central Wirral. It is here that co-operation between the south-west Lancashire and Deeside authorities is essential.

But it is the Ship Canal zone which calls loudest for the fullest possible co-operation. If the hopes of industrial development entertained by the Ship Canal promoters are ever realised, an industrial salient will be thrust forward to separate the rural districts of Lancashire and Cheshire, and the picturesque Keuper escarpment which overlooks the left bank of the middle Mersey will be obliterated by a mass of dwelling houses. Yet this zone at present is divided between the spheres of several distinct regional schemes!

The more Lancastria is studied in all its geographical aspects the greater becomes the impression of unity in diversity. Yet disputes, often bred through fear of absorption by the larger towns of smaller communities, are unfortunately not infrequent in spite of the activities of regional committees. A conception of rivalry between the ports of Liverpool and Manchester indicates but a parochial outlook. Such a conception may loom large in the minds of certain individuals, but both ports depend on the efficient use of the same estuary and can but share the traffic of the common hinterland. There is an immediate and urgent need for the closest possible co-operation between all local authorities, great and small, if the fullest possible use is to be derived from the sum of the factors which grant Lancastria its individuality.

The Potteries

J. Kershaw

The Congleton spur of the Pennines which bounds the Cheshire Plain on the south-east is breached by a series of natural gaps diminishing in altitude from east to west. They are the successive overflow channels of the glacial lakes which formerly covered Cheshire. One of these at a level of 485 ft. (148 m.) leads into a somewhat isolated industrial region known as the Potteries. The pottery industry is here developed to an exceptional extent and employs over 35 per cent. of the total working population of Stoke-on-Trent. This is due to the presence of suitable Carboniferous clays and marls. The Etruria Marl group consists of a great thickness of purple and red unstratified marls with bands of green grit occurring at several horizons. The Newcastle group consists of grey sandstones and grey shales with three or four seams of coal. The Keele group is composed of red sandstones and red marls without any bands of green grit. These strata yield, from open quarries, the bulk of the local supply of clays from which are manufactured common earthenware, drain pipes and tiles for various purposes. When Josiah Wedgwood (1730–95) established his factory at Etruria he "converted a rude and inconsiderable manufactory into an elegant art and an important part of national commerce." To the coarser

POTTERIES

GEOLOGY (CHIEF POTTERIES & COAL MINES)

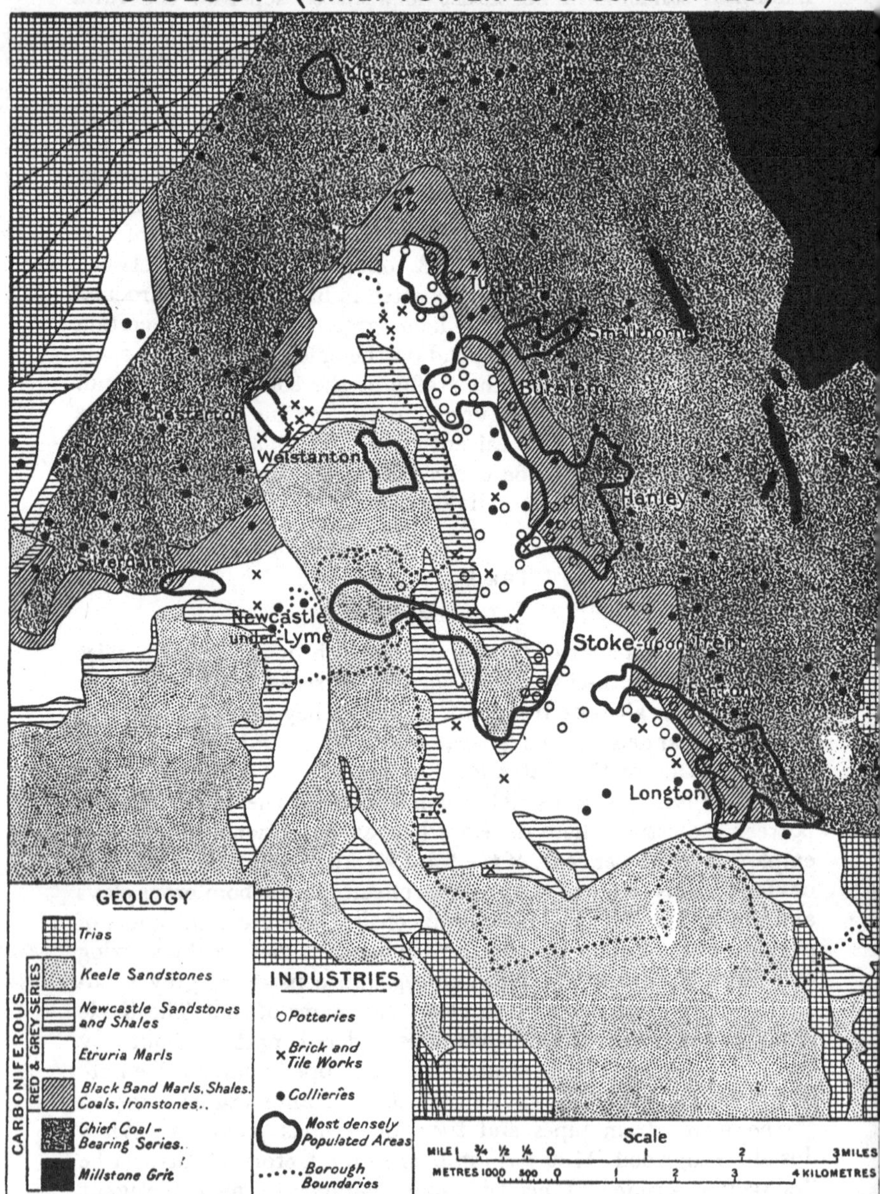

Fig. 37. The Potteries: Geology and Industries.

kinds of earthenware there has been added the still more important china ware for which almost all the requirements, except coal and fire-clay for saggers, come from elsewhere: china clay from Cornwall, ball clay from Dorset, flints from Norfolk and Normandy, feldspar from Derbyshire and Norway, bones from South America. Altogether there are over three hundred pottery works with a number of ancillary trades manufacturing the machinery, brushes and chemicals (colours, glazes, stains and oxides) required in the industry.

Coal-mining employs more than 21 per cent. of the working male population, and serves not only the needs of the pottery industry but also an important iron and steel trade based in part on the valuable beds of "black band ironstones" contained within the North Staffordshire coal seams. The smelting industry and engineering trades employ almost eight thousand men. These industrial activities scattered over definite geological formations have six distinct nuclei—Hanley, Burslem, Stoke, Longton, Tunstall and Fenton—linked by roads and railways. In addition, the Trent and Mersey Canal enables easy transport of raw materials and manufactured goods not only locally but over central England and to the various ports, especially to Liverpool. On the spur to the west stands Newcastle-under-Lyme, an ancient route and market centre, whose interests and outlook have differed radically up to the present from those of the six original industrial towns.

In 1907 an enquiry began, which led, after prolonged negotiations, to the amalgamation of the six pottery towns to form the County Borough of Stoke-on-Trent, which in 1925, after an extension of boundaries, was raised to the status of a city (population 278,900). The usual problems attendant upon a rapidly expanding industrial centre are intensified in Stoke by the fact that the city has not developed from one nucleus, but has been created out of smaller and rapidly growing towns. The development of an urban entity with a well-recognised civic centre is rendered difficult both by these local interests within the borough, and by the shape of the town itself. The six towns which grew up around the headwaters of the Trent and spread along the outcrops of broad bands of marl and clay, and of coal seams used in the pottery trade, give the city its elongated shape, and still more elongated population zones both of which are adverse to civic unity.

Furthermore the city's problems are increased by the difficulties in the way of further expansion to the east and west. On the east is the rising ground of the chief coal-bearing measures, and the high moors of the Millstone Grit. Much of this country, otherwise quite suitable, is rendered unfit for building sites because the widespread coal-mining makes the area liable to subsidence.

On the west the problem is different. Newcastle and Wolstanton, built on Keele Sandstones let in by the Apedale fault, form a residential area amid the mining districts of Silverdale and Chesterton to the west and the Potteries to the east. This is perhaps the ideal area for the expansion of the town, but repeated attempts to incorporate this district have been vigorously opposed by Newcastle, an ancient borough and historic market centre, ranking with Lichfield and Tamworth as one of the oldest Staffordshire towns, and naturally anxious to preserve its individuality.

This clash of interests between the old, static Newcastle, and the modern industrial and rapidly growing Stoke forms one of the most vital problems of the district and affords a striking example of the change of geographic values within the area. Newcastle to-day is only of secondary importance. It is not a Pottery town, but serves largely as a residential suburb of the Potteries.

The difficulties of expansion to the east and west make it probable that the city will tend to grow in a southerly direction, and this, by increasing the length, may accentuate the difficulties of developing an urban entity with a well-marked civic centre. The quarter million population is strung out in a zone of road, rail and canal communications which link the area southwards with the Midlands and northwards with Lancastria. This crowded industrial thoroughfare presents social problems which at present find no remedy, but which sooner or later must react on the economic well-being of Stoke-on-Trent and its non-industrial sister town of Newcastle-under-Lyme.

XVI

THE DON VALLEY AND SOUTH YORKSHIRE COALFIELD

R. N. Rudmose Brown

THE southern Pennines and the Don valley comprise three physical regions, all distinctive in structure and appearance. These cannot be said to form a physical unity, but they are knit together by modern conditions of life, directly and indirectly the outcome of the coal-mining of south Yorkshire.

On the extreme west lies the High Peak and limestone region of the heart of the Pennines. Denudation of the top of the Pennine fold has exposed the older rocks. The Peak of Derbyshire (2080 ft. or 634 m.) belies its name, for it is a flat-topped cap of Millstone Grit which by resistance to erosion has protected the underlying Carboniferous shales that can be seen on its lower flanks and in the wide valleys of Derwent and Ashop on the north and the Vale of Edale on the south. Farther south stretches a plateau of Carboniferous Limestone at an elevation of about 1000 to 1500 ft. (300–460 m.). It is a rolling grass-covered surface trenched in places by deep ravines, famous for their fantastic beauty. Many of these represent subterranean channels of which the roofs have fallen in. Caverns and underground streams abound. The limestone plateau is notable for its mining of lead, barytes and fluor spar. Many parts show deserted mines and heaps of refuse, for the lead industry which dated back to Roman times or earlier is now nearly extinct in Derbyshire. For the rest the region is agricultural with sheep on the limestone uplands and mixed agriculture in the valleys. One or two small cotton mills indicate an extreme outlier of the textile region of Lancashire.

The link between this region and the Yorkshire coalfield is twofold. The upper valley of the Derwent has been dammed to form two reservoirs to augment the Don water supply for Sheffield. These form the first of five reservoirs that are planned in the Derwent valley for the supply of Derby, Leicester and Nottingham as well as Sheffield. The water is carried to Sheffield by a four-mile tunnel through the Millstone Grit on the east. The two reservoirs now completed together provide 15 million gallons (568,000 hl.) a day, and the scheme, if it is ever finished, should provide a total of 88 million gallons (1,250,000 hl.). New reservoirs are now planned for the Burbage, a lower tributary of the Derwent.

A second link between this central Pennine region and Sheffield is the outcome of the piercing of the Dore and Totley tunnel in 1898 on the route of the Midland Railway between Sheffield and Manchester. This has allowed the villages of the Derwent valley, such as Bamford, Hathersage and Grindleford, to become residential or dormitory suburbs of Sheffield well removed from the grime of the town. As a result these villages are growing in size. The later use of motor transport on roads has had a similar effect on villages away from the railway, as Baslow and Bakewell.

Eastward of the limestone region escarpments or edges of the Millstone Grit tower above the Derwent valley marking the beginning of the eastern flank of the Pennines. The structure of the Pennines has been described* and need not be repeated here. The band of impervious gritstone is a poorly inhabited zone occupied mainly by wet moorland on which a few sheep graze. Only in the wide valleys of the east-flowing drainage is there any cultivation. The lower slopes where the coal occurs are among the most densely populated areas of Great Britain, and although the old-time agriculture survives, its importance is masked by mining and manufactures. Still farther east these old rocks are carried beneath the lowlying level Triassic plains which stretch to the Humber and the Trent. They form a region of high fertility which in the past had a purely agricultural population but now is being invaded by the mining interests from the west.

The River System

The River Don to a certain extent binds together the Pennine slopes and the low vale to the east and is the one river of importance wholly within the area. According to Mr Lower Carter the original consequent Don was the present river above Penistone and was continued by a low gap to the east, flowing by what is now the Dove and the lower reaches of the Dearne to the Sheaf about where Mexborough now lies (see Fig. 88). The Sheaf was originally a longer stream than at present. Its course was determined by a synclinal fold and associated faults which carried its water north-eastward in the line of the present small stream. Then a subsequent of the Sheaf cutting backwards into the Wharncliffe plateau beheaded the upper Don. Thus is explained the peculiar sharp turn in the River Don in the heart of Sheffield. The original course of the lower Don is not clear, but a former course can be traced and

* Chapter xiv.

is shown on Fig. 40. The present Rother suggests originally a subsequent tributary of the Moss, at present one of its insignificant tributaries, which in course of time beheaded the Poulter. Its valley thus affords a notable north and south route on the eastern side of the Pennine upland.

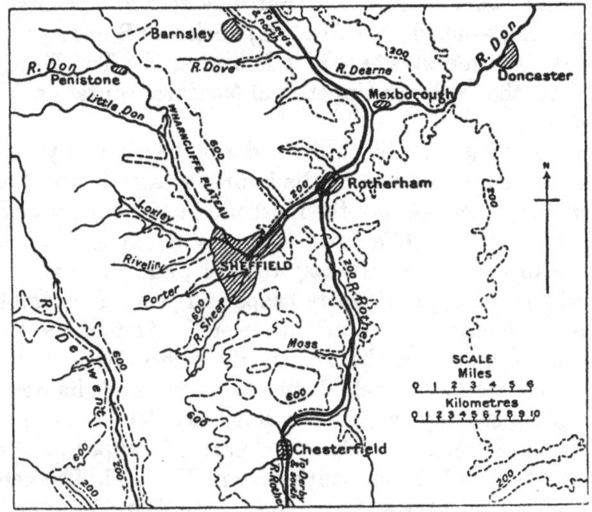

Fig. 38. Diagram to show the isolated hill site of Sheffield, the river originally supplying water-power and the original railway with only a branch to Sheffield.

The Coalfield

On the eastern flank of the southern Pennines extending from the Aire where it ends in the Wharfe anticline, in line with the Market Weighton axis, to the Trent where it thins out beneath the newer rocks of the Triassic plain, lies the greatest of the coalfields of Great Britain. On the west and north the field ends against the older Millstone Grit and on the east it dips beneath the unconformable Magnesian Limestone and the Triassic sandstones of the Vale of Trent. The exposed field is over 70 miles (110 km.) long and from 8 to 20 miles (13 to 32 km.) wide with a total area of 760 sq. miles (1970 km.^{2}). The total thickness of the Coal Measures is over 5000 ft. (1520 m.). The Lower Coal Measures on the east with their beds of massive hard sandstone and relatively great elevation give upland scenery not unlike the moorlands of the Millstone Grit. Economically they are of most value for their ganister fire-clays, building stone and millstones. Sandstone was in the past the one building material used in the Pennine area and layers of sandstone were even used for roofing. Coal seams are not numerous, nor, as a rule, thick. The best are the Beeston seam worked around Leeds and the Ganister or Alton coal. The Middle Coal Measures, with fewer beds of hard sandstone, form lower ground and tamer scenery. They contain the most valuable seams of coal including the well-known Silkstone, Parkgate and Barnsley seams. The best is the Barnsley seam, which is 11 ft. (3·35 m.) thick near Barnsley and thins towards the south-east where it is known as the Top Hard seam. Farther east the lowlying Upper Coal Measures contain a few thin seams of coal.

Outcrop mining on this field no doubt began many centuries ago, but the real expansion of the industry came in the 18th and early 19th centuries. To-day the most productive area is where the upper part of the Middle Coal Measures is exposed. There the valuable seams are relatively easy to reach, and there the hideous colliery villages and gaunt waste heaps and general devastation of the countryside can be seen at their worst. West of the Permian outcrop to the north of the Don is the richest area with about 30 ft. (9 m.) of workable coal. Some of this coal finds its way to the Humber ports for export, some is converted into coke partly for use in the metallurgical industries and partly for export, and a good deal is sent to London and south-eastern England. The centre of the coal-mining industry is the upland town of Barnsley (population 69,000) standing above the flood plain of the Dearne. It was noted for its linen and iron wire, but was of no great size till coal-mining supplied an impetus to growth towards the end of the 18th century. In 1799 it was linked by the Barnsley Canal with the Aire and Calder Navigation at Wakefield and by the Dearne and Dove Canal with the Don Canal (South Yorkshire Navigation). These canals are still of some value, but a modern indication of the importance of the region in the output of coal is the network of railway lines including several built solely for mineral traffic. To the south past Rotherham the zone of mining narrows but widens again in Derbyshire, where a local anticline has brought near the surface supplies of iron-ore, and caused localised iron-smelting around Chesterfield, Clay Cross, and Staveley. Local ore has now been replaced by supplies from Northampton and Lincolnshire. Farther south the nature of the industries changes, and textiles, especially cotton, predominate around Nottingham (Chapter XI).

Eastward of this exposed field lies the concealed coalfield (South Yorkshire) of the Trent and Aire valleys which has a considerably larger extent. It was not until 1859 that a pit was sunk to the coal through the Magnesian Limestone. The valuable Top Hard coal was reached at 1530 ft. (466 m.). Since then so many pits have been sunk that the concealed field produces over a third of the Yorkshire output. There are now pits as far east as Stainforth and Thorne and boreholes within a few miles of the Trent. But the most valuable part of the field is around Doncaster, where a number of large pits are working the Barnsley and other seams. In that area, according to Prof. W. G. Fearnsides*, there are probably 15 to 25 ft. (4·6 to 7·6 m.) of workable coal with a great decrease in number

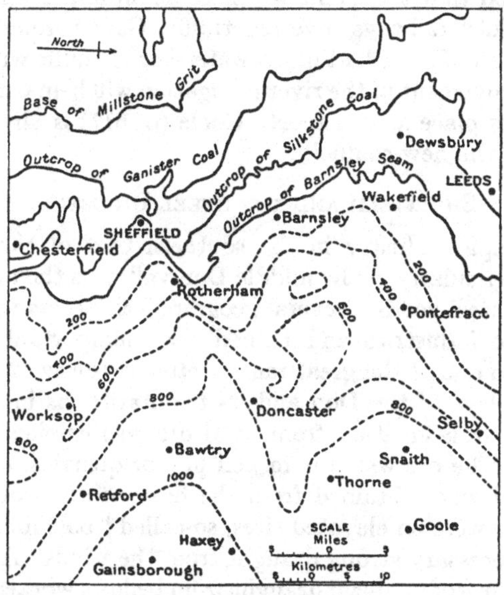

Fig. 39. Sketch-map of the Concealed Coalfield of S. Yorkshire, etc., showing approximate underground contours of the Barnsley Seam (after W. G. Fearnsides). Depths are in feet.

(Note the orientation of this map.)

and thickness of seams to the east and north. The northern edge of this coalfield is probably on the line of the Wharfe anticline. The field seems to extend well to the east beyond the Trent, but east of the clay vale of Lincolnshire coal would be too deep for profitable mining. Possibly the coal continues under the sea since the anticline shown in the Lincoln Wolds may not mark a pre-Cretaceous movement (Fig. 39).

Certain drawbacks impede the development of the coalfield as a whole. The seams vary in value throughout their extent, and they are heavily faulted. Without numerous borings the sinking

* "The Doncaster Coalfield," *The Colliery Guardian*, March 1927.

of a shaft is a problematical venture. Water is often a serious trouble and is very liable to find its way along the Permian unconformity when mining has caused a settlement of the underlying strata. Thus the working costs are frequently high.

Many agricultural hamlets are becoming mining villages, generally with uniform ugliness, although here and there are new growths which do not mar the countryside.

If the new coalfield proves more remunerative in the future there is a likelihood that the Trent will become an important outlet for the coal. Until railways diverted traffic, Gainsborough, 25 miles (40 km.) up the Trent, had much water-borne traffic with London. Recent improvement in the river navigation will help Gainsborough to resume its place as a seaport. Goole (p. 317) is already feeling the effect of the new coalfield.

Sheffield and the Steel Industry

The principal industry in the southern part of the Yorkshire coalfield, particularly in the middle Don valley, is the manufacture of steel and steel goods. Several geographical factors favoured the rise of these industries and turned the unimportant hamlet of Sheffield into one of the great manufacturing towns of Britain.

In the valleys of the Don and its tributaries, as in other parts of England, iron-smelting from local ore was carried on in very early times. The ore was dug in bell pits or quarries, and the fuel and charcoal were obtained from the heavy forests of the times. The furnaces were on elevated sites, so-called "bole hills," in order to get the necessary strong draught from the winds. A later stage entailed the use of artificial draught from bellows worked by water-power. This caused the industry to move to the valley floors utilising particularly the heavy gradients of the upland courses of the streams. The result was the radial distribution of the early scattered furnaces. Dams or hammer ponds were built on the streams and broke them into a series of levels each serving a water-wheel. Along the course of the five upland streams converging on the site of Sheffield as many as 200 water-wheels were at one time working. At a later date these were used to turn grindstones in the cutlery trade. Some of the dams still exist and one mill is in working order.

More than four centuries ago—the exact time is uncertain—the manufacture of steel and steel goods began in the district. But local iron contained too much phosphorus and the cutlery trade throve chiefly on iron imported, at high cost, from Sweden and Spain. The skill of the local craftsmen by this time was an important factor in localising the industry. Other advantages were the water-power and hard sandstones for the grinding wheels. Economic conditions in the steel industry have to some extent survived from the days of small mills, for even in modern Sheffield much of the grinding is done by individual workers renting their premises and power from a landlord.

As the industry grew, foreign iron-ores were imported, but the iron-smelting was on the verge of ruin in the 18th century by the exhaustion of local supplies of timber. The cutlery trade with its imported iron was unaffected. Outcrop coal, worked easily in shallow pits in the deep valleys through the Coal Measures, saved the iron industry and gave it a new lease of life. Coke replaced charcoal; clay and later ganister were required and found locally. Gradually, as water-power ceased to be of value, the steel industry concentrated in the lower parts of the valleys and the town of Sheffield grew rapidly. There was a concentration near the coal and near the easiest transport to and from the district, that is by the Don valley. Thus the modern works, especially those making heavy steel goods, are found chiefly in the Don valley between its elbow bend and the town of Rotherham, while cutlery and lighter metal goods are made on higher sites to south and west. Tradition and social inheritance are practically the sole geographical factors that now give greater value to Sheffield as a steel manufacturing centre than any other place on the coalfields of Britain. But they are strong enough to explain the apparent anomaly of an inland town being engaged in the manufacture of heavy goods, largely from imported ore, and being one of the great centres of ship construction of the country in spite of the lack of access to the sea. "Special" or "alloy" and stainless steels are important new aspects of Sheffield's leading industry. Silver and electro-plate goods are also of importance.

The growth and position of Sheffield account for many of its peculiarities. It is still essentially a great workshop in a nook of the Pennines, essentially an industrial town in which the surface relief, alternation of ridge and valley, led in the period of prosperity in the 19th century to a great crowding in the valleys and somewhat later growth over the ridges. No city has a more beautiful site: scarcely any city has a more dreary appearance. Yet its western suburbs extending on to the gritstone moors and swept by clean winds are in striking contrast to the grimy monotony of its central part and eastern valley area. In this respect it is unique among the great manufacturing cities of England.

Sheffield lies on no great lowland route or natural highway. To north and west and south the valleys that made Sheffield lead only to the Pennine dales and lofty moorlands inhabited solely by sheep and grouse. As long ago as 1732 improvements in the Don navigation were begun in order to help the trade of Sheffield, and in 1793 the Don was joined to the Trent by the Stainforth and Keadby Canal. The original Midland main line passed from Chesterfield by the Rother valley to Rotherham leaving out Sheffield, not for any lack of importance of the town but on account of physical obstacles. For over forty years a branch to Rotherham along the Don valley was Sheffield's link with the main line. Not until 1870 did a main line (L.M.S.) pass through Sheffield after the piercing of the Sheaf-Rother watershed. The Pennines to the west are pierced by long tunnels through which devious heavily graded lines maintain connections between east and west. Even now relatively little of the trunk traffic of England passes through Sheffield. With the exception of Penistone, a small steel centre higher up the Don valley*; Rotherham (population 68,000), a small counterpart of Sheffield, and Doncaster lower down; Barnsley (population 69,000) in a tributary valley of the Don; and Chesterfield (population 61,000), an iron centre over the Sheaf watershed in the Rother valley, there are no contiguous urban areas. Of those mentioned, Rotherham alone coalesces with Sheffield. The Sheffield conurbation is relatively small and isolated and the city is in no sense a great regional capital like Leeds or Birmingham. Its tributary area contains little over 1,000,000 people, of which Sheffield itself has more than half. In recent years the urban area to which Sheffield is the natural capital has grown a little owing to the new Doncaster or South Yorkshire Coalfield to the east.

Between Don and Trent

The Coal Measures end against a narrow belt of Magnesian Limestone which generally shows a marked escarpment face, as high as 100 ft. (30 m.) on the west and a gentle eastern slope down to the low Triassic plain of 25 to 50 ft., the uniformity of which is seldom broken westward of the escarpment of the Lower Lias beyond the Trent.

The Triassic plain nearly everywhere between York and Doncaster is covered with recent alluvial deposits which in some places overlie glacial debris. In a few places islands of rock emerge above the alluvial deposits to a height of 50 or rarely 100 ft. They are probably due to bands of harder marl and gypsum having protected from erosion the softer sandstones and marls. These islands being relatively free from the inundation to which the plain was periodically

* The suggested closing of the English Steel Corporation's works at Penistone will throw most of the population out of employment.

subjected became at an early date valuable sites for settlement. They mark to-day the main centres of population. Thus Doncaster, Thorne, Hatfield, Bawtry, Armthorpe, and other places lie on the Keuper and Bunter sandstones. Crowle, Epworth, Haxey, Gringley, etc. lie on a narrow plateau of Keuper marls farther east. This last is the Isle of Axholme*, so called from its insular position among the alluvial flood plains that surround it (Fig. 40).

In olden days all this lowlying Triassic plain was covered with forest and marsh. Relics of the forest are preserved in Sherwood Forest and other parts of the district known as the "Dukeries." The peat bogs of the Isle of Axholme bear evidence of former forest growth and suggest a subsidence in recent times. In early days agriculture was mainly confined to the limestone belt with its light soil, good drainage and excellent conditions for sheep. Worksop, with its wood manufactures, malting and agricultural implements, was a typical market town on the edge of the firmer ground. The site of Doncaster (Rom. Danum) marked the lowest firm ground on the Don where traffic could cross the river between northern and south-eastern England, avoiding the forests and swamps to the east and the higher ground to the west. The Roman roads from Londinium (London) to Luguvallum (Carlisle) and Eburacum (York) passed through Danum. Much the same route is now followed by the Great North Road and the main railway line (L.N.E.R.) to the north. By the Don valley passes the easiest route between Sheffield and the Humber ports with access by the Dearne tributary to the Barnsley coal-mining area. In early days sheep-rearing on the limestone soil gave Doncaster a textile industry which it lost later in competition with the northern part of the Yorkshire Coalfield. Large mills for the manufacture of artificial silk are now being erected. A favourable factor is the new coalfield employing only male labour and the consequent abundance of female labour.

Doncaster until a few years ago was purely an agricultural centre except for the Great Northern Railway works placed there as a convenient centre near coal and steel. Now it is becoming the hub of a busy mining area with growing colliery villages around it, and the population of the borough has practically doubled in a few years, and is over 60,000.

The drainage of the lowlying plains was not seriously undertaken until the 17th century, although the Romans had cut a few dykes or ditches. The present scheme was planned by Vermuyden, a Dutch engineer. The problem centred on the regulation of the waters of the Don, Torne and Idle, which drained into the district, and meandered over the surface in changing channels. The drain

* Axe (Celt) = water; ey (A.S.) = island; holm (Scan.) = river island.

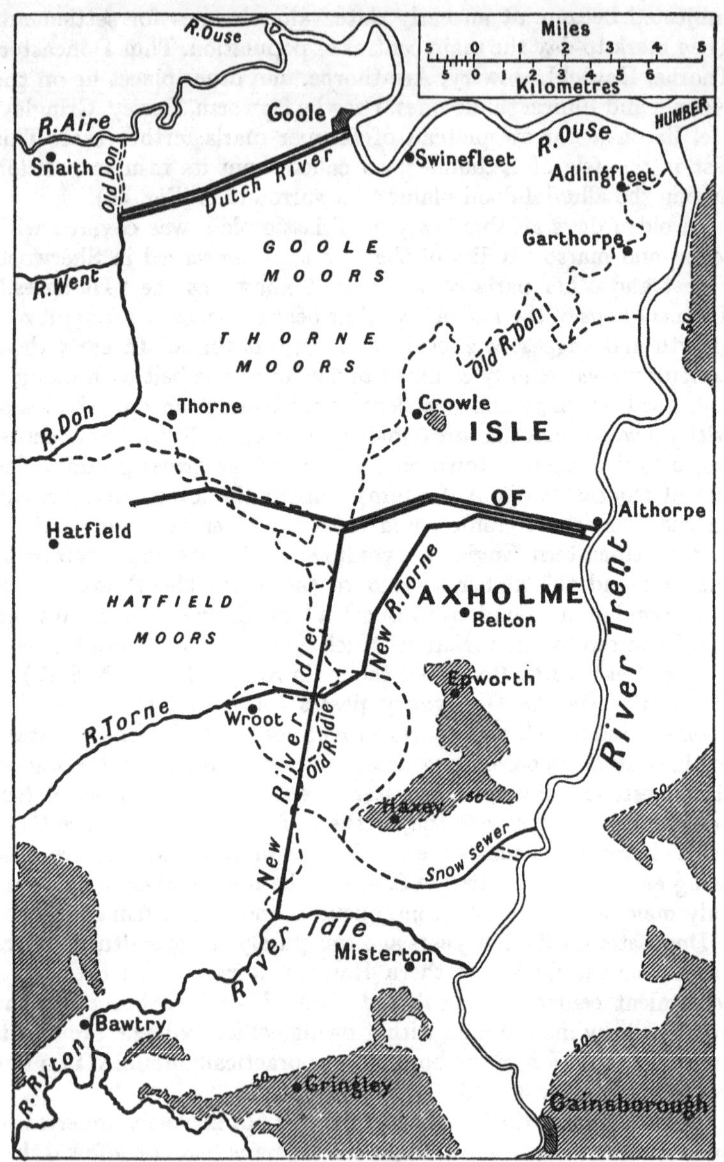

Fig. 40. Isle of Axholme and adjoining lowlands, illustrating Vermuyden's original drainage scheme. Rivers shown in broken lines disappeared after drainage channels were cut. Land over 50 feet is shaded. For canals and present channels, see the Ordnance Map.

called the Dutch River carried the Don waters more swiftly and surely to the Ouse than the Aire could. The Idle waters were turned to the Trent by the Bykersdyke, cut by the Romans, and now called the River Idle. The Torne, which originally flowed by two branches into the Idle, was carried by the New Torne River into the Trent at Althorpe beside the New Idle River which drained part of the eastern arm of the old Don from near Thorne, and by a southern branch, the old course of the Idle. At right angles to these main channels were other drains. Sluices were placed where the main drains entered the rivers and excluded tidal waters. Embankments were built along the Trent and the drains and pumping mills were installed in places where the gradient was too slight to allow a ready flow. In the 18th century the drainage was improved and extended to adjacent lowlands. The mills worked by horse or wind are now replaced by steam pumps. Warping drains, in which the tidal silt is trapped by sluices, are continually increasing the area of fertility. Other improvements are the deepening of the channel of the Trent which allows the water to drain away quickly and prevents the tidal waters of the bore or aegir from overflowing the banks. Several of these drains, notably the Dutch River, are used for barge navigation.

Until these drainage works were made the population on the lowlying plains was sparse and confined to the higher areas of firm rock. It was almost an insular population and derived its livelihood mainly by fishing and wild-fowling. Even now, when most of the region is reclaimed, population is not dense except where collieries are opening. But the drainage works have robbed the inhabitants of their isolation and converted them from semi-nomadic hunters into farmers. The alluvial soil is the most fertile but many parts of it are still liable to inundation. The sandstones form the poorest soil and the Keuper marls are generally the areas of most active cultivation. Agriculture is mixed. On the wet pastures, still prone to flooding, cattle predominate, and cattle farming grows at the expense of crops because of the attraction of the urban growth to the west, and the industrial penetration of the district resulting in depopulation of rural areas and the provision of markets for dairy produce. Wheat and oats are however important especially on the more fertile alluvium. At one time flax and hemp were noteworthy crops of the Isle of Axholme. Potatoes and root crops are extensively cultivated*.

* Acknowledgments are due to two of my students, Miss N. Ward, M.Sc. and Mr J. Corley, M.Sc. whose theses, prepared for their degree work, have been of use to me in writing this essay.

XVII

HOLDERNESS AND THE HUMBER

R. N. Rudmose Brown

THE East Riding of Yorkshire approximately represents the northern end of the Cretaceous lands of the English plain. The hard chalk beds of the Yorkshire Wolds (A.S. weald = waste ground) lying on the east and west fold of an earlier floor, the Market Weighton anticline, extend in a crescentic curve rising to 800 ft. (240 m.) in places from the Humber to the sea where they end in the bold cliff of Flamborough Head. To the north the Wolds slope down fairly steeply to the deposits of the old lake floor of the Vale of Pickering covering over the Kimmeridge Clay of the faulted Pickering syncline. To the west a narrow belt of Lias and Oolite limestone prolongs the slope of the Wolds until the Triassic plains of the Vale of York are reached. Deep vales trench both the escarpment and dip-slopes of the Wolds and mark the sites of villages and routes.

To the south-east the Wolds slope gently down to the plain of Holderness (scan Hallorness = ? promontory of the Hull) which faces the North Sea and the Humber in a long low unindented line of cliffs averaging 30 ft. (9 m.) in height. It is a monotonous but hummocky plain 10 to 30 ft. above sea-level and covering about 160,000 acres (65,000 ha.). Its chalk floor is entirely obscured by glacial material and this in parts is overlain by lacustrine and alluvial deposits. In pre-glacial times an old sea coast extended to the foot of the Wolds from Flamborough Head to Hessle, and all Holderness was a chalk plain below the level of the sea. The cliff continued south along the eastern edge of the Lincoln Wolds. The original swamps and marshes of Holderness, especially in the valley of the Hull, the one river of any note, have given place to cultivation; embankments and drainage channels keep the water in control. The only remaining mere, the name for water-filled hollows due to irregularity in glacial deposits, is at Hornsea. Peat deposits often mark the site of former meres and receive the name of "carrlands."

South of the Humber the same structure is evident. The Wolds, interrupted by the alluvial lands of the Humber estuary, are resumed in the Lincoln Wolds sweeping south-east and increasing in height. The Jurassic beds on the south-western flank of the Yorkshire Wolds, which have only local effect on the topography, increase in thickness south of the Humber as the Market Weighton axis is left to the north, and the limestones of the Lower Oolite form the notable Lincoln Edge while the soft clays of the Middle Oolite are responsible for the low clay vale between the Limestone Ridge and the Chalk Wolds. The limestone at Brough gives a firm footing north of the Humber which the Romans utilised in the continuation of the road since called Ermine Street along the Lincoln Edge to Winteringham.

The Humber, like the Thames estuary, forms a great seaway into the heart of lowland England, but, unlike the Thames, it affords a route also to and from mining and manufacturing England. It collects the drainage of some 9000 sq. miles (23,000 km.^{2}) of its great feeders the Ouse and Trent. Originally determined by a line of fault or a sharp syncline, the Humber was a river channel in pre-glacial days, and its old bed lies some eighty to a hundred feet below the level of the present stream which flows over glacial deposits and river silt. Its pre-glacial course was apparently eastward towards an outlet near the site of Withernsea, but it is now deflected to a south-easterly course by the glacial gravels at Paull. Its greater success than that of other consequent streams of north-east England was largely owing to its having twin source streams in the Aire and Calder, and so a greater volume of water. At the same time the probable line of fault which marks the Humber no doubt facilitated its task of erosion.

The broad estuary of the Humber was in recent geological times of greater width and depth than it is to-day. Much of the area once invaded by tidal waters has now been converted into dry land. Marsh has taken the place of many tidal inlets and dry land has replaced marsh. The slow shallow stream, probably more heavily charged with silt than any other river in Great Britain, is continually obstructing its bed with shifting banks of sand and mud which render navigation difficult and impede access to most parts of the shores. A deep-water channel, swinging towards the left shore at the elbow bend and the right shore farther east, has been largely instrumental in deciding the position of the ports of Hull and Immingham. Hull itself is built on a bank of silt which did not exist in Roman times.

On the northern side, Holderness is growing in area by the deposition of silt and "warp." This warp is augmented by the marine wastage of the coast of Holderness which the tidal currents partly sweep into the estuary. The whole of the Humber bank of the Holderness plain has changed within historic times, and an old seaport like Hedon is now an inland village. Two hundred years ago Sunk Island was a mud bank seven acres (2·8 ha.) in extent separated from the mainland by a wide channel. Gradually the river action, aided by embankment, has led to the reclamation of over 7000 acres (2800 ha.) and the channel between the island and Holderness. The East Riding has gained other land from the Humber and a little has been added to the foreshore of Lincolnshire. Altogether about 15 to 20 sq. miles (40–50 km.^{2}) have been taken from the Humber since Roman times. These changes have apparently not been aided by any change in relative sea-level. (See Fig. 41, p. 315.)

The changes in the Humber have also involved the loss of land between Sunk Island and the peninsula of Spurn. Occasionally inundations still threaten that part of the Humber coast. On the whole, however, there has been a narrowing of the Humber channel, and the so-called lost towns of the Humber, according to Mr T. Shepherd*, were really on the sea coast and not on the Humber as other authorities have maintained. A few small hamlets and farms which lay on the north bank of the estuary higher up than Sunk Island were destroyed by inundation or removed by threat of inundation before the period when Holderness grew southwards into the estuary. On the sea coast of Holderness destruction is rapid, particularly between Kilnsea and Easington. There is probably no section of the British sea coast that is crumbling more quickly. For several centuries the average rate of destruction has been variously estimated at between 4½ and 7 ft. (1·4 and 2·1 m.) a year. This is owing to the shore cliffs being formed of boulder clay, loose sand, and gravels and lacustrine deposits. These overlie the chalk floor which south of Flamborough Head is below sea-level. Many villages and small towns have thus been lost, and the destruction is so rapid as to be visible to-day. Owthorne, Old Withernsea, Sunthorpe and Ravenser, Dimlington (a considerable seaport in the 18th century) have gone; Kilnsea is going. Lanes end abruptly, old routes have to be abandoned, and farms are threatened.

Much of the waste of Holderness and the silt of the Humber have together contributed to Spurn Head. This long hook-shaped lowlying bank of sand and shingle is of modern growth. As its point has moved southward and westward, the lighthouse, first built over two centuries ago, has been frequently moved. In the course of time no doubt the Humber will straighten its course by breaking across the narrow spit and Spurn Point will then be an island

* T. Shepherd, *The Lost Towns of the Yorkshire Coast*, 1912; and *Hull and the East Riding* (*British Association Handbook*), 1922.

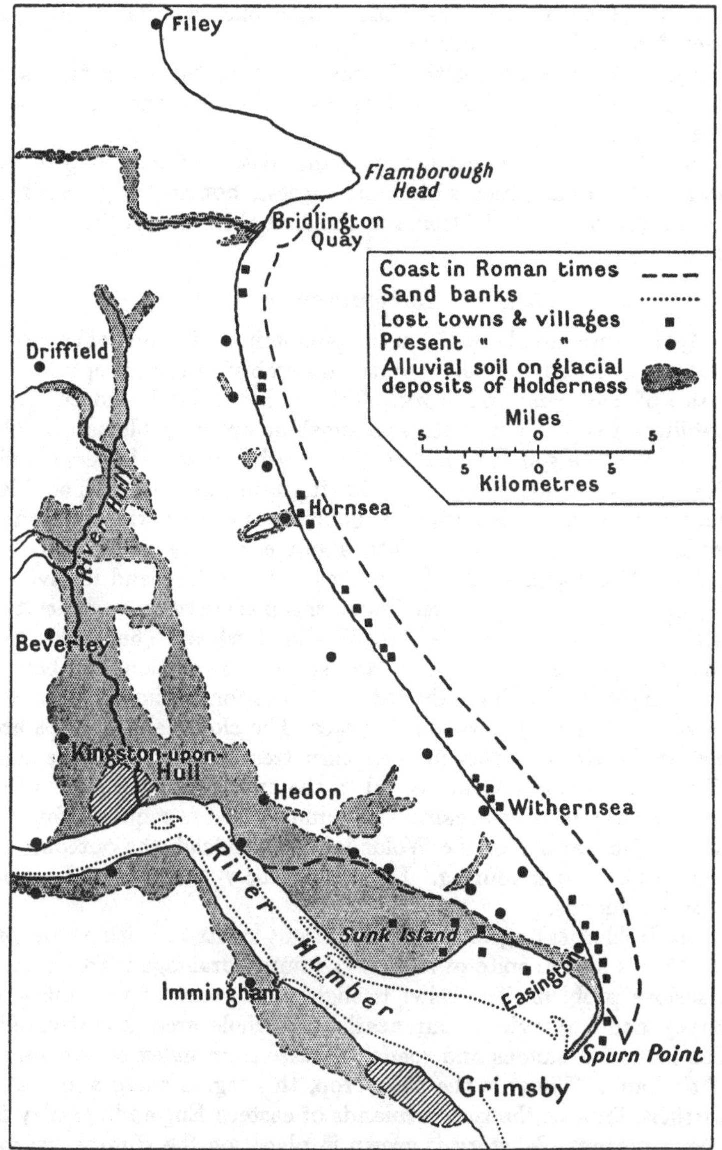

Fig. 41. Holderness and the Humber mouth. Sketch-map showing changes in coast-line, loss of settlements, and distribution of alluvial soil.

which may gradually be linked, in an area of relatively still water, with the Lincolnshire coast. More than once an island has thus been formed from a decapitated Spurn.

Some of the waste of the Yorkshire coast travels farther and is deposited on the coast of Lincolnshire and in the inlet of the Wash.

Efforts to stay the destruction of the coast by the use of groynes and seawalls have met with some success, but on the grounds of expense they have to be localised and thus their aim will eventually be defeated.

Agriculture

Agriculture has always been the predominant interest of the East Riding although the soil is not remarkably fertile except in the Vales of Pickering and York. But the low rainfall and the probability of summer warmth and sunshine are valuable assets. On the Wolds the soil is thin and sandy rather than calcareous, and flints are numerous. Below some 200 ft. (60 m.) there is much boulder clay and gravel. The natural vegetation is one of whin and wiry grass and fertility is low. Until about a century and a half ago the whole expanse of the Wolds was one unfenced and undivided sheep run. But now the breeding of sheep is everywhere associated with the growing of barley, turnips, and wheat. The Wolds are enclosed in fields and there are several plantations of beech. Permanent grass is not abundant: a rotation of turnips, barley, clover and wheat has taken its place. The clover and turnips are mainly for sheep, which in their turn tread and manure the soil. The wheat is thin and not equal to the crops on richer soil to the east. Cattle are few because the summer feed is required for the sheep. The fertility of the Wolds is thus artificial, the outcome of cultivation and manuring. Land left fallow would soon become nearly valueless.

On Holderness agricultural conditions contrast with those on the Wolds and in spite of the lower rainfall drainage is frequently a serious problem. The relief is slight and many of the soils are clayey and impervious, but nearly the whole area is cultivated. The lack of commons and roadside waste is an index to the value of the land. Wheat is the chief crop, this region being about the northern limit of the rich cornlands of eastern England. Barley is also important. Mustard is grown in places on the Humber warp. Sheep are few: cattle are numerous on the wetter lands, but on the warp lands of the Humber brackish water is a serious drawback. The lack of villages on the alluvial soil of the Hull valley is an indication of the former periodic flooding of the area before the waters were controlled.

The natural water supply from springs is not good, for many springs are intermittent, even along the eastern edge of the Wolds where they are most frequent. A good supply can however be obtained from borings through the 70 or 80 ft. (20–25 m.) of drift to the underlying chalk. Most Holderness farms get their water from these sources. Hull has chalk water from wells near Cottingham.

The lack of mineral wealth in the East Riding is reflected in the nature of the few manufactures of the region. Brick, tile and cement works bear close relation to the rock products, farming, brewing and milling to the agricultural wealth, and shipbuilding to the commercial value of the Humber. Oil extraction, seed crushing, cattle cake, paint and soap manufactures originated in whale oil refining and were stimulated by local agricultural requirements. Most of these industries are pursued in Hull, whose wide trade relations add to the scope of raw materials and extend the range of markets.

Towns

The central position and focus of routes account for the importance of such market towns, with minor related manufactures, as Great Driffield (5600), Pocklington (2600) and Market Weighton (1700). The last two have canals, now derelict, to the Derwent and Ouse. But Beverley (13,500), though less central, has always been a more important town. The Beverley Beck, really an old canal, connects it with the River Hull, and allowed Beverley at one time to be a seaport. From the 9th to the 14th century it was more important than Hull, but gradually the greater advantages of the river mouth site told in favour of Hull, especially in view of its widening hinterland and river-borne trade. Beverley to-day retains a little interest in shipbuilding but is no longer a port. Hull is the chief market town for the greater part of the East Riding, but that position alone would not account for its containing about half the population of that region. Hull is not a regional capital of a great area; the population to the north of it is purely rural and so not dense. Around few other English towns of equal size is the population as scattered as around Hull. The coalfields to the west have done what local influences could never have achieved. They have had a similar effect on Grimsby, Immingham and Goole. All these towns are much more seaports of the coalfield than great cities belonging to the counties in which they stand. Their size is not an expression of local influences. The same factors have been responsible for the change from small fishing villages to busy pleasure towns of such places as Scarborough and Bridlington with their largely migratory populations.

Hull (287,000) is an old seaport that owed its origin, like Hedon and Beverley, to a sheltered creek on the harbourless banks of the Humber. Here the Norsemen found a convenient foothold which they called Wyke. But in the days of small ships and poor land transport it served only the immediate neighbourhood. Vessels could ascend the Trent to Burton and via the Fosse Dyke to Lincoln, the Don to Doncaster, the Idle to Bawtry (for Sheffield), the Aire to Castleford (for Leeds), the Wharfe to Tadcaster, and the Ouse and Ure to Ripon. In competition with Doncaster, Lincoln, Gainsborough and York, and Humber ports such as Hedon and Grimsby, the port of Hull had no chance to grow. In 1299 it received its modern name of Kingston-upon-Hull. From then onwards its growth in importance began, as it became the port of trans-shipment between river- and sea-borne traffic. Gradually the Humber became more important than the River Hull to the town. Shipping grew in volume and industries arose of which the more important was whaling, now long since defunct (see p. 319). This phase of Hull's development was largely the outcome of the water-power period in the West Riding woollen industry and the South Yorkshire iron and steel manufactures. These inland industries led to the improvement of water connections with the nearest seaport. Thus, late in the 17th century the Aire and Calder navigation was improved by deepening in some places and making cuts in others so that Hull was linked to both Leeds and Wakefield. Later the Don navigation was similarly improved to Sheffield (1732–1804). Then came the Leeds and Liverpool canal (1770–1816), the Calder and Hebble canal (1786–1829), the Rochdale canal (1794), the Huddersfield canal and others. The 18th century was the great canal era in England, for the art of road making was not understood and the steam engine was not yet in use. Many of these canals are still used for bulky cargoes, but the day of their real importance is past. The cross-Pennine canals in long tunnels and with many locks were the last to be built and had little time to show their value before the successful competition of railways almost robbed them of all traffic. Hull still has a considerable amount of barge traffic, mainly of coal from the west, though this steadily decreases, and of grain inland to the milling centres.

As shipping grew the advantage of the Hull mouth as a sheltered inlet faded in favour of expansion along the Humber frontage. Hull now faces south. Easy railway construction along the plains of Ouse and Trent accounted for early railway connections with the mining and manufacturing regions to the west. Of the imports the most notable are food for man and cattle, timber and wool. The last is a recent innovation in competition with London and Liverpool. Exports are chiefly textiles, coal and metal goods. There is also much coastwise traffic and great fishing interests. Its manufactures are those which require relatively little coal and are related to local needs (p. 317).

The whaling began in the 16th century when London was another Arctic whaling port, and lasted until about sixty years ago when it perished for the lack of right whales in Arctic seas and the competition by the Scottish whalers of Dundee for the few remaining "fish," as much as for the growing competition of mineral oil for lighting. In its time it was of great importance for oil and baleen and to-day its influence may be traced in the active fishing interests and the oil industries of the port. Relics of the old whaling days are still to be seen in occasional gate posts made of whales' mandibles.

Hull, with its 250 acres (100 ha.) of docks and its wharves along the river and in the mouth of the River Hull, is now the fifth port in Great Britain in volume of tonnage entered and cleared.

Hull as a regional capital has little influence beyond the Humber. The ferry is too slender a link to join interests across the river. Only to a small extent has the population of Hull utilised the south of the Humber for residential sites. A tunnel beneath the Humber has been suggested, but there are difficulties. The buried channel lies at least at 120 ft. (37 m.). In order to obtain good rock for boring the tunnel would require to be in the Chalk or better still in the Kimmeridge Clay. The best site would be between Barton and Hessle. But what Hull loses by its isolation from other great cities it gains in other ways. No city in England has developed a higher degree of civic consciousness and pride of place.

Hull has no longer a monopoly of the Humber trade. Twenty-six miles (42 km.) farther up the river, accessible only to smaller steamers, is Goole (20,000), the only seaport of the West Riding. This is a miniature Hull in the nature of its trade, but far below it in volume of tonnage handled. It owes its rise from an insignificant hamlet a century ago to the canal connections with the Aire at Knottingley and the Don navigation with Sheffield. Goole has not the advantage of deep water access, but it lies more central than Hull to a great mining and manufacturing area.

Small inlets near the southern side of the Humber mouth furnished the sites at which originated the fishing villages of Grimsby and Immingham. The rest of that lowlying coast, fringed with alluvial flats or shingle banks, offered little inducement for coastal settlement. The early invaders naturally avoided this stretch of river bank and pushed farther up stream. In fact inundation periodically laid waste these shores until effective embankments were constructed. They now line the shore from Grimsby to Barton. The aftermath of flooding is even now more serious than the flood itself since it leaves a soil saturated with salt and quite unsuited for crop cultivation.

Nearness to the open sea, the Dogger Bank, and other valuable fishing grounds were ample reasons for Grimsby's rise as a fishing port when once railways had solved the problem of finding ampler markets than the immediate vicinity provided. For Grimsby has no densely populated hinterland in which it can establish a monopoly in the supply of fish. Grimsby as a great fishing and commercial port owes more to railways than most ports in Great Britain, while Immingham owes everything. As Hull turns north and west for its hinterland so Grimsby turns south and west. It was the demand for fish to feed the growing mining and manufacturing population to the west that first led the railway to take an interest in Grimsby. In 1846 the improvement of the small haven began and soon after that date, with the era of steam trawlers which migrated originally from Hull, began the prosperity of the port. In 1841 its population was only 3600: in 1921 it was over 83,000. It is the premier fishing port of the world: its vessels go north to Iceland, the Barents and White Seas and Davis Strait, and south to the Bay of Biscay and the coast of Morocco: its fleet numbers about 700 and there is landed annually a sixth to a quarter of all the fish landed in Great Britain. In 1923 Grimsby's catch was 178,000 tons. Most is cod, haddock, and other white fish: herring are comparatively few. Yet Grimsby has no monopoly among Humber fishing ports and its rise did little to injure the interests of Hull. Hull, with about 300 steam fishing vessels, handles about half the annual weight of fish that comes to Grimsby, and it is too far inland to attract many herring boats. But beyond this a great deal of fish caught by Hull trawlers in the North Sea is carried direct to Billingsgate, London, from the fishing grounds by fast "carriers."

In general trade the chief import of Grimsby is timber for the coalfields and manufacturing areas, and the chief export is coal. The Baltic lands that send the one demand the other. Butter and other food products from northern Europe also come in. But trade is not restricted to these commodities. Six miles farther up the Humber lies Immingham Dock, which may be called the newest port of England. It was made by the G.C. Railway (now L.N.E.R.) in 1912 on the site of a long-forgotten haven of olden times. By this means the G.C.R. obtained a port on the Humber suitable for far larger vessels than Grimsby could accommodate and nearer to the South Yorkshire and Derbyshire coalfield. The site is particularly advantageous in relation to the north Lincolnshire iron works at Scunthorpe and Frodingham, and the concealed coalfield of the Trent valley, though for the latter a new port on a dredged and deepened Trent could probably prove a serious rival to Immingham. Immingham is solely a commercial port; its official title of Immingham Dock describes it. There is no town and only a small population resident in this artificial creation. Residentially it turns to Grimsby with which it is connected by electric tramway along the Humber embankment. Coal and textiles are the chief exports: timber, iron-ore and grain the chief imports. In virtue of its capacity to accommodate large ocean-going steamers, too big for Grimsby, its hinterland is all northern England, and its steamer connections are world wide*.

* I am indebted to one of my students, Mr C. Midgley, M.Sc., for much useful information contained in his degree thesis on the East Riding of Yorkshire.

XVIII

THE YORKSHIRE REGION

C. B. Fawcett

THIS Yorkshire Region includes most of the ancient county of York; though the northern edge of that county in the valley of the Tees and Cleveland is part of North-East England, and the area drained by the River Don and its tributaries in south Yorkshire together with the Humber lands is also dealt with in chapters XVI and XVII.

The region obviously falls into three well-marked north-south belts of structure and relief. The western and largest is the Pennine Highland (see chap. XIV), in which only the coalfield remains to be considered. The central belt is the broad lowland of the Vale of York formed by denudation on the less resistant rocks of the Triassic belt which are now for the most part hidden under recent glacial and alluvial deposits. East of this are the alternating uplands and lowlands of east Yorkshire including the upland of the North York Moors and the low Vale of Pickering, respectively on limestones and clays of Jurassic age, the chalk upland of the Yorkshire Wolds and the drift-covered lowland of Holderness.

Central and east Yorkshire is the northern part of the English Lowland and shares its characteristic belted structure and relief. But it is noteworthy that in the southern half of Yorkshire the Jurassic belt is represented only by a very narrow outcrop, and the Cretaceous belt ends in the northern scarp of the Wolds.

Yorkshire is thus in structure and relief, as in many other respects, an epitome of England. It falls into several distinct sub-regions, with only four of which we are concerned in this chapter. These are, in order of increasing populousness, the North York Moors, the Vale of Pickering, the Vale of York, and the West Yorkshire Coalfield, the last two of which overlap to a considerable extent. (Fig. 42, p. 324.)

THE NORTH YORK MOORS

The North York Moorland is a dissected plateau which occupies the north-eastern corner of Yorkshire. It is oval in shape with a length from east to west of some 35 miles (56 km.) and a north-south breadth of about half that distance. A large part of its surface is more than a thousand feet above sea-level though its highest summits do not reach 1500 feet (460 m.). It is formed on impure limestones (Oolitic and Liassic) which have been subjected to slight up-folding along east-west axes, and its north and north-east edges are partly determined by fault-line scarps. This north-east scarp and the eastern edge of the plateau are cut by the sea in a picturesque cliffed coast; while to the west and north-west it is bounded by a steep erosion scarp which overlooks a lowland of Liassic and Triassic clays. To the south the limestones dip under the clays and alluvium of the Vale of Pickering.

This moorland resembles the Pennine Highland in many respects, but owing to the absence of pure limestone no part of it approaches the karst type; there is also much less bog on it than over much of the Pennines. It is for the most part a heather moor with grass-farming on the lower slopes and in the dales merging at the foot of the hills into the richer and more varied agriculture of the lowlands. The population on the moorland is very scanty and the dales contain only a few small villages and scattered farms, with some mining villages to the north-east. The lowland markets at the foot of the hills to the north, west, and south are sufficiently near to serve the dales on those sides; and hence the only town in the region is Whitby on the estuary of the River Esk, market town and seaport and seaside resort in one, with less than 12,000 inhabitants.

Since the middle of the 19th century its deposits of iron-ore have formed the chief wealth of the North York Moorland. This deposit is a phosphoric ore (minette) occurring among the Liassic beds. It was quarried in the Esk valley and exported to the Tyne under the name of "Whitby stone" more than a century ago. But it only became important after the middle of the 19th century, when it was discovered and worked where it outcrops on the sides of the deep-cut valleys and on the northern scarps of the plateau. From this northern edge the main seam of the ironstone dips gently southward, with some slight interruptions by small faults and foldings; but many of the deeper valleys cut into it even on the southern slopes. The ore is at its best near the northern edge where it outcrops. As it recedes southward it becomes deeper and so less accessible, the bed becomes thinner and more broken by interstratified bands of shale, while the ore itself contains a smaller proportion of iron. It is now mined chiefly in the northern area of the Cleveland Hills; and the southernmost mines in Rosedale are abandoned. Mining is almost wholly from drifts or adits which follow the ore in from its outcrop on the hillside rather than from pit-shafts which are used chiefly for ventilation. Almost the whole of the ore is sent to the furnaces of the Teesmouth district.

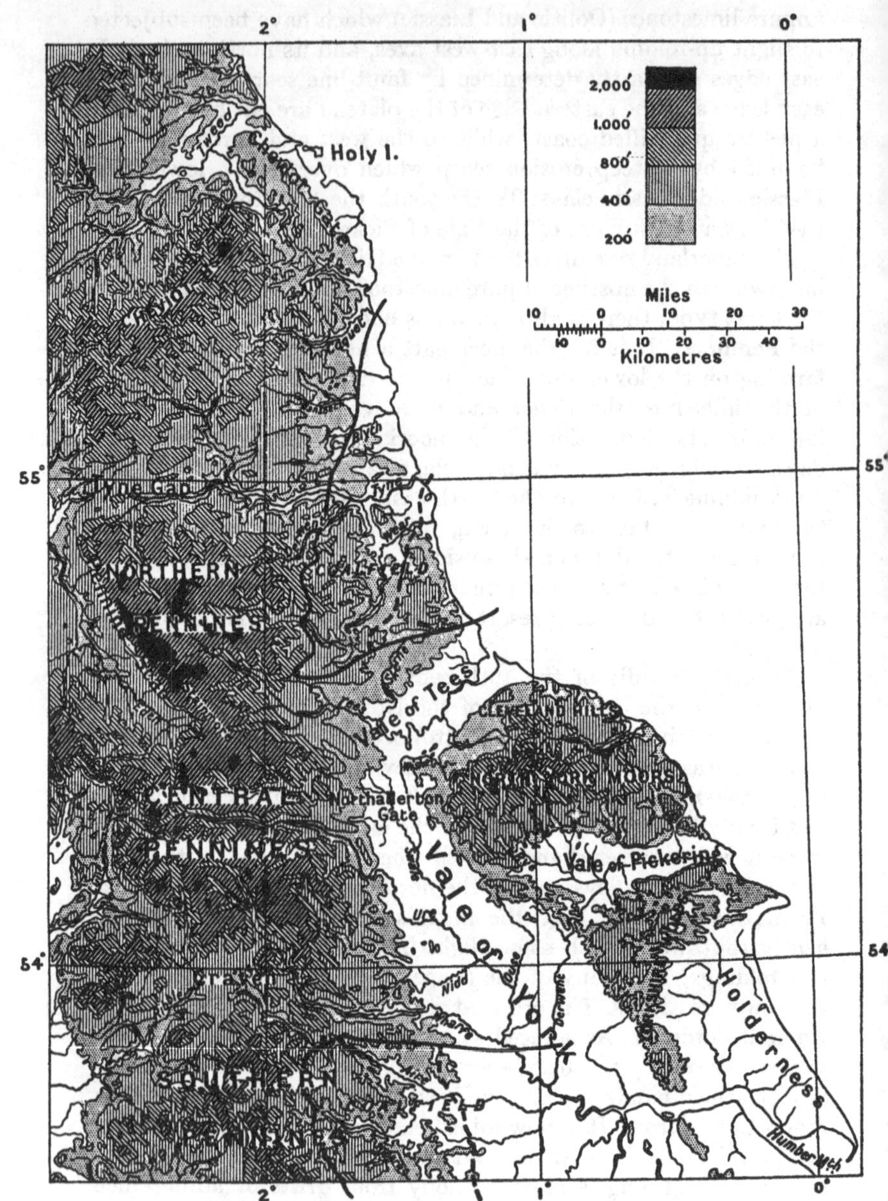

Fig. 42. The Pennines and the North-Eastern Regions: Orographical Map, with coalfields.

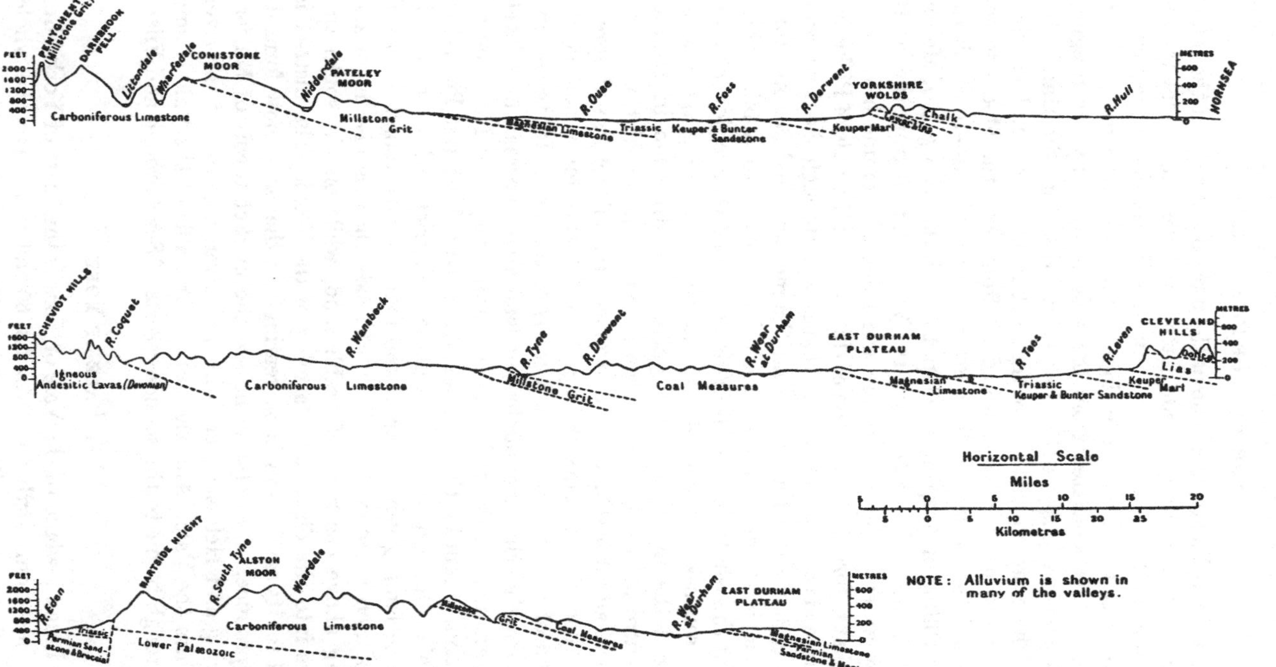

Fig. 43. Geological Sections in North-Eastern England. Upper Section runs nearly WNW.–ESE., about Lat. 54°; Middle Section runs nearly NNW.–SSE.; Lower Section runs W.–E., about Lat. 54° 45′.

The Vale of Pickering

The Vale of Pickering is one of the most distinctive of the minor regions. It is essentially a wide, flat-floored west-east valley formed by erosion on the Jurassic clays between the dip-slope of the Jurassic limestone belt and the scarp-slope of the Cretaceous rocks. At its west end it is separated from the Vale of York by the irregular hills known variously as the Howardian Hills and the Hambleton Hills. These are nowhere more than 600 ft. in height, and less than four miles wide, and are marked off from both the Moors to the north and the Wolds to the south by distinct gaps—to north the "dry" Gilling or Ampleforth Gap, to south the river gap of Malton or Kirkham. At its eastern end the Vale is cut off from the North Sea by an even lower and narrower ridge of moraine hills. It is clear that before the deposition of this moraine the Vale of Pickering was drained eastward by a river that flowed through the depression which is now Filey Bay; and that the moraine, and perhaps for a time the ice-front, dammed back its waters into a lake which covered the whole floor of the Vale and found its overflow channel to the south-west where it cut out the Kirkham gorge and so produced the most extensive glacial river diversion in Britain. The centre of the Vale is on the alluvial deposits of this glacial lake-bed, and it is still a flat, ill-drained area subject to widespread winter floods. An outstanding result of this formation is the location of the settlements and roads of the Vale. Its towns and villages are at the edge of the flat land above normal flood level, and both main roads and railways form a ring round the central flood land, except for the transverse route from Malton to Pickering. The Vale is an agricultural and pastoral district with a broad belt of arable land, given to cereal and fodder crops, round a central area of rich meadow. On the north side there is also moorland grazing above the zone of arable land, while on the south this merges into the distinctive farming of the Wolds. The district is famed for its cattle- and horse-rearing and the cattle now form its chief produce. Its only towns are the market towns of Malton, Pickering and Helmsley, each in the entrance to one of the river ways into the Vale. And the important traffic of the Vale is now the through traffic to the seaside resorts of Scarborough and Filey.

The Vale of York

The broad lowland of the Vale of York is the heart of Yorkshire. It extends from north to south for 60 miles through the whole length of the shire with a width in its southern half of some

30 miles (*c.* 50 km.) between the Pennine foothills and the parallel westward scarp of the Wolds*. Just north of the latitude of York the direction of the scarp changes, and from here the lowland rapidly narrows to a width of some 10 miles in the Northallerton Gate where it merges with the Vale of Tees. The east and west edges of the Vale of York are clearly defined by its bordering hills; but it has no such clear limits to north and south. For ages its southern edge was in fact formed by the fenland about the head of the Humber, which extended almost to the foot of the Pennines and so separated the populous agricultural districts of the Vales of York and Trent by an area of marshland. In the north the River Tees is the historic boundary of Yorkshire; but in modern times the development of communications and industries has brought the whole drainage-area of that river fully into the zone of influence of the North-East Coast region.

It seems probable that the Vale owes its origin to the river captures which have concentrated the drainage of more than a hundred miles of the eastern slopes of the Pennines into the single estuary of the Humber, which cuts through the Jurassic and Cretaceous escarpments where they are narrowest and lowest. But in detail it is not possible to trace these captures, or any direct relation between the Pennine rivers and the gaps in the eastern scarps, because of the vastness of the subsequent erosion and the later modifications of surface detail in the Glacial Period.

Much of the surface of the Vale is occupied by ground moraine; and there are marked lateral moraines along its borders, and at least two important series of terminal moraines stretching across it in addition to that which forms its northern border in the Northallerton Gate. Of these terminal moraines the most continuous and important is the broad low ridge extending westward from the north-east corner of the Wolds to the foot of the Pennine slopes between the Rivers Wharfe and Ure. This ridge is cut in river-gaps by the Derwent above Stamford Bridge and by the Ouse at York. It is in general about 50 ft. (15 m.) above the low, and often marshy, plain at each side of it; and it has long formed the chief transverse route of the Vale. The glacial soils of the Vale are very varied in character and include some almost barren areas of light sands as well as boulder clay, while the irregular morainic topography gave many marshy areas, most of which are now drained. The variety of soils is increased by widespread deposits of river gravels and alluvium on the flood plains of the rivers and in moraine hollows; and as a whole the Vale is a fertile agricultural region.

*** The upper section on Fig. 43 crosses the Vale obliquely.**

Its mean annual rainfall is from 25 to 30 inches and it is the northernmost part of England in which the mean July temperature reaches 60° F. (15°·6 C.). It is thus the most northerly extensive area in which wheat is a main crop: so that in its climatic character and its agricultural development, as well as in its structure and relief, it is essentially a part of the English Lowland.

The population of the Vale is mainly agricultural and is spread over its surface in a large number of small villages and farmsteads. These are distributed primarily in relation to the cultivated lands, and next to that the dominant factor in their location has been the avoidance of flood land. Hence the villages are for the most part either on low eminences in the Vale or towards its edges. The centre of the Vale is in fact less populous than its margins, though the contrast is less marked here than it is in the Vale of Pickering. There are a number of small market towns scattered at intervals of from 8 to 16 miles (13–26 km.) apart at nodal sites; here also, except for the chief market in York, the more important of them are near the edges of the Vale and several are also foothill markets of the upland country, as for instance Ripon, Thirsk and Market Weighton.

The city of York, which has given its name to the shire and the region, was the chief town in the northern half of England for more than a thousand years before the Industrial Revolution, though now with about 85,000 inhabitants it is less populous than the eight largest industrial towns of Yorkshire. Like most towns of long standing it owes its importance to its geographical location. It is situated near the middle of the Vale at the crossing of the two most important natural routes of that region, the north-south waterway formed by the chief river and the east-west landway on the York moraine ridge. The city was at the head of the tidal portion of the Ouse before the regulation of that river in the 18th century* and it was accessible to the small sea-going vessels of the Middle Ages, while river boats could also ascend the Ouse and some of its tributaries to the edge of the Vale. York is on the first site up-stream at which there is high ground, well above flood level, near to both banks, and here is the shortest crossing of the flood land. These facts explain its development as the chief town of the Vale; while the importance of that area as an agricultural district made York one of the largest market towns of England.

The main route between the English and Scottish Lowlands lies along the coastal sill east of the Central Highlands because there

* The tides are now felt as far as Naburn lock, seven miles down-river from York.

is no correspondingly continuous lowland route by the west coast; and until the railway era the west coast route was almost negligible. The location of Yorkshire on this main route and in the northern end of the English Lowland has been the leading factor in its history. It was for many centuries an advanced post of European civilisation towards the less civilised north-western regions of Britain. Here at York (Eburacum) was the military capital of Roman Britain, in the fertile lowland and so an important base of supplies, but within reach of the threatened frontier. Later it became the capital of the Angle kingdom of Northumbria; and when in the 7th century Britain was divided into two archiepiscopal provinces, York was made the capital of the northern one, and it still remains the ecclesiastical capital of a northern province shrunken by the loss of Scotland. During the Viking Period the part of Angle Northumbria to the south of the River Tees became the Danish kingdom of York; but when the Danelaw was conquered and organised into shires by West Saxon kings, the remoteness of this region hindered a similar re-organisation here, and it became first a vassal state and later one shire. Yorkshire owes its position as the largest county of England largely to its remoteness from Wessex and London.

The chief elements in the population and economic life of the city of York are directly due to its position. It is in the first place the central market town of a large agricultural district, and hence a focus of trade and shopping. Next it is an ecclesiastical capital and one of the "show" cities of England and so a centre of attraction for tourists. It is also an important strategic centre and is still a garrison town and the headquarters of the Northern Command. The division of Yorkshire into three separate administrative and registration counties has deprived York of its position as a civil capital, and the growth of the West Yorkshire conurbation has made Leeds the chief centre of the population and economic activities of Yorkshire, so that York is no longer an important regional capital. In the railway era the importance of York as a node of routes was greatly enhanced by the focussing of several routes from the south, south-east and south-west* on it as the principal junction at the southern end of the corridor between the Central Highlands and the North Sea, a position comparable in this respect to that of Edinburgh at the northern end, and the city owes much of its recent growth to the railway system.

* Before the railway amalgamations of 1921 five of the more important companies carried traffic into York from the south, and one carried it on northward.

THE WEST YORKSHIRE COALFIELD AND "WOOLLEN" DISTRICT

The West Yorkshire Coalfield is the northern section of the largest coalfield of Britain. The older part of it, where the Coal Measures form the surface rocks, lies on the lower eastern slopes of the narrowest part of the Pennine Highlands, almost wholly in the drainage-area of the Rivers Aire and Calder; the newer part extends eastward under the younger rocks of the Vale. It is not marked off from the central section of the coalfield, usually known as the South Yorkshire Field, by any decided boundary, though the high ridge which forms the water-parting between their valleys is a partial barrier. The distinction between the west and south Yorkshire industrial regions has its bases in the facts that they are centred in different valleys and that each is specialised on a distinct industry for which its natural resources are suitable. This West Yorkshire industrial area is the "Woollen District." Here are concentrated over four-fifths* of the total number of workers in this industry in Great Britain, in a group of industrial towns along the River Calder and part of the River Aire which have grown towards each other during the last hundred and fifty years so that they now form a continuous urban area which contains about one and a half million inhabitants.

This West Yorkshire conurbation occupies an oval area whose major axis through Leeds and Huddersfield stretches for nearly 25 miles (40 km.) from north-east to south-west, while the minor axis through Bradford and Dewsbury is about half that length. It includes more than forty towns (boroughs and urban districts) of which the four just named, together with Halifax, are the largest and most important. Each of the five has a population of a hundred thousand or more and together they contain more than a million people. Leeds, with nearly half a million inhabitants, is much the largest of these towns; it is the foothill town of the district, situated on its chief river, the Aire, just where the narrow valley of Airedale opens out to the lowland, and since the 17th century it has been the head of river navigation. Next in size is Bradford, with three hundred thousand people, centred in the valley of a small tributary of the Aire south of that river a few miles west of Leeds and well within the highland. Thus, half the conurbation is in the Aire valley, mainly to the south of the river and stretching up towards the watershed where it meets the suburbs of the Calder valley towns which lie for the most part along the northern side of that

* 84 per cent. at the census of 1921.

river from Wakefield to Halifax. Beyond the Calder Huddersfield is in the valley of the Colne, a small tributary from the south-west.

The specialisation of processes in the modern woollen industry has led to some corresponding local specialisation within the "Woollen District" and emphasised the concentration of the whole industry within a small area by increasing the volume and frequency of transport of the material at various stages from one specialised factory to another. Bradford is the centre for wool combing and worsted work, and as such is the chief buyer of the raw wool and exporter of yarn and cloth; it is now the highly specialised focal town of the woollen industry. Dewsbury and its neighbours form the area of the manufacture of cloth from re-worked yarns, usually known as "shoddy." Halifax is prominent in the making of carpets and other heavy woollens; while the Colne valley towns make fine cloths and tweeds.

It should be noted that although Leeds is the largest town in the "Woollen District" it is not now a principal centre of that industry, which occupies only fourth place among the industries now carried on in the city. Here the wholesale clothing manufacture employs a fifth of the insured workers, and is closely followed by the distributive trades and general engineering, which is mainly concerned with the making of locomotives and transport material and factory equipment. Next come the building trades, the woollen industry, and printing, each with somewhat less than half the numbers of those in the first group. The six groups of industries named employ nearly two-thirds of the workers of Leeds, while the rest are distributed among ninety other registered industries. This great variety of industries marks off Leeds from the other towns of West Yorkshire; and the prominence of its distributive trades, which is a result of its position as a chief focus of routes and a commercial centre, indicates its standing as the chief distributing centre of Yorkshire. The outer limits of its influence as a principal distributing centre are almost identical with those of the Yorkshire Region, as it was defined in the opening paragraph of this chapter; but they also include all the East Riding.

XIX

NORTH-EAST ENGLAND

C. B. Fawcett

In the north-west of Yorkshire, south of Stainmore Pass, the Pennine axis is crossed obliquely by the axis of the transverse upfold marked by the highlands of the Lake District and of north-east Yorkshire. These intersecting upfolds have thrown the strata of this central part of Britain into four "basin" areas in each of which the general dip is away from the area of intersection just referred to. The north-easterly basin of these four forms the region with which we are here concerned; its "basin" character is accentuated by the position of the Cheviot section of the Central Highlands, which forms its north-western edge. The centre of the "basin," as indicated by the direction of the general dip of the rocks, lies under the North Sea some miles east or south-east of the mouths of the Rivers Tyne and Wear; so that our region is in fact rather less than half of the "basin," the other and larger portion being submerged. (See Fig. 34 and the middle and lower sections of Fig. 43, pp. 265 and 325.)

This region of North-East England is, in many respects, one of the most clearly defined of the principal geographical regions of Great Britain. It occupies the eastward slopes of the middle of the island, extending along the coast for about 100 miles (*c.* 160 km.) from the high cliffs in which the North York Moors meet the sea to the neighbourhood of the River Tweed, where the igneous dome of Cheviot looks across the narrow strip of coastal lowland which has for ages been the chief passage way between England and Scotland. Nearly midway the coast is breached by the estuary of the Tyne in latitude 55° N., half way between the northern and southern limits of Britain in Lizard Head and the Shetland Isles.

The landward borders of the region are formed by the highland rim of the basin in three principal sections, the Cheviot Hills to the north-west, the Northern Pennines to the west, and the North York Moors to the south-east respectively, separated from each other by two important breaks in the rim which give lowland routes to the rest of Britain by way of the Northallerton Gate to the south and the Tyne Gap to the west. The main mass of each of these highlands rises above the thousand foot contour and thus well above the upper limits of cultivation here; and they are therefore covered by extensive areas of uninhabited moorlands which separate the populous lowlands they enclose from other inhabited land, except in the two breaks just mentioned and along the coastal margins, where alone there is continuously inhabited land.

Of these breaks in the enclosing rim of highlands the Northallerton Gate is by far the most important. It is the widest and lowest, being about ten miles wide below the 200 foot (61 m.) contour level and twice that width between the moorlands: and it offers a direct and easy route to the most important region of Britain, the English Lowland. The Cheviot Gate to the north, between the highland and the sea, is longer, narrower, and more obstructed by low hills, so that it offers more difficult routes; also it forms a way to the smaller and less populous Scottish Lowland which lies beyond the Southern Uplands; so that it has been of much less importance than the southward "Gate." This contrast between the principal ways through its bordering highland was of importance in determining the fact that our region is North-East England rather than South-East Scotland. Midway in the western border the Tyne Gap separates the Cheviot and North Pennine sections of the Central Highlands by a narrow valley, hardly more than a mile wide where it crosses the main divide at an altitude of about 465 ft. (142 m.). This route leads only to the smaller Solway Lowland; and therefore has always been of less importance than those to south and north. But no part of the surrounding highlands is so high as to be impassable in good weather; and there are many tracks and some roads across them, in addition to three railway lines.

Except for superficial deposits of glacial drift, coastal sands, and alluvium, this region contains no strata younger than the Liassic and Oolitic beds of the Jurassic series. And it is possible to give a coherent reasoned account of the present surface features on the assumption that they have been formed by the agents of normal subaerial denudation acting on a peneplane of Tertiary age which intersected the strata at a low angle, with a gentle slope down towards east-south-east, and whose remains still form the plateau surface of the higher moorlands*. This normal erosion was interrupted by the last Ice Age, during which most of the area was submerged by ice; and the surface has been modified in many details by the work of the ice-sheets and the mantle of drift they left on it. There are also features in the topography due to post-

* *Cf.* Ch. XIV.

glacial changes in the relative level of land and water. The most prominent results of these changes are the sunk meanders of the middle courses of the Wear and Tees and breaks of slope on most of the streams, besides extensive river terraces and fragmentary raised beaches. But the larger features of valley and hill are of pre-glacial origin and result from the action of normal denuding agencies.

One of the most important dividing lines in the region is formed by the westward-facing Permian scarp which extends from the coast just south of the Tyne estuary south-south-westward for some 25 miles (40 km.) through the centre of County Durham (see Fig. 43). It is cut by the Wear at Hylton about five miles from its mouth, and from there forms the steep eastern edge of the middle Wear valley with a general altitude of some 300 to 400 ft. above the river and ceases, with the valley it has bordered, at Westerton Hill, east of the bend of the Wear at Bishop Auckland. Immediately south of this, in the Shildon district, which is also the southern end of the coalfield, the Lower Carboniferous and Permian rocks meet across the valley of the Tees without any prominent surface features to mark their junction.

The whole area of North-East England to the west of the Permian scarp is formed on rocks of Carboniferous age, except where they are broken by the igneous rocks of the Cheviot, a few outcrops of Silurian rocks, and numerous basaltic sills and dykes, all of small extent. These older rocks thus underlie about four-fifths of the total surface of the region. The remaining fifth, in the south-eastern corner, is occupied in roughly equal thirds by rocks of Permian, Triassic and Jurassic age.

The Rivers

The rivers of North-East England fall into two distinct groups: those of the northern group, which includes the Tyne and all the streams north of it, have their courses wholly on the Carboniferous rocks, while the Wear and Tees farther south also pass over younger strata. The northern rivers have often been referred to as an example of a system of consequent streams, flowing in parallel courses down the dip-slope from the Cheviot Hills to the sea. In spite of many small irregularities this simple system is complicated by only one considerable modification, where an irregular subsequent valley has been developed by erosion along a belt of relatively less resistant shales at the foot of the Cheviot Hills. This valley is occupied in its eastern half by the Breamish-Till river round the foot of Cheviot and the spreading tributaries of the Coquet above Rothbury, and next by the Elsdon Burn and the lower course of the River Rede to the North Tyne, and it merges into the western end of the Tyne Gap by the valley of the Tipalt Burn. The valley is shut off from the lowland of south-east Northumberland by an irregular range of sandstone hills, breached in river-gaps by the Aln, Coquet and North Tyne, and east of the Till valley by several dry gaps which bear witness to the success of the Till in the struggle for territory. But though this valley is thus continuous from the Tweed to the Tyne Gap, and at its highest points between the Coquet and Rede, and again between the North Tyne and the Tipalt Burn, its floor is barely 800 ft. (244 m.) above the sea, it is not followed by any railway or important road except in the Till valley because the districts which would be connected by such a route have no need of it. The range of hills which separates it from the lowland is not an important feature, though many of its summits rise above the thousand foot line; it has not yet acquired any general name.

The Tyne itself follows a straightforward course eastward from the angle between the Pennine and Cheviot Highlands; and the complex questions of the origin of the Tyne Gap and its relations to the Solway rift and the valleys of the Southern Uplands are outside the scope of this essay.

With slight exceptions, the most important of which have been referred to, all the rivers flow eastward or slightly south of east so long as they are on the Carboniferous rocks: but in the south-eastern part of our region the passage to other rock series has led to important changes in the direction and character of the valleys. The upper Wear, down to its great bend at Bishop Auckland, follows the general "consequent" east-south-east direction. Here it comes to the foot of the Permian scarp and makes a right-angled turn to flow north-north-east along the foot of that scarp as far as Chester-le-Street. This "subsequent" valley is in the heart of the coalfield and the records of numerous borings and shafts sunk in the search for and working of the coal have enabled us to reconstruct the pre-glacial topography. Before the deposition of the glacial material the valley was more than twice as deep, below the crest of the scarp, as it is to-day and continued northward to join the Tyne just west of Gateshead; so that the pre-glacial Wear was a tributary of the Tyne and its waters reached the sea beyond the present northern end of the scarp. The lowest section of the Wear valley, from near Chester-le-Street to the sea, is cut through the northern end of the east Durham plateau. This valley is due to a glacial or post-glacial diversion of the river caused by the blocking up of its northern outlet and the overflow of its waters along this route. Thus the lowest section of the Wear is also the youngest; while the uppermost is the oldest part of this composite river.

Farther south the Tees flows in a course generally parallel to that of the upper Wear across the Lower Carboniferous and Permian strata: though it does not anywhere flow over the Coal Measures, on which the subsequent valley of the middle Wear has been formed. In detail the course of the Tees west of Darlington has been determined by a number of minor fault-lines, and there are two main older valleys, one stretching south-eastward from Cross Fell and one eastward from Stainmore Pass. On the Triassic lowland the Tees turns north-eastward and flows to the sea in a course parallel to that of the middle Wear; but it is worthy of note that this vale of the lower Tees is a broad open lowland in strong contrast to the constricted valley of the middle Wear; and also that the Tees is nowhere close to the Jurassic scarp of the Cleveland Hills which forms the south-eastern edge of its vale.

Between the middle Wear and the lower Tees lies the curious valley of the Skerne. This river flows in a broad, shallow, and often marshy valley, which is almost wholly on the Permian rocks, in a direction directly opposite to that of the larger rivers to join the Tees just below Darlington. Its apparently anomalous course deserves careful investigation. And in Cleveland the upper part of the "obsequent" (?) Leven and the whole course of the Esk lie in a valley between the northern part of the Cleveland Hills and the North York Moors which is a continuation of the line of the upper Tees.

One other feature which should be mentioned here, though its relations to the river system are not yet fully known, is the Ferryhill Gap which connects the valley of the Skerne to that of the Wear across the southern part of the Permian scarp. It is not occupied by any stream; but its position almost in line with the valleys of the River Browney to the north-west and part of the Skerne to the south-east suggests that it may have begun as a river-gap, before the stream which made it was captured by the ancestor of the middle Wear. It has also been suggested, with much probability, that it was an overflow channel of a glacial lake during some part of the Ice Age. It is a very spectacular gap of marked regularity and simplicity of form. Its human importance is wholly due to the fact that it provides an easy route for the main "East Coast" railway; the older Great North Road climbed the hills two miles to the west rather than venture into its marshy and winding bottom levels.

The Coast

The greater part of the coast lies in an almost straight line from north-north-west to south-south-east, ending in slight curves marking respectively the projection of north Northumberland, whose apex is the Farne Islands, and the re-entrant of the south Durham and Cleveland coast with its apex in Tees Bay. It is in general a rocky coast and the simplicity and regularity of these lines, together with the fact that they cut diagonally across the various strata, suggest that it may be in part due to the foundering of the North Sea basin along definite fracture zones; and there is some evidence for this view from the southern part in fault-lines which mark the edges of the Cleveland Hill massif. But most of the present regularity of the coast-line is in fact the result of glacial deposition and coastal erosion. A removal of the glacial drift would transform a large part of south-east Northumberland into a wide shallow bay of very irregular outline, and farther south a similar change would greatly extend the area of Tees Bay; while it would make little difference to the outline of the rest of the coast.

The upthrust of the Tertiary volcanic mass of Cheviot raised the older rocks near it in a low dome, and as a result the coast-line north of the mouth of the Coquet is nearly an arc of a circle drawn round the Cheviot about twenty miles from its summit. The chief interruptions to the regularity of this coastal curve are several small promontories and the outlying rocks of the Farne Islands which are fragments of the resistant intrusive basaltic rocks of the Whin Sill and several dykes, and the somewhat larger Holy Island with the bay between it and the mainland. On projecting points the basaltic rock forms low cliffs, such as that on which Bamburgh Castle stands on a fragment of the Whin Sill, above Carboniferous limestone.

The Holy Island consists of three small rocky areas, each about a mile long and less than half a mile wide, in the east, north-west and south-west respectively, and a central area of sand dune and alluvial deposits which have filled what was once a lagoon. The bay is so shallow that the "island" is an island only at high tide and is reached by a road across the sands at other times; it is almost enclosed by two long sand spits. On the mainland both to north-west and south of the bay, as on the island, alluvial areas shut in by sandbanks show the progress of the process of forming and filling in lagoons which will in time join the Holy Island to the mainland.

The formation of sand bars across the mouths of small inlets, or to inshore islands, and the filling in of the lagoons so formed by silt and blown sand, has already destroyed many of the smaller irregularities of this coast. The rock which was formerly the innermost of the Farne Islands is now joined to the land by sand dunes, as is also Beacon Point near Newbiggin. Just south of this the inner portion of Beadnell Bay is filled up by a small delta behind a sand bar. But the most important of these coastal changes on this "tombolo" coast is that much farther south where the rock on which Hartlepool is built has been changed from island to peninsula. Here the bar of sand and silt is complete on the north side; but half the area of the lagoon remains as the "pool," now transformed into the harbour beside which the two towns of Hartlepool and West Hartlepool have grown.

From Warkworth to South Shields the coast cuts across the Coal Measures. In its main features it is similar to the coast farther north, though it is rather less irregular in detail. There is however a very great contrast in the human relations of the two sections. Between Warkworth and the Tweed there is neither harbour nor dock, and no serious attempt has been made to construct one. But from the Coquet to the Tees the need for transporting the coal has led to the construction of ports at intervals of a few miles. The leading ports are on the estuaries of the Tyne and Tees; but there are also coal ports at Amble, Blyth, Sunderland (where the chief docks are outside and south of the river mouth), Seaham Harbour and Hartlepool; the natural facilities, however, are good only at Hartlepool.

From South Shields to Hartlepool the coast is for the most part a rocky beach backed by the low cliffs (from 20 to 100 ft. high) of the Magnesian Limestone plateau of east Durham, with several stretches of sand where the existence of shallow bays allows it to accumulate. By far the most extensive sand beach is that on the seaward side of the Hartlepool bar.

Between Hartlepool and Saltburn the lowlying Vale of Tees, formed on the easily eroded marls and clays of the Triassic and Liassic series between the Magnesian and Oolitic Limestone heights of east Durham and Cleveland, comes to the sea in the only considerable stretch of really flat coast in our region. Part of this is occupied by the mudflats and sandbanks of the Tees estuary, most of which is now being reclaimed; but south of the river mouth is one of the longest continuous sand beaches of Britain.

The coast of Cleveland south-east of Saltburn well justifies the name of the "Cliff" land. The cliffs are generally much higher than those of Durham or Northumberland, and are frequently interrupted by small gullies and valleys. Here the iron-ore of the Cleveland Hills determined a partial development of a few small ports in the 19th century for the export of the ore to the Tyne and Tees, but the nearness to the Teesmouth furnaces by land has prevented any important development of this kind.

At intervals along the north-east coast there are small seaside resorts which serve the dense industrial population. Beaches of the fine and firm sand beloved of children are not numerous; but nearness to the industrial areas has helped the growth of such resorts as Whitley and Cullercoats just north of the Tyne, Ryhope south of the Wear, and Seaton Carew north of the Tees. The most favoured coast resorts are those south of the Tees, where the sands of Redcar, Marske and Saltburn offer the best and largest playground of this type. To north and south of the industrial areas and the coalfield the coasts of north Northumberland and Cleveland offer more varied scenery in their alternations of cliff and bay, and of rock and shingle and sand beach; and many small fishing villages, such as Bamburgh, Seahouses and Newbiggin in the north, Staithes, Runswick and the larger Whitby in the south, have found a new prosperity as holiday resorts to compensate in part for the decay of their former occupation as fishing villages.

In so small a region as this the climate varies little from place to place and the variations that do occur are chiefly determined by altitude; and the relief is very closely related to the structure of the underlying rocks. Further, the region is now chiefly engaged in the exploitation of its mineral resources and the industries based on them. Its areas of dense population lie on the coalfield and around Teesmouth. These populous areas are so definitely coastal that the region is usually referred to as the North-East Coast. Of the ten towns whose populations exceed 50,000 only one, Darlington, is not a seaport, and three-quarters of the total population are clustered along the estuaries and the coast between them.

A division into geographical sub-regions, based upon the relief, the rock structure, and the distribution of the population, gives four principal sub-regions, with smaller sub-divisions, which may be tabulated as under (and see Fig. 42, p. 324).

I. The Moorlands and Dales on the Lower Carboniferous rocks with sub-divisions into: (*a*) the Cheviot Gate; (*b*) the Cheviot Hills and Rothbury Forest; (*c*) Tynedale and its tributary dales; (*d*) the Northern Pennines; (*e*) Teesdale.

II. The Coalfield: (*a*) the surface coalfield on the Coal Measures; (*b*) the buried coalfield below the Permian rocks of the east Durham plateau.

III. The Vale of Tees—overlying Triassic and Lower Lias and some Permian rocks.

IV. The Cleveland Hill Country—on Jurassic limestones.

The first of our sub-regions is part of the Pennine and Cheviot Highlands. It occupies nearly two-thirds of the total area but contains barely a tenth of the population. Its general features have been discussed in Chapter XIV.

The coalfield which forms our second sub-region offers a striking contrast to the Highlands. With only a quarter of the area it carries more than two million inhabitants or about four-fifths of the population of North-East England. It borders the coast for 50 miles (80 km.) from Amble to Hartlepool, and some of its mines extend below the sea, while no part of it is more than 25 miles from a sea-port. This coastal location is accentuated by the penetration of the estuaries, and the average length of the journey from pit to port is only 10 miles. The field supplies almost all the important types of bituminous coal, including the hard steam coals of Northumberland, the household and gas coals of which Wallsend is a type, and the coking coals of south-west Durham whose special suitability for iron-smelting adds considerably to the local advantages of the Teesmouth area for that industry. Though there are many minor faults the seams are not in general badly faulted, while the thickness of workable seams ranges from 1 ft. 6 in. (0·46 m.) up to nearly 6 ft. (1·8 m.). North of the Tyne the coalfield is a lowland with a mantle of glacial deposits: but in County Durham its area is occupied by two plateaus of moderate elevation separated by the valley of the middle Wear. West of that river is the lower part of the North Pennine Highland where the Coal Measures outcrop at from 300 to 900 ft. (90–270 m.) above O.D. Here many seams outcrop along the sides of the deep valleys which dissect the highland, and in early stages of exploitation the coal was obtained from quarries and adits. East of the Wear is the plateau of east Durham with a surface from 300 to 600 ft. (90–180 m.) above O.D. This is markedly less dissected than the western plateau, probably because its rocks are more porous and surface streams are fewer, while it has a somewhat less rainfall and is not crossed by any considerable stream except for the post-glacial channel of the lower Wear. Here the coal is reached only by deep shafts sunk through the Permian rocks to depths of more than 500 ft. (150 m.) and the conditions of mining differ widely from those of west Durham.

This coalfield has been continuously worked since the 12th century, when the export of coal from the Tyne to London was already important, and it has been one of the chief areas of evolution of modern mining and its related transport and industries. Many of the older mines of the surface coalfield have been exhausted; but those of the buried coalfield are comparatively new and there are also extensive proven deposits of coal in the Lower Carboniferous rocks to the north-west of the present coalfield, some seams of which are already worked at Scremerston, Morpeth, Hexham and Haltwhistle, so that the region may continue to produce for a long time to come.

The third of our sub-regions, the Vale of Tees, is a broad lowland formed by denudation on the band of less resistant Triassic rocks and lower Liassic clays and extending on to the lower dip-slope of the Permians. Nearly the whole of its surface is covered either by glacial drifts or by more recent alluvial deposits. Towards the south-west it merges into the Vale of York through the Northallerton Gate, and here a hummocky area of moraine material reduces the fertility, and so the population, of the agricultural lowland. This morainic drift forms the southern edge of the valley and diverts two important pre-glacial tributaries of the Tees, the Wiske and the Swale, southwards to the Yorkshire Ouse.

The Vale of Tees is the most important agricultural area of the North-East, but its agriculture is now dominated by the demands of the nearby industrial towns and dairy farming is the most prosperous branch. Its ancient market towns of Stockton and Darlington are both industrialised, and the banks of the estuary are occupied by a dense industrial agglomeration which contains some 300,000 people.

The last of our four sub-regions, the hill-country of Cleveland, is similar in many respects to the first. Its higher areas, above 1000 ft., are similar open moorlands. But its lower dales have more fertile soil and a somewhat better climate than those of the Cheviots and Pennines. And also it contained large deposits of iron-ore which have been mined extensively since the decade 1850–60 and are now nearing exhaustion. This iron-mining has added to the rural population nearly 20,000 people directly dependent on the mines, grouped in small mining towns and villages, and a smaller group of workers round the furnaces of Skinningrove. But most of the ore is sent directly down to the furnaces by the Tees estuary. The population is also influenced by the spread of the outer suburbs of Teesmouth into the accessible parts of the hill country, so that Cleveland retains very little of its ancient isolation and from the view-point of human geography is almost a part of the industrial region.

INDUSTRIAL GEOGRAPHY

The industrial development of North-East England has been based on its coalfield. As has been noted this coalfield is essentially coastal and from the beginning of mining here a large proportion of the coal has been sent out by sea. The sea-borne traffic to other parts of Britain, especially to London and the east and south coasts of England, is still active; and in addition over half of the normal output of about 40 million tons per year is exported. This field supplies about a third of the coal exported from Britain. Shipbuilding began with the building of keels and colliers to carry away the coal. The exhaustion of the local supplies of timber aided the growth of a trade with Norway and the Baltic Lands in the export of coal and the import of timber and ores. The iron and steel industries, which are now second in importance only to coal-mining, became prominent with the discovery of the "black-band" iron-ores in the Coal Measures; and down to the middle of last century these ores formed a principal source of supply. The location of the Consett Iron Works in north-west Durham away from the coast was due to their initial use of such ores; and they are the chief survivors of a number of inland iron works of that period. The transport of the coal down to the waterside for shipment was easier here than in most other coalfields because of the short distances and the fact that the earlier mines were often on the higher land and so the transport was downhill: but the high relief effectually prevented the building of canals and the problem was solved by the development of the wagon ways, the precursors of the railway. The replacement of the first wooden rails by iron (later steel) rails and the whole of the railway development formed a great stimulus to the iron and steel industries. The steam engine came in here primarily as a pumping machine to aid coal-mining and was developed into a locomotive to haul coal to the ports. The chief aim of the first public railway, the Stockton and Darlington, was to carry the coal of south-west Durham to the sea at Teesmouth.

The coal-mining population is necessarily spread over the whole of the coalfield; but its local distribution differs considerably on the two sections. The great depth of the mines in the newer "buried" coalfield of east Durham has limited them to a comparatively small number of large pits, each of which provides employment for some hundreds of men. The greater area of coal worked from such deep shafts gives them more permanence, and their workers are housed in colliery towns with populations of from

5000 upwards. These large pits are in striking contrast to many of the shallow older pits on the surface coalfield to the west. Very few of these older collieries employ as many as 500 men and a large number have less than 100 men each; while the shallow shaft is easily supplemented or replaced by another at some distance away if the underground workings are sufficiently extensive, so that the population here is spread out more widely and in smaller groups. It is an area of pit villages rather than of colliery towns.

The other industries of the North-East are all concentrated round the three large estuaries of the Tyne, Wear and Tees if we include the Hartlepools with Teesmouth. The communications between the estuary towns are made by efficient roads and railways and a busy service of coasting vessels, amongst which there is keen competition; and most of the more important industrial establishments are placed by the waterside.

TYNESIDE

For fourteen miles inland the lower Tyne is the axis of a cluster of industrial towns whose total population is nearly 900,000. The river has been made navigable for the whole of this length, and up to Scotswood bridge there is now 30 ft. (9·1 m.) of water at high tide. The valley here is markedly U-shaped, with a flat floor from half a mile to two miles wide, bordered by steep banks rising from 50 to 150 ft. above the river. This relief has determined the form of the urban area and its chief divisions. The strips of low ground by the river are occupied by docks, staithes, wharves, shipyards, furnaces and other industrial establishments; while nearly all the residential areas and the shopping and business centre are on the high ground. The city of Newcastle-on-Tyne, the focus of this conurbation and the unquestioned regional capital of North-East England, is on the north bank at the lowest point where the river is bridged, some ten miles from the sea. Its site was originally determined by a narrowing of the valley here and the position of a high bluff on the north bank bordered by two small tributary valleys. These valleys strengthened the site for defence by forming obstacles to east and west so that the site was approached from three sides by a steep climb and only on the north was there easy access. The eastern tributary also entered the Tyne almost opposite a similar small tributary on the south bank, and the two valleys led the road down to a ford by more gradual slopes than elsewhere. To add to its value this crossing is in line with the valley of the middle Wear and the Team so that the route from the south came directly to the Tyne at this point. The site was occupied by the Roman station of Pons Elii, round the north of which the Wall was built to reach the estuary at Wallsend. It was later the site of the Angle Monkchester; and it only assumed its present name when the Normans built their "new" castle here. Here the East Coast Route between England and Scotland is joined by the transverse route across the island by the Tyne Gap, and the junction forms the chief node of natural routes in our region. When we note also that Newcastle is nearly central in the coalfield and in North-East England as a whole its predominance as the regional capital is explained.

The Tyne is the chief seaport of the North-East Coast and its trade reflects the life of the region. Measured by the net tonnage of vessels entered and cleared it ranks third or fourth among British ports with some 20,000,000 tons, closely rivalled by Cardiff but far behind London and the Mersey ports. Measured by the value of goods passing through the port, however, the Tyne falls to seventh place, little above Bristol and Harwich and well below Southampton. By far the most important export is coal, which accounts for more than half the value of the exports and nine-tenths of the weight of the outward cargoes. Next in value among its exports are the ships built on Tyneside, followed by the products of its engineering works, and chemicals. The imports are chiefly foodstuffs for the dense population of its hinterland, which is almost limited to North-East England, and raw material, chiefly ores and timber, for its industries and the coalfield.

The principal industry on Tyneside is shipbuilding, in which the district rivals the Clyde for first place. The yards produce vessels of all types from the largest liners and warships down to the colliers. The iron and steel works are largely devoted to producing material for the ship yards, and large quantities are also obtained from the Tees. The engineering industries began early in the Industrial Era to provide machinery and equipment for the collieries and for the transport of the coal, and now produce large quantities of mining machinery, marine engines and ship's fittings, locomotives and railway and tramway equipment both for local use and for export.

In addition to its coal and iron industries Tyneside has long been an important seat of the lead-working industries. It still obtains lead from the Northern Pennines; but most of the metal is now imported, as are also the ores used in its comparatively small copper works.

The cheapness of coal here and the production of waste gas from the iron furnaces have aided the development of electrical engineering on a large scale as well as a chemical industry and glass-making. And the tendency of some industries to concentrate near a port is illustrated by the flour mills and food factories of the Co-operative Wholesale Society at Pelaw.

The advantages of Tyneside as a centre of industries are chiefly its favourable location, the abundance and cheapness of the local supplies of coal and therefore of power, and the accumulation here of an industrial momentum and experience which dates back to the very beginnings of the Industrial Era.

Sunderland, at the mouth of the Wear, is only 15 miles (24 km.) from Newcastle. The estuary of the Wear is a rocky channel and hence it has not been practicable to make it a deep-water harbour by dredging as was done in the glacial drift of the Tyne; so that it is accessible only to vessels of moderate size for a distance of less than two miles*. The trade is almost confined to a large export of coal and the import of timber for the mines. The hinterland of the port covers a very small area because the continuation of the valley of the middle Wear northward by the Team valley to the Tyne leads the coal and other trade of the western and central parts of Durham towards the Tyne ports, and leaves to Sunderland only the special trade of a small but very active part of the coalfield. Its second activity is the building of cargo vessels, largely for the coal trade, and marine engineering. With its immediate suburbs the town counts over 160,000 inhabitants; but it is essentially dependent on the greater regional focus on Tyneside and is not a distinct centre of urban life.

The third of the important estuaries is not on the coalfield, whose southern boundary is near the northern watershed of the Tees, and the development of large towns and an industrial region here dates back only to the mid-19th century. Its beginning may be placed in 1825 when the first public railway brought coal down to Stockton for export; but at that time the lower course of the Tees was shallow and winding. Furthermore, at low tide the river reached the sea by four separate channels no one of which was navigable even for small boats. These difficulties led to the extension of the railway to the site round which Middlesbrough has grown and the foundation of that town as a port for the export of coal. The meeting here of coal with iron-ore imported for inland furnaces to supplement the diminishing local supplies of the coalfield gave rise to a small iron industry comparable to that of other north-east coast ports at the time. But it was not until after the discovery of the Cleveland ores about 1850 and of the Thomas and Gilchrist process for ridding the iron of the phosphoric impurities in 1869–71 that

* *Cf.* p. 338.

Middlesbrough became a great centre of the iron industry, for which the district has a very favourable location. To produce a ton of pig iron requires on the average the collection in the furnace of 2½ tons of ore, 1½ tons of coke produced from some 2 to 3 tons of coal and ½ a ton of limestone for fluxing purposes. If we add the heavy and bulky materials used in the construction and lining of the furnaces we get some measure of the extent to which the success of an iron industry depends on a favourable location for the collection of these materials. Middlesbrough is from 15 to 25 miles away from the mines of south and south-west Durham which produce a suitable coal: it is a similar distance from the great limestone quarries of Ferryhill and Teesdale, while the iron mines of Cleveland are from three to twenty miles away; and the fact that a large proportion of the furnaces is close to the estuary facilitates the use of imported ore. In the latter decades of the 19th century the production of Cleveland ore was at its maximum and Middlesbrough flourished. Though more than half of its ore is now imported, the momentum of that period and the more lasting advantages of the nearby supplies of coal and flux keep it among the best locations for the industry. Since the War this district has produced from one and a half to two and a half million tons of iron per year, a fourth to a third of the total made in this country.

The iron industry naturally gives rise to many related industries. These may be classed in two groups, both of which are carried on here. The processes of coking the coal and smelting the ore produce more heat than is necessary for the smelting as well as vast quantities of impure coal gas. This surplus heat and fuel is used in subsidiary processes of which the most important are steel making and the working up of the steel and iron into various products beginning with rails, plates and girders, and extending to ships and bridges. An important recent development is the production of corrugated iron goods on a large scale. The other and less important group of industries utilises what were formerly waste products and also use, in some cases, the surplus power referred to. The slag, which was formerly got rid of as cheaply as possible, now forms the basis not only of slag wool, but of an important industry in the production of tar macadam and other road-making material, of cement, and of concrete and slag bricks on a large scale. This newer industry has given a value to the formerly waste heaps of slag and is now quarrying them away more rapidly than they are being renewed from the blast furnaces.

The chemical industries of Teesmouth are mainly situated to the north of the river. Here borings have revealed the presence in the lower layers of the Triassic rocks of thick beds of salt and of gypsum. The former is obtained by forcing down water and pumping up the brine, and the latter may be reached by mining. To these resources are added the supplies of fuel from the coalfield just to the north and by-products of coking and the iron industry. This area has developed considerably since the War and seems likely to become a principal seat of the heavy chemical industries. At present its output consists chiefly of salt and of artificial manures, the nitrogen in which is obtained from the atmosphere. The iron and shipbuilding industries are carried on on both banks of the Tees and at the Hartlepools.

Unlike the Tyne ports, which form one port for Customs purposes and records, the Teesmouth towns are grouped in the three ports of Middlesbrough, Stockton and the Hartlepools. Of these the Hartlepools at the south-east corner of the Durham coalfield exports some 2,000,000 tons of coal and some iron goods. It is also one of the chief timber-importing ports and obtains some iron-ore for use in the local furnaces. It is second to Tynemouth as the fishing port of the North-East Coast. The ports of Stockton and Middlesbrough on the Tees estuary are essentially ports of the iron industry, dependent almost wholly on the immediate hinterland. There is a widespread export of iron and steel goods to many countries of Europe, Africa, Asia and South America; and the regular sailings to the Far East have drawn to Middlesbrough for export some of the cotton goods of Lancashire, and the woollens of West Yorkshire. By far the principal import is iron-ore; and for the supplies of imported foodstuffs and general goods Teesmouth is part of the hinterland of the Tyne ports.

Of the other towns only the city of Durham can be referred to here. It is one of the few examples in Britain of the mediaeval fortress city. The site is a rock peninsula in a sunk meander of the River Wear some three-quarters of a mile long, with a flat surface nearly 70 ft. (21 m.) above the river, reached by a low isthmus about 300 yards wide. This peninsula was settled by the monks of St Cuthbert during the troubled centuries of the Viking Period. It became the seat of the Bishop and the capital of the territory granted to him as "The Patrimony of St Cuthbert." After the Norman Conquest William I erected the Bishopric into a County Palatine, and from that time the city was the capital of an ecclesiastical border march. As the capital and stronghold of the Prince-Bishops it was during centuries of border warfare the chief city of the North-East; but in times of peace it was of less importance than the rival city of Newcastle, since the very difficulty of access which made it a natural fortress hindered its commercial development. Before the palatine privileges of its Bishop disappeared in the 19th century he founded here a university; and to-day the city owes its importance to its rank as the capital of the county, a cathedral and university city, and a local market of the mining population. The cathedral is the greatest architectural monument of the region and helps to make Durham a centre of attraction for tourists: but to-day, with less than 20,000 inhabitants, it is one of the smaller towns of the North-East.

XX

CUMBRIA

F. J. Campbell

JUTTING out to the west from the side of the long upland mass of the Northern Pennines is the dissected dome of the Cumbrian Mountains. This dome is thrust out at the end of a wide and short ridge into an area of lowlands and seas, so that the whole upland mass, which has been aptly likened to a short-handled spoon upturned, has the distinctness of a promontory to the west, abruptly ending in St Bees Head. It lies across the main West-Coast Route, and a geographical effect can at once be noted in the comparative difficulty of the alternative railway routes to the north; the lowland route winding far round by the coast, the direct one climbing 1000 ft. over Shap Summit.

It rises from the surrounding Eden Vale, Carlisle or Solway Plain, the Solway Firth, the Irish Sea, Morecambe Bay, and the Kent estuarine lowland of south Westmorland.

These uplands, together with their marginal lowlands, clearly demarcated from the remainder of Britain by the Pennines to the east, by the Solway—and in earlier days also by mosses or marshes—to the north, and by Morecambe Bay to the south, make up the region of Cumbria.

The region, and more particularly the subdivision of the dissected Cumbrian plateau, is of extraordinary interest to the geographically minded. The rugged dome-shaped massif, with its unique radial finger-lakes in a deeply incised drainage-system which cuts it up into radiating plateau-blocks, and with its striking diversity of scenery, carved by erosion and glaciation out of a complex mass of strata, is both a natural "museum" of physical geography and a thing of beauty.

The human reactions to the physical basis, both now and in the past, afford very definite and interesting illustrations of the effect of geographical factors. The Dome of the Lakes Mountains with its cloudiness and high rainfall is predominantly pastoral. But in the Carlisle and Eden Lowlands there is a population whose life is dominated by agriculture based on the richness of the red soils there. And in almost violent contrast to both there are, on the west coast, coal- and iron-mining areas with modern large-scale industry.

With all its distinctness, the region is not now isolated, for it is entered at several points by modern lines of communication—the main railways and roads. This occurs where the highland barriers are pierced by "through" depressions or where a way is found

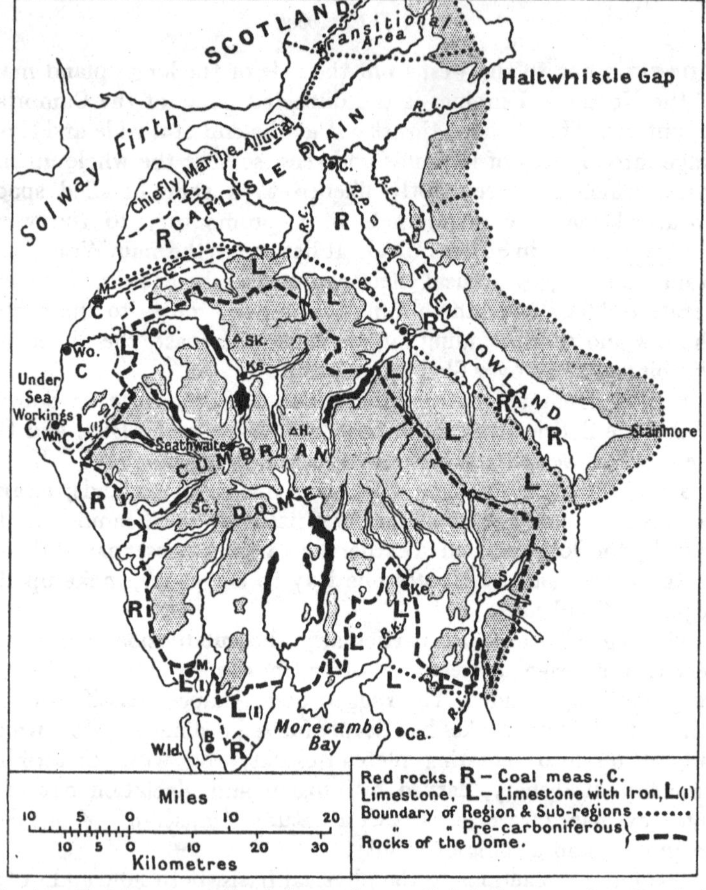

Fig. 44. Sketch-map of Cumbria, to show radial drainage, and rocks of the "frame" around the Dome. (Land above 600 ft. or 183 m. is shaded.)

round the heads of the isolating arms of the sea. The main crossings of the Southern Uplands necessarily focus on the northern limit of the Carlisle Plain where the Solway most nearly approaches the Pennines. Into the same plain debouches the Haltwhistle gap, continuing the trough of the Tyne. The headwaters of the Eden have worked back towards the Tees basin (River Greta) with the result that the Stainmore Pass enters the Eden Lowland. The southern gate lies between Morecambe Bay and the Pennines, and has a subsidiary entry up the Lune valley. All are utilised by railways. There are also two road entries of some importance, a difficult one from Alston, and one from Wensleydale head.

Facing these entries are the points enjoying nodal importance, which command the ways that link the region with the remainder of Britain, themselves joined together by connecting routes and sending ways into the interior. They find the material basis of importance in the productiveness of the lowlands about them, and are thus withdrawn from the boundaries of the region. They have long been the chief market towns, and are now railway junctions and the headquarters of auction mart companies. The northern gate and the Haltwhistle gap focus upon Carlisle. Penrith receives the Alston road-entry, but depends mainly on an exceedingly important local nodality for its significance. Stainmore Pass leads to Appleby, just as the upper Wensleydale pass leads to Sedbergh. Kendal is the southern counterpart of Carlisle, for the roads; but the main railway avoids it, and takes advantage of local relief*, the junctions being outside the region at Lancaster and Carnforth.

Carlisle's complete monopoly of the northern communications makes it the most important nodal point in the whole region, with marked effect on the features of that city. All towns other than the above have only local nodality, with the exception of the seaports, which make their extra-regional contacts by sea or actually originate traffic of importance by manufacturing. These are Barrow-in-Furness, Whitehaven, Harrington†, Workington, Maryport and Silloth.

Definition of the Regional Boundaries

The region has been roughly defined by description. Since much of its unity and individuality follows from its very definite physical separation from the neighbouring regions—Northumbria, Yorkshire, and Lancastria—by barriers (to the reality of which the definite nature of the passages draws attention), it would be well to state the precise limits. The problem always arises in regional studies, and deserves a respect which it does not always receive.

In the north the influence of Cumbrian geographical conditions is neutralised by that of the Southern Uplands in the threshold zone of the northern gate about the River Esk. It so happens that the International Boundary is in or near this zone. Right down

* Local opposition kept the main line away from Kendal.

† Now defunct.

the east of the region the special Cumbrian influences vanish when the water-parting on the top of the western escarpment of the Pennines is passed. All the parishes end with or near the crest, with the exception of a piece of Cumberland which is really Northumbrian, Alston Parish. Viewed from within the region the Pennine Scar is indeed the boundary wall.

The region must include the lowland debouchures of the gaps, and thus it ends near Gilsland in the Tyne Gap—actually with the Northumberland boundary. It ends near the Rere Cross and the Yorkshire boundary in the Stainmore Pass, and so includes the Yorkshire parish of Sedbergh, where this entry opens out. Thus the Howgill or Langdale Fells are all included. The "frontier" follows the most westerly (secondary) water-parting of the Craven-Lonsdale Pennines and includes all Westmorland parishes. Part of three of these are in Dentdale and other dales, which are transitional in nature to Yorkshire conditions. The southern boundary again lies across a gate, where Lancastrian influence begins to be felt in the geography, in a belt including a little of Westmorland. Hence the region includes practically the whole of Cumberland and Westmorland, Furness—or Lancashire-North-of-the-Sands—and one parish of Yorkshire.

The region, thus defined, falls naturally into three main sub-regions (Fig. 44). They are:

(*a*) The Carlisle Plain, together with the slopes forming the semi-circular arena to the north-east entered by the Haltwhistle gap.

(*b*) The Eden Lowland, together with the long slope flanking it to the east, bearing sheep pastures.

(*c*) The Cumbrian Dome of the Lake District proper, the heart of the whole region. Much of the lower slopes has an economic orientation towards the two former sub-regions and towards Lancastria, but yet belongs to the unit of the Dome.

The Physical Basis*

The physical geography of Cumbria outlines the basis of the distinctness of region and sub-region, and, further, of the whole superstructure of the human geography. The facts of the materials and structure are most eloquent when they are given in their place in the history of the region, the phases of which have stamped their imprint directly upon its geography.

The oldest rocks of the district are the sediments known as the Skiddavian rocks, of Ordovician, and in part perhaps even of

* *The Geology of the Lake District*, by J. E. Marr, is indispensable here.

Cambrian age. In a sea trough floored with these arose a line of volcanic islands in later Ordovician times. These were worn down, depressed, and covered with sediments of Silurian age. The whole was later folded along ENE.–WSW. axes by the Caledonian movements, with molecular packing, giving good cleavage in the finer rocks. These mountains in their turn were reduced to a peneplane, sank, and were covered with the limestones and shales of the Carboniferous Limestone, and, finally, upon a foundation of Millstone Grit, with Coal Measures. Again there was uplift with axes in the same "Caledonian" trend, and the centre was worn off down to the Silurian and older rocks, with loss of the coal. Gradual depression followed, and the area was covered, first with desert sands red with iron oxides and inland seas possibly depositing more iron salts, then by the Lower Lias and perhaps later deposits. It may be, as suggested by Marr (*op. cit.*), that now vanished Chalk or even Tertiary rocks covered the district. The red sands form rocks like the St Bees Sandstone, of which that headland is built. The low plateau in the centre of the Carlisle Plain is of Lower Lias.

At last the present dome* was raised up and enormous amounts of the newer rocks removed. This event took place in post-Lias time, probably during the great movements of the mid-Tertiary. All later than Silurian was removed from the central part, and any Cretaceous and later deposits from the whole area. The great Crossfell-Craven series of faults separated the Dome from the limestone plateau of the Pennines, and so its lower slopes on the east form one side of the faulted trough of the Eden Lowland. Farther south there is no such break in the surface relief between Cumbria and Pennines, but the fault brings up the (Silurian) Coniston Grit of the Howgill Fells, typical of the regional rock-type, against the Carboniferous rocks and so marks the limit of the region. The exact causes of the uplift into dome-form are at present a matter for speculation and research.

In the Eden trough have been preserved the red Permian and Triassic rocks—sandstones and breccias—which run like a great tongue from the main red rock lowland of the Carlisle Plain between the Dome and Pennines.

The Carlisle Plain is a shallow basin, left low between the Southern Uplands, the North Pennines and the Lakes Dome. In its centre are preserved the youngest rocks in Cumbria, the Lower Lias, already noted, necessarily left in the centre of such a topographical

* Where printed as "Dome" the sub-region is referred to; as "dome" the structural and physical feature. The two roughly coincide.

plain with basin-structure. It is very definitely cut off from the Dome of older rocks by the long Maryport fault, running east-north-east from that seaport, with downthrow on the northern side.

The distinctness of the two sub-regions from each other and from the Dome is as clear in structure as in relief.

The Dome presents a quite marvellous picture of complex unity. A core of ancient rocks, themselves bent into arches, contorted, cleaved, and fractured, protrudes, as the result of post-Tertiary denudation, through a garment, or, more correctly, a series of skins of newer rock. The remnants of these now make a frame round the exposed core. Everything dips radially from the central region of the dome. This "frame" of younger rocks is found to change its nature as the dome is circled, according as denudation has or has not been able to remove particular layers. On the south-west coastal flank the Triassic rocks lie directly upon the ancient core. But to the north-west are the remnants of inward-facing scarps of Carboniferous Limestone, with the edges of the overlying sheets showing in turn—Millstone Grit, Coal Measures, Triassic rocks. To the north-east and south the "frame" is composed of limestone alone. In the east its place is taken by the Howgill anticline ("the spoon-handle").

Application of each of these chief facts of the geological history to human geography will be made in its place.

As a result of the Caledonian flexing, the outcrops of the ancient rocks of the core run in parallel belts from east-north-east to west-south-west. The oldest rocks, the Skiddavian slates and grits, naturally outcrop over the main axis; to the north and south appear the ashes and lavas of the Borrowdale Volcanic rocks, the northern exposure being the narrower; the Silurian grits, flags, and roughly cleaved "slates" form the third great belt to the south.

This rock variety leads to great local variety in scenery. There are usually smooth slopes in the Skiddavian fells, as the rock weathers into small fragments, and a striking contrast to these is seen in facing round to the Volcanic fells, with mixed crag and slope due to mixed lava and ash. South again are the smoother outlines of the Silurian grits and shales of the Low Fells. In the latter we find the "park-like" scenery of Lake Windermere, while in the Volcanics are the wild and even savage surroundings of Wastwater. Then, on the flanks, are limestone moors to the east, isolated karsts to the south, and low rolling plateaus on the Coal Measures to the north-west.

The foregoing summary stresses diversity and contrast and scarcely seems to uphold the geographical unity of the Dome sub-region. One of the chief reasons for considering it to be a sub-unit is found in the effect of the glaciation of the region in the Quaternary Ice Age. The dome had its own ice-cap, fed from the high central region, so that the ice moved out in all directions towards the lowlands. The central fells are made of slatey or igneous rocks yielding a siliceous, lime-deficient soil. Ice-movement has spread this over the "richer" rocks of the lower slopes as boulder clay, masking them, and has thus impressed a great measure of similarity on the soil conditions of the whole dome, including part of the limestone ring or foreland to the east. Stony and clayey in nature, this soil type is far more suitable for pasture than for crops, and fundamentally different from the red soils of the agricultural lowlands.

A map of soils shows an exceptional area, the salient of red soil which runs up the Eamont valley from Penrith through the limestone. This is to be expected, as the ice has passed outwards over an important tongue of red basement conglomerate, with much sand, which reaches to Great Mellfell. Crop maps show the effect of the richer soil strikingly.

Returning to the physical history of the region, the drainage features should next be considered. The system of drainage naturally initiated on the newly-risen dome was radial, and upon a surface of young rocks. Under the comparatively homogeneous blanket of Chalk and younger rocks was the mass of ancient rocks, interlocked, mixed up, and broken. When the streams eroded their valleys down to this old surface they had already etched their permanent courses. Thus, despite the intricate strata over which they flow, the valleys are radial—a beautiful example of superimposed drainage discordant with the strike. Wordsworth has it: "eight vallies...like spokes from the nave of a wheel" radiate from the Scafell area, the summit-mass of the dome, with peculiar symmetry.

Another great factor in the relief is the extensive faulting that has taken place. In the older rocks more adjustment was needed than folding and cleaving afforded in the Devonian or earlier movements. Faults sliced the rocks almost horizontally and packed them up. These look like thrusts, but may be "lag-faults." The sheets moved, and as some parts moved more quickly than others, they tore away, making "tear-faults" at right angles to the moving front*. Such long, narrow shattered bands often run for great distances and are important elements of relief. The excavation of some main valleys has been helped by these, where they happen to coincide with radial valleys. They have interfered

* Harker and Marr in *Proc. Geol. Asscn.* 1900 and 1906.

a little with perfect radiality elsewhere. With such help the cols between valley heads are much lower than the stage of the valley erosion promises, and thus many of the passes used by shepherds and visitors have originated. One striking series runs across the radial ridges obliquely from Windermere head over Wrynose and Hardknott towards Ravenglass in a straight line—a Roman route of importance. The passes were much used by the pack-horse traffic. But a much more important shattered line provides the only easy route through the dome, difficult to cross from south to north because of the radial dissection. This is followed by the road to the north from Windermere over Dunmail Raise to link Kendal and Keswick, a real artery with all-year-round motor services. The building of a light railway was suggested in recent years. The height of the next lowest pass (Kirkstone) is 1469 ft. or 448 m., to be compared with 782 ft. or 238 m. of the one discussed.

The function of this same weak belt is also to cut the Lakes Mountains into two distinct masses: a western one, sloping to the sea; and an eastern half-dome articulating with the Pennines through the Howgill uplift*. As the Skiddaw area is separated from the other highlands by the synclinal trough of the Glenderamackin valley, there are three distinct divisions of the massif, each rising over the 3000-ft. (914 m.) contour line. The northern one has Skiddaw, 3053 ft. (930 m.), the western one the Scafell Range (Scafell Pike at 3210 ft. (978 m.) being the highest point in all Cumbria), and the eastern one has Helvellyn, 3118 ft. (950 m.).

Later fracturing has its effects. Carboniferous rocks were involved, breaking along the old trends. In this way arose two disadvantages of industry in the west—the expensive working of faulted coal-seams, and the devious land communications due to the breaking of the southern limestone "frame" of the dome into blocks running north to south, with consequent alternation of hilly peninsula and estuary. Long viaducts fail to make the route at all direct.

Reference has been made to only one result of the glaciation of the region. This phase in its history has given it most of the minor and visible features which are most striking in its relief. Dissection had reached the stage when valleys were deeply cut while yet the ridges were little lowered, so that the relief was very strongly marked. But the slopes were smoothed by the vast accumulation of detritus. The ice came and went, and rivers returned to the valleys. But these had been over-deepened, especially where helped by faulting, and in places plugged with morainic material. Differing combinations of these factors produced the English Lakes—long,

* Cut off by the Lune, older and antecedent to the uplift.

narrow, usually almost straight, with their steep valley sides rising almost sheer from the water, often partly in rock basin, partly dammed by drift. Pairs of lakes, as Derwentwater and Bassenthwaite, are one basin divided by deltaic material. In the combes are tarns, either true rock basins or ponded up. From the tributary valleys, left high in the walls of the main valley by over-deepening, cascades tumble down. The valley sides have been stripped of their old detritus and undercut by glacial erosion, leaving them unstable; while frost action has split up the bared rock. The result is a terrain exceedingly rugged, with much crag and scree. The unique scenery, by attracting cragsmen, fell-walkers, and beauty lovers, is the basis of one of the chief occupations of the region—the "tourist industry," reacting alike on dalesman and shopkeeper. The contrast of lake or green alluvial meadow with craggy valley walls is typical. The low altitude of the valleys gives the hills the maximum of height-effect. And all this variety is packed into an area not three dozen miles across.

The ice was not able to reach the lowlands of north Cumbria unhindered—movement was free only to the south. The Scottish ice moving southward contended with it for possession. The prevailing condition was one in which the ice-cap of the dome was almost confined to its slopes. It was able to wedge the northern ice into two branches, one of which skirted the dome to the east, following the Eden Lowland before crossing the Pennines, while the other passed down the western coastal plain. These are both lowlands of red rock and, as northern ice was confined within them, the drift, and therefore the soil, has remained red and sandy. Numberless drumlins arrange themselves parallel to the circumference of the dome, showing how the ice swept round. The parishes and even the villages have this arrangement with their long axes parallel to the circumference of the dome.

Glaciation is a main geographical factor not only in the central uplands. The surface of the Carlisle Plain is almost entirely moulded in glacial and post-glacial deposits, and all the most distinct features are to be found in the great area covered by the former. Along the southern tract the drumlins give the effect of "ripple-marking" on a gigantic scale, in sharp contrast to the smoother slopes of the Dome. Eskers wind over the plain, and there is much evidence, in the form of overflow channels and strand-lines, of temporary glacier-lakes, including a great "Lake Carlisle*." A wide tract, up to 4 miles (*c.* 6 km.) across, fringing the Solway, is of marine alluvium or warp (uplifted tidal mudflats).

* See the important Geol. Surv. Memoir—*The Geol. of the Carlisle, etc., District.* 1926.

Near the sea, the present warp, lower than the terraces of the warp of the Plain and occasionally covered by sea-water, is the source of the Solway turf, used on famous tennis-courts, and pasture for sheep from the Dome in winter. Upon the flats, of both glacial and marine deposits, are the bogs ("mosses"), with slightly convex surfaces. A contrast between the Plain and the Dome is seen in the use of wells dug in the glacial sands or the warp of the former, and of springs by settlements in the latter.

Climate

The region lies on the side of England windward to the Westerlies, and is the part nearest to the most frequented track of storms. Hence the amount of cloud and rainfall tends to be high over the whole area. A minor "winter gulf of warmth" reaches up the Irish Sea over which the prevailing and strongest winds blow, so that winters are mild.

Naturally the rainfall is highest in the Dome sub-region, where it increases with the altitude, roughly. Some low valleys, which lie across the track of the south-west winds, have very heavy fall, as the air currents have to cross them at a great height from ridge to ridge; Great Langdale, at 400 ft. (122 m.), has over 100 in. (2500 mm.) a year.

The maximum is not on the crest of the Dome, but immediately beyond, to leeward, where the air currents rise still higher. So Styhead at the head of Borrowdale has the record fall in England, over 150 in. (3750 mm.) a year. The special condition present is the convergence of radial valleys in the south-west quadrant. The amount everywhere is usually proportional directly to the velocity of the wind from the south-west quarter. The rainfall of the coast generally is just over 40 in. (Whitehaven, 42 in. or 1050 mm.), except along the lowest shorelands. The two lowland sub-regions are in the rain-shadow of the Dome (Penrith 33 in. or 825 mm., Carlisle 32 in. or 800 mm.)*. A peculiar coincidence is that where

* Whether in the areas of maximum fall, or of rain-shadow, there is a quite pronounced maximum in August to be remembered when considering the farming. There is another maximum from Oct. to Jan., usually with the peak in December, and most strongly marked in the fells.

Rainfall, 1881–1915 (in inches and millimetres):

	Jan.	Feb.	Mar.	Apr.	May	June	July	Aug.	Sep.	Oct.	Nov.	Dec.
Carlisle (Spital.) (in.)	2·4	2·1	2·3	1·9	2·2	2·4	3·1	3·9	2·5	3·2	2·8	3·0
" " (mm.)	61	53	58	48	56	61	79	99	64	81	71	76
Seathwaite in (in.)	13·3	11·9	11·1	7·4	7·4	6·5	8·5	11·6	9·9	12·0	13·6	16·3
Borrowdale (mm.)	338	302	282	188	188	165	216	295	252	305	345	414

the soil is permeable its lowland nature, mentioned previously, produces a comparatively low rainfall.

With increasing altitude temperature falls, rainfall increases, as does exposure, while the soil is leached and is thin and stony because of the slope. As the soil throughout the Dome is of the same general nature, the vegetation is in general distributed in concentric zones. The botanist Hodgson suggests the following classification:

4. Mid-Arctic Region (39° F. or 3°·9 C.): area over 2700 ft. (*c.* 820 m.).
3. Infer-Arctic Region (42° F. or 5°·4 C.): area from 1800–2700 ft. (*c.* 550–820 m.).
2. Super-Agrarian Region (45° F. or 7°·2 C.): area from 900–1800 ft. (*c.* 270–550 m.).
1. Mid-Agrarian Region (roughly mean annual temp. at sea-level, *i.e.* 48° F. or 8°·9 C.): up to 900 ft.

(The region is too far north to have an Infer-Agrarian Region.)

The limits of these are very irregular, altered by aspect, etc. To view these zones geographically, there is in the mountainous central region *felsenmeer*, mountain and moorland, where the conditions of extreme exposure are reflected in the size of the Least Willow—½ inch in height. Around and below this is a ring of rough sheep pasture*. Farther out still is good hill pasture and some cultivation (*e.g.* oats), and, finally, in the lowland borders meadow and arable land.

But here must be noticed again that the radial valleys have been very deeply excavated, so that narrow "inlets" of lowland penetrate to the heart of the Dome. Borrowdale is less than 400 ft. (120 m.) above the sea within two miles of Great Gable, almost reaching to 3000 ft. (914 m.). As screes and debris carry moorland conditions down to the edge of the valley-bottom, the valleys, floored with the alluvium of filled-up lakes and less advanced deltas, cut sheer into the zonal arrangement with their lowland meadows.

Agricultural Geography

The primary division of the region is into two quite different agricultural units, the Dome and the Eden-Carlisle Lowlands. In the former the conditions of high rainfall and clay soils are against crop-farming on any but the smallest scale; what alluvium there is, is moist. The lowland conditions of deep, warm, red soils, or of alluvium, just where the rainfall is lowest, allow of a large percentage of arable land (Fig. 45).

* This would appear to have been followed round by pre-historic migrants, of Neolithic, Bronze and Iron Age times.

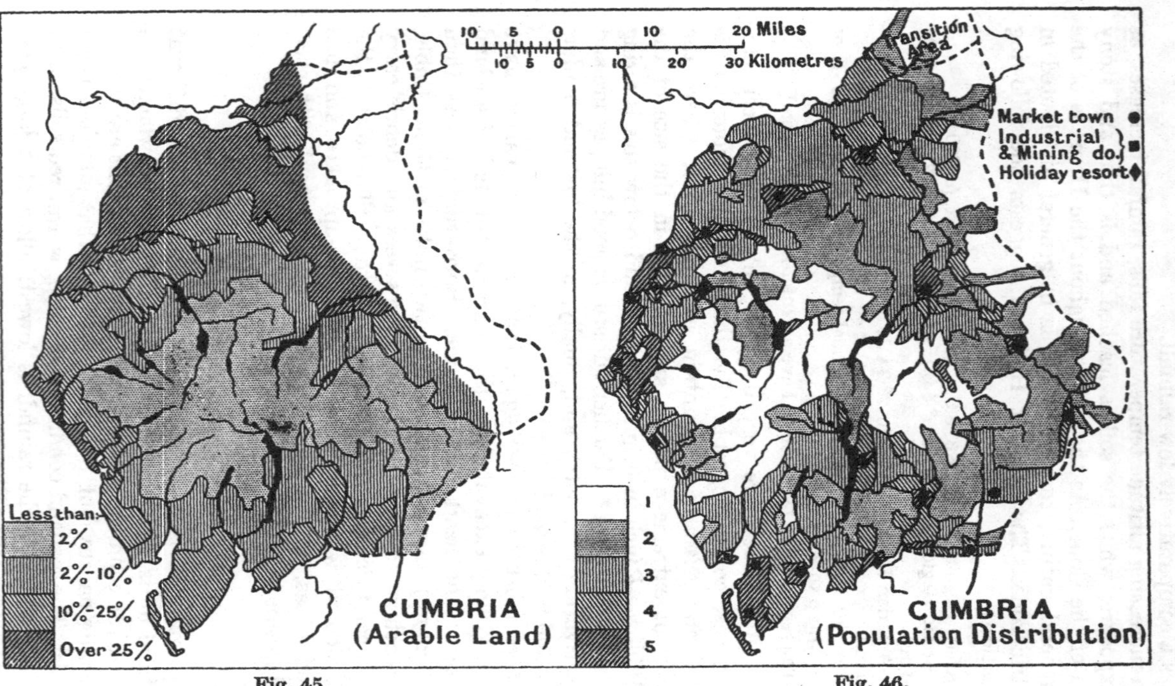

Fig. 45.

Fig. 46.

Fig. 45. Cumbria: arable land, percentage of the total area under the plough.

Fig. 46. The three Divisions of Cumbria according to density of population by parishes. 1, under 30 persons per square mile; 2, 30 to 60; 3, 60 to 100; 4, 100 to 500; 5, over 500. Note the relation of 1 to the fell areas, of 3 to the lowlands with much arable, of 5 to the areas with much coal and iron. (*Cf.* Figs. 44 and 45.)

In the central fell parishes but 2 per cent. of the land is arable. Working outwards from here the parishes of the Dome show the concentric zoning in having more arable as the average altitude falls, a belt with under 10 per cent., then one with 10–20 per cent., and finally the transition belt on the very edge of the Dome with over 25 per cent. of the surface arable. Even taking meadowland with ploughland more than half the parishes cannot show 20 per cent. of good land. Yet in the lowlands at least 75 per cent. of the surface is farm land, and ploughland usually occupies about 40 per cent. or more of the total area.

In the central upland region, then, the environment is unsuitable for crop-farming, except in small favoured patches*, where, in the main, roots and oats are grown, largely for winter feed. In some valleys the late summer rains cause even oats to be too risky a crop; but grass for pasture need not ripen and thrives in a moist soil, and grass for mowing ripens in diffused light and gleams of sun. With the mild winters in mind, it is obvious that the area is suited to pastoral farming. Sheep are the animals adapted to make the most of the rough well-drained slopes, while cattle thrive in the moist meadows of the alluvium in the valley-bottoms.

Sheep-farming is the main occupation of man in the fell regions—leaving out the tourist centres. The beautiful silver-faced Herdwick sheep is sturdy enough to live through the winter on the exposed seaward fells in the west, but it has a fleece of only 3½ lb. (roughly 1·6 kg.). In the east and now spreading over the centre is the black-faced Rough Fell sheep, horned, and more slender than the Herdwick. In parts of the Low Fells of the south are the Scotch Black Faces. The Rough Fell sheep area is extending westwards, invading the Herdwick area, while from the limestone of the east the Swaledales are moving into the fells, after having sustained great losses in the first years in the new environment. All round the lower slopes of the fells are farms which breed fat lambs by drafting in the small fell ewes and crossing them with bigger breeds, such as Wensleydales and Border Leicesters. The mountain strain ensures a good flesh, the other a fair wool, and the cross rapid fattening qualities.

Examination of the distribution of cattle in the sub-region shows the importance of dairying here. Practically everywhere over 20 per cent. of the total cattle are in milk. Most important, however, are restricted areas in the immediate neighbourhood of the industrial towns of the west and of the tourist centres. The lowlands

* Frequently smoothly curved, lenticular masses of boulder clay, on the valley-walls and therefore well-drained.

fringing the Dome export milk, those in the north to Newcastle, those in the south to Liverpool, etc. Cattle feeding or meat production is most important in the north-west (especially around Cockermouth) where extensive good grassland has rail communication with the industrial west and to the markets of Carlisle and Penrith.

On the small proportion of land that is ploughed the typical crops are those thriving in cloudy and moist conditions; oats and sown grass take up to 60 per cent., and in places up to 80 per cent., of the area. Swedes also are grown in almost every valley, but especially on the north-east flanks of the Dome, where clay soil and the lower rainfall favour them. The greatest percentage of ploughland is given to potatoes, first, near the towns, and again in the thinly peopled and isolated areas. Wheat is grown in small quantities, but only near to the limit of the Dome, except in the case of the western coastal "enclaves of red soil."

In the Lowlands the farming is of the true mixed type, large production of oats*, turnips and swedes and potatoes (in order of acreage) as well as sown grasses, together with the raising and fattening of stock. With the exception of the upper Eden valley arable land occupies at least 30 per cent. of the area. The percentage rises to 50 per cent. or more for many parishes in the warp and on the Lias plateau, which is covered with the thinnest drift on the Plain. The many drumlins look curiously alike, as the fields run in parallel strips across the "whale-backs," the larger of which also have a hedge (dividing the strips each into two fields) which looks like a dorsal fin.

Mining and Manufactures

While the occupations of the Cumbrian lowlands have always been primarily agricultural, the predominance of sheep pasture in the fells area gave early rise to wool manufactures. Some of the processes were carried on in the cottages and farms; the smallness of the holdings of the "statesmen" (as copyholders under the Border military tenure were called) calling for supplementary sources of wealth even for a low standard of living. Carding and spinning were done for the scattered mills driven by water-power, or for the centres of the industry, the biggest of which was Kendal†. Stocking-knitting was also important. The American War of Independence and other events ended the export of the coarse

* In most parishes about one-third is under oats—over half in some.

† The motto under Kendal's Coat of Arms, translated, is: "Wool is my bread."

cloths made, and no recovery was possible after the rise of steam-driven processes in Yorkshire. However, some woollens are still made in Kendal—of high quality or heavy weight.

On the west coast another industry naturally rose to great importance before the canal and railway era, that of coal-mining. Here were the most perfect conditions of the time for marketing, a coalfield on the only easy means of general communication—the sea—with Ireland opposite and near, and poor in coal. On the shallow Solway natural harbours were absent, except in the mouths of the bigger streams for the smallest of vessels. Artificial harbours were dug to accommodate the 100-ton sailing vessels carrying the coal. Later the railways brought the great inland fields, more economical to work, into competition with the west coast; but they also made possible the development of the iron industry, by enabling coke to be brought from Northumberland by way of the Tyne Gap, and by allowing the cheap assembly of the local raw materials by means of mineral lines.

Here was the coal for power (now used for coking also, as technique has developed), and also fire clay; here were harbours, and, because along the north-west flank of the Dome all the layers of the "frame"* of younger rocks are exposed, here also was limestone containing very pure hematite iron ore† and material for use as flux. The tortuous route by rail southward is disadvantageous, and an effort to minimise the effect is seen in the digging of the large Prince of Wales Dock at Workington, the iron-centre of the north-west, to avoid trans-shipment.

In the south-west is the peninsula of Low Furness, which has had and probably still has extraordinarily rich hematite deposits, but no coal, having a sequence in the rocks of the "frame" different from that of the north-west. Here, simply because of the iron, in a portless agricultural region, is a growth almost miraculous, the greatest city and the greatest port of Cumbria—Barrow-in-Furness. It is on the site of a few seaside cottages, facing the narrow and protected channel behind Walney Island. Its harbour is artificial, made up of docks excavated between the mainland and Barrow Island. The largest town in Cumbria, a region with towns of long history, is younger than many new Western towns in America. The Bessemer works were deliberately placed upon a site for a

* *Vide* p. 354.

† The roots of the iron industry are to be found in the geological story—the iron oxide once uselessly distributed among sandgrains has been dissolved and, as the underlying layer was either limestone or the faulted rocks of the core, concentrated into great rich masses in the potholes and caverns in the former and along the faults of the latter.

seaport near ore supplies rather than on a coalfield. There are two similar but comparatively tiny centres, Askam and Millom.

The Western industrial districts thus give to the Dome of the Lakes a seaward fringe of smoke and tip-heaps.

Other regional manufactures are small in comparison with the foregoing. But note may be made of the paper mills of the Kent valley in Westmorland, using the soft water of a siliceous catchment basin. Carlisle* possesses in her wonderful position a great advantage which has been well exploited; raw materials are assembled for the manufacture of fadeless fabrics, furnishing cloths, biscuits, woollens, and for engineering.

Truly regional in character is the slate-quarrying in the fells, where the volcanic ash cleaves to yield a beautiful green slate, tough and weatherproof, if rather heavy. The line of quarries runs across the area over ridge and valley along the main ENE.–WSW. strike.

Batholiths and other intrusions are quarried, where they are near a railway line, for so-called "granites," either as monoliths of highly ornamental stone, as at Shap, or for road metal, as at Threlkeld and near Muncaster Fell.

There has been much mining in the past, for lead, copper, graphite, and even gold, with probably more gain than loss on the whole.

Historical Geography†

Some of the influences which have come down from the past are too powerful to-day to be ignored, even in a brief study. One is the tradition of its Border position. Cumbria is the part of England farthest from the "Continental angle" and nearest to Celtic Scotland of the past—indeed the northern half was outside Domesday England and within Strathclyde. The permanent Roman frontier passed through the north of the region—Hadrian's Wall, and the associated works. Up to the Union, and, it is to be feared, afterwards, most of Cumbria has been a "zone of friction." Included in England in 1072, it was alternately recovered by Scots and English, with usual accompaniments of pillage. The northern area suffered most, but everywhere architecture is influenced. Manor houses and often churches also have peel towers for defence. Most farm buildings are fairly modern (post-Union); for, while beautiful half-timbered houses were built in the south, here in the "North-West Frontier" there was wattle and daub, untempting

* Population in 1921 a little over 50,000, *i.e.* about four times as large as Kendal, but about 10,000 less than Barrow.

† W. G. Collingwood's *Lake District History* (Titus Wilson, Kendal) well summarises much local work.

and easily rebuilt, for the mass of the people. The centuries of functioning of the region as the Western Marches with Scotland perhaps account for certain traits in the character of the modern Cumbrian.

The racial composition of the population is strongly affected by the position of the region, facing the historic Norse seaways. There is a very large Norse element, which came largely via Ireland and Man, possibly attracted by the familiar aspect of the fells from the sea, offering sheep-pastures. The distribution of the place-names sketches out the movements of the Angles and later the Danes as they encircled the Dome from the east, occupying the land, while the Norse penetrated through this agricultural ring to the dales. Physical characteristics, family names, and dialect—especially where the terms refer to sheep-farming—afford striking proofs of the importance of this Norse infiltration in the earlier half of the 10th century.

Carlisle is the "Border City," a true if hackneyed phrase. It is built on a bluff facing northwards across the Eden, with both flanks protected by water, on one side the River Caldew, on the other the Petterill. There is even to-day no road-bridge between it and the sea. Moreover, the nearest bridge up the Eden cannot be reached without first crossing the Irthing and leaving the Plain. Carlisle was isolated in the old days from the rest of the Kingdom by wide stretches of mountain and moor while yet facing the critical gateway on which the western exits from Scotland converged, and having behind it the easiest route into England for a Scottish invading army. Thus it was a stronghold in Norman times, and up to the Union of the two Kingdoms was besieged six times by the Scots, but only once taken. The strategic position of the city again became of importance during the Jacobite Rebellions of "the Fifteen" and "the Forty-Five," with their bases in Scotland. The city fell in the later rising, after its last siege.

Distribution of Population

When the density of population for the whole region is plotted on the map by parishes, all the facts of human geography examined before are, in a sense, summarised. An abrupt change is seen where the dome-shaped plateau ends and the agricultural lowlands of the north and north-east begin, and where the minerals of the seaward fringes of the dome are exploited on a large scale (Fig. 46).

The parishes of the plain have about 70 or 80 persons to the square mile, whilst those adjoining them on the very edge of the

Dome have between 40 and 50. The great bulk of the fell parishes have under 80, many from 20 to 10, while two have but five persons to the square mile*. Any parish in the sub-region of the plateau showing an increase on these figures usually is influenced by the tourist "industry." Such are the concentrations around Keswick, Bowness and Windermere, Ambleside, and other places, and on the sea coast at Grange-over-Sands. Of the industrial belt on the north-west Coal Measures and of the south-west iron region almost a third of the area is made up of parishes with over 1000 to the square mile—several with many more—and practically all carry over 500.

The nodality of Kendal and the richness of the pasture land of the wide vale of the River Kent where it enters the lowest part of the limestone "frame" also have their reflection in the population map, as does the salient of red soil in the River Eamont valley in the north-east, with a density four to five times that of the surrounding area. The population of the plain is also increased far above the average for the sub-region for miles around Carlisle, as well as around the other nodal points mentioned as facing the entries.

The settlements in the fell valleys, where the best land is in long narrow strips, with many springs of water, are usually of the dispersed type, while in the agricultural lowlands, with wells for water, are farm-villages, usually of the "straight-line" type.

The census figures for the 19th century show a very gradual diminution of population in the agricultural parishes, with a rapid increase in the industrial ones. The change in the population of Low Furness since 1851 has been most marked—a rich agricultural area becoming an industrial one. In the fell area the tourist centres have steadily expanded.

The Relation of the Region to Britain

The population of Cumbria, about 435,000, is not large. Its coalfield is a minor one. The whole region has fallen behind in the modern economic struggle; first, because of the smallness of the coal area and the difficulty of working cheaply seams that are often thin and faulted; and secondly because of the poor communications by land, both within the region and, as far as the west is concerned, with the remainder of Britain.

Yet the region makes its contribution to the agricultural wealth of the country in crops and animal products. Its iron and coal

* 10, 20, 30, 40, 50, 70, 80, 500, 1000 per sq. mile = 26, 52, 78, 104, 130, 181, 207, 1295, 2590 per km.^{2}

industry is of far more than local importance, in spite of physical difficulties. It performs the function of supplying the huge Manchester agglomeration with water, taken from the Atlantic winds by the mountain core*. Slopes so rough as to be of little use even as sheep-pasture are being devoted to afforestation by the Government and by individuals; Douglas firs, Corsican pines and other trees are being planted, and the fullest use is thus being made of this land.

In the future increasing use is sure to be made of the water supplies by great centres far south, and afforestation will be extended. The high rainfall will never give rise to great industrial developments based on hydro-electric power, even although the desiderata of high relief and glacial interference with gradients are present, because, after all, it is but a small mountain area, and the catchment basins are not large. Moreover, the water-power available is very variable. Nevertheless increasing use is being made of this source of power as a supplement to coal. The Solway tides may, some day, be harnessed. There may be renewed activity at a future period in the mining of metals, when prices and technique have both advanced far enough.

The enduring importance of the area of the fells would seem to lie in its enormous value as a National Park—in the beauty of its scenery, and in the scientific interest of its geography and geology. What is more, the scenery is accessible to all—whether viewed from the quiet of the valley, or with exhilaration after strenuous walking or climbing. Some of it does belong to the Nation now, being under the control of the National Trust. An array of mountain tops, given to the people, is the Memorial to cragsmen killed in the late War. The importance of this function of the Lakes Mountains can hardly be overestimated, and the region must be jealously protected from short-sighted economic aggression†.

* Thirlmere has now to be aided by Haweswater to meet the growing consumption.

† A section on the Isle of Man was to have followed this chapter; but it was found impossible in the space available to add anything material to the chapter by G. W. Lamplugh in the *Oxford Survey of the Empire*, Vol. I. ED.

XXI

THE HIGHLANDS AND HEBRIDES

A. Stevens

THE HIGHLAND PENEPLANE

THE Highlands of Scotland form an area of clearly marked individuality. On all sides but the south-east they are bounded by the sea, save where, as in Caithness and the shores of the Moray Firth, they are bordered by a more or less narrow coastal plain. The south-eastern boundary is no less clearly marked. It is a conspicuous topographic feature, a scarp so slightly gashed that from Strathmore or a point of vantage on one of the heights, which form the southern boundary of that furrow, it seems in the rather unfavourable lighting characteristic of the region to present a wall scarcely more broken than those of the ancient fortalices which guard its edge.

This highland block is built mainly of rocks of Archaean age, or of rocks so much altered and disturbed that their exact degree of antiquity, undoubtedly high, has not been determined. From the Kyle of Tongue to Schiehallion and from Morven to the Moray Firth these "Moine Gneisses" or "Eastern Schists" have a uniformity as remarkable as that of the scenery they build. For the most part they are flaggy or granulitic gneisses, compact and hard; though they are associated with limestone bands and streaks of schist of different nature. Similar uniformity of geological composition and landscape characterise the Long Island, the archipelago which stretches from the Butt of Lewis to Barra Head. There the dominant rock is gneiss of coarse Laurentian or Lewisian type, and the landscape is sketched in subdued curves: rounded hills on the higher ground, mammillated bosses of naked rock thickly scattered among silver lakes and sullen bogs on the lower. The mainland block shows on all its margins certain differences of composition. Between Sleat in Skye and Cape Wrath is a band of country 25 miles broad at its widest which inland is bounded by a remarkably even line from Loch Eriboll to the Sound of Sleat, and seaward is deeply fretted by sea lochs and bold headlands. Here great fantastic pyramids of pre-Cambrian sandstones rise steep-sided and barren above the no less barren rock shagreen of a Lewisian landscape. To the east Torridonian glacis and scarp are replaced by curtain and ramp of fossiliferous Lower Palaeozoic rocks thrust over, and over-thrust, by gneisses and schists from the east. In the south the Torridonian cover is more continuous, in the middle a bastion of Palaeozoics pushes out eastward in Assynt. Nevertheless there is a sameness of aspect all over: deep glen and rock-walled massif in the field, even skyline furrowed in the distance from the heights, narrow ridge and narrow glen set out north-westward on the map. Again, from Lorne and Cowall to Buchan and Kincardine run parallel bands of limestone, slate and grit, pebbly or smooth, quartzite and grauwacke, alternated and repeated in a fashion which has not yet been unravelled to the minds of all geologists. On the whole this builds a landscape very similar to that of the north-west, and of the Highlands generally, though with altitude distinctly decreasing in the south-west and north-east. In most of the Inner Hebrides there is a considerable development of sandstones, conglomerates, shales and limestones ranging from the Trias to the Cretaceous. These, however, are to a great extent capped by so-called Tertiary plateau basalts or intruded by a wide range of alkaline igneous rocks, and in consequence also present the appearance of an elevated land-surface deeply gashed, and are not conspicuously distinct from the landscape type of the Highlands proper. Finally, the Old Red Sandstone areas of Caithness, Orkney and Shetland, Easter Sutherland and Ross and the inner angle and southern shore of the Moray Firth, together with the Mesozoic patches on the coast of eastern Sutherland and Moray only show themselves in a widening of the more kindly coastal plain, which is all but wanting on the north and west sides of the Highlands.

The surface of the Highlands has been developed upon material of such antiquity, and bears the traces of so many vicissitudes, that it exhibits remarkably great and uniform resistance to the agents of erosion. In its general aspect the landscape has the rounded and massive character associated with the scenery of granitic country. Most marked in the mammillations of the Lewisian lowlands and in the lumpy masses of the Cairngorm group, this is conspicuous in massive shoulders and rounded flanks throughout the Highlands. The variations produced by differences in geological constitution may be locally striking, but are relatively unimportant. Even the limestone bands of the south-eastern (Dalradian) series only affect the landscape indirectly through the vegetation, and the grand white cliffs of the Durness Limestone in the north-west are dependent rather on structure than on material. Such differences as one might be inclined to ascribe to material are not consistent. The granite landscapes of the Cairngorms contrast less with the scenery of the neighbouring schists than with the more rugged granite country of Lochnagar or Cruachan or Arran. The gabbros of Skye build country whose forms belong to the same topographic family. Distinction is to be found mainly in detail: the rounded shoulders about Loch Lomond and Loch Tay have a knobbed appearance, a shagreen finish. Extensive areas of the Lewisian platform are pustulated, now with rocky knobs, now with morainic mounds. The Torridonian pyramids are much less contrasted with the general "Highland aspect" than the all but uniquely sharp pyramid of Schiehallion. The flat capping alike of the Tertiary basalts of the Isles and of the lower Palaeozoics of the north-west only makes the extension of plateau features more ubiquitous.

Extensive deforestation has taken place in the Highlands, and vestiges of the Caledonian Forest are rare and dubious. Such stands of timber as remain, even on high ground and rather inaccessible peaks, are artificial; and many of them which were removed during the Great War, have, for various reasons, never been replaced. In consequence a certain monotony of colouring emphasises the general uniformity of landscape: brown for the greater part of the year, but dappled in the summer with the green of bracken and in the autumn with the purple of heather. It is in this connection that the influence of geological composition shows itself most. It is often, but by no means constantly, possible to trace from a distance the outcrops of the Dalradian limestones in Perth and Argyll by the greenery of the vegetation upon them; and the striking nature of this phenomenon seems to have led observers to assume it is more general than the facts warrant. Everywhere the only constant, and the most important, contrast which vegetation shows is between recent alluvia and sedentary soil or local drift. Where well-developed moraines occur they may run through the brown heath as green streaks, but only here and there. The contrast between the green vegetation of the alluvia and the dun heath of the general landscape is as marked when it occurs near the 1000-ft. (305 m.) contour in Strath Bran (eastern Perthshire) as near the 500-ft. contour in the valley of the Tay. Elsewhere puzzling contrasts occur; as, for example, between the south-eastern aspect of Glen Roy (western Inverness), which is strikingly brown and forbidding, and the green and apparently less favourable north-western aspect.

The Highlands then build a plateau of fairly homogeneous structure. The surface of this plateau has been described as a plain of marine denudation, but the general opinion now is that it represents a peneplane. The evenness of its skyline from any judiciously selected point of view has been insisted upon in many places by Sir Archibald Geikie and many others. It is, moreover, generally agreed that the peneplane surface of the plateau is of some antiquity. For this the main evidence is found in the numerous basic igneous dykes which traverse it. These cross indifferently hill and valley of the present stage of dissection. Had they intruded a land cut up as at present their material would undoubtedly have gushed out at the lower levels and flooded the valley system with lavas. These dykes are, perhaps without more than specious accuracy, associated with the Tertiary phase of volcanic activity which produced the plateau basalts over a wide province extending from Ireland to Greenland, and this correlation assigns a maximum antiquity to the final stage of physiographic evolution evidenced by the plain or peneplane nature of the Scottish plateau*.

The Dissection of the Highland Peneplane

The existing valleys of the Highlands, which, apart from the scenic attraction of the mountains, are their most interesting feature, have had a complicated history to which their form must be related. It is not the intention to enter here on matters of evolution, which have been described and discussed in literature by no means obsolete or inaccessible†. Peneplanation had been accomplished in later Tertiary times. Rejuvenated drainage, partly along old *thalwegs*, partly along new lines, began to excavate valleys now submerged in their lower parts by the sea. The deep silt-filled pre-glacial valleys of the Tay and the Forth have been proved by boring. The contemporaneous valley of the Clyde has its floor north of Glasgow more than 200 ft. (60 m.) below sea-level. The shore-line probably coincided with the present 600-ft. (180 m.) isobath, and the modern valleys are the truncated remnants of the drainage system of a larger area. Ice invaded a system of immature drainage, arrested its development, and interfered with its direction. The precise influence of the ice is a matter of considerable controversy, but at least it left smoothed outlines, morainic barriers, and a surface swept all but free of pre-glacial surface deposits and soils. Since the departure of the ice normal drainage has resumed its development, but along modified lines. The truncation of the river system is itself competent to account for its rejuvenation. This

* An interesting summary of Highland Geology will be found in the posthumous work of B. N. Peach and J. Horne, *Chapters on the Geology of Scotland*, 1930.

† Mackinder's *Britain and the British Seas*; B. N. Peach and J. Horne in *Report of the Bathymetrical Survey of the Scottish Fresh-water Lochs*, vol. I; Geikie's *Scenery of Scotland*; etc.

drainage is still immature, and has not trenched to any marked extent the post-glacial surfaces*.

THE MORPHOLOGY OF THE HIGHLAND VALLEY

There is a dearth of detailed study of the forms of the Highland valleys, and many of their features await explanation. Quite generally it may be said that the characteristic transverse profile is convex above. The dissection is deep, but the valleys are narrow on account of their immaturity, and in spite of abundant surface water. The residual uplands are consequently lumpy and broad-shouldered. Hence the low ground of the Highlands is usually remarkably low, but it is widely scattered and relatively small in total amount. It is also in great part relatively remote and inaccessible. A most interesting and, at present, puzzling feature throughout the greater part of the Highlands is the stepped nature of the upper slopes. In the Lochaber country many of the higher masses show skylines which suggest the normal profile of the upper part of a mature *thalweg*, but which is abruptly truncated by steep slopes just under the level of 3000 ft. (900 m.) In western Perthshire a similar feature is fairly conspicuous, but the elevation is in the neighbourhood of 2000 ft. (600 m.). In Arran, F. Mort has described a "thousand foot platform,"† and extensive level tracts at about that height are to be seen in Rannoch, Cowal, Kintyre, and the south-west of the Highlands generally. These various planes or platforms at different elevations are certainly to be connected with definite stages in the topographic development of the country. They must be independent of each other, because their differences of level are so great that they must be steps rather than slopes. They could only be connected by slopes of considerable steepness which would contrast forcibly with the slopes it was sought to connect. The most competent authorities, Drs Peach and Horne, believe that platforms between 2000 and 3000 ft. belong to a high plateau or peneplane above which the highest peaks rose as monadnocks. Below that are surfaces belonging to an intermediate plateau. The *upper* limit of this is said to be about 1000 ft. Mort, however, believes his thousand-foot platform is sea worn in Arran. The lowest surfaces of all are said to have been developed in relation to a base level corresponding approximately to the 100-fathom isobath off the west coast at the margin of the continental shelf.

* The valleys of many streams of second and third order in Central Scotland and bordering areas are almost entirely developed in till. *Cf. infra*, p. 429.

† *Scot. Geog. Mag.* vol. XXVII, 1911, pp. 632 *et seq.*

On many mountain slopes structural terraces can be seen at various other levels. In the valley of the Spey below Aviemore three are particularly noticeable below 1500 ft. (450 m.).

In longitudinal profile also the Highland valleys as a rule exhibit a stepped formation. The valleys include one or more intermont areas of accumulation. Where there are more than one of these intermont flats in the same valley they are separated by sections of active downward erosion. It often happens that the stream on leaving the alluvial tract flows in a shallow bed continuing the built-up flat before it plunges over a fall or cataract into a gorge, as in the case of the Treig issuing from the northern end of Loch Treig. Post-glacial erosion has been so slight that the gorge developed in the river section where trenching is active has scarcely invaded the basins of accumulation; or if it has invaded them it has scarcely developed beyond the trough stage within them. This is the case in the transverse portion of the Tay valley. The existence of lochs in many of the valleys is evidence in the same direction, especially where, as in the case of Loch Laggan, an extensive body of water lies high above sea-level. Of main streams and of considerable torrents alike it may be said that their upper courses lie in shallow beds, often meandering, while lower down they have entrenched themselves by means of waterfalls and cataracts; and in not a few cases this is repeated several times in the case of one stream.

The areas of accumulation are sometimes flood plains, sometimes lake plains. In the valley of the Tay below Struan several terraces of accumulation, especially on the east bank, account for the agglomeration of a relatively dense population engaged in agriculture. These terraces are fragmentary and occur at two, and perhaps at three, levels, if not more. Their composition suggests that they are accumulations in water. Below their level the river has formed a shingly flood plain which is liable to be overflowed in time of spate, and which carries rough pasture. Many of these basin alluvia must have collected in shallow moraine-dammed lakes, as in Strath Fillan and Glen Dochart, where the remnants of the moraines persist. Such basins may still be poorly drained on account of the failure of the river gorge to reach them, when they are damp and sedgy, or even peaty. Where the natural drainage is moderately favourable they may be cultivated up to considerable elevations, as at Tomintoul (1200 ft. or 360 m.) in Banffshire, and Amulree (900 ft. or 270 m.) in Perthshire; whereas the upper limit of arable agriculture in Scotland generally is considerably lower.

The most interesting, as well as the most puzzling, of these areas of accumulation within the valleys are of course the typical long, narrow and deep valley lochs. In no essential feature does the scenery of their banks differ from that of the unflooded valleys. In most cases they lie deep in the glens within rock basins, and where they are closely hemmed in by high land sloping steeply from their shores they have a gloomy forbidding aspect. No villages dot their shores, and an occasional cottage on a lateral torrent fan or on a limited flat at one end, inhabited by a shepherd or a keeper, only heightens the desolate aspect by suggesting contrast. Such are Lochs Treig and Ericht. Their high level and complete investment within barren highland moorland contrast them with such lochs as Venacher, Tay and Earn whose lower ends are in contact with low ground in which alluvia have accumulated and which have a more verdant and pleasing appearance, or with the exceptional Loch Lomond whose southern end broadens out on the edge of the Midland Valley. In a word, the lochs may represent upper or lower members of the suite of basins of accumulation which are the Highland valleys.

A very peculiar feature of these valley lochs is the fact that many of them have been silted up at both ends. In the case of Loch Tay a possible explanation is obvious*. The Lyon which comes in just below the loch must, from the absence of lakes, have a much more irregular regime than the Tay at the confluence, and might easily pond back the waters of that stream in time of flood and so distribute sediment eastwards and westwards. Such an explanation will not answer in the case of Loch Rannoch. Moreover the accumulation seems to have gone on *from* both ends, to judge from its deltaic outline; and vegetation and human occupation suggest that the alluvia at the lower end of the lochs are drier and perhaps older than at the upper.

In relation to the scale of the country these lochs are of remarkable depth. The depth of the water may be as much as 1017 ft. (310 m.), as in the case of Loch Morar. Loch Lomond has an extreme depth of 620 ft. (189 m.) and Loch Ness of 750 ft. (229 m.). The trough-like form of the basin is brought out by the high value of the mean depth in relation to the extreme. In the case of Loch Ness the mean depth is 433 ft. (132 m.), 57 per cent. of the extreme. Frequently the cross profile shows subaqueous benches similar to those observable on the valley sides. These, of course, reduce considerably the mean depth of the lake, and in this way the relation of the mean depth to the extreme masks the steep-sided,

* Investigation by J. S. Thoms discounts this suggestion. Rather general reversal of flow through the lochs seems to be the only alternative explanation.

flat-floored nature of many of the basins. The depth of some of the lakes in relation to the level of their surface is such that their floors lie below sea-level; and in the case of the deepest basin, Loch Morar, its surface being only 30 ft. (9 m.) above the sea, its floor lies nearly 1000 ft. below sea-level, while the actual sea bottom does not attain this depth until it leaves the continental shelf west of St Kilda.

The loch basins are merely exaggerated examples of the rock basins represented in so many of the glens by marshy flats and reed-fringed lochans, which are easily explained by the erosive power of ice. But the great depth and abrupt ends of the loch basins are much harder to explain with perhaps the majority of thinkers as due to ice excavation. Professor J. W. Gregory regards them as tectonic depressions in the development of which ordinary destructive and transporting agents have had little part, though they may have added the final touches to their form. He sees in their western distribution an essential connection with the sea lochs or fjords of the west coast, while the more orthodox uniformitarians explain it by the greater precipitation due to the western aspect, and a greater speed of ice novement due in part to steeper slopes, in part to less obstruction to flow. The arguments of the rival schools cannot be examined further in this place.

The western sea lochs present in many respects the same features as the fresh-water lochs. In many cases they represent closed depressions, so that any considerable lowering of the sea-level would convert some of them into valley lakes like those farther inland. The difference between Loch Nevis with its 30-foot (9 m.) threshold and its basin descending to 300 odd feet and Loch Morar close by, discharging its overflow across a bare mile of land by a drop of 30 feet is a difference merely of degree, and a small one.

In the stepped longitudinal profile of the valleys the Highlands have a most valuable asset, the possibility of developing a modern hydro-electric system. For the purposes of small beginnings or for the supply of power in the constructional stage the valley lochs are available as natural regulators of a water supply which, in a wet climate with little seasonal variation of rainfall, is not inconsiderable, and the application of which is facilitated by the numerous breaks of slope. For development to the maximum the steepness of the valley sides are very favourable, reducing the length of barrages, offering stable foundations on solid rock, a large volume of storage per unit of height of dam works, and a very slightly increased evaporation surface. Should serious attempts to make use of tidal energy succeed, the sea lochs, for similar reasons,

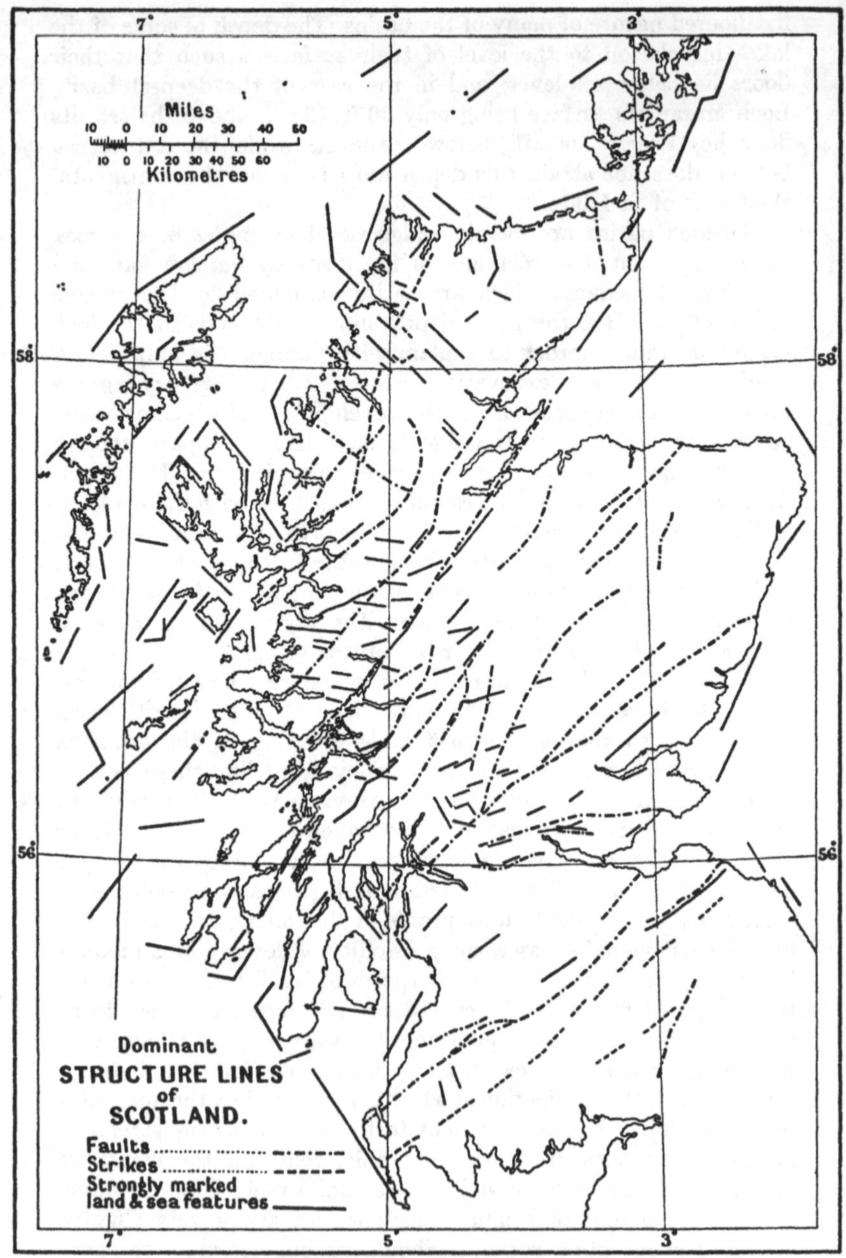

Fig. 47. Scotland: dominant structure lines.

would be very favourable for such projects, but they represent, like the water-power resources of the valleys, assets which have remained at a heavy discount, but which in future may occasionally have a value. As harbours merely they are often ideal. Their zig-zag courses mean deep water with the maximum of shelter from all directions, and their steep-to shores offer the possibility of wharf accommodation with the minimum of construction. In many cases land communication along their shores might present problems too costly for solution, but their unused state is due to their situation in unproductive country. On account of the establishment of aluminium refining at Kinlochleven, Loch Linnhe has attained a very moderate usefulness which will grow with the development of the industry by means of the Lochaber hydro-electric project now in execution; and it is not impossible that the latent value as harbours of some of the sea lochs farther north will be of considerable importance as the harnessing of the water-powers north of the Great Glen is undertaken.

The Pattern of the Valley System

The pattern of the river, loch and glen system differs in different parts. In general it follows several directions which are well marked throughout the country in the alignment of coastal stretches and land features. So far as the structural grain of the country is concerned, in the Central Highlands, south of the Great Glen, the longitudinal (strike) trend is approximately south-west to north-east, the transverse north-west to south-east (Fig. 47). Geographically these terms are adequate and, in the uncertain and incomplete state of knowledge, preferable to the commonly used alternatives, subsequent and consequent respectively. In the North-Western Highlands these directions are somewhat different and somewhat variable. In the Lewisian or fundamental complex, both on the mainland and in the isles, the strikes swing about between east and west through north to north-east and south-west. It is not easy to see justification for the statement by some authorities that the dominant grain in the North-West Highlands is from north-north-east to south-south-west.

Besides these fundamental directions several others are well marked in the pattern of the land features and of the sections of coast-line. These lie from north to south, north-north-west to south-south-east, north-north-east to south-south-west, east to west. In part at least they are due to tectonic lines which are often fault-lines. Among these a series running generally slightly obliquely to the north-easterly strike is very conspicuous in the

Central Highlands. As a rule they cut across to the north of the strikes, and swing to the left until they run about north and south. The fault diverging from the Highland line at Aberfoyle to pass across Loch Vennacher and Loch Earn at an angle of about 70° coincides very nearly with the trend of the middle section of Loch Tay, and probably runs north down Glen Feshie. There are thus in the Highlands lineaments following several very definite directions, and these lineaments are connected with structural features of different type. It follows that the land features are boldly drawn, whether by tectonic or erosional agencies, that they follow a pattern of some complexity, and that the complexity is somewhat increased by the tendency of lineaments of different nature to follow nearly like directions. The dominance of this or that trend, the considerable development of several trends in the same neighbourhood, these are criteria which are of importance in considering the division of the area into regions for detailed geographical study.

The Tay Valley System

An examination of the pattern of the Tay, the greatest drainage system of the Highlands, will serve to bring out the dominant directions of the Highland drainage and certain of the varying characters and episodes of its development.

The main transverse trunk of the Tay originates under the level of 1500 ft. in the Pass of Drumochter*, where, in an area of singularly indeterminate drainage, there collect the waters of several corrie torrents. These feed temporary pools and marshes, which in flood time drain both northward into the Truim and southward into the upper transverse Tay, the Garry. The pass itself is a wind-gap whose sides rise up left and right to over 3000 ft.; that is, the valley is here 1500 ft. deep. If the transverse Tay is regarded as a consequent stream, and if the tops of the Monadhliath represent the local level of the Highland peneplane, then, judging by the gradients of main streams above the level of 1500 ft., the original source of Tay must have been somewhere about that of the modern Markie Burn, which enters the Spey from the north above Laggan. The Markie may have flowed up lower Strath Mashie and over a col at 1500 ft. to Dalwhinnie; this upper part of the ancient Tay or Garry Valley being one of the flats characteristic of these high-

* The Highland "passes" are rather defiles than mountain saddles. The low level at which some of them lie, *e.g.* the Pass of Brander, emphasises the deep dissection of the land. The rich descriptive value of the native Gaelic names is usually lost in the common English substitutes.

land valleys. The Garry itself swings rather eastward from the transverse gradient, and perhaps the Bruar represents the true transverse (consequent) main trunk. Whichever way it is, the development of the Tay has been troubled by the rapid growth of the, relatively, immensely long Spey along the Highland grain,

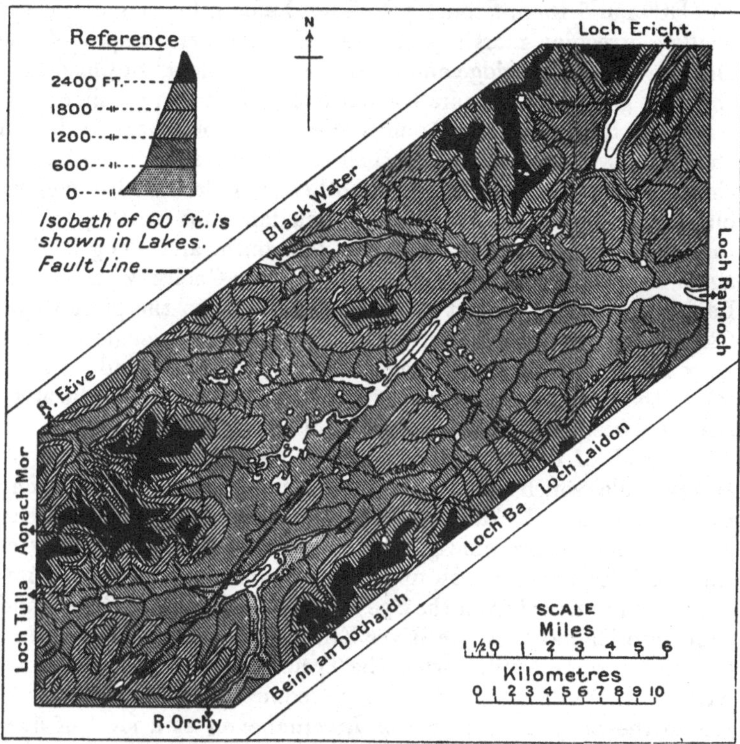

Fig. 48. Sketch-map of the Moor of Rannoch. Most of the lochs in the Black Water valley are now merged in the reservoir which supplies the Kinlochleven hydro-electric plant.

and this growth of the Spey has been one of the chief disturbing elements in the gradual occupation of their drainage basins by the streams of the Central Highlands. Between the Pass of Killiecrankie and Struan lies one of the conspicuous flats of the transverse Tay.

The first great longitudinal stream of the Tay on the right is the Tummel, which enters a couple of miles below Killiecrankie. This is a remarkable valley. It includes three flats at different levels. The highest includes Loch Laidon, 924 ft. (282 m.); the middle Loch Rannoch, 668 ft. (204 m.); and the lowest Loch Tummel,

458 ft. (138 m.). Loch Laidon is a deep gash in the floor of the Moor of Rannoch (Fig. 48). This moor is an undulating "flat," about the level of 1000 ft., forming the bottom of a bowl excavated out of granite. Its northern side is a granite ridge running up to 2500 ft. (760 m.), at the foot of which collect the headwaters of the Etive and those tributary to Loch Laidon, forming a maze of shallow irregular marsh pools. On the southern slopes of the Beinn a' Chrulaiste ridge and on the northern slopes of Clach Leathad the gutter streams are trained like close-grown espaliers on a wall. Clach Leathad and Aonach Mòr with Stob Ghabhar shut in the moor on the west and send their eastern waters to Loch Laidon through Loch Bà and numerous other irregular flooded hollows on the flat floor. From Beinn Achaladair runs north-eastward the edge which closes the basin to the south, and along its foot again drainage divides, eastward to Amhainn Gaoire, which drains Laidon to Rannoch, westward by Loch Tulla to the picturesque Glen Orchy and Loch Awe. The basin is a long triangle with its apex stretching eastward almost to Loch Rannoch, and its sides follow the dominant directions of longitudinal drainage in the area: east and west, and north-east and south-west.

The gash of Laidon is exceptional in the moor. It alone, if we except Tulla, which is marginal to the basin, is long and deep among the numerous lochs of the moor; and its north-eastward trend points over definite wind-gaps to Loch Ericht and the valley of the Spey. Ericht forms an upper ante-chamber in the Tay (Tummel) basin annexed from the Spey. It lies at 1150 ft. (350 m.) and spills into Rannoch by the River Ericht, which falls nearly 500 ft. (150 m.) in some 5 miles, while the Truim, which at one time must have drained Ericht for the Spey, requires 9 miles (15 km.) to fall to the Spey by the field of Invernahavon at a level of over 800 ft. (240 m.). The lusty vigour of the pirate stream is convincing. The line of the depression runs more northerly than the longitudinal (strike) direction, and it is that of a great fault whose course is clear across the granite from near Cruachan to the foot of Loch Ericht at least. This fault is a tangent to Loch Tulla, and its trend shows in part of the extreme western shore of the loch, the main run of which, however, follows the rock strike. A shallow western extension of Loch Laidon exhibits the eastward line of the consequent Tummel valley, which, in spite of the "side-step" of the Laidon trough is maintained with great consistency to the confluence with the transverse Tay. This is the trend of the so-called east and west dolerite dykes, which, curiously enough, are not conspicuous so far north in the Central Highlands.

The upper Tay itself forms the next longitudinal member of the Tay system. The glen on the whole approximates to the strike direction, though the middle section of Loch Tay is deflected along a fault running more northerly, and the lower end has the east and west trend. This trend is conspicuous in the valley of the main tributary valley, Glen Lyon. Loch Lyon, however, lies along a fault-line with the usual course north of that of the strike. The Fillan, almost the extreme main head-stream of the Tay, is transverse. The main longitudinal valley continues beyond the bend of the Fillan at Crianlarich, but westward of this the drainage is to the west. Here is one of those remarkable low-level through valleys which frequently form both convenient lines of communication and debatable grounds that are perhaps the true boundaries within the Highlands; for the natural regions are the mountain fastnesses with their glens, and the low stretches, too narrow to be concourses, were in the past no man's land, and are now links.

In the present instance the line of communication is, as it were, short-circuited by reason of the rather difficult transverse Glen Ogle, which rises 600 ft. (180 m.) in some 3 miles (5 km.), but connects with Stirling and the south. At the western end of the Tay valley, Glen Falloch, which continues the longitudinal depression to the head of Loch Lomond, makes connection with the valley of that loch, and admits traffic from the western centre of the country. Crianlarich is thus a railway junction of some importance in spite of the desolation of its site, and it consists of two railway stations and an hotel, a few railwaymen's dwellings constituting the village. In barren Strath Fillan two railways run within sight of each other on either side of the valley for some 6 miles until they diverge north and west by Tyndrum, a petty lead-mining centre which also has two railway stations. The Tay valley shows a very close analogy to the Great Glen in relation to communications. A spur from the Oban line reaches Killin at the western end of Loch Tay, just as a spur taps Fort Augustus on the Great Glen. Aberfeldy, 6 miles east of the Loch, with the difference that it is served also by a spur from Ballinluig, corresponds to Inverness.

This valley, of course, is also stepped. Lochs Dochart and Iubhair represent a stage in process of change from lake to boggy flat which will be closed to settlement until in the course of time the gorge below shall have invaded it and provided adequate drainage. The former lake represents the upper part of a longer water body split by the delta of the Benmore Burn, which, with its adequate drainage, carries several farmlets. That lake, shallow and weed-bottomed, according to the Scottish Lake Survey, serves as a settling tank for 85 per cent. of the drainage area of the two lochs, and so, for the present, enables Loch Iubhair to maintain its superior depth.

The Earn valley is again east and west in its trend. It has been truncated by the Balvag, which drains Loch Voil by Loch Lubnaig for the Forth, and which is also stepped. Its valley is virtually a lowland extension, and not even a corridor into the Highlands; for the gradients at its head are bad. In relation to communication it is a kind of "circle" traversed clockwise, and it is therefore a dependency of lower Strathearn.

The left bank tributary of the Tay, the Isla, is definitely lowland, but it gathers the waters which spill down from the Mounth along valleys which for the Highlands are unusually productive, with their southerly aspect, but no longer thoroughfares for the most part. They are truncated transversals.

Subsequent or merely longitudinal, it is natural that these longitudinal through valleys should cut more deeply into the country than the transverse; though, in the matter of modern communication the transverse valleys are perhaps of the greater importance. Glen Spey, however, is an exception in this matter, and deserves a glance. It is decidedly the largest longitudinal river of the Highlands, and its development has interfered with that of several other systems. Partly it runs in a fault direction, partly along the rock strike. With the ancient highway into the Highlands through Drumochter Pass it forms a T, and to this Kingussie owes its importance as a coaching centre; for until the West Highland Railway penetrated the Moor of Rannoch the road by Loch Laggan was the chief means of access by land to the Lochaber country and the west end of the Great Glen. The headwaters of the Spey to-day are found close by the fountain head of the westward drainage of Glen Roy, right on the brink of the Great Glen. The direct up-stream line of the upper Spey, however, is continued by Loch Laggan and the Spean, which has captured the main fountain head of the earlier Spey. Near the great elbow of capture made by the Pattack, the chief affluent to Loch Laggan, as it turns west from the Spey, the drainage is as indefinite at a lower level as it is in Drumochter Pass. Laggan with its vigorous drainage lies in a kind of oasis in the midst of the moory waste among the still waters. The Pattack is wild and tumultuous, and there is no uncertainty about the westward outflow from Loch Laggan. There is presented here convincing evidence of one of the most interesting cases of diversion of drainage to be found in Scotland.

The pattern north of the Great Glen is rather different north and south of the through valley of Strath Carron and Strath Bran, which carries the most recent railway in the Highlands, the Skye line of the L.M.S. Railway. South of this is a country strikingly furrowed from east to west, the most noticeable exceptions being Glen Moriston, an ante-chamber to the Great Glen, which follows the strike of the Central Highlands; and the gloomy trough of Loch Shiel on the west, and Strath Glass on the east, which lie in the same line parallel to the Great Glen. To the north the dominant furrowing is transverse, running from north-west to south-east, a direction which is conspicuous in great dip-faults and in swarms of dykes in the Lewisian area of the Mainland.

Human Settlement in the Highlands

Apart from a few small workings of metalliferous minerals there is little industry in the Highlands. The coal of Brora and the granite of Aberdeenshire belong to areas which are treated elsewhere. In most parts of the Highlands, in spite of the ubiquity of bare rock, there is a dearth even of building stone, and the older structures bear witness to the uncouthness of the material of which they are constructed. Perhaps one of the most interesting occurrences of mineral which have been exploited, although on an insignificant scale, is the alluvial gold of the Suisgill Burn in east Sutherland. Zinc, lead and copper have at various times been worked in different parts of the Highlands, but the projects have never been lasting on account of the smallness of their scale, and are now extinct. Barytes is a mineral of frequent occurrence, and it was from the allied mineral, strontianite, which occurs at Strontian in Morvern, that the metal strontium took its name. Even there lead is more important than the heavy earths.

It follows that the only possible basis for the human occupation of the land is agriculture, and the population of the Highlands is primarily agricultural. The particular mode of agricultural occupation is essentially a product of the geographical environment, for though it is connected with the evolution of the Clan as it occurred in Scotland, that evolution itself has its fundamental basis in the physical geography.

Human settlements in the Highlands are conspicuous for their compactness, small size, peripheral distribution and sparsity. Agrarian settlements as they are, they speak of the small extent and scattered distribution of ground fit for cultivation. Their situations are littoral or in glens, on level patches at a low altitude widely scattered in a land of bulky uplands of moderate elevation. Even within the glens and the littoral region their distribution is limited; mainly by the nature of the soil in a land of moderate climate whose greatest drawback is the moderation, which is no moderation, of the warmth of summer, coupled with a heavy rainfall and an abundance of soil water. Much of the valley ground is marsh or bog. The difficulties in the way of its cultivation are the acidity of the peaty soil, and the poorness of natural drainage to which in part that is due. The former is hard to cure artificially on account of the poverty in limestone of the Highland rocks over broad areas, the latter on account of the difficulty in ensuring an adequate fall for artificial drainage. The smallness of the areas concerned and their wide dispersion mean inability to justify expense on any considerable scale. Within their littoral or valley limits "townships"* have found a footing in areas naturally drained and in those in which the bog acids have been killed naturally. On the margins of such favoured spots forlorn attempts have been made to colonise the bog land, attempts whose persistence only a depressed standard of well-being could explain. The typical sites (Fig. 49) are on haugh lands in valleys; in coastal situations where blown sand has ameliorated both the drainage and the acidity of the soil, such sand containing a greater or less proportion of comminuted shell; by steep shore-lines, either on raised-beach terraces, or on shelves interrupting steep hill-sides; on flat land bounded seaward by cliffs. In extent and productivity the sites vary in the order in which they are named, the last mentioned being the poorest.

Of the haugh lands the most important are naturally tongues of good farm land pushed across the boundary of the Highlands, and continuous with the Lowlands; and on such ground the arable holdings are much larger than those typically Highland. The Tay valley above Killiecrankie, the valley of the South Esk, where wheat is grown above the constriction at Glenarm at elevations over 700 ft. (210 m.); Glen Moriston and Glen Urquhart, alcoves off the Great Glen; Strath Halladale, a valley corridor leading south from a dune-fringed sandy bay in the east of Sutherland; these are examples of the valley-haughland type of site, in order of diminishing productiveness, but not of decreasing relative populousness. Considerable stretches of the west coast of the Moray Firth present examples of the coastal situation in which improved drainage and

* The *township* was the area held jointly by a community of crofter tenants, either directly or through an intermediary, from the landowner. It included the arable patch, divided each year afresh among the tenants between seedtime and harvest, and the outrun or grazing, which with the arable in winter was used jointly. The arable lots are now generally permanently allotted, and the term, township, has a looser sense.

soil are due to drift sand. For various reasons these are best regarded as areas of lowland type, but as a rule they send tongues up the valleys which are definitely highland. At Strathy and Bettyhill on the north coast of Sutherland and at Barvas on the west of Lewis the type is repeated; while the island of Tyree with its notably coastal population, and Iona in its productive eastern part show instances on a scale relatively large. The use of shell sand as a "manure" was at one time common in coastal parts of the Highlands, and is still carried on in Islay and Tyree at least. Perhaps it is worth noting the shell sands of John o' Groats. The coast there is rocky, but shelving, and in hollows of the foreshore there have accumulated deep wreaths of a silvery sand consisting almost exclusively of comminuted shelly material of the coarseness of barley. Inland this has been piled into a raised beach of ten or twelve feet in thickness, and shows a tendency to become coherent. Although the area covered is very small, only a few acres, it is very noticeable on account of the calciphilous nature of the vegetation, whose sclerophyllous character contrasts strongly with that of the xerophytic vegetation of the nearby moorland. Shell sand, nearly as exclusively calcareous, occurs at other points on the Pentland Firth, at Uig in Lewis and on the shores of Tyree and Iona. It sometimes happens that poorness of natural drainage makes the influence of limey sand of little avail in rendering tractable the soils of the Highlands. In eastern Lewis the inner shores of the Broad Bay are sandy. To this is due the possession by the town of Stornoway of a golf course on the damp flat links of Melbost; but the township of Melbost itself is rather a miserable settlement in spite of its sandy soil, which, lying low, is water-logged and loath to yield a return to cultivation.

The raised-beach sites also vary considerably. Many of the coastal sandy areas are raised-beach, but their dunes and sand beds are largely formed by drift off the present beach. Apart from these are the terraces on which pebbly loam has developed on the marine sediment of the beach, reinforced by the deltaic accumulation and soil wash from the steep slopes above. Sometimes these terraces are relatively broad; excellent examples are the Moine Mor at the root of the peninsula of Kintyre and the Laggan towards its southern end. Here again, however, defective drainage counteracts the natural porosity and favourable composition of the mineral basis of the soil, and there is much damp meadow, which produces pasture but is difficult to cultivate. On the shores of the sea lochs, and around islands sheltered by the skerry-guard of the outer fringes of the two Hebridean groups the terrace lands of the beaches form long continuous strips, as much as two furlongs (400 m.) wide where the old shore recedes into bays relative to the modern coast-line. It is in such situations that the crofting townships are least compact. Elsewhere they tend to consist of a row of cottages, single or double, fronting the thoroughfare, and huddling almost as closely as the houses of the modern "colliers' rows" of the industrial lowland. On the inner beaches they have formed open order, like a platoon of infantry deployed, on the broader stretches of the beach; and from a soil which looks like an unconsolidated pudding-stone they produce wonderful crops of oats and potatoes, especially the latter. No stretch of the coast shows this phase better than the inner part of Loch Linnhe, Loch Aber, and its *annexe*, the shores of salt-water Loch Eil. On the exposed coasts of the Minch the bossy surface of the land stoops ever more steeply until it plunges headlong to the sea. The waves are everywhere undermining the cliffs, and only in patches does a narrow ledge of the 25-foot beach remain. On such one finds the most forlorn of the Highland hamlets, a single row of white-washed cottages nestling under the cliff which cuts them off from all communication by land with the interior. Perhaps the most accessible of such hamlets is Applecross on the Inner Sound of Skye; but it has its Srath Maol Chaluim, with church, manse, and "big house," which, however, ends in the dreary gloom of Applecross Forest; and its highroad leading to the Carron corridor at sea-level over moors at 2100 ft. (640 m.). Another is Mallaig, which boasts an old and a new "town." The old is a "street" of low grey "black houses" pressed against the ledge at the bottom of Port Faochagach, which ends with the promontory on the east side of the inlet. The new clusters round the terminus of the West Highland Railway, which broke trail from Arisaig to Mallaig to establish a rail-head for the Inner Hebrides and the herring fishing on the banks of the north-west. Its modern slated houses, its shops and its hotel clamber uncomfortably among the *roches moutonnées*, and its fishing bothies cumber the beach (*cf.* Fig. 49, No. 4).

Here and there a hill-side terrace below the limit of agriculture provides a cultivable strip of the drier slopes on which a man who does not demand too much of life may expend much labour and ingenuity and snatch a crop from nature in her most niggardly mood. Often these are haugh lands whose natural vegetation is not grass but heather. In a brown or purple landscape they hang above sodden valley bottoms with a semi-tundra cover of peat-hag and heather bush. Sometimes they are rock terraces formed in the distant history of the drainage systems. Excellent examples are

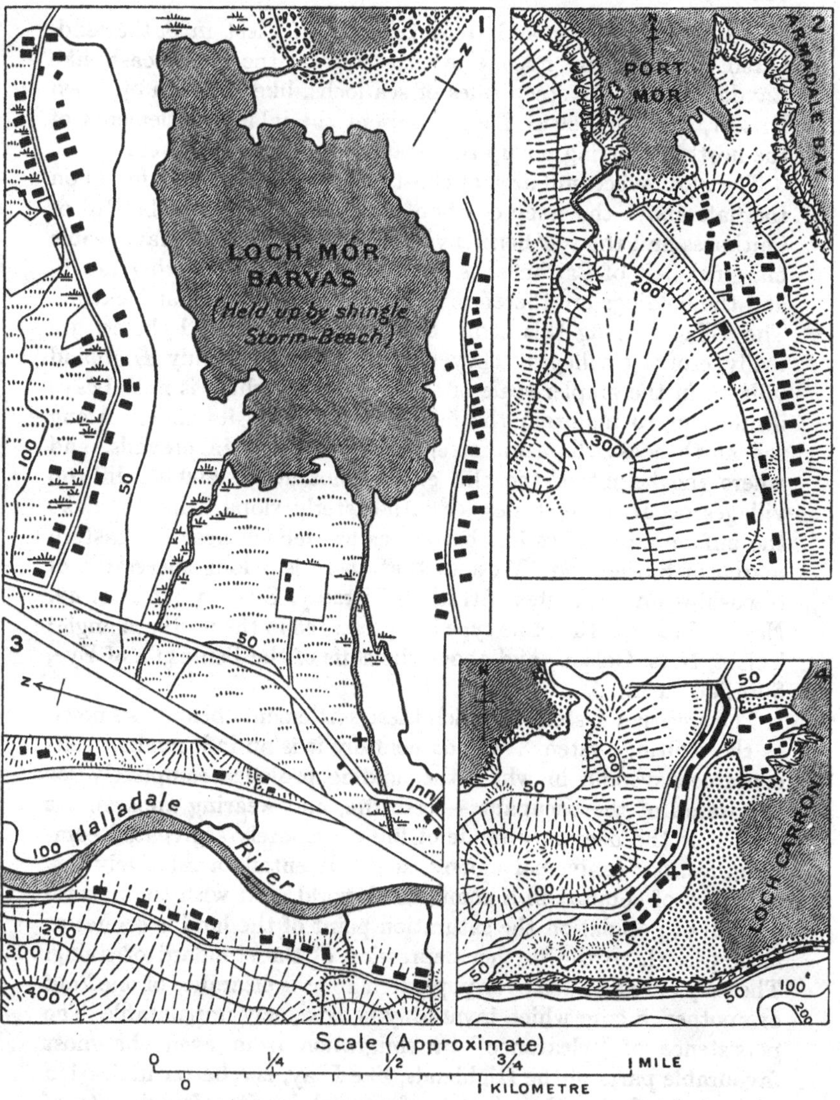

Fig. 49. Sketch-plans of Highland village sites. 1, Barvas, Lewis: sandy coastal type; 2, Armadale, Sutherland: cliff-top type; 3, Trantlemore, Sutherland: valley-slope type; 4, Plockton, Wester Ross: raised-beach type, a modern village, less "Highland" in aspect.

found between Lairg and Rogart in Strath Fleet in Sutherland. Often they also are marine in situation, on the open coast, like Laxdale, or on the steep sides of sea lochs, like Balallan on Loch Erisort, both in Lewis. They represent the inland settlements of the north, occurring far up the straths, as in Strath Oykell.

The most successful area of cliff-top settlement is to be found on the flats above the steep coasts of Caithness. Between the Ord of Caithness and the southern environs of Wick the more favourable marginal belt of Caithness is much narrower than farther north, and the holdings are therefore more Highland in scale and nature. The villages along this strip, Dunbeath, Latheron, Lybster, are really crofting villages. They contrast with the truly Highland villages in the ampler scale of their industry, which is not always insufficient to maintain the family and occupy its time. Along the north coast the same steepness of sea margin prevails, and where the height of the cliff edge is not above 200 ft. cliff-top villages occur. Sometimes a south-easterly slope helps to make the unfavourable sites less barren, as at Melvich near the eastern march of Sutherland. West of that there is a long succession of townships on such sites: Strathy, Kirtomy, Skerray (*cf.* Fig. 49, No. 2). In Lewis the same type is represented in the northern angle: Tolsta, Ness, Galson; and above the cliffs of the Peninsula of Eye, Suordal, Aird.

The agrarian system to which these settlements belong is known as the crofting system*. Fundamentally it is an independent self-contained system, in which the holding provides completely for the wants of the occupants—food, fire, and wearing apparel. As such it is the necessary mode of life of remotely segregated communities, and universal among them. It entails a relatively low standard of comfort in a commercial world, and with the natural growth of population the saturation point of the land is sooner or later passed, to the further depression of the standard of living. The only cure for such over-populated areas is clearance of one kind or another, a cure which invariably is extremely unpalatable. The persistence of "clearance" or emigration from even the most favourable parts of the Highlands, like Islay, is a better indication of over-population than figures of general density. The density of population is low in persons per square mile over a parish or a county, but high when expressed as arable area per person. Within the crofting townships 50 souls per square mile (19 per km.^{2}) is a high value of density, for it means average holdings of less than

* See a valuable article, with bibliography, by I. F. Grant, "The Highland Openfield System," *Geog. Teacher*, vol. XIII, 1926, pp. 480 *et seq.*

5 acres or 2 ha. arable per holder, and numerous crofts of less than an acre. So much is "outrun" or rough hill grazing, productive only in summer, and not ploughable at a profit. The crofter therefore finds his time only partly employed, and seeks other employment such as fishing. This is a consideration which favours the littoral habitat; but the modern fisherman is a migrant with the power fishing fleets, and his daughters migrate with him as fish cleaners. The wife and the younger members of the family must do much of the farm work, and they too migrate after the fashion of the Swiss to make use in summer of the distant grazings*.

The crofting community is a village community, collected on the fertile spots on account of the sterility of the land. Isolated dwellings do not belong to the native exploitation of the land. They house those engaged in the large scale modern sheep raising industry, the farmer, and his shepherds in their lonely cots; or gamekeepers associated with the equally modern systematic exploitation of the game resources.

The Regions of the Highlands

Focussing attention on the areas presently open to human occupation, we may describe the Highlands as a moist habitat without definite summers consisting of a large number of stations limited in area, scattered widely and to a considerable degree isolated from each other. It passes eastwards into a habitat which is less moist, has a more definite summer, and is much less discontinuous. In this volume the latter region is therefore definitely separated from the Highlands. Insular regions, being insular, are isolated, but they may, and do, as in the case of Lewis, Islay and others show greater continuity within themselves in regard to the human superstratum. There is here scope for more minute subdivision than the present essay can contemplate, as between Skye and Harris, which are more definitely Highland on the one hand, and Lewis and the archipelagoes of Orkney and Shetland, which last on account of their composition and situation are outliers of Caithness.

The attempt to divide the Highlands into homogeneous subregions such as are required in modern geographical study is handicapped by the absence of detailed regional surveys. The most obvious dividing line, the Great Glen, ought itself at first sight to be a region and a corridor in one. One is brought to face the fact that

* For an amplification of this in relation to Lewis, see A. Stevens in *Scottish Geog. Mag.* vol. XLI, 1925, p. 75.

in mountainous country, even if man is confined to low tracts, the natural boundaries must be hollows.

The significance of this line is the greater in that it separates the nearer Highlands from the more remote. To a large extent the settlements of the Central Highlands are protrusions from the lowlands of the south or east, or outliers therefrom. They are organised upon lowland centres or centres so marginal as to be really lowland; and reference is made to them elsewhere. It is proposed here to say a little about the remoter regions of the north*.

The Great Glen rift stretches from the Firth of Lorne to the Moray Firth, and is a narrow flat-bottomed gash even when under water, as in the Lochs Lochy, Oich and Ness. It follows a remarkably rectilinear course often described as bearing north-east, but really running N. 37° E. Especially on its western side the Glen is steep, rising abruptly to elevations of the order of 3000 ft. This western rim is more broken than the eastern, opening into populated alcoves like Glen Urquhart and Glen Moriston, which share the more prosperous aspect of the Great Glen as a Highland region. The eastern side of the rift, which is the northern flank of the Monadhliath, if lower, is only slightly slashed by torrent courses. The whole of the floor of the rift lies under the level of 250 ft. (76 m.), and the highest of the lochs, Loch Oich, has its water-level scarcely above 100 ft. (30 m.). Especially on its northern side the Great Glen is cut off by lines as emphatic as the Highland line, but its very limited extent, even including the ante-chambers of the eastward-running glens which open into it from the west, reduces its importance to insignificance, while leaving it a very peculiar individuality and interest. It ought obviously to be a corridor of communication; but a roadway in a desert depends for its importance on the desert being an obstacle between complementary productive regions, and the very minor importance of the Caledonian Canal throughout its history demonstrates the uselessness of this corridor as a highway under existing economic conditions. The Glen is but one of the greater Highland oases. Its significance has lain in separation of the Central from the Northern Highlands. It has been judged in this connection to be a strategic line of sufficient importance to carry the 18th century forts of Fort George, Fort Augustus and Fort William in their function of controlling the

* The following recent articles cover areas not treated specially here: M. I. Newbigin, "The Kingussie District: a Geographical Study," *Scot. Geog. Mag.* vol. XXII, 1906, p. 285; J. M'Clement, "The Distribution of Agriculture in Kintyre," *ibid.* vol. XLIII, 1927, p. 20; J. Holmes, "The Upper Clyde Estuary," *ibid.* vol. XLIII, 1927, p. 321; I. D. Duff, "The Human Geography of South-western Ross-shire," *ibid.* vol. XLV, 1929, p. 277.

remoter Highlands. To the idea of a corridor of communication the towns at its ends, Inverness and Fort William, lend a specious support, with their nodality expressed in the number of their hotels. But this nodality is due to a position on low routes over the Highland Plateau from the south, coupled with a focal situation in relation to routes leading farther north and seaward; and the double rail-head on the Great Glen is an expression of the relative unimportance of the corridor.

A boundary between east and west is best followed in Hardy's vegetation map*. Fully aware of the arbitrary nature of any such line, we draw it from the Great Glen down the valley of the upper Spey to Dalwhinnie, up the Truim and along the divide on the east of Loch Ericht and the north of Loch Rannoch to the meridian of Schiehallion, along that meridian across Loch Tay, and thence by Strath Bran to the Highland Border. Perhaps there is more to be said for a continuation by the Pass of Drumochter to the Garry and so to the border. The western Highlands so separated are characterised by more minute dissection; that is smaller human stations and more of them, greater dampness which may have a connection with the greater fragmentation, cooler summers. The vegetation map presents to some extent a summation of these features: high pastures and marshy grass moors contrasted with dry grass moor and broad heathy uplands, birch and even oak woods in place of the coniferous woods more abundant in the east. In the west a deeper dissection goes with its minuteness, and the jagged fjord-gashed coast is topographically related to the zig-zag plan of the over-deepened flooded valleys of the western lochland. If for the sake of having a definite line, as well as for other reasons, one is inclined to pitch upon depressions and *thalwegs* in separating east from west in the Central Highlands, these do not run far from water-partings, and on the whole the two sub-regions drain one to the North Sea, the other to the Atlantic.

North of the Great Glen the parting between east and west must cut definitely across the drainage. Moreover the width of the land from east to west is much less. There are no important transverse routes leading off from the Great Glen, which has the same sort of relation to the North-West Highlands that Strathmore has to the Central. The latter are penetrated by two trunk lines of communication, the Highland and the West Highland. The more massive eastern half is served by spurs from the east coast route. Only this route persists north of Inverness, sending a branch to Kyle of Lochalsh, which may be compared with the Oban line in

* *Scot. Geog. Mag.* vol. xx, 1906, facing p. 341.

the Central Highlands, and which establishes a third rail-head on the west coast. Perhaps one might draw a line from Loch Laxford along the Shin Valley to Lairg, which is virtually a post on the border. The line would continue to Ardgay, pass up Strath Carron, down Strath Vaich and Glen Garve, across the foot of Loch Luichart, southward to the fork of the Glass and the Farrar, and along the Beauly to the sea. The areas separated are virtually Sutherland and Easter Ross on the east, and the more remote western Highlands, which are divided into two by the Carron-Bran corridor. The northern part includes Lewis, the southern Harris and the rest of the Long Island and the Hebrides as far south as Ardnamurchan. The modern communications of these island groups centre on Kyle and Mallaig respectively, Skye being lengthwise divided between them.

These western divisions one might call Wester Ross and Clan Ranald, in order to secure a short name without doing more violence than necessary to current or historical terminology. In the former one finds quite a considerable coastal fringe on both sides of the Minch, settled in agricultural hamlets on the inlets of the fretted coast by a considerable population, which on the outer coast of Lewis is repelled a few hundred yards inland by a less indented shore. Fishing on the inshore banks for local consumption, and "low ground" herring-fishing by small boats serves to eke out the meagre yield of the fields. The "cash" crop, the only surplus exportable, is stock. Each crofter sells at a few months or a year the young beasts which are the condition of his meagre milk supply and which his holding will not feed through the winter. The trade is mainly local: the quey* of one man goes to replace the superannuated cow of his neighbour; the stirk* to stock the broader acres of some more fortunate holder. But there is always a movement of stock by way of local "markets" (fairs), which occur annually towards the beginning of winter, in the direction of organised sales at marginal villages, outposts of the "south." These outposts arrange themselves in relation to the area radially towards the rail-head in order of increasing importance, and at the foci of roads. The essential equipment of these centres for their function is the inn and the market ground. The function of the inn was rather to facilitate the bargain than to provide rest for man or beast, and this remains the most lucrative part of the business, though the premises have often been modified to cope with the accessory trade brought first by the sportsman and later by the motorist. The clachan (hamlet) is more or less secondary to the

* *heifer* and *yearling ox* respectively.

focus. Altnaharra, at the upper end of Loch Naver, probably the settlement of Sutherland farthest from the sea, is such. Forsinard, a rail-head overshadowed by Lairg and Thurso, because it lies too near the Caithness border, is even more so. Lairg is the main focus of the mainland region. Here the coastal railway makes a detour into the Highlands, and reaches its remotest point where the converging glens lead paths together to a broader croft-studded hollow at the lower end of Loch Shin. Lairg to-day consists essentially of a market place, a railway station, a hotel and a garage, with a clachan somewhere in the background. It collects the surplus and distributes the wants and mails of the country from Lochinver by Scourie to Tongue, in virtue of the fact that here rail transport from the progressive east meets the primitive road transport (though now motor-borne on second- or third-hand vehicles) of the over-populated stagnant west. If there is the rudiment of coastwise communication and a certain amount of tangential road which acts as a sympathetic system, in relation to the essential modern communication, Lairg is the ganglion, and the coastal villages mere peripheral nerve-endings of the central nervous system.

The north coast as far as Skerray is served from Caithness, and Thurso is the centre. Caithness itself is more definitely lowland than the east coastal strip of Sutherland, where the narrowness of the strip encourages fragmentation in land tenure. On the flats of Caithness the size of the farms expands. Even the sheep farms, which may be regarded as the product of an exotic type of agricultural activity, and which in the Highlands are purely sheep walks, in Caithness are often to a considerable extent arable and feed horned cattle. This is a general characteristic of the sheep farms of the northern borderlands. Caithness, however, like the Highlands is a boggy or heathy moss, and its settlements which do not share in the characters of crofting settlements are confined more or less to the margin. The Ord of Caithness, permitting the passage of the main road, excludes the railway, and until the low ground widens out north of Sarclet, which is one of a series of small farming townships strung along the precipitous coast south of Wick, true Caithness with its treeless sloping flats hardly begins. The railway enters Caithness after its most marked detour into the Highlands north of Inverness, dropping down the monotonous slope of the Caithness moors from the height of land at Forsinard, and not by Strath Halladale and the north coast as was originally intended. The coastal strip from Melvich eastwards is so dotted with settlements demanding service from the rail-head at Thurso that it was natural to continue the service by road westward;

hence the curious traffic divide near Tongue, itself a lone settlement at the end of a long barren trail, is reminiscent of a very asymmetrical water divide.

Lewis again is tributary to Stornoway, and Stornoway is to its hinterland almost as foreign as Caithness to its own remoter glens and to the west of Sutherland. As a port Stornoway has seen better days. Even the glory of its herring fishing, at the end of the year and at the beginning of summer, has been sapped by the advent of the power fishing vessel with its longer radius of action, which enables it to land its best catches at a rail-fed port such as Kyle, Mallaig, or Oban. The cosmopolitan aspect of the town remains, and it serves as metropolis to a primitive Highland hinterland, with its direct sea and rail connection through its county town, Dingwall, of which isolation makes it semi-independent, and past which the railway enables it to look to the Eastern Lowlands, with which are its most intimate connections. It is a curious symptom that Stornoway has received part of its daily supply of fresh milk from the farms of Moray.

The mainland part of Wester Ross becomes progressively deserted southward. Lairg still gathers up the threads of the communications of the northern part, the southern is brought into closer contact with Dingwall by the Skye railway and its string of nodes like Garve for Loch Broom and Achnasheen for Lochmaree. Since the extension of the line to Kyle, Strome Ferry has fallen back into mere local importance in relation to the shores of Loch Carron. As to Skye, neither its volcanic rocks nor its Mesozoics have redeemed it from a purely Highland condition, not even the ironstone of Raasay. Its eastern margin, tributary to Wester Ross, is the more populous, but no town comparable with Stornoway in size or importance has grown up upon it. Portree with its few hundreds of people stands midway along the northerly stretch of coast which is beyond the reach of the ferry at the narrows of Kyleakin, and Broadford at the southern end of this strip is not of a lower order of importance.

In these remote and insular regions the only industry of note is the production of "homespun" tweed. Even this has succumbed to the modern move towards mass production. The vulnerable point is the slowness of spinning relative to weaving. So the home grown wool, which, with the labour expended on it, to some degree replaces cattle as exportable material, is largely spun in local factories and returned to be woven on the croft. Herein the antithesis between town and country, merchant and crofter, is strengthened; for the merchants of the town buy the wool crop wholesale, spin or have it spun, and sell it back retail to the crofters, purchase the fabric again wholesale to retail it locally or by post. No money passes: only "credit"; credit which means much indebtedness on the part of the crofter and many bad debts which he has to liquidate by means of swollen prices, and the merchant makes retail profit on the yarn, on the groceries he gives for it, and on the fabrics he has to sell to the south at luxury prices.

In Clan Ranald country dissection and isolation are at their maximum. They are seen in the mainland glens of Knoidart, Morar and Moidart, in the deep cut coasts, and in the scattered small isles. Even the Long Island has a continuity which is only apparent in spite of the road, which, with low-water extensions across tidal sandbanks, runs from Stornoway to the tip of Harris and from the north of North Uist to the tip of South Uist. In the low islands of the Outer Hebrides the land is cut up, not by mountain masses but by deeper lake-flooded hollows in the low damp surface. Neither on the mainland nor in the islands is there a centre comparable with Stornoway, and the population is small and very scattered. The outer rail-head is Mallaig, an outport of Fort William; but the rail journey from the heart of the south is little shorter than by the Skye line; and Fort William has not the semi-terminal importance of Dingwall or Inverness. The surplus of the land is still woollens and cattle, but the mainland portion was in early times a source of supply of hardy ponies exposed in the market of Callander and later at the Trysts of Falkirk.

South of Ardnamurchan there is a progressive increase of prosperity and diminution of pressure on the land. The effect of remoteness is less marked, connections with the southern centres are closer, produce is more easily marketed, and the islands and loch shores become more and more a kind of playground annexed to the teeming western industrial districts of the Lowlands. The southern lochs and isles are in more or less direct connection with Glasgow by the Crinan Canal, the more northerly are in the sphere of Oban, which is naturally the busiest of the three western rail-heads.

XXII

NORTH-EAST SCOTLAND

John McFarlane

THE north-east of Scotland is a region which in many respects has a well-marked individuality. It may be regarded as beginning where the last outliers of the Grampians approach the coast near Stonehaven, and as including the lower valleys of those rivers by which the Highland region is drained, together with the plain of Caithness, the lowlands around the Moray Firth, and the worn-down plateau of Buchan. The region as a whole possesses few geographical advantages, its soils, if not infertile, are seldom rich, its climate is generally harsh, and its mineral resources are of comparatively little value. Situated as it is between mountain and sea, the occupations of the people are in the main confined to the cultivation of root crops and the hardier cereals, to stock raising, and to fishing. With relatively few exceptions industry does not provide for more than the immediate wants of the population. On the other hand the many attractions of the region have led to the development of holiday resorts, and the growth of a new if subsidiary occupation for a large number of people.

But if the north-east of Scotland is not richly endowed by nature, it must not be regarded as a poverty-stricken land. An energetic people has done much for the improvement of its environment. The glaciated soils have been cleared of their boulders—the huge consumption dykes in which the latter were stored are still to be found in many places—the land has been drained, the peat mosses by which much of the surface was at one time covered have been reduced in area, the soil has been cultivated, and one of the most important cattle-rearing industries in Britain has been developed. With equal vigour the resources of the sea have been exploited, and the towns and villages along the coast are the local centres of one of the chief fishing areas in the country. If these pursuits have not brought great wealth to the region they have at least rendered it moderately prosperous, and Aberdeen, which may be regarded as its capital, is one of the finest cities in the country.

Before proceeding to a regional survey of the north-east of Scotland it may be well to discuss briefly one or two matters which affect the development of the region as a whole. In some important aspects the soils in this part of Scotland differ greatly from those found in England, more especially in the south-eastern counties. In the former region they are the product of the glacial period, and usually consist of the crushed and ground-up remains of igneous and metamorphic rock. But the period which has elapsed since their formation has not been sufficient to permit of profound chemical weathering, whereas the rocks in south-east England are themselves largely composed of the results of such weathering. The result is that in many of the Scottish soils "the great reserves of phosphate, potash, lime, magnesia and soda which are present in the form of slightly weathered minerals do much to account for the surprising fertility of what appear to be poor light thin soils, and the great reserves of bases contained in their undecomposed silicates enable them in the absence of liming to produce gradually a supply of lime which enables them to stave off indefinitely the evil effects of extreme lime hunger."* Much of this soil tends to be acid in reaction, but the main crops of the North-East, oats, turnips and grass are suited to a moderate degree of acidity. For the cultivation of barley it may be necessary to supply lime, which is also of value in checking finger-and-toe disease in turnips.

Fishing for demersal and pelagic fish is the second important industry of the north-east of Scotland. The former are caught mainly by trawlers and steam liners, and for these Aberdeen is the chief centre. Line fishing by small boats is practised along the coast northwards, and as far as Shetland, but, except in a few places, it is decreasing in importance. The region in which the trawlers operate extends from the Great Fisher Bank in the south round the north of Scotland as far west as the 100-fathom line and south to the Flannan Islands; it also includes the Faroes, which are growing in importance. The trawlers may go out twice weekly, or they may be absent for a period of seven to ten days at a time as distance and weather conditions necessitate. In the northern part of the North Sea haddock is the most important fish, and Aberdeen is the chief port to which it is brought, over 50 per cent. of the demersal fish landed there consisting of haddock. Other fish brought by trawlers include cod, ling, plaice and lemon soles. German boats which fish mainly in Icelandic waters land large quantities of cod, part of which are split and dried before being exported chiefly to Catholic countries. Steam liners fish in much the same area as trawlers, but they extend their operations into the deeper water on the edge of the continental shelf where much halibut is found. Along with halibut they land cod, ling, and skate, all of which are sold fresh.

* J. Hendrick, "Some characters of Scottish soils," *Trans. Highland and Agricultural Society of Scotland*, 1925.

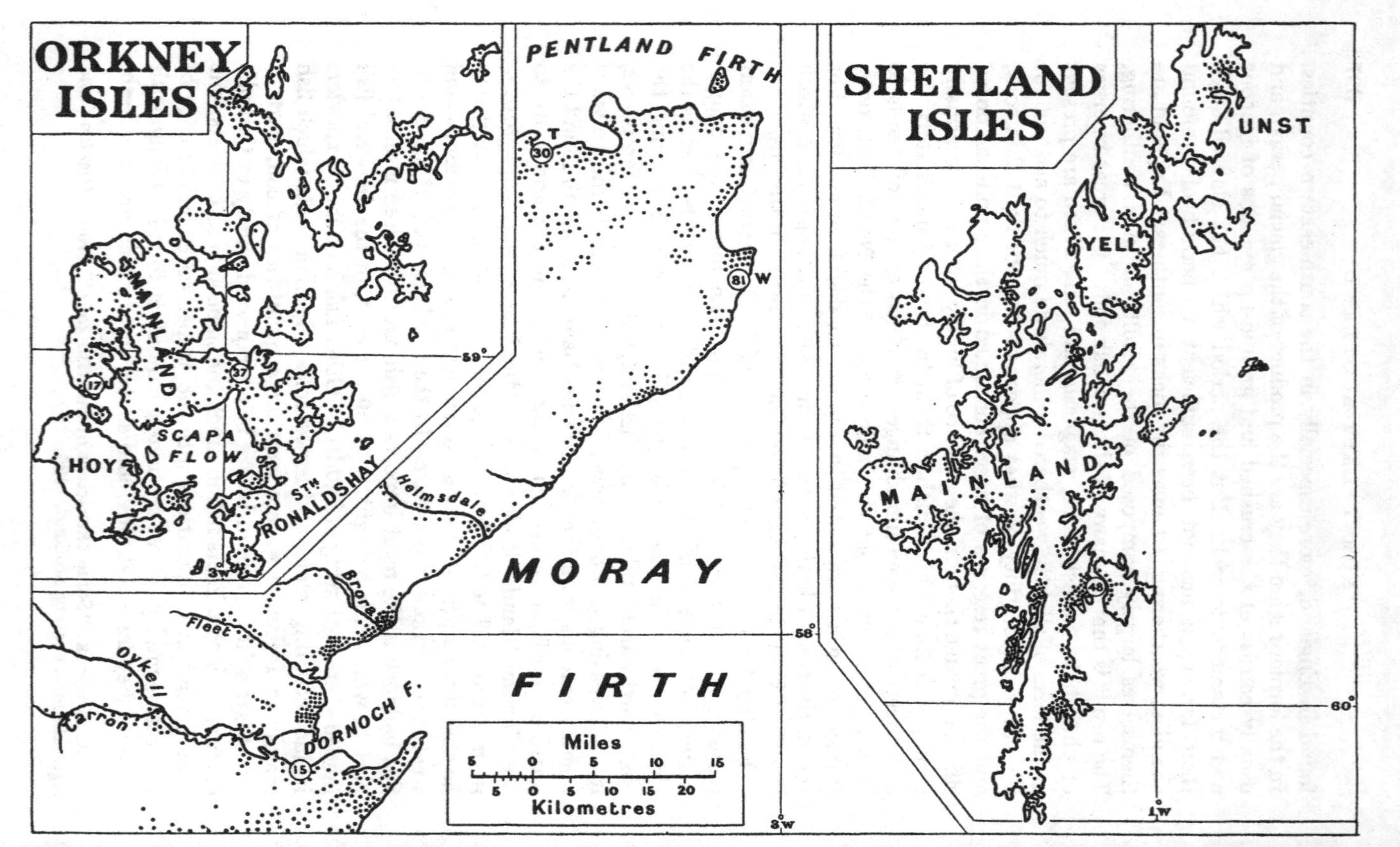
ORKNEY ISLES
MAINLAND
HOY
SCAPA FLOW
STH RONALDSHAY
17
37
59°
3°W
PENTLAND FIRTH
T
30
81
W
Helmsdale
Brora
Fleet
Oykell
Carron
DORNOCH F.
15
MORAY FIRTH
58°
3°W
Miles
5 0 5 10 15
5 0 5 10 15 20
Kilometres
SHETLAND ISLES
UNST
YELL
MAINLAND
48
60°
1°W

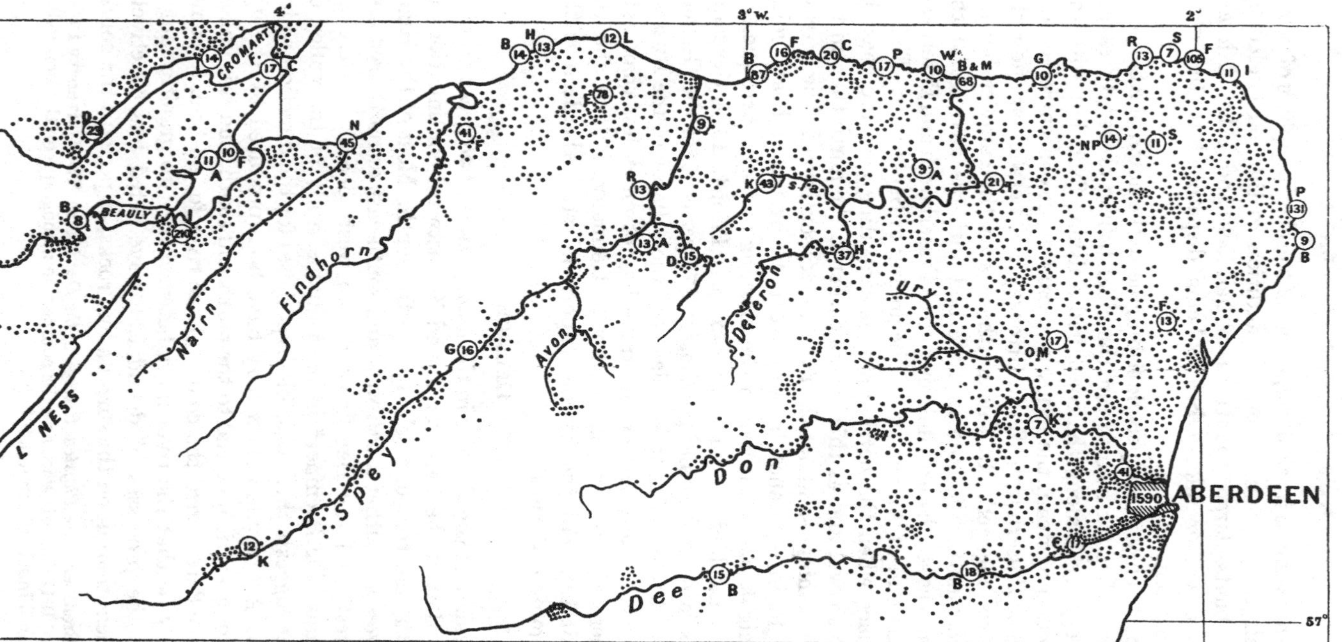

Fig. 50. North-eastern Scotland with the Orkney and Shetland Isles; distribution of population (1921). Each dot represents 50 people. Towns are shown by circles, the enclosed figures giving their population *in hundreds*.

Aberdeen, owing to its position, is one of the chief suppliers of prime fresh fish in the country, but, in addition to the cod landed by German boats, large quantities of haddock and other fish are smoked, dried, or filleted before being dispatched to the chief consuming centres.

The ports engaged in pelagic fishing lie to the north of Aberdeen, which is rather far away from the chief fishing areas. In earlier days every creek along the coast had a fishing fleet, but with the introduction of steam the industry became more concentrated, and Peterhead, Fraserburgh, Wick, Stronsay and Lerwick are now the chief ports engaged. In this northern area the summer herring fishing begins in Shetland in June and with the advance of the season moves southward, finishing up at Fraserburgh and Peterhead about the end of August or the beginning of September. Fishing takes place when the herring are concentrating near the coast in order to spawn, and as this occurs earlier in the north than in the south the fishing—and not the fish—moves southward. The herring are caught in nets shot by steam drifters or motor boats going out at night, and returning in the morning to port where the fish are cleaned, salted and packed for export. Prior to 1914, Germany and Russia were the chief purchasers—75 per cent. of the German share eventually going to Russia—and the collapse of the Russian market has reduced the export trade by not much less than half. At present the exports are mainly to Germany, Poland and the Baltic States. For the herring export industry Aberdeen is the chief commercial centre.

Deeside

The valley of the Dee lies in the main between two great granitic masses, that of the Cairngorms and its eastward continuations in the north, and that of Lochnagar in the south. Most of the Highland rivers run either north-west and south-east or north-east and south-west, and the Dee flowing from west to east is therefore anomalous. Dr Bremner* in his valuable study of the valley of the river suggests three possible hypotheses for the direction which it takes. A tectonic hollow may have been produced by the upwelling of granitic material to the north and south, and this hollow may have determined the course of the river for all time to come, or it may be that the river flows between the two granitic ranges because the processes of denudation have acted less effectively upon them than upon the surrounding rocks, or perhaps the course

* A. Bremner, *The Physical Geology of the Dee Valley*, The University Press, Aberdeen, 1912. To this work and the companion monograph on the Don valley the writer is indebted for much help.

of the river was marked out by the conditions which existed when the Old Red Sandstone covered the land. Dr Bremner, while evidently inclining to the first of these hypotheses admits its uncertainty.

From its source high up on the slopes of Braeriach the Dee falls nearly 3000 ft. to reach the Linn of Dee, a few miles above Braemar, where it crosses the 1200-foot (365 m.) contour. During this part of its course it flows through wild Highland country abandoned to deer and grouse, though in some of the glens there are indications of past cultivation. Below the Linn of Dee the river enters a large alluvial flat which extends as far as Braemar and may be an old lake basin. Its fertility probably contributed to the growth of Braemar, though the site of that village first assumed importance as a meeting place of two routes from the south, one by Glen Clova, and the east bank of the Clunie, and the other by the Spittal of Glenshee, and the west bank of the Clunie. Here in the 14th century was built the castle of Kindrochit, the remains of which have recently been revealed.

Between Braemar and Ballater the alluvial strip along the river again becomes narrow while the well-forested hills rise steeply on either side. The population is small, and is mainly, though not entirely, dependent upon certain great residences of which Balmoral is the most important.

Ballater stands on the edge of an alluvial plain, once the floor of a large lake which was drained when the water escaping from it succeeded in cutting through the morainic bar at Cambus o' May. The village itself only came into existence about the end of the 18th century when a bridge across the Dee was built and a church erected nearby to serve the three neighbouring parishes*. A little later a loop on the Deeside road, which then went through the Pass of Ballater, an old overflow channel, its slopes now covered with granitic scree, put the new village on the main route from east to west. With the extension of the railway to it in 1866, it rapidly grew into the popular holiday resort which it now is.

Ballater is typical of many small villages in the north which are rising into importance on account of their popularity as health resorts. Quite apart from their summer residents the population of these villages has special characteristics. Married couples who have retired from active work, widows, and spinsters add to a limited income either by providing for lodgers or by letting their houses during the season; in the latter case they themselves frequently retire to some temporary abode in their gardens. In places

* G. M. Fraser, *The Old Deeside Road*, The University Press, Aberdeen, 1921.

like Ballater this has a definite effect upon the population. In the county of Aberdeen the female population numbers just over 53 per cent. of the total, but in Ballater it is 60 per cent. Children under 15 years of age number 31 per cent. of the whole population of the county, in Ballater less than 21 per cent. The percentages for Grantown in Moray are almost the same as for Ballater.

Below Ballater the Dee valley begins to open out as the hills become lower and retire from the river; consequently arable farming becomes more important and sheep farming less so. The Muir of Dinnet, in autumn a beautiful expanse of heather, consists of outwash gravels from the valley glacier, while the plain round Tarland is an old lake basin, the waters of which were held up by glacial debris at Coull. Aboyne stands on the edge of another alluvial basin; the limestone hills by which it is surrounded are relatively low, and the village is an agricultural centre as well as a summer resort. A little lower down, from Kincardine O'Neil to Banchory, the valley of the Dee becomes more constricted and the railway makes a detour to the north, which enables it to tap the alluvial basin of Auchlossan, the site of a former loch not yet completely drained, as well as a considerable stretch of agricultural country lying between the Hill of Fare and the Dee. On the other side of the river and separated from it by low but somewhat steep hills lies the erosion basin of the lower Feugh and its tributaries, in the formation of which the hard rocks which obstruct that river near Banchory have probably been the chief factor.

Situated just above the confluence of the Dee and the Feugh and at the point where the Deeside road was joined by one from Strathmore by the Cairn o' Mounth and by another from Stonehaven still known as the Slug, Banchory was in the days of rural fairs an important market centre. If its importance as such has declined it has gained in other ways. Built on a southward facing slope and protected from cold northerly winds by the Hill of Fare it has became a noted health resort. The surrounding district is well wooded, especially to the south of the Dee, and timber industries are of some importance.

From Banchory to the coast the basin of the Dee gradually becomes more contracted. On the south it is bordered by the last extension of the Grampians as they run towards the north-east, while on the north the divide between the basins of the Dee and the Don runs across a low platform of granite and gneiss which seldom exceeds 500 ft. (150 m.) in height. The region is in the main agricultural, and the population is somewhat more evenly distributed than it is farther west. In the valley of the Dee itself a number of suburban villages are strung along the southward-facing slope.

Agriculture, the chief occupation of the people of Deeside, varies in character from place to place. To the west of Ballater, there is very little cultivable land, and of what there is much is kept in permanent grass. Oats is the principal crop, the acreage under turnips is below the average, cattle are few, and sheep are numerous. In the middle portion of the valley where the lowlands are opening out and where climatic conditions are better, oats and turnips are more extensively grown, and cattle-rearing becomes important. In the last stretch from Banchory to Aberdeen the rainfall is lower and the temperature higher, and barley takes the place of oats to some extent.

Donside

According to Dr Bremner* the Don has been developed on an easterly sloping peneplane. On the undulating moorland where they rise, at a height of over 2000 ft., the head-streams of the river flow in shallow valleys; but, as it descends, the main stream becomes more deeply entrenched and by the time that it has reached Strathdon it is flowing in a valley which is narrow and steep-sided. As far as Towie population is very sparse and is confined almost entirely to the alluvial strips along the main river and its chief tributaries, but from Towie, what is perhaps the most characteristic feature of the Don valley, the alternation "of close gorge and open reach," becomes well-marked, and exercises a very important influence upon human activities. In the Towie and Kildrummy basins which have been excavated out of Old Red Sandstone, agriculture on a more extensive scale becomes possible; the gravel-covered country in the basin of Kildrummy, for example, produces good corn, and the importance of the region in former times is indicated by the fact that on its margin stood Kildrummy Castle, the largest castle in the north of Scotland.

The Alford basin is even more striking; it measures twelve miles from north-east to south-west and nine miles from east to west, and its general character is well indicated by the fact that in addition to the gaps by which the Don enters and leaves the basin there is only one other point—the gap at Tillyfourie—which is well below 1000 ft. Thus the basin is sheltered on all sides and especially so in that part known as the Howe of Alford which lies below the 500-ft. contour. The boulder clay with which the greater part of

* A. Bremner, *The Physical Geology of the Don Basin*, The University Press, Aberdeen, 1921.

the surface is covered provides a fertile loam, and the whole basin is one of considerable agricultural importance. Lower down lies the Kemnay basin, on the border of which a fine-grained light-grey or nearly white granite, much in demand, is obtained, and the less well-defined basin of Kintore. Generally speaking, indeed, after the Don has left the Alford basin it flows through country of a lowland type, the valley being broad and undulating and the hills low and wooded.

At Inverurie the Don is joined by the Urie which, with its tributaries, drains the Garioch or "rough country" lying to the north of Bennachie. Except to the south of the Gadie, where the soil appears to consist mainly of granitic debris of relatively little value, the country is fertile. Especially is this the case between Inverurie and the eastern slopes of Bennachie where the stiff clays which prevail generally over the region have been mixed with granitic material carried by surface water from the hills above. The fertility of the region led to the growth of Inverurie as a manufacturing centre from early times.

From a point a few miles above Inverurie to the neighbourhood of Dyce the Don flows between the contours of 200 ft. and 100 ft. with an average fall of 6·1 ft. per mile (1·1 m. per km.). Below Dyce the gradient suddenly rises to 14·5 ft. per mile (2·4 m. per km.). This Dr Bremner ascribes to glacial action. Here the Don flowed in a narrow valley across which the ice streamed almost at right angles; "the nature of the valley, the ground rising sharply on the west side, and its direction favoured the accumulation of glacial debris." The result is that "the river now flows for the most part between high banks or stepping terraces cut in boulder clay or coarse glacial gravels." The special interest of this to the geographer lies in the fact that the water-power developed by the rapidly flowing river was a very important factor in the early industrial development of lower Donside.

Economic activities are somewhat more varied in Donside than in Deeside, but throughout the greater part of the region agriculture is the chief industry. Above Towie it is generally confined to the alluvial strips along the river and its tributaries, but in the successive basins already referred to, the area under cultivation rapidly increases, and in the parish of Alford nearly two-thirds of the land is cropped. In the western part of the Urie basin conditions are somewhat similar, oats and turnips being the principal crops and the type of agriculture being very much like that of the Buchan district to be discussed later. In the Garioch proper and in the Don valley below the basin of Alford crops rather approximate to those in lower Deeside where barley to some extent takes the place of oats.

In the basin of the Don there is more agricultural land than in that of the Dee and much of it is more fertile. The fluvio-glacial grounds which occupy so large an area in the valley of the latter river absorb water easily, and the crops grown upon them often suffer in dry summers. A local saying that one day's rain will do for the Don what it takes two days' rain to do for the Dee illustrates this point. On the other hand the Don flows through country the scenery of which is much less interesting than that of Deeside; accordingly, while many of its villages have an influx of visitors during the summer months there are no holiday resorts like Ballater and Braemar. Finally, it may be noted that while the valley of the lower Don has been rejuvenated by the events of the glacial period and has as a result a certain amount of water-power, that of the lower Dee is a well-graded river and is almost without industries. The agricultural importance of the Don valley in prehistoric times, and especially of that fertile region of which Kintore is the centre, is indicated by the number of stone circles, sculptured stones, and "short cist" burials, and in mediaeval times by the many ruined castles scattered over the region.

Aberdeen

Aberdeen at the mouth of the Dee and close to that of the Don is the chief and indeed the only city in the north-east of Scotland. Situated as it is to the south-east of the most productive part of that region and just north of that gap between the Mounth and the sea through which pass all the main routes from it to the south, it is in an almost ideal position to act as its capital. To its development land and sea have each contributed. It probably originated as a Teutonic settlement, and about 1180 when it received its charter it was a trading centre and port, while the Church of St Nicholas, the patron saint of traders, is of even earlier date. For centuries it exported the products of the surrounding country—hides, furs, wool and salmon—to Flanders, first to Bruges and later to Campvere, and when the power of the Hanseatic League began to decline it took advantage of its position as the nearest British port to the Baltic to establish an important trade with Danzig. Meanwhile the industrial development of the town had been proceeding. Flemish weavers had introduced the woollen industry, and in the 17th century Aberdeen and the surrounding country were unrivalled in Scotland for the manufacture of cloth, just as they were in the 18th century for the manufacture of hosiery. These domestic pursuits were, however, unable to survive the Industrial Revolution, and Aberdeen without either coal or iron was compelled to reorient its activities in a world in which these were all important. In the middle of the 19th century its position opposite to the wood-producing Scandinavian countries led to a brilliant if short-lived period of shipbuilding, and some of the most famous of the China clippers were launched from its yards. But the most important feature of recent years has been the development of the fishing industry. Industries closely connected with trawling such as fish-curing, shipbuilding, and marine engineering have naturally followed in its train. In the early stages of its growth paper-making probably owed more to the abundant water-power in the lower reaches of the Don than to any other single cause, but to-day when coal is largely used for power, and when, in addition to esparto, wood pulp is also used, the ease with which the latter can be imported from Scandinavia is a factor in the success of the industry. Although often called the "granite city," Aberdeen did not use local supplies of that stone for building purposes until within the last two centuries, important buildings of earlier date being usually of sandstone brought from a distance. Granite polishing, now an important industry, was only introduced in the 19th century. The raw material is only partly of local origin, Kemnay, Peterhead and even Norway providing additional supplies.

In earlier times, as already pointed out, Aberdeen carried on through its port a considerable direct trade with various foreign countries. The building of larger ships and the development of internal communications led to the abandonment of much of this direct trade and to a very great increase in the coasting trade. Even so, for every cargo ship that arrives in its docks there are six trawlers. The principal imports from abroad are now mineral oils, esparto and wood pulp, wheat and flour, phosphates, granite and oil-seeds, while coasting vessels bring coal, fish, cement, beer, and miscellaneous goods. Exports abroad are not large, but the export of herrings has revived the old interest of Aberdeen in the Baltic countries.

Aberdeen is one of the most northerly cities of its size in the world, and situated in an environment none too friendly is typical of the whole region of which it is the capital, a region in which the obstacles to human development have been great. Its influence extends as far as the Shetlands, though Inverness is to some extent the legal centre, and the collecting and distributing point, for the districts round the Moray Firth. As the intellectual capital of the north-east of Scotland Aberdeen's position has long been recognised, and for nearly three centuries prior to 1860 it was the seat of two independent universities which are now united.

BUCHAN

Buchan, in some respects as distinctive a region of Aberdeenshire as Deeside itself, is generally regarded as being bounded on the west by the Deveron and on the south by the Ythan. It is in the main a lowlying platform of ancient rock, and only in comparatively few places does it exceed 500 ft. (150 m.) in height, while in the east and north-east there is a large area which is less than 250 ft. above sea-level. The most important line of elevation in the region is the series of low flat-topped heights which separates the basins of the Ugie and the Ythan; they run west from Buchan Ness and then bear round to the north-west. Farther to the north, where they pass into the Old Red Sandstone area, the Braclemore and Windyheads Hills rise to over 700 ft. (210 m.). But if the surface features of Buchan are comparatively tame and uninteresting some compensation is to be found in parts of the coast-line, notably between Peterhead and the mouth of the Ythan where the granites and schists have been worn into a picturesque line of cliffs, stacks, caves and skerries.

The soils of the region vary in character. The best are probably in the basins of the Ugie and the Ythan where they consist either of alluvium, good fertile loam, or a light clay. On the northern slope between Rosehearty and Fraserburgh there is also good soil, while east of a line running south from Rathen to Lonmay the soil is generally either a light loam or a light retentive clay. From a short distance north of Peterhead to Ellon the lower slopes near the sea are covered with red boulder clay, which, in the north at least, is difficult to work and much of which is kept under grass. Over the greater part of the remainder of the region there is lower boulder clay which on the uplands between the Ugie and the Ythan is poor and cold, and generally speaking consists largely of granitic debris.

But whatever the diversity of the soils there is in some respects a singular uniformity of agricultural conditions. Over four-fifths of the land is either cultivated or in permanent grass, and only in a few exceptional cases does the percentage for individual parishes fall below seventy-five. When the various crops are examined the same uniformity is apparent. Oats, turnips and swedes, and rotation grass occupy the larger part of the cultivated area, and the percentage of the cultivated land occupied by each crop remains remarkably constant. For example, out of the nineteen parishes under consideration all but two devote from 11 to 13 per cent. of the farmed area to turnips and swedes, all but two have between one-fourth and one-third of the same area under oats, and in all but one the percentage of rotation grass varies from 40 to 50 per cent., in most cases being between 43 and 47. It is evident then that a very distinctive type of agriculture has been developed in the region; the climate is the governing factor as far as the choice of crops is concerned, though the character of the soil may help to determine their yield. The cool cloudy summer with its frequent showers but not too heavy rainfall, and the mild autumn which allows the plant to mature, are well adapted to the cultivation of turnips, and it is said that these grow here under conditions closely approaching the optimum. For oats too conditions are very favourable, though it may be true that the yield is greatest when the rainfall is above the normal and when temperature is also high*. For wheat the temperature is too low, and for potatoes the soil is too cold. Barley is cultivated to a slight extent, mainly in the extreme north-east and in the south-east, in both of which regions the rainfall is low. But the great agricultural industry of Buchan is cattle-raising and it is to it that the agricultural development of that region is largely due. Until the middle of the 18th century the bleak and forbidding country with its boulder-strewn soil and harsh climate had proved too much for its inhabitants. Progress began when the value of the turnip for cleaning the land and for feeding cattle was realised, and to-day the agricultural economy of the region is specially adapted to the cattle industry as is indicated by the fact that a very common rotation of crops is a six years' one—oats, turnips, oats, grass, grass, grass. The two most important breeds are the Aberdeen Angus and the Aberdeen Shorthorn. For years some of the best stock has been exported to the Argentine, and the improvement in the herds of that country has been so marked that there is more than a hint that Buchan is beginning to suffer from their competition.

Except on the coast there are no towns of any size in Buchan. The villages are essentially agricultural centres, some such as Maud, where railways diverge for Peterhead and Fraserburgh, having important cattle markets. Peterhead and Fraserburgh owe much to a position which allows them to command alike the North Sea and the Moray Firth. In the 19th century the former was the chief centre of the whaling industry, and when that declined its

* A. E. M. Geddes, "Weather and the crop-yield in the north-east counties of Scotland," *Quart. Jour. Roy. Met. Soc.* July, 1922.

place was taken by herring fisheries. The quarrying of red granite in the neighbourhood of the town has led to the growth of a granite polishing industry of minor importance. In the years immediately preceding the War Fraserburgh was the chief centre for the herring fisheries of the east of Scotland. Its proximity to the fishing grounds of the Moray Firth and the shelter afforded by its harbour made it in the days of sail-boats the rendezvous of all the fisher folk of the north-east coast. At the beginning of the season boats used to arrive from all quarters bringing men, women and children, often with their household goods, who resided in the town during June, July and August. With the introduction of steam drifters and motor boats which have a greater freedom of movement and in many cases return to their home port for the week-end this seasonal migration has greatly declined.

The Deveron and the Spey

To the west of the Deveron and upper Ythan, and to the north of the hills which border Strathbogie in the south, the land is at first merely a continuation of the Buchan platform, and although its elevation is somewhat greater it is seldom that it much exceeds 700 ft. (210 m.). The Deveron itself flows through the region in a series of west to east and south to north stretches, the former being mainly transverse, and the latter mainly longitudinal. Strathbogie, through which passes the railway route to the north, has been formed by the Bogie working back in an easily eroded trough in the Old Red Sandstone. Farther to the west, in the region drained by the upper Deveron and the Isla, the character of the land alters as the Highland ranges advance to the north. Aultmore, in the north, does not exceed 1000 ft., but to the south of the Isla the country increases in height, and south of the Deveron it passes into the Highland area. The rivers present some features of interest. It is regarded as probable that the Dullan Water and the upper Fiddich at one time flowed into the Isla, but were tapped by what is now the lower Fiddich working its way back from the Spey; it is also suggested "that the Upper Spey once went through the Mulben Gap to the Isla, and that a stream, now the Lower Spey, working back along the softer Old Red Sandstone rocks, has tapped this Upper Spey-Isla river at Orton."* In any case the arrangement of the rivers has influenced the communications of this region very considerably.

The surface in general is covered with boulder clay which varies with the underlying rocks. As these run generally in a south-south-

* Geological Survey Memoir, Sheet 86, p. 3.

west direction, more or less at right angles to the direction of the ice-flow, the boulder clay is, broadly speaking, divided into several main belts following roughly the direction of the solid rock. To the south-west of Cullen and Sandend the clay is red and forms a poor, stony loam. The second belt extends along the coast from Sandend to Banff and inland as far as Huntly; as it includes various kinds of schists, the boulder clay which lies upon it, and the soils derived therefrom, are likewise varied. The latter are, however, generally stony loams. The third belt lies upon the Macduff group of grits and slates, and the soil derived therefrom is a stony clay loam. The area within which this lies is defined by lines joining Banff, Aberchirder, Huntly, Gartly, Fyvie, Turriff, and Macduff. Lastly, in the east of the region the drift lies upon the Old Red Sandstone and has developed into a stony sandy loam of fairly good quality*.

All this region may be further divided when economic conditions are considered. In the eastern part, which has been described as an extension of the Buchan platform, and in the coastal districts as far west as the Spey, from 75 to 80 per cent. of the land is cultivated or in permanent grass. But the type of agriculture practised here differs appreciably from that of Buchan, and in the cultivation of barley approaches that of Moray and Nairn to be discussed later. Along the coast seed potatoes and sugar beet are successfully cultivated.

In the western part of the region, which includes much of the hill country of Banffshire, less than 50 per cent. of the land is as a rule in crops or under grass. Cultivation is frequently carried to a height of 1000 ft. (300 m.) or more, but at an elevation of over 600 ft. (180 m.) the danger of crops not ripening makes grain somewhat uncertain. Accordingly the acreage under barley is low and that under permanent grass high. Cattle are reared in the more fertile districts and sheep in the less fertile. For some reason which appears to be unexplained the higher reaches of Banffshire and Speyside are pre-eminently suited for the rearing of Aberdeen Angus and Aberdeen Angus cross-breeds, a fact to which the high prices obtained at local sales bear ample testimony.

The fishing industry along the coast provides another type of economic activity the importance of which may be gauged by the fact that the coastal parishes of Banffshire have a density of about 300 people to the square mile, while in the purely agricultural region the density is usually about 60, and only in exceptional cases does it exceed 100 (say 115, 23, and 39 per km.^{2} respectively).

* Geological Survey Memoir, Sheet 86, p. 219.

The position of some of the fishing villages on this coast has been concisely described in Memoir 86 of the Geological Survey: "The harbours of Findochty and Portknockie are constructed in breached anticlines. Cullen harbour lies in the shelter of Cullen Bay and Sandend in Sandend Bay: Portsoy uses a cleft in the igneous rocks; Whitehills is protected from the east by the promontory of Knock Head; Banff and Macduff lie on opposite sides of Banff Bay at the mouth of the River Deveron: the villages of Gardenstown and Crovie use as their harbour Gamrie Bay, carved out of Old Red Sandstone rocks and protected from all but northerly gales by the slate headlands of More and Troup Heads." One or two further points may be noted. Some of the places just mentioned had in the past quite a considerable foreign trade—a trade which has been lost since railways were made and ships increased in size. In the case of Banff, which used to send passenger ships abroad, the harbour has been so silted up by the Deveron that its place as a port has been usurped by Macduff. As regards the larger places also it is necessary to distinguish between the town engaged in the fishing industry and that section of it in which the fisher folk themselves live. All along the coast the separation of the latter from the rest of the community is noticeable. "Banff," says the *Statistical Account* (1791–9), "consists of two parts completely separated, of which the first lies partly on the lower extremity of the plain on the river side, partly on the declivity; the other part (called the sea-town) on an elevated level which generally terminates abruptly within a small distance of the sea." At Buckie the fishermen's houses are dotted about on a raised-beach locally known as the Yardie while the remainder of the town lies at a considerably higher level. At Cullen also the sea-town is built on a low raised-beach. The cause of this segregation of the fishing community was the necessity for easy access to the sea, but it was accompanied by a social separation which resulted very largely from the fact that the hours of labour and of leisure did not coincide with those of the landsmen. Thus until lately, when the increased profits made during the War frequently enabled them to move from the sea-town, they remained a secluded people, simple-minded and manly, but without much knowledge of the outside world, and rather inclined, when moved thereto by untoward circumstances such as a disaster at sea or an unfavourable fishing season, to indulge in periods of religious emotionalism.

The chief inland towns of the region are Turriff, Huntly and Keith. Turriff is situated where the main road from Aberdeen to Banff crosses the Deveron; while Huntly, the capital of Strathbogie, stands on a stretch of alluvium near the confluence of the

Deveron and the Bogie. Both are agricultural centres, and the latter, which is the largest inland town in Aberdeenshire, has also woollen manufactures. Of more interest as a route centre is Keith whence there are railways to the coast at Buckie, down the Isla and up the Deveron to Huntly, up the Isla and down the Fiddich to Craigellachie in Speyside, and through the Mulben Gap towards Elgin. The position of Keith makes it the chief agricultural centre of Banffshire.

For the concentration of the distilling industry here and in certain other districts in the north of Scotland it is rather difficult to account, but a variety of circumstances appear to have favoured its growth. The water flowing over the ancient rocks of the region contains little mineral matter in solution; pure water makes malting easier, and at the same time does not leave a residuum in the still. Moreover, many of the Scottish distillers use malt only, and the spirit has imparted to it a flavour derived partly from the peat used in drying the malt. Other considerations are the extensive cultivation of barley in the neighbourhood, though much is now imported from abroad, and the relative cheapness of land and labour. Finally, when Highland whisky had once acquired a reputation it was but natural that distilleries should increase in number.

The Spey to the Beauly Firth

The Spey, the Findhorn, and the Nairn "may be regarded as belonging to a system of longitudinal north-easterly flowing streams developed subsequently to, and to a large extent at the expense of an earlier system of eastward, or south-eastward running consequent rivers."* The Spey may be divided into three parts. As far as Laggan Bridge it has a rapid descent, but from there to Grantown the slope is gentle, and in places almost horizontal. Below Grantown the gradient again increases, and it has been suggested that a possible cause of this may be found in an uplift of the land crossing the Spey valley somewhere in the vicinity of Grantown†. The Nairn is a swift stream flowing for the last part of its course within the Old Red Sandstone area which in the west rises to form Drummossie Moor. Like the Spey and the Findhorn in their lower courses it is bordered by alluvial flats; below Cawdor these develop into a considerable flood plain.

The whole of the coastal region of this part of the Moray Firth is bordered by an upland region of ancient rock which descends to a lowland where Old Red Sandstone and later formations are

* Geological Survey Memoir, Sheet 74, p. 3. † *Ibid.* p. 4.

found. In Moray this plain has a breadth of ten or twelve miles, but it narrows towards the west and in Nairn is only a few miles broad. To the north of the Old Red Sandstone there are raised-beaches often covered on their seaward margin with large deposits of blown sand. A belt of false-bedded sandstone of Triassic age runs along the coast from Burghead to near Lossiemouth*.

From the agricultural point of view the region between the Spey and Inverness Firth may be divided into a northern lowland and a southern upland. In the former region the soils derived from the more recent formations are often very fertile, and the "laigh of Moray" has a long established reputation. On the other hand, the percentage of land under crops and grass is much lower than in northern Banff and Buchan, and in only a few parishes does it exceed 60 per cent., a result at least partly due to the large extent of coastal sand, and partly to the more extensive woodlands. Barley is a still more important crop than farther east and in a number of parishes occupies at least 10 per cent. of the farmed area. For this climatic conditions appear to be in the main responsible. This is not at first apparent if the mean temperature of Nairn, say, is compared with that of Aberdeen. From May to August the figures (°F.) for the former are 48·8°, 54·2°, 56·7° and 56·2°, and for the latter 48·3°, 53·6°, 56·7° and 56·3° (9·3, 12·3, 13·7, 13·4; 9·1, 12·0, 13·7, 13·5 degrees C.). But when maximum temperatures are compared, Nairn has 56·8°, 62·1°, 64·3° and 63·7°, as against 54·9°, 60·2°, 63·1° and 62·7° (13·8, 16·7, 17·9, 17·6; 12·7, 15·7, 17·3, 17·1 degrees C.) for Aberdeen. Account must also be taken of the low rainfall of this region and of the long daylight of the summer months. Anyone who has witnessed the slow ripening of the crops in Aberdeenshire during the late autumn of an unfavourable year will appreciate the advantages possessed by the lands round the Moray Firth. Even in the Highland area in the south barley is a fairly important crop. Throughout the whole region oats are not so extensively grown as in Buchan but turnips are more widely distributed, and cattle-rearing is important. In the valley of the upper Spey there is relatively little arable farming and sheep farming is the chief industry.

Owing to the large extent of sand the coastal districts are not so densely populated as farther east. The only coastal parishes with a high density in Moray lie upon the Triassic sandstones between Lossiemouth and Burghead, and in Nairn near the county town

* For a detailed study of this and neighbouring regions see A. G. Ogilvie, "The Physiography of the Moray Firth Coast," *Trans. Roy. Soc. Edin.* vol. LIII, pt. II, 1923, pp. 377–404.

where the Old Red Sandstone approaches the coast. In Moray the most important town is Elgin, which grew up as a cathedral city but owes much to its position. Situated in a fertile district it lies on the direct route from Aberdeen which crosses the Spey at Fochabers, the lowest bridge town on that river, and runs to Inverness. This route is crossed at Elgin by another which runs from Craigellachie in Speyside to Lossiemouth by the Glen of Rothes through which the Spey probably flowed when its waters were held back by the quartzite rocks of Ben Aigan. Nairn and Lossiemouth have become holiday resorts, the fishing folk of the latter having removed to Branderburgh. Forres, Garmouth and Kingston have all lost their importance as harbours, the last two partly owing to the vagaries of the Spey. Rothes is an important distilling centre.

Inverness Firth to Dornoch Firth

To the west of Drummossie Moor lies the valley of the Ness. A raised-beach sweeps up it as far as the 50-ft. (15 m.) contour, beyond which it is covered with alluvium. A narrow strip of this beach, continued along the southern shore of the Beauly Firth, expands at the western end of the Firth and partly fills it up. To the south of the raised-beach lies a prominent ridge of schistose upland, the northern part of which, known as the Aird, is heavily wooded. The lowland tract is continued to the north by the littoral belts round the Cromarty and Dornoch Firths, the lower parts of Strathconon and Strathpeffer, the Black Isle, and the peninsular part of Easter Ross. The whole region lies within the Old Red Sandstone area, and is generally lowlying except in the Black Isle where a line of elevation running to the north-east rises to a height of over 800 ft. (240 m.). The most interesting physical features are, firstly, the continuation of the fault traversing the Great Glen in the straight coast-lines of the Black Isle and Easter Ross; secondly, the drowned valleys of the Conon and the Oykell now represented by the Cromarty and Dornoch Firths respectively; and thirdly, the very large extent of raised-beach around the coast.

The soils on the Old Red Sandstone are often very fertile, and the same is true of some of those on the raised-beaches. Considering the high latitude the climate is favourable, and during the summer months Strathpeffer has a higher mean temperature than Aberdeen. Barley continues to be an important crop especially on the south coast of Inverness Firth and in the Black Isle. It is also of interest to note that the small area under wheat in Easter Ross gives one of the largest yields per acre in Scotland. In other respects, however, agricultural conditions are much the same as in the preceding region.

The population is in the main a rural one, and the bulk of it is concentrated in the arable areas, that is, in the lower valleys of the larger rivers and along the coasts. Of the towns Inverness with a population of 21,000 is the most important. It first emerges into the light of history in the 12th century when it appears to have been both a royal castle and a local trading centre. Situated near the mouth of the River Ness and almost surrounded by the fertile lands which lie along the Moray Firth, it was well adapted to be the centre of a not unimportant agricultural area, while it owed its political and strategic importance to the various routes which diverge from it, eastwards along the coastal plain into Moray, northwards to Caithness and the Orkneys, westwards by Glencarron to Skye and the Hebrides, south-westwards by the Great Glen to Argyll, and southwards by the Pass of Drumochter to the towns of the Central Lowlands. But since the fertile region around Inverness is limited, while the routes which diverge from it do not link up highly productive areas, the town has never developed to any great extent. Nevertheless the fact that by rail it is almost as near to Edinburgh and Glasgow as it is to Aberdeen has made it more or less independent of the latter town, and justifies its being regarded as a local capital.

No other town of this region has a population exceeding 2500. Cromarty, at the north-east extremity of the Black Isle, is a place of some strategic importance as was shown during the War. It guards the narrow entrance to the Cromarty Firth, the depth of which is sufficiently great to allow it to be used as a naval base. Dingwall, at the head of the same firth, is the most important railway junction north of the Caledonian Canal; lines diverge from it to Thurso in the north, and by way of Strathpeffer, Strathban, and Strathcarron, to Kyle of Lochalsh in the west. Its central position has given it a certain amount of importance since Norse times at least. Strathpeffer owes its importance as a spa to the distinctive character of its sulphur waters though it also possesses chalybeate springs. Tain, situated opposite Dornoch on the firth of that name, was formerly a port, but the silting up of the river on which it stands, together with the shifting sands in the firth, and the bar across, have led to its decline as such, and it is now the market town for the prosperous agricultural region lying to the south of it.

All these towns are affected to some extent by their proximity to the Highland line. Inverness by its position and history is perhaps rather a lowland outpost in the Highlands than a Highland town proper. But farther to the north the Celtic influence is still strong, and a certain detachment of the people from material matters is perhaps shown by their leisurely life, while their attachment to things which are immaterial is indicated by their interest in all that concerns their Celtic past.

Sutherland and Caithness

From Helmsdale to Strath Fleet extends a narrow coastal plain, which is backed inland by an almost continuous line of steep hills. The plain is underlain by rocks of Old Red Sandstone, Triassic and Jurassic age, while its surface from the hills seaward is covered successively by morainic deposits, raised-beach, sand links, and sand dunes. South of Strath Fleet the coastal region becomes broader as the raised-beaches around Dornoch become wider. The coastal range, the heights of which vary from 1000 to 2000 ft., is marked off from the coastal plain by a fault; in the north it consists of granite and in the south of Old Red Sandstone; but south of the Fleet the latter rocks become lower and the coastal range disappears. To the west of the escarpment of the coastal range lies a dreary plateau region where peat-clad hills alternate with wide stretches of peat moss. The larger rivers flowing across it, such as the Helmsdale, the Brora and the Fleet, have developed broad alluvial straths in which a certain amount of settlement is possible.

The common type of glacial deposit in the south-east of Sutherland is a loose morainic drift probably with areas of boulder clay over the coastal plain. The soils vary from a clay loam to a light sandy loam, the most common types being medium loam and sandy loam. The great part of the cultivated land lies below the 200-ft. (60 m.) contour. Only in the parish of Rogart, where the bulk of the crofter population of the Moine plateau is situated, does the cultivated area rise to 750 ft. (230 m.).

Of the whole county of Sutherland less than 50 sq. miles (130 km.^{2}) or about one-fortieth of the surface is cultivated or in grass. Of that about two-thirds lies in the south-east coastal parishes, the most extensive area being around Dornoch. A little barley is grown but the principal crops are oats, turnips and rotation grass. A few cattle are kept in the lowlands and a large number of sheep on the hills.

The plateau which occupies the south-western part of Caithness seldom exceeds 1500 ft. (460 m.) in height, and falls towards the north, north-east, and east to form a gently undulating plain, the greater part of which is less than 500 ft. (150 m.) above sea-level. The plateau consists of crystalline schists and granitic rocks, but some of the prominent hills which rise above its general level, such as Morven and Maiden Pap, consist of outliers of the Old Red Sandstone. The plain belongs almost entirely to the Middle or Orcadian Old Red Sandstone. In the south and west the rocks are breccias, conglomerates and red mudstones, while in the north and east the typical grey Caithness flagstones cover large areas*. The coast scenery is often very striking. To the south of Berriedale, where granite prevails, the almost vertical cliffs are sometimes 400 ft. (120 m.) in height, while farther north in the Caithness greystones are the "goes" or "long inlets with more or less parallel and vertical side walls," which with the detached stacks are so characteristic of the region.

Of the superficial deposits shelly boulder clay and peat are the most interesting. The former, which lies east of a line running approximately from Berriedale to Reay, is a "dark grey or blue stony clay with fragments of shells scattered through it," and is believed to have been carried from the region round the Moray Firth by ice moving in a north-west direction across the Caithness plain. The lime which this clay contains gives it a special value for agriculture. Peat is widespread and it is estimated that it still covers two-thirds of the whole surface of the country, notwithstanding the fact that much of the cultivated area now lies on land from which the peat has been removed.

In contrast to Sutherland about one-fourth of Caithness is either under crops or in permanent grass. The distinctive feature of its agriculture when compared with the counties south of the Moray Firth is the high percentage—about 25—of the land in permanent grass, with a corresponding decrease in the areas under oats, turnips and rotation grass. For this the climate is in the main responsible as is indicated by the fact that the yield per acre of oats is low while that of turnips remains high. The most productive region is the extreme north-east.

The fact that Caithness is a plain of which a considerable part is capable of cultivation while the south-east of Sutherland is at the best "a riviera of pastoral links and fertile ploughland," has naturally a great effect upon the distribution of population. Even in Pictish times this was the case as is evidenced by the distribution of brochs, the so-called Pictish castles. In Caithness they are widely distributed over the agricultural areas of the north-east, while in the south-east of Sutherland they are less numerous and are

* Geological Survey Memoir, Sheets 110 and 116, p. 1.

confined in the main to the fertile lands of the seaboard and to the broader straths.

Later on when the Norsemen, driven from home by the poverty of their own country, invaded the Pictish province of Cat it was in these same fertile lowlands that many of them settled. It is an interesting commentary on this fact that to-day not only are Norse place-names numerous in this region, but they are almost entirely confined to the land below the 500-foot contour.

Wick with a population of 8000 and Thurso with one of 3000 are the only towns. Thurso with the most fertile part of Caithness as its hinterland was in Norse times and for long thereafter the chief port in the north of Scotland. It was only when the fishing industry developed that Wick with its more favourable position on the east coast rose into importance. Since it ceased to be the centre of the herring fishing its population has begun to decline. In Sutherland the only centres of population are small villages situated on the raised-beaches along the coast at or near a point where there is easy communication with the interior. Dornoch the county town, for example, with less than 1000 inhabitants, is situated on the 25-foot beach and commands routes by Strath Fleet and Strath Oykell to the north and west coasts*.

Orkney and Shetland

Not because of their similarity but rather because of their dissimilarity the Orkney and Shetland groups may be considered together. The former is physically and geologically a continuation of the Caithness plain, while the latter in many respects resembles the Highlands. With one or two exceptions the whole of Orkney consists of rocks of Old Red Sandstone age, and the surface generally forms low undulating tablelands terminating seaward in a bluff cliff or sloping downward to a sandy beach. In the Mainland, however, there are large stretches of hill and moorland, while Hoy, where the Upper Old Red Sandstone predominates, is trenched by deep valleys occasionally flanked by conical hills rising to over 1400 ft. (430 m.) high. In Shetland Old Red Sandstone is limited to a few districts and the region consists in the main of igneous and metamorphic rocks. In the larger islands, at least, the interior is hilly, though heights of over 1000 ft. are exceptional. The coast-line is also more irregular; the voes or sea-lochs sometimes run for miles into the heart of the country, and appear to have an origin analogous to that of the fjords of

* For Sutherland, see A. B. Lennie, "Geographical Description of the County of Sutherland," *Scot. Geog. Mag.* vol. XXVII, 1911, pp. 18–34, 128–142, 188–195.

Norway. In Orkney the typical rocks are the flagstones of the Middle Old Red Sandstone, and from these are derived the prevalent fine-grained silty soils, free from stones but somewhat difficult to work on account of their close, stiff nature. The soils of Shetland are more varied owing to the more varied geological structure of the region; some are fertile but large areas are covered with heather and peat moss. In climate, also, Orkney has the advantage, the mean temperature for the four months from June to September being 52·25° F. as against 51·37° for Lerwick.

The combination of soil and climate has made Orkney the more important agricultural region. In the North Isles over one-half of the whole area is in crops and grass; conditions are less favourable in the Mainland which in places is hilly, but nearly one-half of the land is farmed; in South Ronaldshay and Walls the proportion falls to one-fourth. In the Mainland of Shetland one acre in thirteen is in crops and grass and in the North Isles one acre in eighteen. Nowhere in Shetland has farming advanced much beyond the crofting stage, 55 per cent. of the holdings do not exceed 5 acres (2·02 ha.) in extent, and 88 per cent. do not exceed 15 acres. In Orkney, on the other hand, 60 per cent. of the holdings are over 15 acres and 16 per cent. over 50 acres. As a result agriculture in Shetland has to be supplemented by fishing and fish-curing. In Orkney over 50 per cent. of the people are engaged in agriculture and 4 per cent. in the fishing industries; for Shetland the corresponding percentages are approximately 30 and 24.

The population of Shetland (25,000) is somewhat greater than that of Orkney, but the density of the latter is 64 per sq. mile, while that of the former is about 45 (25 and 17 per km.^{2}). Kirkwall is the chief town of Orkney; it is situated on a narrow isthmus which divides the Mainland into two parts, and has thus a double harbour. To the south is Scapa Flow, a great naval base during the War. Stronsay, an important fishing centre, has a population almost wholly migratory. In Shetland, Lerwick is the chief town.

XXIII

CENTRAL SCOTLAND

Alan G. Ogilvie

PART I. CERTAIN ASPECTS OF THE REGION*

CENTRAL SCOTLAND, consisting of the so-called Midland Valley between the Highlands and Southern Uplands, contains the greater part of the Scottish Lowlands. Here the Scottish nation was evolved, and the region has always supported by far the greater part of the population of the Kingdom; to-day this is over three-quarters of the whole. It has thus long been and still is the nucleus, economic, political and cultural, of Scotland. From the glens which open abruptly on its north-western border there poured at frequent intervals the "Highland host" to take part in civil wars and to ravage the land. But from these glens mid-Scotland first received Christianity afresh after the torch lit by St Ninian in the Roman period had been wellnigh extinguished by the raiding Norsemen; and commercial relations with the Highlands as a source of cattle and skins and of wood were always maintained. The continuous independence of Scotland to the 17th century may be ascribed in part to the difficulty which English armies experienced in successfully invading this central heartland. For the Southern Uplands offer few openings that are not easily defended. The brunt of the smaller frontier wars, then, fell upon the dales of southern Scotland nearer to the "auld enemy"; when, however, great armies such as those of Edward I and of Cromwell did penetrate these natural defences, usually by way of the east coast, they left marks upon the landscape that remain to this day—ruined abbeys and castles. But these same ruins and other ancient buildings serve to remind us of the foreign sources that contributed to Scottish culture.

From England or through England came, above all, the civilising influence of the Church with a power of material benefit to the land—*e.g.* in farming—much greater than that exerted by the older Celtic Church. And in periods of peace with England, Scotland traded with the south. The "Scottish baronial" architecture of the castles still recalls the "auld alliance" with France against England. On the other hand, the style of humbler dwellings,

* In view of the limited space available, treatment of select aspects seemed to be preferable to a too-brief synthesis of *all* the geographical factors.

especially in east coast villages with their crow-stepped gables, white-washed walls, outside stairs, and roofs of curved red pantiles, tells of long maintained trade connections with the Baltic and North Sea ports; whereas from the early centuries of our era down to the 18th, Scotland derived little advantage from her Atlantic seaboard. The West remained a backwater from the decline of Celtic Christianity until the development of American trade which was greatly promoted by the deepening of the Clyde estuary. Nowhere in the world can there be found a better example of the effects of position and of inherent resources in altering completely and rapidly the aspect of a country, directly a new vista is opened in the geographical and economic outlook.

Structure and Minerals

Geologically, Central Scotland is a tectonic trough some 50 miles wide whose axis has the "Caledonian" trend (north-east to south-west); and its limits are the nearly rectilineal edges of the Highland metamorphic rocks and of the tough greywackes, shales and slates of the Southern Uplands. These must have presented in-facing scarps—intermittently at least—for a very long period; they represent lines of structural weakness. The Highland edge coincides with a fault, or faults, overthrust from north-west, in its entire length; the south-eastern limit is also marked by almost continuous and nearly straight faults with the downthrow to north-west. The rocks of the intervening region consist of sediments, the Old Red Sandstone series occupying a broad continuous belt on the north-west and less regular areas on the south-east; while Carboniferous rocks lie generally in the central zone. These sediments are greatly disordered, and moreover, associated with the strata of both periods there are large sheets of igneous rocks—lavas, intrusive dykes, sills and plugs. The igneous rocks have resisted denudation better than the sediments, so that their distribution is of prime importance as determining the orographical features of the region. Thus of their two greatest expanses the one following the Caledonian alignment forms successively the Renfrew, Kilpatrick, Campsie and Gargunnock, Ochil and Sidlaw Hills, with smaller features reaching to the Kincardine coast. The other, lying nearly perpendicular to this and across the trough, forms the watershed between upper Clydesdale and the Firth of Clyde. Between the Highland edge and the first named igneous belt lies a lowland zone of rather uniform structure, known from the east coast to the River Earn as Strathmore. The south-eastern part of the Midland Valley displays no such structural and orographical simplicity; and its limit on this side is consequently much less sharply defined. One special feature may be noted from this zone, namely that the Pentland Hills, a horst somewhat complex in structure, follow a trend line not otherwise evident in this region, but paralleled by the Great Glen of the Highlands.

With this introduction, the relationship of surface features to geological structure must be left. For there is much accessible literature on this subject, notably Sir Archibald Geikie's fundamental work*.

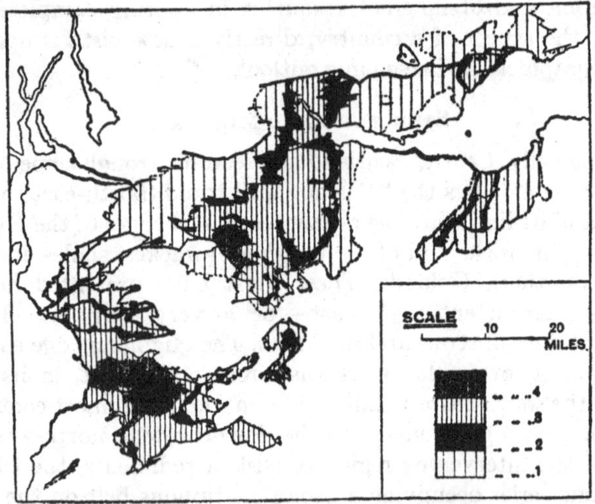

Fig. 51. Central Scotland: distribution of coal-bearing (and younger) rocks. Reference: 1, Carboniferous Limestone series; 2, Millstone Grit; 3, Coal Measures; 4, Barren Red Measures and younger rocks. (After P. R. Crowe.)

But there is one other aspect of geological structure that deeply affects the geography of Scotland; and this must be dealt with briefly. The economic minerals of the region other than building stone and brick clays consist of coal, iron-ore, fire clay and oil shale contained in the Carboniferous rocks. The present distribution of these minerals depends mainly upon three geological factors. First, that the floor upon which the strata were deposited was uneven. This has affected the distribution in detail; a fluctuating shore-line at the time of deposit having here and there caused the omission of certain strata from the series; *e.g.* the oil-bearing

* *The Scenery of Scotland, etc.*, 3rd ed., 1901. See also the various Memoirs of H.M. Geological Survey and the "1 inch" maps appertaining to them, as well as B. N. Peach and J. Horne: *Chapters on the Geology of Scotland*, 1930, Chapter I.

shales of the Lower Carboniferous are present only in the Lothians and east Lanark. But we cannot here consider this factor further. Secondly, the region was subjected to crustal stresses at a later geological period, not definitely determined, so that the Carboniferous rocks, including the Volcanics of like age, were folded and much broken. While the nature of these movements appears not to have been studied as a whole, the result is clear, namely that four synclinal basins formed (Fig. 51). The axes of the easterly pair of these run from north to south and are both now crossed by the estuary and Firth of Forth. The other two axes trend approximately east-south-east to west-north-west and lie respectively in the Clyde valley and in northern Ayrshire, passing through Kilmarnock.

Thirdly, immense post-Carboniferous denudation has removed the Coal Measures and much of the coal-bearing limestones from the upfolded belts between these synclinals.

Thus there are three mineral basins in the region: (1) the Lothian-Fife Basin, bisected by the Firth; (2) the Ayrshire Basin; (3) the Central Basin in which two of the synclines merge, so that this, the largest mineral area, comprises the coalfields of Lanark, Dumbarton, Stirling and Clackmannan counties.

The reserves of coal as estimated in 1913 are as follows (in millions of tons): Lothians-Fife Basin 12,000, of which 4000 lie under the Firth of Forth; Central Basin 6000; Ayrshire Basin 2000. Thus while it will be seen below that hitherto the Central Basin has been the most important, the eastern field has the greatest future, even if much of the sub-marine coal be inaccessible.

The relative importance of the chief Scottish coalfields in recent years may be discerned by reviewing the quantities of coal mined in the three years 1922 to 1924, when there were few labour troubles. The average annual yield expressed roundly in thousands of tons was as follows:

Ayrshire Basin	Lanarkshire Coalfield	Dumbarton-Stirling Linlithgow-Clackmannan Coalfield	Lothians Coalfield	Fife Coalfield
3,819	14,391	5,337	4,366	8,113
11·4 %	39·2 %	14·5 %	11·9 %	22·1 %

The percentages refer to the average Scottish production for the period which amounted to 36,710,702 tons; and if the fields be grouped naturally, in geological basins, the division is as follows: Western, 11·4 per cent.; Central, 53·7 per cent.; Eastern, 34·0 per cent. = 99·1 per cent. This subject is further illustrated by Mr Crowe's map showing the distribution of mine employees (Fig. 52).

The coalfields contain valuable bedded iron-ores, the so-called "clay-band" and "black-band" ironstones; and it is on these that a great iron industry has been based, the ore having been mined for the most part with the coal. So long as wrought iron was the chief material for construction the Scottish ores, which are phosphoric, fully met the demand. But since 1860 acid steel has been the predominant product in Scotland, made chiefly from imported non-phosphoric ore, with the result that the local ironstones supply less than one-seventh of the ore used. Nevertheless at any time in the future Scottish ores may again be more actively mined; and we may note the reserves as estimated in 1919 by H.M. Geological

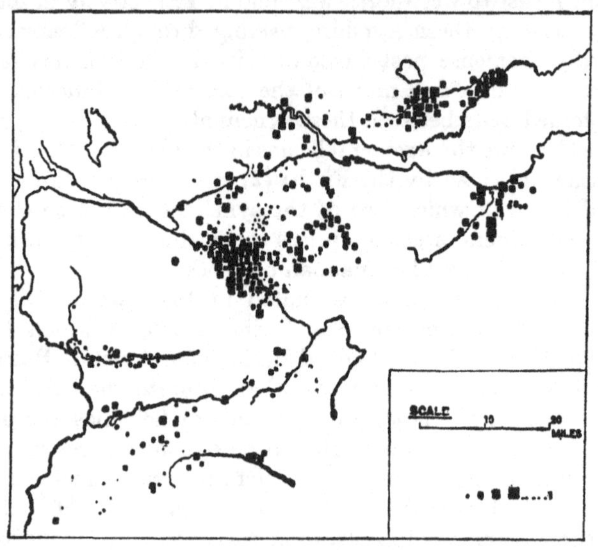

Fig. 52. Central Scotland: coal mines in 1925. Reference: 1, mines employing 20, 100, 500, 2000 men respectively. (After P. R. Crowe.)

Survey. The figures, in millions of tons are: Ayrshire Basin—probable 62·1, possible 69·1; Central Basin—probable 18·6, possible 342·6; Fife—probable 1·4, possible 6·9; Midlothian—probable 2·3, possible 4·6. The proportion however of the present output is quite different. For the period 1922–24 the average was (in thousands of tons): Ayrshire 1·4; Central 68·6; Midlothian 10·4.

Associated with coal and ironstone is a valuable fire clay useful for many industrial purposes. This is now mined in large quantities in all the fields, but especially in Lanarkshire.

The oil shale of the Lower Carboniferous rocks, occurring almost solely on the eastern fringe of the Central Basin and formerly in the Lothian field, gave Scotland in the 19th century a well-nigh unique oil industry which has now however seriously declined.

The industrial consequences of this mineral wealth will be touched upon later. Meanwhile we may return to a more detailed discussion of the surface features apart from their direct relation to geological structure*.

Land-forms

Central Scotland bears the traces of several cycles of normal erosion which may perhaps be considered as marking successive broad uplifts that took place in the Tertiary Period. The physiographic history of the pre-glacial surface of Scotland has still to be worked out. But for the present purposes it is sufficient to draw attention to an element of the land-form not hitherto broadly discussed. The differentiation here suggested will be found to have real geographical importance, even if the genetic classification and the altitudinal limits described be found faulty on subsequent investigation.

The highest hills of the Midland Valley, shown as black patches on Fig. 53, correspond for the most part to the distribution of strong igneous rocks. The smoothness of their summits is in some cases partly accounted for by their structure; but the comparative uniformity of their altitudes, in spite of separation, mostly between 1500 and 2000 ft. (say 450 and 600 m.) indicates that they are the residuals of a surface once continuous from the lower Highland plateau to that of the Southern Uplands. In the Pentland Hills the sixteen principal summits, of varied structure, are between 1600 and 1900 ft. (490 and 580 m.) in height. In this range there is also a well-marked erosion surface at about 1300 ft. (400 m.). Most of these residual blocks are bounded by steep slopes, and the scarps which are precipitous for long stretches are indicated on the map. Some are crossed by profound valleys, forming through routes often now dry, that seem to belong to the old transverse drainage system from the Highlands, and to which the dissection of the ancient plateau may be ascribed.

Of greater human importance, from its lower altitude, is the higher Lowland peneplane of Scotland (Fig. 53, 2)†. This feature is

* See further: P. R. Crowe, "The Scottish Coalfields," *Scot. Geog. Mag.* vol. XLV, 1929, pp. 321–337, from which Figs. 51 and 52 are reprinted by permission of the Royal Scottish Geographical Society.

† This feature and the "lower peneplane" are here tentatively named for the first time, in the hope that subsequent investigation will either demonstrate their separate individuality, or explain otherwise the usual break between the two surfaces.

everywhere characterised by smoothness, apart from minor dissecting valleys; and it truncates various rocks. Its altitudes differ considerably from district to district, but without any marked break of slope. The greater area stands between 500 and 750 ft. (say 150 and 280 m.), but near to the east coast it falls in places below 400 ft. (120 m.); and around the hills it rises slightly above 1000 ft. (300 m.). The variation of height may prove to represent simply the slope of a peneplane or it may indicate that warping has taken place since the surface was formed. This peneplane is least broken south of the Forth-Clyde furrow. Its preservation is clearly related to the position and relative power of river systems, and later of Quaternary glaciers. Most of the streams that now cross it flow in immature valleys incised in its surface, the chief exception to this being the Clyde and its tributaries, above the Falls. The higher peneplane is to be correlated with many of the valley floors in the Highlands and Southern Uplands and with the short through valleys referred to above.

A somewhat greater area of the region is occupied by a lower surface (Fig. 53, 3) which reaches the coast on the inner edges of raised-beach, and which in most parts is separated from the higher peneplane by a well-marked break of slope. This may perhaps be termed the "lower peneplane" to indicate that it is not a plain. But its general appearance is rather that of maturity; and its relief, though slight, is varied on a small scale. This variety is however mainly the result of glaciation, for it is pre-eminently the area of glacial deposit. The seaward edge of the higher peneplane is least well-marked in eastern Ayrshire and in the lower Clyde valley. Elsewhere the lower surface lies in general between the 100- and 500-foot contours. The relationship of this surface to that of the surrounding sea floor is of course masked by the sea and by glacial detritus. It would appear that much of this "lower peneplane" represents land that remained undissected during the pre-glacial period of maximum emergence to, say, 600 ft. above the existing level. The greatest river, the Forth, has been shown to have then still flowed in a narrow canyon whose floor at Stirling is probably nearly 300 ft. (90 m.) below sea-level, and descends to some 460 ft. (140 m.) at Queensferry*. Under such conditions normal lateral dissection could not have proceeded very far before the Ice Age, for the buried lateral canyons of the tributaries have steep gradients.

True "plains" (Fig. 53, 4) are restricted in area. They consist

* Midway between the depth is at least 570 ft. (174 m.), possibly owing to glacial scour. For a detailed examination, see H. M. Cadell, "The Story of the Forth," *Scot. Geog. Mag.* vol. XXVIII, 1912, pp. 225 *et seq.*

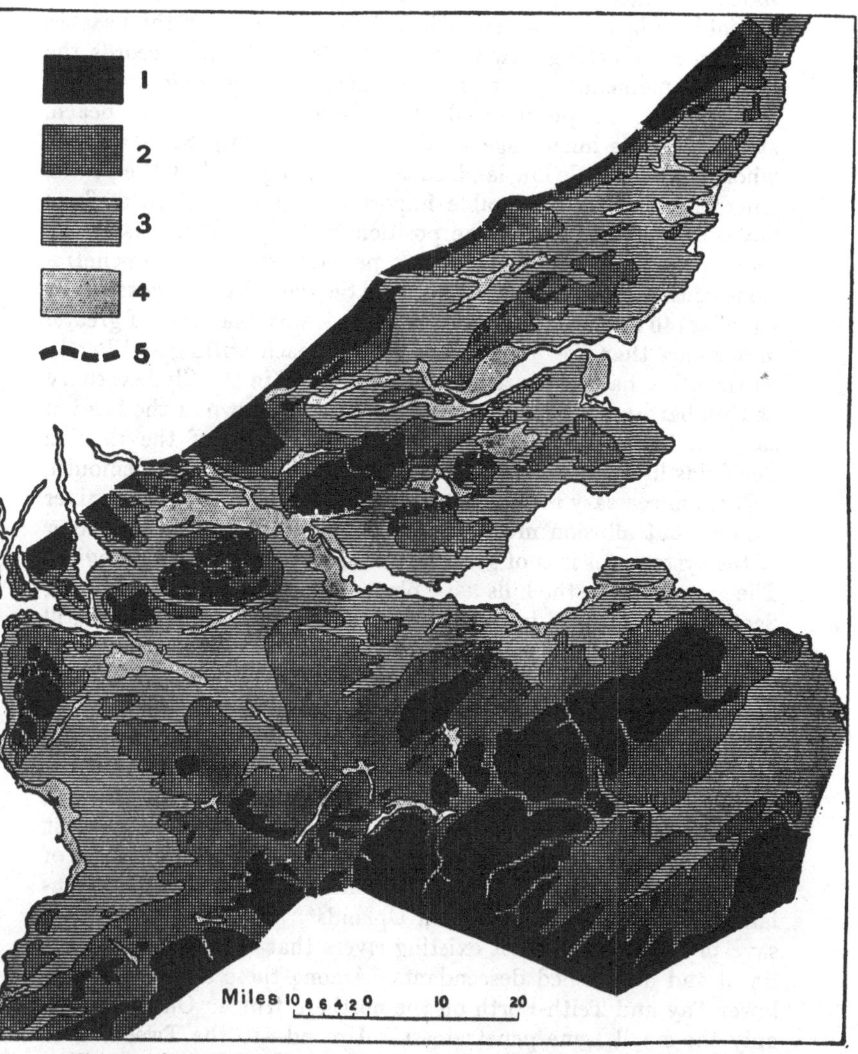

Fig. 53. Central Scotland and the Tweed Basin: Physiographic Divisions. 1, Plateaus of the Highlands and Southern Uplands and the higher surfaces of the Midland hills; 2, Remnants of the higher Lowland peneplane; 3, the lower Lowland peneplane; 4, Plains of raised-beach and alluvium; 5, Abrupt scarps on the Midland hills.

chiefly of the raised-beaches—the floors of estuaries or of coastal shoals, emerged since the Ice Age—with smaller patches of post-glacial lake beds and several flood plains of rivers. Of the beaches the highest, reaching now up to about 100 ft. (30 m.), records the great submergence of late-glacial times, while among the lower terraces the most pronounced, the "Mesolithic" or Azilian beach, stands with its inner edge at about 25 ft. (7·5 m.). Scotland as a whole includes little low land, so that the soil gained by the partial emergence is of considerable importance to her. More striking however is the fact that the position of the shore-line as it has remained at least since the Roman period gives the deep penetrations of the three Firths and narrows Scotland by a "waist" only 25 miles (40 km.) wide. In former times it was relatively of greater importance than now that shipping could reach Stirling and Perth; on the other hand the presence of a rocky sill in the Clyde estuary at Dumbarton gave special importance to that town at the head of easy navigation. To-day the modern industries of the Central Coalfields have ports at their doors in Glasgow and Grangemouth.

It is unnecessary to attempt to trace here the history of the river valleys; but allusion must be made to the pattern and character of the system, for it is of great importance in the life of the region. The gaps between the hills have played a great part in determining Scottish history, and in modern economic conditions they are still vital. In all cases these gaps are the work of rivers, often following lines of structural weakness, though many have been widened and deepened by glaciers, and some, like those occupied by the estuaries of Forth and Tay, have been rendered swampy by the great submergence. The present "grain" of the country on the whole has the Caledonian trend; and of the two eastern firths the outer Forth and the inner Tay are so determined. But this is not the earliest "grain" of which we can see traces to-day. The ancient system of transverse rivers, flowing generally south-eastward from the Highlands and crossing the Southern Uplands*, has nowhere persisted save in short stretches of existing rivers that can be regarded as lineal and diminished descendants. Among these are North Esk, lower Tay and Teith-Forth on the northern fringe. On the south, only the small Lyne penetrates the Upland—to the Tweed. But there the great valleys remain with low heads in the escarpment—Nith, upper Clyde, Biggar-Tweed gap, Eddleston and Gala Waters. Within the region there are several rivers flowing from south-east to north-west, which may in part be following, in reverse direction,

* This is the generally accepted view, but see in regard to penetration of the S. Uplands that expressed by Mr Stevens in Chap. XXIV, pp. 471–2.

lines that were formerly taken by rivers of the group just mentioned. The greatest of these is the Clyde from its elbow above Lanark, while Doon (in the south-west), Gore (from Gala head) and Blane (north of Glasgow) are in this group. Then there are gaps, as already mentioned, through the Midland hills, such as Glen Devon and Glen Farg in the Ochils, the Sidlaw gap north-west of Dundee, the broad gap between the Lomonds and the east Fife plateau, and Glencorse in the Pentlands.

At a later stage the main drainage ran generally eastward, much of it carried to the old "North Sea" river system by the ancient Forth. To this stage may be assigned the formation or deepening of such valleys as the Lochwinnoch and Loch Libo gaps (south-west of Glasgow), continued by the deep furrow south of the Campsies and now followed by the Kelvin, and the mid-Fife valley drained now by Devon and Eden. In the south-west the valley watersheds respectively between the Rivers Irvine and Avon, Ayr and Douglas, together with many minor valleys, bear witness to the beheading of this former eastward drainage system. And it is important to note that a journey may be made from Ayr to Dunbar without crossing the 1000-foot (305 m.) contour by following the Ayr and Douglas valleys and by passing south of the Pentlands. The corresponding route along the north-western fringe, from Dumbarton to Stonehaven, rises above 250 ft. in only three short sections.

To sum up, the drainage system, whatever be the age of its component parts, may be conveniently resolved into a net of rivers of all sizes, either transverse or longitudinal for the most part; and it may be correct to apply the terms "subsequent" to the latter group and "consequent" and "obsequent" to the members of the former. The pattern of Central Scotland is thus satisfactorily accounted for; and it bears a very close relationship to the railway map and even to the road map of the country.

Most of the rivers have a long history behind them, but there have been so many interruptions in erosion cycles and so many captures that probably few have normal profiles. On the other hand, waterfalls are rare. The three Falls of Clyde near Lanark, however, marking the passage of the river from a post-mature section to a young section of its valley, constitute the greatest waterfall in Britain, if volume and height be considered together.

A striking feature of Scottish topography is the small amount of denudation that has been accomplished since the Ice Age except on the actual beds and banks of the greater rivers. At the same time it must be remembered that the rivers have almost everywhere a considerable fall. Incised meanders occur only locally in the present valleys; and there is no case of a stream of any size meandering freely on its flood plain. Even the lower Forth has its great "links" slightly incised in the clays of the Stirling Carse*.

It has been mentioned above that the lower and the flatter ground of the region above 100 ft. is in large part a surface of glacial deposit. This consists chiefly of boulder clay apparently deposited by ice-sheets that radiated from the Highlands and from the western part of the Southern Uplands respectively. This deposit of till is somewhat variable in texture and composition; and the drumlin characterises its surface, especially on the lower land. Moraines are of very local occurrence, generally associated with small corrie glaciers of a late glaciation. Fluvio-glacial deposits on the other hand are widespread both as sheets and as kames, and are connected with the recession of the ice-sheets or with later valley glaciers. Again, there are numerous overflow channels incising the peneplanes above described or cut in the sides of the higher escarpments. Many of these are still swampy, and are not linked with the present drainage system.

The coast-line cannot be described here in detail. Its general form is accounted for by the great submergence followed by relatively slight subsequent emergence. And if we disregard possible tectonic considerations in the Firth of Clyde, we may add that large embayments coincide with the distribution of weaker rocks, as illustrated for instance in the Ayrshire coast and in the Firth of Forth, where the sea penetrates farthest along the synclinal axes of relatively weak Carboniferous rocks. As to the details we may generalise by saying that cliffs alternate with sandy beaches in response to exposure and to tidal currents. Only in the extreme north and south of the east coast are the cliffs sufficiently imposing to prevent a railway or a road from running near the shore. In the sea areas of Clyde above Greenock, of Forth above Queensferry, and in the Firth of Tay, tidal deposit is going on as it did when the carses of the raised-beach were formed. In the Clyde estuary constant dredging is necessary from Port Glasgow to Glasgow; likewise in the Firth of Tay and at the Forth ports above Queensferry. Land reclamation has been begun in the inner Forth by the building of dykes and allowing the tidal deposit to fill them; much more land remains to be gained in this way.

* "Carse" is the Scottish name for a level tract of alluvial land bordering an estuary.

Rivers and Water Supply

We know comparatively little of the regimen of Scottish rivers except of those which have been used or are to be used for hydro-electric schemes. It may be said in general that the period of low water is the summer and that the maximum flow occurs at varying times in the winter six months. Serious flooding takes place only where streams with considerable fall emerge upon alluvial plains of little slope. And in such places in Central Scotland embankments are generally to be found which are sufficient to restrain the ordinary spate. It is commonly stated that the rocks of Scotland are impermeable. But this is a relative term. It is true that even in the Midland Valley pure limestones are uncommon; but Carboniferous limestones and some of the sandstones must have a fair degree of porosity. However, the rocks are for the most part cloaked by glacial deposits of variable permeability. The fact remains that in Central Scotland there is no excessive stream density even in the west where the rainfall is high and evaporation is low. It follows that there must be a large amount of drainage to the streams via the soil and subsoil if not also into the rocks below. A tentative and approximate measurement (made on the O.S. map, 1 : 63,360) of the total length of streams within two areas, somewhat similar in regard to relief but differing in the amount of rainfall, proves interesting while it cannot be taken as conclusive. Ayrshire, north of the River Ayr and below the 1000-foot contour, an area of some 345 sq. miles, has about 800 miles of streams. Lothian, east of the Pentland Hills and also below 1000 ft., an area of some 390 miles, has about 540 stream miles. Thus the mean stream density for north Ayrshire is 2·3 stream miles per sq. mile, and that for eastern Lothian is 1·4, or about 1·4 km. and 0·8 km. respectively per square kilometre. The difference must be the result of a variety of causes. These figures, however, suggest that there is at least some correspondence between rainfall and river density; for the rainfall of the Ayrshire area is between 35 and 60 in. (900 and 1500 mm.), while that in the Lothian area is between 25 and 40 in. (625 and 1000 mm.).

Central Scotland certainly does not suffer from insufficient water supply. The character of the agriculture will be seen to be greatly affected by the fact that the rainfall is usually excessive; and much labour and money have to be spent upon drainage of fields and of hill pastures. The domestic water supply of rural districts—villages and individual farms—presents few difficulties, small but permanent springs and streams being the usual source. On the other hand, the growth of towns and of water-using industries has led to rapid extensions of catchment areas reserved for these agglomerations, and sometimes to keen competition between towns on this account. The provision of water would have been much more difficult without the hill groups of the Midland Valley. In all of these and along the escarpment fringes on both sides artificial reservoirs are now a feature of the landscape. Thus the numerous towns of the Ayrshire lowland and those of the Clyde valley are supplied from such sources constructed on the plateaus separating these areas. Clydesdale also draws upon reservoirs in the Upland edge to the south and on the peneplane surface to the east, where it competes for water with the Lothian towns. In a lesser degree the Campsie, Ochil and Sidlaw and Fife Hills are sprinkled with artificial lakes. Dundee, Arbroath and Montrose now take most of their supply from the Grampian valleys, and water which Dundee gathers in the enlarged Lintrathen Loch near the mouth of Glen Isla and leads across Strathmore and the Sidlaws is carried over the Tay Bridge to the north shore of Fife. Perth still uses the abundant water of the lower Tay, and is now the only large town in the region to rely upon such a source; adequate storage and purification have just been effected.

The two chief cities alone have had to go far beyond the Midland Valley for their water. In Loch Katrine, with its level artificially raised, Glasgow has a magnificent supply carried in an aqueduct south-eastward across the head-streams of the Forth basin and so along Strath Blane. Edinburgh's growth has been marked by successive extensions of its waterworks first to the northern Pentlands, then to the Moorfoot edge, and lastly, to the reservoir of Talla, a head-stream of the Tweed, the supply being led thence down the Tweed to its eastward bend and then along the east foot of the Pentlands to Edinburgh.

Underground water, which plays so great a part in south-eastern England, is scarcely used in Scotland. In the great brewing industry of Edinburgh, however, specially suitable water reached by deep wells, determined the brewery sites in the city and more recently in a southern suburb.

Soils and Vegetation

Study of the soil and subsoil as well as of the place-names of the Midland Valley leads definitely to the conclusion that at least towards the end of the pre-historic period it was covered with forest, mainly deciduous, except where the ground was not sufficiently drained to support such woods. But on land with little slope, at whatever altitude, there lay a thick mantle of peat; and in the multitude of enclosed hollows among the drumlins were lakes or swamps. What a change has been wrought by Man upon the landscape since then! Nearly all of the small lakes have been drained, and of the original forests nothing has been left. Historical references, however, and the place-names tell us that these were composed of such trees as oak, birch, alder, hazel, pine, willow, yew and thorn. While there are now no remnants of primeval forest the remains of the older vegetation, that of the earlier post-glacial phases, are still with us in the peat, of which there are large tracts not merely on the flatter parts of the higher hills, but covering great expanses of the higher Lowland peneplane, especially in its western and central areas. The oscillations of post-glacial climate account for the succession of tundra by forest, increasingly varied in its trees, but later invaded by peat. Subsequent and more rapid changes again led to forest growth which, in the early centuries of history, suffered from renewed extension of peat*. The latter stages are illustrated in the low Vale of Menteith (between the Rivers Forth and Teith), a large part of which is a peat bog covering the stumps of a great oak forest, some of which was cut by the Romans. In the 18th century, laborious clearance of this peat from a section of the Vale laid bare for agriculture the fertile soil of the carse.

There can be no doubt that conditions of temperature and of rainfall throughout the region to-day are those of a natural forest climate†. It is probable that the original forest varied in character from the wetter west to the drier east; and it is most likely that the soils when examined systematically will be found to exhibit a progressive increase in the amount of leaching from east to west. In the parts of the region that are not used for agriculture (see Fig. 54) the woodlands are practically all plantations—deciduous on the lower ground that would otherwise be agricultural, and consisting of Scots pine (*P. sylvestris*) or of non-native conifers on the hill slopes. But a very much greater area is treeless. The vegetation is of sub-Alpine moorland type. It is of low stature and consists of various associations with numerous gradations between *Sphagnum* and cotton grass (*Eriophorum*) on very wet peat, through heather moor also on peat, drier heath, grass heath

* See, *e.g.*, T. W. Woodhead, "The Forests of Europe and their development in Early Post-Glacial Times," *Brit. Asscn. Report*, Glasgow (1928) Meeting, p. 618.

† Owing to lack of space no discussion of climate is included in this essay save as incidental to this and succeeding sections.

and grass, especially *Nardus*. Moreover, it cannot be regarded as natural, since it has been profoundly modified by long use as sheep pasture and more recently by the practice of preserving grouse moors. The sheep by their tramping, manuring and selective grazing have their own function in the process; while the shepherd by drain cutting strives to increase the proportion of edible grasses. The modification on account of the grouse is produced by frequent burning of the heather, to give the plants full opportunity to produce young shoots, and this generally strengthens the heath association. Thus in the interests of sheep and grouse there is a constant war between the chief elements in the hill vegetation—grass and heather. When the western areas have been examined with the care shown by R. and W. G. Smith in the east* the effects of rainfall upon the associations will doubtless appear. But in general all these hill slopes present in the landscape a similar alternation between greens on the one hand—of intensity varying with season and with composition—and the brown or purple of the heather on the other hand; and the more assiduous the attention of sheep farmer and gamekeeper, the more patchy and abrupt the alternation†.

Population and Historical Changes

The people of Central Scotland like those of most of Britain are of mixed origin. Indeed there must have been many strains in their blood by the opening of the historical period. But anthropological surveys do not seem to have progressed sufficiently to allow of any clear statement of variation in physical type throughout the region. Archaeological remains have yielded as yet a somewhat fragmentary picture of the spread of mankind in the region and of his use of the land. Among the place-names are some that puzzle and are assumed to be pre-Celtic, are said indeed to have affinities with the Basque language. But the names of physical features and of villages are in overwhelming majority Celtic—either Brythonic or Goidelic, although many have been anglicised so as to lose completely their original meaning. Indeed, Gaelic was the language of all Scotland in the 11th century; and it probably gave place very slowly in a northerly and westerly direction to English spreading from a centre in the Lothians. Gaelic was still spoken in parts of Galloway in the 16th century.

Since pre-historic finds have not yet been scientifically mapped in Scotland we may turn to the more obvious relics of early cultures,

* "Botanical Survey of Scotland," *Scot. Geog. Mag.* vols. XVI and XXI.

† An account, by Dr W. G. Smith, of the hill pasture of the Southern Uplands, which resembles that of the Midland hills, will be found below, p. 491.

remains that are still a feature of the landscape. But before attempting to outline their distribution it is necessary to state that agricultural operations in the last two hundred years have been very thorough on all the fertile land; and it is therefore likely that many of the less imposing evidences of ancient settlement have been removed. This probably applies to tumuli, cairns and stone circles and especially to hut circles, the relics of dwellings. Fortifications, on the other hand, are to be looked for on the hills that have never been cultivated, and so the distribution of that characteristic Scottish monument the *dun* or rude hill fort is worthy of study*. There are two remarkable concentrations of these on the southern border of the region†. One of these is in the area where the Clyde enters the Midland Valley and in the adjoining Tweed basin. A circle centred on the heights of Broughton and with radius of 12 miles includes fifty forts as well as some cairns and tumuli. There can be little doubt that this group controlled three routes of entry into the central lowlands via Clyde, Lyne and Eddlestone. The second group, which is more scattered, consists of some thirty forts in the Lammermuir Hills between the Leader valley and the coast. But half of these are concentrated on the plateaus astride the ancient coastal road of entry from the south and also controlling the deep valley of the Eye which has since carried the modern road and railway. Nine forts stand widely spaced along the Highland edge, and a remarkable series of thirteen faces the Grampians from the hills across Strathmore, extending in a line from the northern Ochils to the Kincardine plateau. Five of these are evenly spaced astride the Tay gap. In Lothian and in Fife the volcanic necks with their characteristic ice-ground precipices generally facing west have ever offered excellent sites for strongholds, and each district has a dozen such or similar sites clearly capped by fortifications. If it is useless at present to attempt a full interpretation of this distribution, it is still harder to account for the distribution of stone circles and of monoliths known locally as standing stones, large numbers of which have been respected even in the highly farmed land. Again the remains of actual habitations of these early peoples are rare in this region; a few subterranean earth houses, a few wooden piles of *crannogs* or lake dwellings, the "kitchen middens" of hunters and shell gatherers on raised-beaches or on the islands, and a very few undoubted patches of cultivation terraces on hillsides, chiefly in

* It is perhaps well to bear in mind the possibility that some of the so-called forts may not merit this name, but may be structures of some religious significance. The most important work on this subject is *Early Fortifications in Scotland* by D. Christison, 1898.

† *Vide infra*, p. 484 and Fig. 58.

Lothian and therefore perhaps of Anglian origin*—such are the scattered relics with which archaeology has still to deal in a comprehensive manner.

The term "historical geography" may be taken to cover two distinct ideas. It connotes in the first place a review of history from the view-point of the geographer; and most of the literature of the subject seems to fall into this category. Secondly, the study may be regarded as having for its aim the investigation of the physical features and landscape of a country as they were at different periods, and of the ordinary life of the inhabitants as related to these. There is a recent brief but excellent summary of the historical geography of Scotland mainly in the first sense†. But most of the research needed for a work of the second type has still to be done, save perhaps for that relating to the Roman occupation.

Among the materials for such study that merit special mention here we may select from the mediaeval period the distribution of castles and of monasteries; the first as illustrative of the country's organisation for local or national defence, and the second as indicating the spread of cultural and economic progress in periods of peace. We may also note the situation and character of mediaeval towns as related to the development of external trade.

The mediaeval castles of feudal Scotland, unlike the prehistoric *duns*, are placed not upon hills withdrawn from the lowland, but rather in the midst of the more habitable agricultural land. Reliance in defence was placed in the building itself; but generally the castle site is some small crag or drumlin or the neck of an incised meander. Some of the greater fortresses however stand on the prominent igneous rocks, notably the royal Castles of Dumbarton, Stirling and Edinburgh, which were strategic points of the first order. As the grouping of the very numerous castles is too large a subject to enter upon we may pass to the religious houses.

It seems probable that throughout the mediaeval period in Scotland, from the 12th to the 16th centuries, which is marked by the great influence of the monasteries, the various religious orders usually settled in places already hallowed before the 8th century by the Celtic monks in their great missionary crusade. But the founders of the new establishments did not by preference occupy

* *Cf.* A. Raistrick and S. E. Chapman, "Lynchet Groups of Upper Wharfedale," *Antiquity*, June, 1924.

† W. R. Kermack, *Historical Geography of Scotland*, 2nd ed., 1926. Much matter that has geographical bearing will also be found in I. F. Grant, *The Social and Economic History of Scotland*, 1930, an exceptionally well documented work.

the secluded "deserts" of the Culdee communities. They chose the fat lands. And they seem fully to have justified their possession of such lands by the attention which they devoted to agriculture. Before the foundation of the numerous abbeys and priories farming must have been very primitive; but the monks with their continental connections were in a position to introduce new crops and new methods. From what is written below, it will be seen that the monastic farms lay in districts where the farming of to-day is of a high order, and where the yields are very large. Thus in the Midland Valley we find important houses at Scone, and several priories in Strathmore; a large convent at Aberbrothock (Arbroath) on the rich coastal belt; great abbeys or priories at Dunfermline, Culross, St Andrews and Lindores on the fertile fringe of Fife; Abbeys at Cambuskenneth on the Stirling Carse, at Holyrood on the edge of the Lothian raised-beach, and Newbattle in the nearby Esk valley; at Paisley on the best soil of lower Clydesdale, and Kilwinning near the north Ayrshire coast. From these and other centres the knowledge of more enlightened agriculture must have spread, as it did also from the great convents of Tweeddale, the Solway region and the Moray Firth coast. To the monks Scotland owed much also in the realm of the arts and crafts; and when the monasteries were abolished at the Reformation the torch of learning was already firmly held by two universities in mid-Scotland, St Andrews and Glasgow*, while that of the Capital was soon to bear its share.

In Scotland the trading fraternity grew up under the protection of the sovereign who granted charters to the so-called Royal Burghs in which mediaeval commerce was concentrated. From the situation of these burghs it is evident that by far the greater part of the country's trade, both domestic and foreign, was carried on in the east. Of the forty Scottish towns which received their charters before 1400 twenty-six are in the Midland Valley, and of these, twenty lie east of the main watershed. Of the thirty-three granted charters between 1400 and the Union with England (1707) sixteen are in our region, and of these only four are in the west. Twenty-nine of the old burghs of mid-Scotland are either on the coast or they controlled nearby seaports†. The great predominance of the eastern part of the country in trade is also made clear by the records of revenue derived from commerce at each place. This state of affairs was of course the result largely of the fact that the

* As at Aberdeen, see p. 407.

† For a table showing the rise and decline in importance of burghs from the 14th to the 16th centuries, see I. F. Grant, *op. cit.* p. 353.

foreign trade, while at first chiefly with England, was from the 13th century mainly with the opposite shores of the North Sea and with the Baltic ports, and also that undoubtedly the main body of the population was in eastern Scotland. It is to be noted that the monasteries benefited by trade on account of their possessions within the burghs, and from the fact that the king also created church burghs. The exports consisted for the most part of wool, hides, skins and fish—chiefly salted herring and salmon, and salt evaporated from seawater over coal fires in the coastal burghs; the imports were manufactured articles. Historical records and study of the plans of the burghs reveal the fact that the burghers in mediaeval Scotland were very largely occupied with agriculture in their immediate vicinity.

The great changes wrought upon the face of Central Scotland and still more upon its population, as a result of the opening of commerce with America and of the modern industrial developments based on coal, have been admirably treated in a recent paper by Mr Crowe. His study, containing four very instructive maps of the population distribution in 1801, 1841, 1881 and 1921, may be regarded as if it were an integral part of this essay, and should be read along with it*. Geographical influences upon industry and upon commerce are often most obvious in the early stages of development, and as readers of Mr Crowe's paper will have followed the evolutionary stages at least in outline, discussion of the economic aspects of human geography will here be restricted to a statement of the present conditions.

In 1921 there were about 1·8 million people in Central Scotland whom the Census describes as "occupied." Of these the vast majority were occupied in manufacturing, mining, trading and transporting. Only about five per cent. were engaged in agriculture. But the figures do not adequately represent the importance of agriculture in the geography of the country. This subject requires some general treatment in relation to physical geography; while the other aspects of human geography will be dealt with in the sections devoted to description of the sub-regions.

* P. R. Crowe, "The Population of the Scottish Lowlands," *Scot. Geog. Mag.* vol. XLIII, 1927, pp. 147–67. It had been intended that the present essay should be in large part devoted to this subject, but Mr Crowe's recent paper, which must be accessible to readers of this volume, deals with it most effectively. Reference may also be made to "The Distribution of the Towns and Villages of Scotland" by James Cossar, *idem*, vol. XXVI, 1910, pp. 183–91 and 291–318; and further to the recent interesting and well-illustrated volume, *The Settlements and Roads of Scotland, a Study in Human Geography*, by Grace Meiklejohn, 1927.

Agriculture

Everyone who knows Central Scotland is aware in a general way that in the use of the land there is a marked contrast between east and west, that in the east arable farming prevails while pastoral

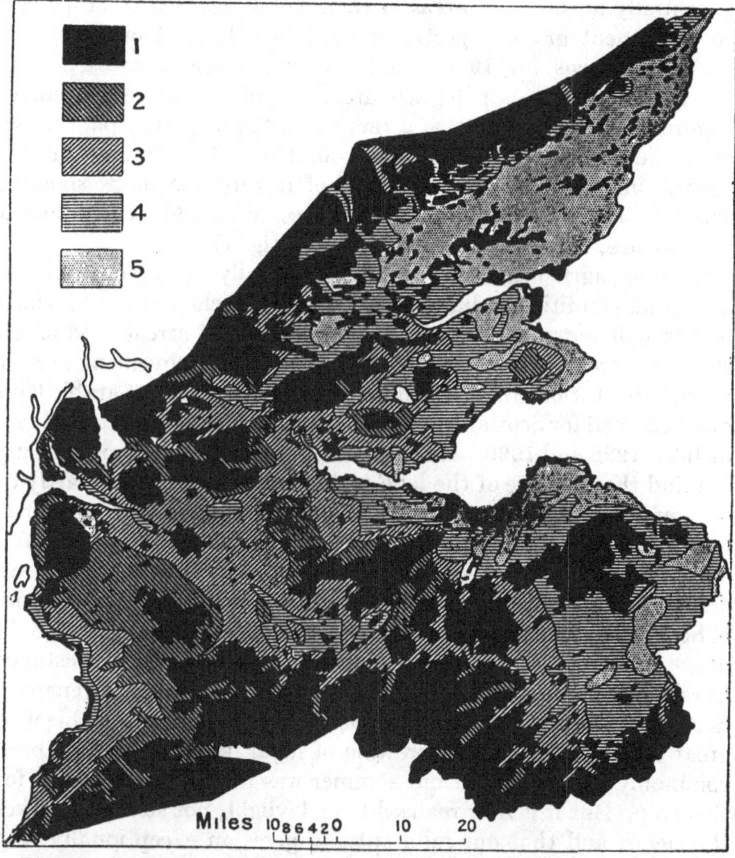

Fig. 54. Central Scotland and the Tweed Basin: Proportion of farm land under permanent grass (1926): 1, uncultivated land—hill pasture, woods, towns, dunes, etc.; 2, 3, 4 and 5, farm land: 2, over 60 per cent. in grass; 3, 60 to 40 per cent.; 4, 40 to 20 per cent.; 5, less than 20 per cent.

farming predominates in the west, with cattle as the chief stock; and that the region is flanked north and south by sheep walks. It seemed interesting to examine in what degree this contrast is real; and a map has been prepared that may be regarded as illustrating this aspect of rural economy. Fig. 54 is a reduction, somewhat simplified, of this map. In making the original, the limits of enclosed land, *i.e.* of the actual farm land, were first plotted from the O.S. maps (1 : 63,360). The area thus excluded and shown in black represents the hill pasture of the region as well as small patches of woodland, towns and coastal sand dunes. All other land is potentially arable. The areas of each parish devoted to crops and to permanent grass respectively were then taken from the agricultural returns for 1926*, and the grass area, expressed as a percentage of the crop + grass area, was plotted for each parish. Figures were found to be so arranged on the map that percentage lines could be drawn with some confidence. The interest of this map is increased by a comparison of it with the maps showing the distribution of cattle and sheep (Figs. 55 and 56), each plotted by parishes; and with the rainfall map (Fig. 1).

Scottish agriculture by 1926 had virtually emerged from the abnormal condition induced by the War, the chief effect of which was to halt temporarily the steady regression of cereal production that had proceeded since the late 1870's. This movement is now again in operation, and 1926 showed the smallest area of arable land ever recorded for Scotland as a whole, but further decreases occurred in 1927, 1928 and 1929. We may therefore study the maps, bearing in mind that the use of the land is due partly to physical, partly to economic circumstances. The chief physical factors here are, first, relief associated with air and soil temperature, and secondly rainfall and drainage. Among the local economic factors the chief are density of population and facility of transport. The climatic factors in Scottish agriculture have been little studied, although they would prove an interesting and important subject for research, for instance in relation to the limit of wheat cultivation which passes generally northward through the Highlands. Or again, to take a subject of great topical interest—the growing of sugar beet—it has long been commonly supposed that our summer was too cold and sunless for this crop. But it is now realised that daylight, not sunlight, makes the sugar, and that our mild autumn gives an exceptionally long period for growth and sugar formation. It seems likely, however, that economic considerations will decide the future of the crop.

When the acreage devoted to wheat cultivation is plotted on a map of this region the effect of climate is at once apparent. In 1919 when the greatest war-time effort was made to raise this grain, it appears as a very scattered crop in Ayrshire, somewhat more

* Kindly supplied by the Scottish Board of Agriculture. The author further wishes to thank Mr J. M. Ramsay, O.B.E., M.A. of the Board for help and criticism in regard to his treatment of agricultural matters.

cultivated in lower Clydesdale, of much greater density in east Fife and southern Forfar and Perth, while by far the greatest area was in Mid- and East Lothian. This arrangement is mainly a response to the summer and autumn distribution of rainfall and sunshine. The temperature factor also chiefly accounts for the restriction of the wheat to land mainly below 400 ft. (120 m.).

In regard to the effect of relief, it will be recalled that there is a peneplane surface rising in places to over 1000 ft. above sea-level. It follows that only in small areas below the upper edge of this are slopes so steep that ploughing is difficult or soil erosion excessive. Figs. 53 and 54 illustrate in some measure the fact that arable farming in the eastern area is carried to the upper edge of this peneplane and in places above it. Oats may be found on land over 1100 ft. high (330 m.) in Mid-Lothian, thus, for instance, giving continuous arable farming from the Esk to the Gala and Tweed valleys. On the other hand, in the western quarter the 500-foot (say 150 m.) contour commonly marks the upper limit of arable farming; and this contrast is almost certainly the result of climatic difference. The data are not yet available, however, for fixing the average altitude of this limit in all districts, but 700 ft. (say 210 m.) may be given provisionally as its average position in the east. It may be added that exposure is of some importance; other conditions being equal, southern slopes permit of higher cultivation in the region than do the opposite slopes. This is well marked, for instance, on the Pentland and Ochil Hills; and differences due to the same cause have been noted by botanists in the character of the hill pasture above the fields.

A third physical factor is of course the type of soil. The arable land consists almost everywhere of soils derived from superficial deposits—glacial, fluvio-glacial, alluvial, or marine. The character of these "parent materials" in turn varies with their origin, and this is especially true of the most important, the boulder clay. This is usually of a red colour where derived from the Old Red Sandstone or from most of the igneous rocks, and greyish when overlying or to leeward of Carboniferous rocks. From the fact that the soil in the fields generally resembles in colour the parent deposit it might be assumed that a soil profile corresponding to the climate has not had time to form. But it must be remembered that in the farm land deep ploughing has thoroughly mixed the upper horizons and has destroyed the profile.

In general it may be stated that on the boulder clays soil texture varies abruptly in the region over small areas, even within single fields. But on the other deposits the texture may be more broadly classified; the fluvio-glacial sands and gravels and the raised-beaches other than carses giving light soils, while the carse clays are among the heaviest. The alluvial haughs, though very fertile,

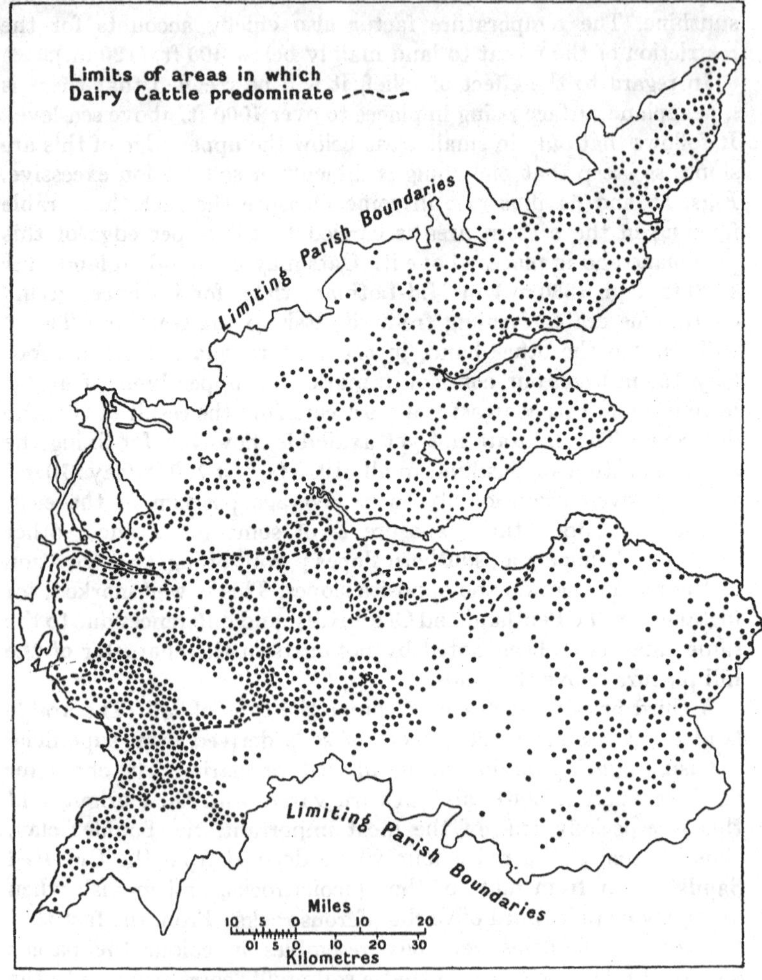

Fig. 55. Central Scotland and the Tweed Basin: Distribution of Cattle (1926). Each dot represents 250 cattle.

are left largely in pasture from risk of flooding or from dampness of soil, or because they are often included in the parks of estates. They are the Scottish equivalent of English pastures that are too good to break up.

We may now review briefly the features of the maps. The distributions on the western half of the Midland Valley are accounted for by the presence of the great industrial population of Clydesdale with its demand for milk, and by the physical factors of high

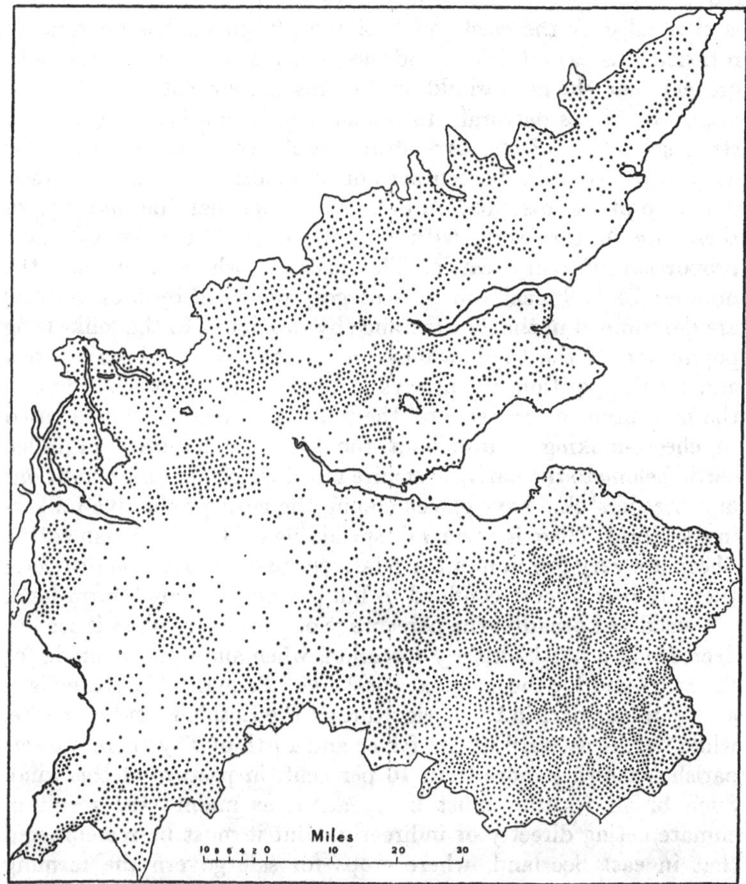

Fig. 56. Central Scotland and the Tweed Basin. Distribution of Sheep (1926). Each dot represents 500 sheep. The lines limiting the region on north and south are parish boundaries.

rainfall, low sunshine, and a soil often heavy, cold and ill-drained. Most of the land that receives over 40 in. of rain lies permanently under grass to the extent of over 60 per cent. The remaining area, with from 35 to 40 in. of rain, shows grass to over 40 per cent. in all but a few places, such as the raised-beach strip of Ayrshire and in the upper Clyde district where soil or economic factors are the cause. The high cultivation in the upland Renfrew parish of Kilmacolm (only 9 per cent. in grass) where the rainfall is about 50 in. should be mentioned here; for by far the greater part of the parish is devoted to hay, which gives a very large yield. The fact is that all over the west much of the plough land is devoted to rotation grasses and clover, and the countryside consequently looks greener than the map would imply; this of course stamps the land unmistakably as pastoral. In this and in some other areas a more striking picture of the agriculture would be presented were the map to be based on percentages of permanent *and rotation* grass. A glance at the distribution of cattle shows that the majority of these are in the west, with almost everywhere a considerable proportion of dairy cattle. The districts where over half the number fall technically in this category (enclosed by broken lines) are determined in the first instance by proximity to the milk-using population of the Clyde valley and Forth-Clyde industrial area, and by the position of railways leading to it. In south Ayrshire the percentage decreases, and the milk to a greater extent is used in cheese-making. Throughout the west the majority of milch cattle belong to the native Ayrshire breed which climatic conditions and breeders' skill have placed among the most productive milkers in the world. There is also a fair sprinkling of British Frisian cattle. Sheep are very few in number relatively to other Scottish districts; they are here almost confined to hill pastures and high farm land.

Considered broadly, the percentage of permanent grass is seen to decrease in a north-easterly direction, when allowance is made for the relief and for the irregular coast-line. Permanent pasture plays a very minor part in the agriculture of the Lothians and east Fife, while north-eastward of the River and Firth of Tay there are few parishes showing more than 10 per cent. in pasture of this kind. Such broad features must be regarded as mainly the result of climate acting directly or indirectly. But it must be remembered that in east Scotland, where crops for sale govern the farming system, cattle are essential. Manure is spread in vast quantities. It must be obtained, and the making of prime beef is incidental. Milk production is of first importance only around Edinburgh and Dundee. On the other hand, where sheep are folded on arable land, they are there to be fed with a view to mutton production. The hill pastures in the Midland Valley, where they are attended to, can carry about one sheep per acre if lambs are included. The map (Fig. 56) represents the maximum number of sheep just after the lambing season.

Prior to the Agrarian Revolution Scottish agriculture was carried on in openfield farms divided as "in-field" and "out-field," with no proper rotation and with oats and bere as the chief crops*. Small amounts of peas and beans and more broom were grown, at least by the 15th century, for winter fodder, but these were insufficient for maintaining the herds alive. For about two centuries, however, agricultural practice in the eastern districts has been steadily improving. Tremendous labour was expended at the instance of the landlords in ridding the boulder-clay soil of stones and in drainage of the farm land. There was then also a great advance in the use of manure, that of the farms themselves having been increased and used over all the land instead of merely on the nearer "in-fields" as formerly. To-day artificial fertilisers and lime are employed universally as required. Of these the region produces phosphates from the smelters and sulphate of ammonia from coke ovens and oil-shale retorts; moreover, the farms are not far from ports where foreign manures are landed. The great extension of the road system in the 18th century and of the railway net in the 19th was obviously a great boon to farmers, more particularly since it revolutionised the marketing of produce, the movement of stock, and the import of fertilisers and of implements. All this progress which began in the east spread to the west, so that at the present time farming methods almost everywhere compare very favourably with those of most foreign countries and of many English districts. The high standard of agriculture is generally obvious not merely to an expert but to all who can appreciate the substantial well-built and well-kept farm "steadings," the exceedingly clean fields and the absence of wasted land between them.

Part II. The Sub-regions

An attempt will now be made to characterise the several more or less natural sub-regions into which Central Scotland may be divided. The landscape will be described, and the human occupation indicated; but relatively little mention will be made of the towns as such†.

The Fringing Corridor of the Highland Line

Stonehaven is the north-eastern point of entry to the region. The route approaching it from the north is hemmed in between the gneissic Highland plateau and an uninterrupted line of high sea cliffs.

* *Cf.* I. F. Grant, "The Highland Openfield System," *The Geog. Teacher*, vol. XIII, 1926, pp. 480–488, with a full bibliography.

† See footnote on p. 438.

The narrow shelf above these carries the road and the single railway route used by two competing companies; it is a strategic defile in any age, and a concentration of prehistoric remains here apparently marks its early importance, comparable to that of the coastwise entrance to mid-Scotland from the south-east about St Abbs. Stonehaven stands on the only large embayment in a rock-bound coast extending from Aberdeen to Montrose. From the haven, the ascent, inland, to the south-west along a deeply entrenched valley brings the traveller to a pass about Drumlithie, and from here he may continue his journey in the same direction for nearly fifty miles (80 km.) to the River Tay with but little change in the view. Overlooking the Howe (lowland) of the Mearns and Strathmore alike is the Highland escarpment, rising abruptly on every spur to a break of slope, and beyond this the rolling and rising crown of the truncated Grampian plateau; every mile, a steep scarp valley, and every six or eight miles, a deep Highland glen bearing the marks of its former glacier. On this face the sheep of the Strath intermingle with the grouse, and the grass merges into the heather.

Fronting the scarp on the ancient fault-line is a strip of the higher peneplane preserved by conglomerate and hard sandstone, but gashed by the cross valleys and ribbed by many short furrows—both "consequent" and "subsequent" scars of earlier dissection. These give access to secluded crofts and sheep farms that occupy this upland belt. On the forward edge of this stands the outer line of towns of the Midland Valley and many smaller centres of life. Then comes the great belt of Strathmore, perhaps once swept from end to end by a single river, but now the corridor of several—Bervie, both Esks, Isla and Tay—which cross or follow it in leisurely loops, before plunging through defiles to the next parallel trench. The Howe of Mearns in the northern part of the corridor, for structural reasons, shows a less geometric pattern in its topography. To south of this there are several long ice-scraped whale backs trending eastward, and the parallel valleys of South Esk and Lunan; but here, though with other trend, are already the smooth contours of the Strath. These however are most characteristic beyond Forfar. The smooth red plain of a well-fed soil, with the meandering haugh-edged rivers slightly incised in it, has some variety on its fringe. South-west of Kirriemuir are several miles of hummocky drumlins, and south of Blairgowrie is a gravelly sheet studded with lakes and woods and elsewhere largely given up to the precious jam-making raspberry. At a distance a traveller might even take this zone for a fringe of the Lombardy plain with raspberry canes for vine stocks! Apart from this specialisation in a native fruit and another in exotic bulbs by an English grower at Newtyle on the eastern edge, Strathmore and the Howe are given over to a fairly uniform type of well-managed mixed farming, in which rotation grasses and oats have the largest areas, followed by barley and turnips (especially in the Howe), potatoes and wheat. The fine Angus cattle are bred here and fattened along with Irish and other store cattle for market; likewise the prime mutton sheep—the Cheviot—and the native "blackface." That basic geographical feature the "Caledonian trend" is reflected not only in streams and drumlins, but in the diceboard fields, in the roads, and even in the set of the farms; these are built of "Old Red" stone with slate roofs*.

For 25 miles (40 km.) the Strath is walled in by the Sidlaws with a narrow moorland summit. This, the dip-slope of the lavas, is smooth and gentle compared to the short, craggy scarps overlooking the Firth. But there are several gaps that give access to Dundee and the Carse of Gowrie. Westward, the River Tay has trenched the Strath, and beyond it from Caputh to Crieff a bulging zone of higher plateau presses forward, hummocky with glacial debris, bearing woods or heath on the poorer soils and farms elsewhere. Deep in this surface run the Almond and other tributaries of Tay in narrow, winding valleys. The smooth axis of the Strath, however, continues westward as a narrow flat-bottomed funnel—artificially drained—to meet the Earn at Crieff. From this town up to Comrie the scenery and the life of the Earn valley is semi-Highland in type, while below Crieff Strathearn resembles Strathmore in its smoothness and highly farmed character. West of the Tay all these districts are to a considerable extent occupied by small estates with parks and woods of fine and varied trees.

South-westward lies the important threshold between the Tay and the Forth country, narrow and higher than the straths, since the foothills of a Highland deer-forest (Glen Artney) in this part take the form of a smooth dome-like moorland of sandstone that pushes out toward the Ochils. The threshold itself is well drained by the Earn's affluents, and farms of the strath type flourish on the red soil. But the change from the raspberry fields about Auchterarder is abrupt after the deep gulch of Glen Eagles is passed. Strath Allan falls only 75 ft. in the 9 miles (23 m. in 14·5 km.) to Kinbuck where the winding Allan breaks through a barrier of kames and enters on its rapid descent to the Forth. The threshold, then, is narrow, has a wet and acid soil, a high

* On Strathmore, etc., see B. Hosgood, "Southern Forfarshire: a Regional Study," *Scot. Geog. Mag.* vol. xxv, pp. 15–29 and 55–71.

rainfall (over 40 in.) and a cloudier sky. In this bleaker and greener zone a thinner population is mainly interested in sheep. The great bulk of these feed however on the broad plateaus of the Ochils which are grassy up to most of their summits and deeply cut by a geometric grid of valleys.

The triangular basin of Forth and Teith is all but enclosed by high heathery or grassy hills. The frame includes precipitous escarpments; on the north-west that of the corrugated plateau of uptilted conglomerate, the Menteith Hills; and on the south the bold front of the Gargunnock and Fintry lavas. West of the latter the moor-clad upper peneplane is broken by a single gap (at 220 ft. or 67 m.). The two rivers thread different types of country. The Teith valley from Callander to Doune is broad, mature, and flanked by drumlins and kames; it has woods and parks and scattered farms. Near Doune the profile steepens and at Deanstone, as at Dunblane on the Allan, the power drives small textile mills. The Forth traverses the Vale of Menteith, nearly flat and in large part still a dark peat bog; for the clearing of Flanders Moss was never seriously attempted in the western half. In this moist district less than half of the farmland is in crops; its stock is its main feature, sheep and especially cattle, including some that contribute to the milk supply of the towns to the south. It will be seen that this thinly peopled basin, while structurally like the north-eastern straths, is also akin to the western pastoral districts. The whitewashed and slated houses resemble those of the west.

On passing south to the Endrick vale we meet a rolling landscape of hay and grass and small woods of fine trees that resembles Ayrshire. Moreover, Strath Blane, of similar aspect but steep-walled, already contains the outer suburbs of Glasgow, and beyond the Kilpatrick Hills in the wider Vale of Leven, draining Loch Lomond, residential districts are interspersed among dairy farms, while dye works and calico-printing works, using the pure Loch Lomond water, line the river to the tide mark above Dumbarton.

The North-east Coast and Firth of Tay

No large barriers interfere with free intercourse in this region. But the coast in general is transverse, and cliff-rimmed bluffs alternate with sandy bays. The landward uplands are cultivated save in small areas—the Sidlaw summits and the gravelly belt of wood and heath at Montreathmont. Elsewhere arable farming occupies the land to the cliff edges. In its general character the agriculture resembles that of Strathmore, save that near Dundee the demand of the city induces dairy farming on a considerable scale. The rainfall is below 30 in. and a feature of the climate, which prevails also in east Fife and Lothian, is the predominance of east winds and frequent sea fog in the late spring and early summer. This delays the maximum temperatures, but the autumn is seldom so wet as to interfere seriously with the heavy harvests of oats, barley and wheat. Where the coast is low and has raised-beaches (Arbroath to Dundee) sand dunes are found and are used as golf courses. The Firth entrance is flanked by two wave-built forelands of this character. The cuspate promontory of Buddon Ness is similar in type to Dungeness in Kent and is used for the same purpose; Buddon is the Lydd of Scottish artillery. Montrose Basin, which is dry at low tide and presumably is reclaimable, is enclosed by a broad "bay bar" strip built from the north. This is the site of Montrose. The Carse of Gowrie is a district apart. This belt, salved from the sheltered estuarine clays by emergence, has one of the most fertile heavy loams in Scotland, and is sheltered by the steep Sidlaw scarp on the north. The damper inner edge is largely pasture on which stock are fattened. The rest is famous for the high yields of its cereals, for its marvellous bean crops and for its fruit orchards. The land is admirably farmed on a long rotation, and Dundee is the chief market.

The urban links of these coast-lands with Strathmore are Perth, Forfar and Brechin. These and also various smaller places in both zones continue the traditional textile manufactures of the east with some leather-making and distilling, and in Perth dyeing as well. In Perth and Arbroath the industrial flavour is preponderant; but Dundee alone stands quite outside the genus "country town." Dundee has nearly half of its earning population in mills, in a large proportion women and girls. Its social problems are numerous, and are partly perhaps related to the separation of the wealthier elements in Broughty Ferry. The interesting industrial history of this, the third city of Scotland, must be read elsewhere*. But Dundee may be noted as an outstanding example of a town with several great industries—textiles (jute, hemp, flax, etc.), engineering, preserves (largely marmalade)—all the materials of which have to be imported. Dundee is the only port in the zone that has much sea-borne trade. It ranks fifth in Scotland in foreign shipping tonnage, its heaviest import being jute from India, while the largest export tonnage sails for the United States.

* *E.g. British Association Handbook, Dundee*, 1912. Dundee's population in 1921 was 168,217, about 9000 more than that of Aberdeen.

The Fife Peninsula

The old title "Kingdom of Fife" recalls the ancient independence of the peninsula based on its isolation; and the metaphor applied to it by King James VI, "a beggar's mantle fringed with gold," was a good one at the time when foreign trade was distributed among the string of burghs along the south and east coasts. The map of wheat distribution to-day shows this as essentially a fringing crop. Its three great abbeys were on the edges, south, east and north. The economic resources are now enormously enhanced by the coal—again near the south coast. But hard work, long continued, has wrought wonders in the interior, and to-day good mixed farming is found everywhere. Nowhere in Scotland has this result been more directly aided by railways and, above all, by the two bridges spanning the Firths.

The peninsula is shut off on the north-west by the Ochil rampart and by its more broken prong extending to Tayport. But east of Glen Farg this less formidable barrier is farmed in most places and intensively so along the raised-beach. The medial corridor rising eastward up the Devon leads by a threshold to the Kinross basin, a beautiful hollow shelving to its residual lake in the corner, whence it is drained by the deep incision of the Leven. Another broad threshold, overlooked by the lava cliffs of the Lomonds, leads to the Howe of Fife, flat and mostly heavy land, but well tilled save on gravel patches that are heath or woodland. Farms dot its surface, but the old villages are on the rim. Two are royal burghs, for this was long the King's hunting forest, and the palace still stands at Falkland. These and other inner Fife villages have small factories devoted to linen weaving, distilling, paper and floorcloth making—socially a healthy distribution. The Howe is drained eastward by the Eden which, after flowing in a rather narrow valley controlled by the market town of Cupar, emerges to an estuary. And this is flanked by the wide raised-beach of Tents Muir, largely a heathery expanse of fixed sand dunes, but with a zone of younger dunes on its edge. The much smaller grassy sandspit on the south is the Mecca of golfers. Indeed, it is safe to say it is more widely known to-day than the ancient cathedral city on the cliff behind it. The mediaeval importance of St Andrews can still be visualised in its buildings and streets whose greyness is offset by old red roofs and the still redder undergraduate gowns. St Andrews is the seat of the oldest Scottish University and it is perhaps the only ancient British academic centre where industry has not intruded. Such is peninsular isolation.

The hilly land north of Eden is unlike that to the south, which forms a secluded oval of the peneplane still much more intact than the northern part. This southern plateau has small picturesque villages and scattered farms that keep about half their fields in grass, feeding sheep and cattle. In contrast, its lower coastal fringe of rich and lighter soil is largely given to crops of cereals and potatoes. This is the Neuk of Fife, where the red-roofed fishing villages, including several royal burghs, are most distinctive and the fisher-folk most characteristic.

South-western Fife is mostly a plateau, with a well-marked grain east-north-east to west-south-west, the hollows of which, clogged with boulder clay, are ill drained and several of them still hold lakes or peat bogs. But the bulk of the area has been won for tillage, and the soil where drained is good, being derived largely from the underlying basalts. The plateau presents a bold scarp to the Kinross basin, and from Dunfermline to Kinghorn it falls steeply toward the Firth. The triangle of the Coal Measure outcrop is lower and much smoother, but the Leven valley notches the surface. Although the germ of industry in this region is old, as elsewhere in Fife, the country is now stamped as a coalfield. Not only are the old coastal villages from Dysart to Leven hemmed in by mines, but throughout the plateau to the west, where the Lower Carboniferous rocks have coal seams, large sprawling villages of sordid mien alternate with pit-head dumps; and these even press upon the ancient royal city of Dunfermline itself.

The Fife Coalfield stands second in production. About three-fifths of the coal raised is exported either coastwise or overseas, the docks at Methil having been specially built for this trade. This little port has acquired a remarkable position, ranking fourth in Scotland in regard to tonnage of overseas shipping. Burntisland ships much of the coal from the western mines of the field. Fife coal has also an important local market, apart from the domestic demand, in the great linoleum industry of Kirkcaldy which is pre-eminent in Britain in the making of the highest class of this material. The raw materials are not local, but Kirkcaldy draws from Dundee most of the jute Hessian cloth that forms the base. The "Lang Toon," as it is called, stretches with its jumble of factories for three miles along and behind the raised-beach, and might be taken as a warning to all future town planners! Among the newer industries of the area is one at Burntisland, the making of alumina from imported bauxite. The alumina passes thence to Foyers or Kinlochleven for its reduction to the metal.

The dolerite intrusions of west Fife and Lothian account for the narrows of Queensferry, a crossing important from prehistoric times, and now much more so because of the Forth Bridge. Inverkeithing, the old port of Dunfermline, is a centre of foundries and paper-making, and the secluded anchorage west of the hills has been converted into the great naval port of Rosyth, which however has been little used since the War*.

The Central Node

A narrow zone west of the Forth estuary forms the premier crossways of Scotland; and the setting of this surely constitutes a combination of land and water as striking as any in Europe. Westward, the high lava table of Gargunnock descends by gentle and well-wooded steps to the flat Carse. Northward the imposing fault scarp of the Ochils rises abruptly to 1200 ft. (360 m.), cut as with a knife along a ruler, and gashed by deep glens which, however, leave between them the triangular facets of the scarp; at each glen mouth, a cone of deposit on which stands a village with woollen mills and distilleries using the water of the burns. These streams, dissecting their fans, are collected by the Devon which itself trenches deeply the older alluvial valley of the scarp foot. South of the Devon valley a triangle of low hilly land with an east-west grain forms the agricultural part of Clackmannan, on the shore of which the ancient burgh and abbey of Culross nestle in a bay that seems to mark a meander of the pre-glacial Forth canyon†, and Kincardine is the ferry point where the estuary widens. In the jaws of the gap threaded by the Forth is Stirling Castle on its crag, guarding the most critical bridge in Scotland, and overlooking the winding and widening tidal estuary which becomes the inner Firth near the great modern port of Grangemouth, where the flat Carse ends. South of the Carse the land rises by wooded slopes to the moors of the smooth upper peneplane. At the foot of these slopes and on the "100 foot" beach runs the Forth and Clyde Canal, and just above it is the Roman Wall. Both pass westward to enter the narrow funnel leading to the Clyde. Under all this low land and water lies the Stirling and Clackmannan section of the central coalfields. The mining operations in Clackmannan scarcely mar its woods and farmland. Its other industries, such as worsted, brewing and iron founding, are chiefly centred in Alloa. The northern Carse is clear of pits; these appear, however,

* For the above region see L. J. Saunders, "A Geographical Description of Fife, Kinross and Clackmannan," *Scot. Geog. Mag.* vol. XXIX, 1913, pp. 67–87 and 133–148.

† *Cf.* p. 426 and footnote.

on the undulations of raised-beach behind; the battlefield of Bannockburn is in view of them, and the dozen square miles about Falkirk and north of the Roman Wall present an industrial concentration unparalleled outside of the Clyde basin. The significance of the position of Falkirk is emphasised by the importance of its cattle markets ("Trysts") in pre-railway times. This is one of the great iron-working districts, where some smelting is done, but foundries, mostly devoted to light castings, are the chief feature. The supply of pig iron is mainly received at Grangemouth from the Tees. Since Grangemouth is also the eastern shipment port for the heavy iron and steel products of the Clyde valley it ranks high among Scottish seaports, especially in coasting trade, and it handles the largest oil imports into Scotland. Modern Falkirk, an unlovely and over-crowded product of the Industrial Revolution, is now anxious about its future, for amalgamation of many of its foundries with similar industries in England threatens to withdraw work south of the border. The soil and the farms of the Stirling Carse resemble closely those of the Carse of Gowrie. As there, the yields are heavy where drainage is attended to, and again we find beans added to the usual crops; but fruit is little grown, and nearly half the fields are left in grass. On the slopes to west and south a heavy soil needs both draining and liming; but it yields much fodder, and many fields are in Timothy grass for several successive years. Cattle are numerous, dairy cows here beginning to predominate. The coal workings extend eastward along—and under—the Firth to the ancient port of Borrowstounness (Bo'ness) which ships coal and imports vast quantities of pit props from the Baltic. Mining villages now dot the bleak uplands over a broad belt extending southward to within a few miles of the Clyde at Carstairs. Without these this moorland would have only the thin population of the widely scattered sheep farms. The plateau is crossed by several railways, all with an important coal traffic.

Clydesdale

The lower basin of the River Clyde seen from Campsie Fells in north-west or from Tinto in south-east—when smoke will permit the view—presents a shallow saucer-like profile, symmetrical save that the western rim is higher and includes scarps of andesite. This is the profile of an ancient valley. But near the middle is the straight gutter of the Clyde, with steep but not precipitous slopes of 400 or 500 ft. and a flat floor half-a-mile wide. The Clyde below Crossford has almost attained maturity; not so its tributaries. The down-cutting of the Clyde has left terraces on its valley sides, and the existing meanders have all but removed the spurs between them. From Hamilton upwards the valley, with its white-washed and thatched or slated villages, presents a delightful rural picture. Yet apart from timber strips and parks, every yard of slope and haugh is intensively used. Cattle and cereals there are, but it is the fruit orchards, the strawberry fields and the acres of new glass tomato-houses that impress one; for such a combination on a large scale is unique in Scotland. In connection with it is the jam factory of Carluke.

Above Crossford the Clyde flows in a trench that narrows and steepens up-stream. At the Stonebyres Fall the river drops 100 ft., and three miles beyond at the two great Linns in a red sandstone gorge the combined fall is 190 ft. (58 m.). Water-power has long driven the cotton mills at New Lanark in the gorge, but these three waterfalls, recently harnessed, are now yielding electric energy to the Clyde industries and are linked to the Scottish "grid."

Lanark, the county and market town on the plateau above the gorge, advertises by its stock yards the importance of cattle throughout the rolling upland. This is evident along broad belts on both sides of the valley. Oats and roots are much grown, but the land is mostly green, and milk is the most marketable commodity. This district is the home of the Clydesdale, the heavy draught horse that has spread throughout Scotland and beyond, and has played a great rôle in the progress of agriculture.

Because of the coalfield the valley slopes have been invaded by industry. West of the Clyde, collieries are not obtrusive though they occur at intervals along a zone reaching south-eastward to near Stonehouse, and they reappear in the small Douglas basin. But the richest part of the coalfield lies across the Clyde and is in touch on the east with the many upland mines already mentioned. The Clyde valley contains, besides collieries, the great coal-using industries originally based entirely on the local "black-band" iron-ore. It is difficult to think of iron and steel objects that are not made to-day in Airdrie or Coatbridge, Motherwell or Wishaw, or in the smaller towns between them. Blast furnaces, iron foundries and steel mills are here assembled to the number of over 70 separate establishments dealing with iron and steel in one way or another, while others handle brass and tin. On the left bank of the Clyde above old Bothwell Bridge is Hamilton, a town which, though now surrounded by rich coal mines, and indeed to some extent undermined by these, does not participate in the iron industry. Coal-getting was delayed here, partly owing to the presence of Permian sandstone—long taken for "Old Red"—overlying the measures.

When pits were opened the iron industry was already established. Hamilton as a market town corresponds to Lanark in the higher district. Below Hamilton the valley widens its haugh lands, the plateau edges recede and slope more gently; the river sweeps enlarge and then, where it ceases to wind, the Clyde has reached the heart of Glasgow. The present form of the river bed is however here artificial and replaces a wider channel with sandbanks*.

This part of the present valley is a pre-glacial channel deeply clogged with glacial and marine detritus such as to present no great obstacle to the excavation of docks and the clearing of a waterway deep enough to accommodate the largest ships. Glasgow, growing from a centre about its cathedral on a ridge north of the Clyde, has spread upon the "100 foot" beach, and down over the land reclaimed by restriction of the river. It has extended northwards over many drumlins and is pushing westwards on a lower raised-beach. On the north-east alone does the city rest on solid rock which forms obstacles to approach; and the railways here enter in tunnels. Such, briefly, is the site. The nodality of the position in relation to natural routes is obvious; and engineering enterprise has led ocean shipping to the heart of an exceedingly productive district.

The industries of Glasgow and surrounding towns are vast and multifarious, but some geographical classification of them is possible. Out of nearly 244,000 industrial workers in Glasgow†, 101,000 are metal workers of some kind—from smelters to instrument makers—and engineering works are the most widely spread. But from Rutherglen on the east—a town that missed the chance of deepening the river and retaining its former leadership as a port—the relationship of engineering to shipbuilding seems to become gradually closer westward toward the great shipyards of Clydebank. The industries dependent upon heavy foreign imports tend to congregate on the west near the river, or north, along the canal. This group includes the factories handling timber and grain, or making paper, vegetable oils and paints, and those dealing with heavy chemicals among which the alkali and soap industries are notable. These taken together easily come after the metal group in number of workers. The modern economic history of Glasgow has been aptly summarised—"tobacco lords yielded place to cotton lords, these to iron and steel lords." Tobacco and textile

* On the site and growth of Glasgow see *Scot. Geog. Mag.* vol. XXXVII, 1921, pp. 1–79. See also C. A. Scott, "The County of Renfrew," *ibid.* vol. XXXI, pp. 225–240, and *British Association Handbook, Glasgow*, 1928, edited by J. Graham Kerr.

† In 1921 the total population was 1,034,069.

factories are still there, but are relatively of small importance. A few miles out, however, in Paisley, we find cotton as king. This is almost a self-contained industry in which many factories combine. Cotton thread, with 10,000 workers, is the chief product; but some weaving is still done, though Paisley shawls went out in spite of Queen Victoria's fostering effort. Chemical works, potteries, bleach works, dye works and many engineering shops take a hand in the common task. But here again the metal workers are the largest group, as they are at Johnstone nearby.

This must suffice here to indicate that the Clyde Valley region is of vast economic importance to Scotland and to the world. The foregoing facts account for its very large population (about half that of Scotland), for Glasgow's lead in Britain as a builder of ships, and for her rank as a port—fifth in Britain in the value of its total trade, but third in the value of its exports.

As the tidal Clyde widens below Glasgow the hills approach the water; and roads, railways and towns are perforce crowded on the narrow raised-beaches. The ancient port of Dumbarton crouching on the Leven estuary behind its fort on the rock now has the single great industry of shipbuilding, involving, of course, many trades. The north point of Renfrewshire would have remained a *bout du monde*, were it not that America lay beyond! Because of trans-Atlantic trade and the industries arising from it—sugar refining, shipbuilding and marine engineering—Greenock has risen from a herring-fishing village to a large hill-climbing city, while her little sisters Gourock and Port Glasgow have spread along the beach so as to give the trio a length of seven miles and a population of 113,000. Twenty years ago Greenock was doing a quarter of British sugar refining, but in view of increased foreign competition the industry has greatly declined. A host of other works tributary to the shipyards however are in operation.

AYRSHIRE

This well-defined district, resembling on the map the crescent of the waxing moon, was divided of old into three parts: from north to south, Cunningham, Kyle and Carrick. The traditional rhyme

> Kyle for a man; Carrick for a coo;
> Cunningham for butter and cheese;
> And Galloway for 'oo—

is still partly true as Figs. 55 and 56 will show. The uplands of Carrick*, where they touch Galloway, still form fine pastures for

* They form part of the Southern Uplands.

black-faced sheep; but the wool is now of less account than the mutton. Cunningham is still the prime dairy land of Scotland, but fresh milk is the chief product sold, and sold by nine of the most modern co-operative "creameries" placed at good collecting points. Kyle, which apparently had no product to offer but men—potential soldiers—is now equally a dairy country; but its southern part, with the Carrick lowland, being farther from the milk market, makes the butter and cheese. The roughly concentric contours of Ayrshire, which give centripetal rivers, are reflected in the isohyets and in the zones on Figs. 53 and 54; and, had maps of the distribution of population, mines and industries been added, the essentials of the shire's geography would have been set out. To the tourist the district appears markedly rural and "pretty" with its many winding streams, its low hills where fields are not interrupted by moors but only by small woods. The quiet villages, the white-washed farms frequently bearing new thatch now so rare in Scotland, the neat cattle—white or splashed with red or black, the luxuriant growth whether of hay or of trees that carry ferns on the trunks; such are the details of the landscape in lower Ayrshire. It is a country of distant views; from every higher hill one can see the encircling moors and the sea-girt mountains of Cowal and Arran. Much of this land was first won from bog in the 18th century. This is especially true of Kyle where the Carboniferous drift is little affected by limestone as in the north and gives a heavier soil than the red sandstone of Carrick.

The Census shows that only one-tenth of the workers of Ayrshire live by agriculture, a third by industries and an eighth by mining. This surprises the tourist if he does not visit Kilmarnock and a few northern towns. True, he will see coal pits throughout the Irvine valley, and occasionally south-east of it, and amid the hill pastures at Muirkirk and Dalmellington will come upon iron mines and furnaces*; but mining is so scattered that it does not disfigure. Much of the Ayrshire coal like that of Fife can be spared for export; and a ready and accessible market lies in Ireland.

What may be called the native industries, wool and leather working, survive, the former in carpet and "highland bonnet"-making at Kilmarnock, the latter as a shoe industry at this town and again at Maybole in Carrick. But the modern industries that represent an overflow from the Clyde overshadow these. Cotton mills are widely scattered chiefly along the hill foot, and the upper Irvine with its three cotton towns recalls a Lancashire valley. Kilmarnock, while making lace, is predominantly an engineering town with its railway works and many others.

* Out of blast, 1929.

The sweeping curves of the coast from Ardrossan to Ayr call other interests to mind. The two estuaries mark the two ancient ports and royal burghs of Irvine and Ayr. Of these Irvine has been the better able to maintain its trade, having the better access to sea and inland markets. It was largely used by Glasgow before Port Glasgow developed. It lies nearest the main coal basin, and has its own chemical works. Ayr, owing to its position, has long been the capital. Architecturally and socially it suggests an English county town more than does any other in Scotland. The sites of Troon and Ardrossan reproduce those of Tyre and Sidon. Yet as ports they are modern and are due to the initiative respectively of a duke and an earl, whose foresight on the whole has been justified. The old process of getting salt from seawater was long practised at Saltcoats. Industrially this may be said to have as a successor, across the sand dunes at Ardeer, the more complex manufacture of explosives on a vast scale. The raised-beach of these great bays is mainly given over to golf, and the links are unsurpassed. Thus dunes acquire a rateable value. But in economic value the flat sandy strip south of Ayr, especially near Girvan, though very narrow, far surpasses these. Given abundance of kelp (a seaweed), it yields potatoes year after year, and the mild climate puts them on the table before any others in Scotland, with much profit to the growers*.

The South-eastern Fringe

There is a certain uniformity of landscape and life throughout the zone immediately overlooked by the Southern Uplands all the way from Ayrshire to Midlothian. This belt consists of portions of the upper peneplane with hill groups rising abruptly from it. The inhabited land has small relief save that in Carrick the zone is dissected, giving a topography resembling that on the north edge of Strathmore; and this character is repeated in East Lothian, where, however, the prevalence of arable farming modifies the life. Elsewhere the population, nowhere dense, is mainly interested in stock, for the season of growth is shorter than elsewhere in the Midland Valley and tillage is mainly practised for the stock, oats, hay and turnips being the leading crops. It is a zone marked by arable farms of moderate size, carrying cattle for feeding, as well as by large hill sheep farms with a small proportion of arable land on which sheep may be folded. Except for a few coal mines in the Douglas valley and the iron-smelting already mentioned, industries

* For Ayrshire, see T. M. Steven, "A Geographical Description of the County of Ayr," *Scot. Geog. Mag.* vol. XVIII, 1902, pp. 393–413.

are negligible. Lanark and Biggar are the two nodal market towns, the railway node being Carstairs between them.

Lothian

The land below this upland and drained to the Firth or sea is divided by the Pentlands into a smaller western and a larger eastern lowland linked by Edinburgh which fills the six miles between the Pentland scarp and the Firth, and thus controls all passage. Ridges from west to east, ranging from igneous crags to drumlins, characterise western Lothian; and the shale mines with their large brick-red dumps dominate the landscape. For an equal distance east of the Capital "bings"* are scarcely less obvious. This coalfield however underlies a smoother surface than that of West Lothian. The smooth plateau is seamed by the rocky or woody canyons of the Esk system which conceal from the plateau all traces—save the tall chimneys—of a great paper industry, the need for the river water keeping the mill villages crowded in the deep valleys. East and Mid-Lothian are separated by a long finger of plateau that narrows the coast way at Prestonpans, and the coalfield extends but little east of this. New coal pits are a feature of Lothian as they are in Fife, a fact of significance for the future; and already the Firth is undermined from both sides. Scenery throughout the belt is attractive, for, apart from natural features, small estates with old castles and fine woodlands are typical. The farming is unsurpassed in Britain; holdings of over 300 acres (120 ha.) prevail with large fields, massive buildings, amounting almost to hamlets, each with a chimney indicating steam power often now replaced by oil and soon perhaps by electricity. A large rural population is so housed; and in East Lothian are a number of old and pleasant villages. Haddington is representative of the Scottish country town. Fine fields of almost all British crops are to be seen, and potatoes perhaps stand out, reaching perfection between East Linton and Dunbar. Except in the central industrial zone of Lothian where dairy cows preponderate, cattle are kept as fertilisers and sheep to eat the turnips. Ploughed fields are in a majority nearly up to the moors of the escarpment, all the way eastward to the cliffs of St Abbs. The parish boundaries significantly run from the lowland up over the scarp so as to include portions of the hill grazings, where formerly the herdsmen had their summer shielings. The East Lothian coast is famous for its golf links almost continuously from Edinburgh to North Berwick. But from Cockenzie westward to Granton the shore is mainly urban. Ancient red-roofed

* The local name for mine dumps.

fishing villages have been engulfed in modern extensions of Edinburgh. Alongside pitheads and a great power station there are still to be seen the salt pans that have been worked since mediaeval monks first mined coal for the purpose of evaporating the seawater.

The genesis and development of Edinburgh have been geographically treated in a recent collection of papers, and the subject need not be referred to here*. The city now covers the raised-beaches and five of the six parallel ridges (trending E. to W.) that are overlooked by the Pentlands, and is pushing west and east. Edinburgh as the Capital and headquarters of legal, banking, insurance and administrative groups has an unusually large proportion of "professional classes" in its population†. It must not be forgotten however that industry has great importance, two-fifths of its workers being engaged in manufacture. Engineering claims the largest group; and then follow in order: textiles and clothing, paper-making and printing, wood, beverages and milling‡. The engineering is largely tributary to many smaller manufactures. The presence of the port accounts for milling and wood-working, for the making of chemical manures from imported raw materials to serve the large agricultural demand of mid-Scotland. The barley fields of Lothian are related to the distilleries and breweries; Edinburgh is the Burton of Scotland with a large export of beer to England. The paper-making of Midlothian is also closely linked with the city and its highly developed printing trade. Leith is the second port of Scotland; but the trade of Glasgow measured either by ship-tonnage or value of cargoes is between three and four times as great. More than half the imports are foodstuffs from the Continent which are distributed throughout Central Scotland. About a fifth of the exports are coal, part of which goes to London, but most to the Continent. The remainder includes manufactures from all southern Scotland. Leith, together with the subsidiary harbours of Granton and Newhaven, is the third fishing port in Scotland in respect of the quantity of fish landed.

* See *Scot. Geog. Mag.* vol. XXXV, 1919, pp. 281–329. See also J. Cossar, "Notes on the Geography of the Edinburgh District," *ibid.* vol. XXVII, 1911, pp. 643–654, and XXVIII, 1912, pp. 10–30; and C. M. Ewing, "A Geographical Description of East Lothian," *ibid.* vol. XXVIII, pp. 624–641, and XXIX, 1913, pp. 23–35.

† Pop. in 1921, 420,281.

‡ The largest single factory, however, is a rubber-works located here for no known geographical reason.

Conclusion

While geographers, geologists and many historians have long recognised and discussed the essential unity of the Midland Valley, it is only in the present century that the close economic interdependence of all parts of the region has been realised by administrative authorities. Among the influences that have helped to accelerate this appreciation of unity are the advent of the motor car and the post-War economic depression. The first has greatly increased the road traffic throughout the region. It has also led to the extension of urban communities into rural districts, so that city authorities have had to widen their schemes of improvement, building and public utilities to embrace ever-extending zones of land formerly agricultural. Thus city is being brought into contact and even conflict with city. Regional planning, with all it connotes, has become a necessity. Amenities must be preserved and improved: the component towns must be served as economically as possible with heat, light, water, means of transport and recreation grounds; and their waste, whether carried by chimney or sewer, must be scientifically dealt with. A fair future in these respects has been provided for Edinburgh by a recent administrative expansion to the limiting rivers Esk and Almond, so as to enclose some 50 sq. miles (130 km.^{2}). In the west, however, the town cluster extending from Wishaw to Gourock is faced with the difficulty of interlocking communities. This can be solved only by regional co-operation. The problem is similar in the central traffic node of the Falkirk industrial district, with the growing port of Grangemouth, a district whose commercial importance must be enhanced by any further economic developments east or west of it.

The modern road system of mid-Scotland which, apart from the coach roads, had grown up largely to meet the needs of an enlightened agriculture, was generally adequate for the new motor traffic. The Scottish roads as a rule are as straight as the relief will allow them to be. New arterial roads are being built to carry the greatest traffic, *e.g.* from Glasgow to Edinburgh and Berwick. But in the east of Scotland the new traffic is confronted afresh by the sinuosity of the coast. The clamant need to-day is for road bridges or at least adequate ferries over the two Firths. The obvious sites—at Queensferry and west of Dundee—require high level bridges; and the great expense of these may lead to the building, instead, of low bridges farther up the Firths.

The railway system, in view of road traffic, will call for little extension. The coast-line and relief and the distribution of population and industries are such that there is great variation in the intensity of rail traffic. To give some idea of this, a count has been made of the number of scheduled trains passing daily over certain stretches of line*. It must be noted, however, that this neglects "special," *i.e.* non-scheduled, trains; and these carry most of the coal and some of the agricultural traffic; the numbers also are those of the winter service which is somewhat smaller than that of the summer. Let us look first at the points of entry and exit and thus get some measure of the exchange with southern Scotland and England on the one hand, and with the Highlands and the North-East on the other.

Six railways enter the Midland Valley from the south, and the total number of in- and out-going trains crossing the threshold of the region daily is about 330. These lines, given in order of the approximate percentage of the total trains carried, are those following respectively: (*a*) the upper Clyde valley (the "West Coast" route from England) with 36 per cent.; (*b*) the east coast, with 22 per cent.; (*c*) the Gala valley (the "Waverley" route), with 18 per cent.; (*d*) upper Nithsdale, with 16 per cent.; (*e*) the Ayrshire coast south of Girvan, with 4 per cent.; (*f*) the Eddleston valley, with 4 per cent.

Four railways cross the Highland threshold. These carry a total of about 170 in- and out-going trains daily. The routes, similarly arranged, are: (*a*) by the east coast (beyond Stonehaven), 55 per cent.; (*b*) by the Tay valley (beyond Stanley, north of Perth), 26 per cent.; (*c*) by the Gareloch (beyond Helensburgh), 10 per cent.†; (*d*) by the Teith valley (beyond Callander), 9 per cent.

In comparing the two aggregates it must be remembered that while the traffic with the north is over half of that with the south as measured by trains, its importance would be considerably less if it could be represented by train and freight weights, by number of passengers or by value of goods carried.

Within the Midland Valley itself the greatest traffic density over considerable stretches of rail is undoubtedly to be found where the railway net is also most closely meshed—in the industrial Clyde Valley. But the volume of through traffic using the various lines between Forth and Clyde and between Clydesdale and Ayrshire is very great; and if its nature were analysed it would throw much light on the internal commerce of the region. The volume may be briefly stated. The four gaps linking Ayrshire with the

* By courtesy of the London, Midland and Scottish Railway Coy. and of the London and North-Eastern Railway Coy. The data refer to 1928.

† Some of this traffic is with Loch Lomond shore only.

Clyde Valley together carry over 250 trains daily. Of these the two northern routes are by far the most important. Thus the Lochwinnoch Gap—the lowest in altitude—with its two railways has 58 per cent. of the traffic; and the Loch Libo Gap, also with two railways, has 34 per cent. of the trains. The percentages for the two southern gaps are comparatively insignificant, *viz.* Irvine valley to Avon valley, 3 per cent.; Ayr valley to Douglas valley, 5 per cent.

Passing east of the Clyde Valley, we may first note that of the railways mentioned above (p. 453) as crossing the sparsely peopled plateau south of the Forth-Clyde furrow, only two are important lines for ordinary traffic, *viz.* that between Airdrie and Bathgate, and, farther south, the Glasgow-Edinburgh line via Shotts. These together carry over 100 trains in about equal proportions. The eastern end of the Forth-Clyde furrow, with three railways (two being main lines), furnishes a very striking example of traffic concentration from geographical causes. Over 310 trains pass daily; nearly half represent Forth-Clyde traffic, but the remainder are trains connecting north with south, for these also must traverse the furrow.

In considering the north-south railways it is interesting to note that about 175 trains use the Forth Bridge and a similar number pass through Alloa Junction, south of Stirling. About 150 trains use the Tay Bridge, while the average number running on the various sections of the main line along Strathmore is probably between 50 and 60.

In contrast to the heavy traffic of the Forth-Clyde furrow less than twenty trains use the line following the River Forth north of the Campsies (between Stirling and Balfron). This gives food for thought. What will happen if a mid-Scotland ship canal is ever constructed? Two routes for this have long been discussed. The one following the existing barge canal; the other passing from the Firth, either north or south of Stirling, up the Forth and over (at 260 ft. or 79 m.) to the Endrick; up Loch Lomond to Tarbert and thence through the neck, $1\frac{3}{4}$ miles (2·8 km.) long and 160 ft. (49 m.) in altitude, to Loch Long. If the second route were chosen and the canal were to be largely used, it seems likely that the upper Forth basin would become the centre of considerable industrial and commercial life. The distances to be saved on voyages from Glasgow to North Sea and eastern Channel ports by using such a canal would be considerable*.

* *E.g.* in nautical miles: to Le Havre, 80; Antwerp, 345; Rotterdam, 390; Bremen, 371; Hamburg, 369; the Skaw (and so to Baltic Sea), 330.

The post-War depression has resulted in widespread determination that in the future coal shall be used more economically. To this end Central Scotland—an area nearly the same as that here discussed—is soon to be a unit in the "super-power" system of Britain. Electricity is now generated at 36 stations. This task is to be given in the first instance to ten selected stations, *viz.* two at Falls of Clyde (both hydro-electric) and three about Glasgow; Kilmarnock; Bonnybridge (near Falkirk); Edinburgh; Dunfermline; Dundee. The geographical reasons for such a choice are obvious.

There is a further suggestion to impound nine square miles (23 km.^{2}) of the Firth of Tay on the north side west of Dundee and thus to harness the tide to the extent of 11,700 continuous horse-power. But this is doubtless for the distant future.

On the whole the Midland Valley is a compact region with a well integrated economic life. Lowland Scotsmen, though varying somewhat in dialect and widely in intonation according to district, have the same sterling characters on the average throughout the region. Unfortunately many of the cream of the people are emigrating every year; some 300,000 left Scotland between 1921 and 1926.

The industrial and commercial success of Central Scotland has been built mainly on its coal, and this was exploited by methods now largely out of date; and others among the older industries dependent upon the coalfields see themselves in need of modernisation. Moreover, Scottish industries as a whole, faced after the War with heavily increased costs of production and of freight, are finding their northern location in Britain a more serious disadvantage than they have done since the introduction of railways. This is definitely a consequence of geographical circumstances. Thoughtful Scotsmen are forced to realise that a great change is taking place in their nation; they see with regret the integrity of the national type menaced socially and economically from two directions—Ireland and London.

In this part of the Kingdom the deficit due to emigration is being replaced chiefly by recruits for unskilled labour from Southern Ireland (the Irish element in the population of Scotland is computed at about 700,000). "The evidence is overwhelming that the Irish in Scotland will increase while the Scottish race decreases." This quotation is from a Report to the General Assembly of the Church of Scotland (1923). In most Scottish parishes there are two Protestant churches, and often more. In many mining and industrial villages of Central Scotland each recent year has seen the addition of a new one—usually distinctive in its architecture and in the colour of its new brick; and a similar development appears in the cities. This sign of the penetration of Protestant Scotland by a Roman Catholic community is viewed with alarm by many of her people, not so much on account of religious prejudice, as because of the social implications. The Irish immigrants come out of a quite different background, and they can have no appreciation of the influences that have built up the Scottish nation. Moreover their average standard of living is distinctly lower than that of the equivalent Scottish class.

The other influence at work emanates from London, an influence long exerted over the rest of England and in recent decades permeating Scottish life at an increasing rate. It is perhaps symptomatic that the number of residents of English birth in Scotland rose from 131,350 in 1901 to 189,385 in 1921, representing a 44 per cent. increase as compared with the 9 per cent. growth of the total population. This English element is mainly in the population of the towns and is to be found in all walks of life. It represents a movement reciprocal to that well-known and longer established flow of Scots across the border—an interchange that is wholesome for Britain. The presence of English people in Scotland is by no means necessarily connected with an economic change that is taking place and which is noted with apprehension by Scotsmen. The absorption of Scottish railways and banks and newspapers by greater English concerns centred in London, and the multiplication in Scottish towns of branches of London shops, to mention only outstanding examples, mark a stage in the amalgamation of Britain, begun at the Union, which is perhaps inevitable and doubtless is for the ultimate good of the commonwealth. And yet the change makes for loss of individuality that is to be regretted on many grounds; and the fact remains that Scotsmen who value their nationality and traditions are—rightly or wrongly—watching with some disfavour the growing dominance of London in the affairs of their homeland.

XXIV

THE SOUTHERN UPLANDS

William A. Gauld, A. Stevens and Alan G. Ogilvie, with a contribution by the late William G. Smith

A. THE LAND FORMS

A. Stevens

THE Southern Uplands form the second of the higher level non-industrial divisions of Scotland. In many respects they have points of difference from the Highlands ensuring them an individuality apart from their isolation. Yet it is generally, and surely reasonably, assumed that many of the chapters of evolutionary history are common to the two areas, and that therefore they will show such similarities as are to be expected between two regions of resistant, but different, material, having each its proper structure, and exposed to the action of epigene forces in the same latitude and aspect. It is of some interest to consider what these similarities may be.

In the latitude of the Southern Uplands the western land margin of Great Britain has swept westwards, and so the fjord coast of Scotland is continued by the west coast of Ireland, where the fjord form has passed to the fiard, a modification which is seen in early stages in the lower grounds of Lorne. Towards the south the division between eastern and western Highlands is somewhat complicated by considerations other than topographic: the topographic boundary separating the country of scenery allied to the fjord west coast might be taken to be a line tangent to the lower ends of Lochs Tay, Earn and Lomond, a line which is nearly parallel to the Great Glen, and which approximately coincides with the lie of the coast between Ballantrae and the Heads of Ayr*. The division between east and west in the Highlands, then, passes out to sea west of the Southern Uplands, which are to be compared with the eastern Highlands. In a very general sense this comparison holds. With the exception of Loch Doon, Loch Ken and St Mary's Loch, together with a few relatively insignificant small lochs, narrow lakes of fjord type are absent from the Southern Uplands, and the hill masses are of the lumpy coherent type of the Monadhliath and the Cairngorm, rather than of the deeply dissected chequered type of the western Highlands.

* See Fig. 47 (p. 376).

Except where the sea puts it beyond discussion, the boundary of the Southern Uplands has not the clear-cut distinction of the Highland Line. The break is not spectacular to the observer approaching it from the Midland Valley, nor is it, for the most part, so clearly defined on a topographic map. Exception must be made of the northern face of the Upland above the flats of eastern Lothian. Elsewhere, even in the south-east of Ayrshire, the upland platform of the Midland Valley, in which the lowlands of Ayrshire, Clydesdale and the Lothians are but topographic hollows, gives, if not a gradual, at least a gently stepped approach to the higher southland. The traveller, however, penetrating along the avenues of approach from north or south, has no doubt as to when he has passed the limit of rolling upland pasture—here damp and sedgy, there heathy, so much gashed rather than excavated by narrow gorges that the deeper valleys disappear from the broader landscape—and entered different country. He is in the lower passes of a country of more marked relief, with less sluggish waters and drier subsoil, less wood about the flatter ground above, but less heath about the more undulating upper levels, a country which has been carved on more generous lines out of a higher platform still, whose levels are preserved on the ample shoulders of the hills. Approaches from the south lead in from marginal areas, whose aspect, different in respect of richness and variety of vegetation, more kindly with its evidence of human activity in productive farms and quiet villages, is due generally to a different solid substratum, but in the west only to lower level and modified history in recent times.

The Southern Uplands owe their general characters and their unity in great measure to the uniformity of their material and the nature of their structure. Broadly, they have been modelled out of the southern limb of a great anticlinorium of Ordovician and Silurian rocks. The former occupies a belt of country which on the west is 20 miles broad and extends from south of Girvan to south of Port Patrick, but narrows more or less gradually eastward so that it has disappeared at Dunbar. The northern boundary separating these rocks from the later materials of the Midland Valley is a fault-line, which is more or less simple, and which is traceable for the greater part of the distance between a point on the coast some little distance south of Ballantrae to the neighbourhood of Dunbar. Step-faulting on the west advances the Ordovician boundary as far as Girvan, while farther east there is a salient in the basin of the Eddleston Water, a tributary of the Tweed. North of the boundary there are faulted inliers of the uppermost Silurian members rising through the upper platform of the Midland Valley near Girvan, at Douglas, between Muirkirk and Lesmahagow, in the bend of the Clyde above its confluence with the Douglas, and in the Pentland Hills. These, which must represent parts of the northern Silurian limb, strengthen the evidence of the asymmetrical drainage that the Uplands form, topographically, a sort of "cuesta," due to the submergence of their northward continuation beneath younger material, and to depression.

Lithologically, variations in the solid material of the Southern Uplands are confined to a group of fine, even-textured rocks resistant to weathering: cherts, grits, greywackes, hard flags, indurated agglomerates, and igneous material. Limestones are relatively subordinate, and the black shales in which the graptolites have been found form very thin bands. The whole series has been intensely and minutely plicated, so that the various beds present their edges to the surface, and, like a well-grained cabinet wood, exhibit a definite striped pattern, but take and maintain a pattern engraved and finished by the hand of nature as if on homogeneous material.

Later sedimentaries occur on the southern margin and show themselves in the topography. Quite generally the 250-foot (75 m.) contour, which forms a squat isosceles triangle based on the Tweed, bounds the Scottish extension of the Carboniferous of the Northumberland coast, which laps up on the rising ground of the Cheviot porphyries. This distinctive part of the Tweed basin, dependent for its level, perhaps, on its soft "cementstones," rural because its thin coal-seams and its fire-clays are not profitably workable, and because its waters run gently, nevertheless owes its general aspect mainly to its drifts. Disposed in an endless series of drumlins, these drifts give a gently furrowed appearance to the ground, and this furrowed surface runs inland over the Old Red Sandstone on to the margin of the Silurian as high as 800–900 ft. (240–270 m.). The drumlins run north-eastward, along the Silurian strike, but towards the east they swing to the right to take an eastward course where they meet the Tweed. Coincidence with the strike direction seems to be fortuitous, for the drumlins do not appear to have solid cores. In this area the ancient ice trend has already swung southwards, to judge by the striated rocks, and takes the drumlins on their flank. It is here that Professor J. W. Gregory finds his strongest evidence for aeolian sculpturing of the drumlins*; but this evidence seems unconvincing, partly because it takes no account of the swing of the drumlins, as it were about the flanks of Cheviot, and

* *Trans. Roy. Soc. Edin.* vol. LIV, 1926, p. 433.

partly because it seems to fail to distinguish aeolian action, which is inhibited by damp conditions, from rain-wash.

On the west again Carboniferous and Trias series, continuous with those of the Eden valley, build an agricultural lowland extending on the Scottish side to the mouth of the Nith; and the Solway Moss like the Solway sands makes an indefinite sort of boundary which geographical description can hardly recognise. In the Nith valley the Coal Measures of Ayrshire push a V up as far as New Cumnock and advance the industrial area to the very margin of the Uplands, while a little lower in the same valley an inlier of the same rocks has been faulted down to give, within the borders of the Southern Uplands, the coal-producing coign of Kirkconnel and Sanquhar, which is very conspicuous as an upland depression lying between 400 and 500 ft. (120–150 m.) in country which rises to 1500 ft. (460 m.) round about. Lastly the measures, conspicuously red sandstones, provisionally classed as "Permian," of the middle Nith and Annan form level upper "treads" divided by low "steps" from the still lower mouths of the valleys, which, in the same material, continue the Plain of Carlisle along the Solway shore to the Criffel promontory. This ground, with its favourable circumstances in respect of altitude and shelter, and its generous cover of drifts, till and outwash material* forms the eastern part of the rich farming and dairying lowland of Solway side. The western part represents the lowest part of the Silurian wedge, but shares the ample cover of loose material of the eastern plain. But in this case the ice swung out across part of the Solway floor and its burden with its shelly content made a richer raw material of soil; and besides, the flat includes broad raised-beach terraces, often stony, always fertile. This is the farm land of Galloway (*v.* Fig. 57, p. 474).

Finally there are several granitic areas mainly in the western half of the Uplands, of which three of considerable extent make their mark on the landscape, notably by their reaction on the encircling Silurians. These form metamorphic aureoles about them, more resistant than either the granite or the normal country rock. They are the Loch Dee mass, extending from the head of Loch Doon to south of Loch Dee, the elliptical mass on the right bank of the River Dee which culminates in Cairnsmore of Fleet, and the mass between Criffel and Dalbeattie, with extensions west of the Urr Water. All of these areas are so characteristic in their topography that their limits can easily be traced with some accuracy

* For a full account of this glaciation, see J. K. Charlesworth, *Trans. Roy. Soc. Edin.* vol. LV, 1927, pp. 1 and 25.

on a topographic map, but the first, perhaps on account of its central position in relation to the drainage system, is the most striking. The core has been excavated to form an irregularly floored cauldron at about the level of 1000 ft., whose walls rise towards the contact rim which has a maximum elevation of 2764 ft. (842 m.) in the Merrick. The floor of the cauldron is flooded in its deepest irregularities by ragged lakes. Indeed in and about this depression are concentrated one of the valley lakes and most of the other lakes of the Southern Uplands. The divide between the northward drainage of the Doon and the southward drainage of the Dee is very indefinite. The medial ridge separating Dee and Trool, with its deep notching by glacial overflows, is also an indefinite water-parting. This is the wildest part of the Uplands. The bold dark background of the hills and the sodden peat bog of the bottoms contrast almost as much with the greener slopes of the Southern Uplands as with the country of the Central Valley, and the land between the Doon and the Stinchar forcibly recalls the remote parts of the Cairngorms. Perhaps the extent of these greater masses is necessary for the formation of cauldron-like depressions in order to provide lodgment for the peaty waters so destructive of granite, and so characteristic of these granite hollows. In any case, less extensive granite areas, such as that of Cairnsmore of Carsphairn, are corrie-sided masses, reminiscent of Snowdon, and an aureole of contact-hardened Silurian is not conspicuous.

The Drainage System

A very striking feature of the Southern Uplands is the coincidence between its northern boundary and the divide separating the drainage towards the Firths of Clyde and Forth on the one hand from the water-flow into the Solway and the Tweed on the other. The divide swings to and fro across the fault just as a normal water-parting zig-zags on either side of the crest line of an ideal mountain range. The only river which pushes far into the Southern Uplands is the Clyde, and one wonders that a fault-line scarp should show so little of the so-called usual tendency to migrate up-stream.

For the most part the valleys trenching the Southern Uplands are steep-sided and narrow, carrying rapid streams of irregular regime. The valleys are stepped, and the staged appearance is emphasised in such as the Nith and Annan whose lower levels are cut out of younger less resistant sediments. Valley lochs, however, are not remarkably developed, and hence in the absence of natural regulation the streams in a country of no unusual permeability rise to every heavy rainy spell. The valleys show the usual obvious traces of glacial finish at least, and, as in the Highlands, excavation below the smoothed glacial levels is gorge-like, and, in the solid, remarkably small in amount. Moreover it is more or less confined to the lower levels, many hill streams meandering in rounded valley bottoms at upper levels almost on the surface of the ground; as in the case of the Stinchar above Barr. The head-streams of the same river, however, have made considerable excavations in unconsolidated glacial material; and indeed valley cliffs of such material, resembling normal boulder clay much more closely than similar deposits in the Highlands, and occurring up to levels between 800 and 900 ft., besides valley terraces of the same composition, are a feature of many valleys in the Uplands.

The modern valley system, as has often been remarked, seems to have been carved, mainly in post-Tertiary time, out of a higher surface which slopes rather gently from a marginal height of 1000 ft. up to the highest elevations, over 2500 ft. (760 m.), and the pre-glacial drainage system, itself in no very mature condition, has been succeeded by a modern system still decidedly juvenile. Traces of an older drainage system have been recognised, for example, in the "Permian"-filled hollows, but on the whole the present drainage appears to be a new system dating from Tertiary times. According to many authorities the Southern Upland drainage developed as an integral part of a general Scottish, or even British, system on a great slab sloping generally across the strike of the rocks towards the south-east; but this view does not seem to have been supported by any systematic study of transport of the materials of the alluvia. To Mackinder is due the suggestion, that the Clyde may originally have been one of the headwater streams of the Tweed, which has frequently been repeated as a fact established by sound evidence. In the wide peaty hollow of the Biggar Gap, and in its relations to the Clyde valley, we can see no positive evidence for this episode in the history of the drainage*. Moreover, if the original upper surface into which the modern streams began to cut was of the nature of a peneplane, very considerable tilting must have occurred; for the residual transverse slopes, as indicated by the upper spurs, are of order not inferior to 1 in 50, usually greater, and suggest complete separation from the

* This gap has a "tread" position in relation to the stepped form of the valleys it connects, and may be analogous to the "Permian" treads in the Nith and the Annan, drainage having been modified on the removal of the less resistant younger material. Its width and general form lend colour to this suggestion.

Highland tableland. In the view of the present writer the drainage of the Southern Uplands was developed independently on a roll of the ground whose core was Silurian, and whose crest lay along the line of the present fault which runs inland from Dunbar; that is, just inside the northern boundary of the area. Except on the right bank of the Tweed, the drainage has remained dominantly consequent and transverse. On the whole the longitudinal direction is not well marked. Except where the granite masses form a disturbing influence it comes out discontinuously in the ridges, especially in the north centre between Nith and Yarrow; but by no means conspicuously. A manifestation of strike influence for which we have no explanation to offer is a curious apparent decrease in stream density south of a line joining Port Patrick and Thornhill (a similar decrease distinguishes the greater part of the Tweed valley). But the relative unimportance of longitudinal features within an area constructed on such a close grain of very uniform material is surely not a matter of surprise. Just beyond the boundary in the Clyde valley, and again in the step-faulted outwork of the Girvan district, where the faults are strike faults, longitudinal stretches in the streams are strongly developed.

Less resistant material on the flanks of the Uplands seems to have exercised the dominant influence in directing the development of the drainage of the area along its present channels down these flanks. This is very clearly so in the case of the Tweed system, and accounts for the contrast between this basin with its discharge eastward and its strongly marked longitudinal trends, and the remainder of the Southern Uplands, where the outflow is transverse; but in Nithsdale, Annandale, and Liddesdale, the last longitudinal, marginal, and of a lower order of importance than Tweeddale, directing influence of the same nature is obvious. Farther west a contrast of the kind one expects in normal circumstances as between one belt and another lying along the strike due to different powers of resistance occurs in the transverse direction. For the resistant baked frames of the granite massifs of Loch Dee and Loch Fleet merging produce a continuous upland running across the strike, and the Dee and the Cree, largely fed from their enduring slopes, naturally enough take a consequent course along the less resistant material beyond their influence.

Physically there are three sub-regions of the Uplands clearly distinguished. In the heart is an area of maximum elevation, and drainage ultimately radial, of which the upper Clyde basin is the core. The boundaries of this are the Cumnock-Sanquhar gap on the west and the Beattock gap on the east. It is the smallest division; and its southern margin belongs to the plain of Solway, including Liddesdale, an integral part of the basin of Carlisle. Westward is Galloway, slab-like, with an irregular coastal plain on the south, a southern half with a dry surface, curiously so in the Rhinns and the Machers of Whithorn, in spite of a westerly situation and a considerable rainfall. Here consequent (transverse) drainage is the rule. Here also the topography is at its wildest, and dissection is so deep and minute as to produce a gloomy interior country. In the east is Tweeddale. Wide open in contrast with the rest of the Uplands, it stretches to the ultimate margin of the Silurian area, the scarped north face of the Lammermuirs. The Carboniferous plain, continuing the Northumbrian coast, embayed deep between Siluria on the north and the Cheviot on the south, is more completely enclosed in the Uplands than the western plain. A larger proportion of Upper Palaeozoics, and consequently lower elevation, etching on broader lines and a curiously drumlin-furrowed surface make its aspect at once characteristic and more pleasing. An easterly situation combines with other factors to give a drier soil and less abundant stream density.

B. GALLOWAY AND THE DALES

William A. Gauld

The dominant physical feature in the hill mass of the Southern Uplands is the system of river valley corridors which dissects it from north-west to south-east. The province of Galloway lies to the west of the Nith valley, and stretches as a peninsula between the Solway and Clyde firths. Nithsdale, Annandale, Eskdale and Liddesdale open the ways from the Plain of Carlisle in England to the industrial midlands of Scotland. In Galloway there are no "dales" in the true sense, but there are important river valleys linking Ayrshire with the Solway plain. These have influenced the whole history of the region, and to-day the location of population and towns, social and economic intercourse, and the variety of occupation are all bound up with them. The major consequent valleys, Ryan-Luce, Cree, Ken-Dee, and the "dales" proper make transverse communications easy. For the most part they are wide with even slopes and low gradients. The alluvial "holms" of the valley floors and the cultivated slopes often coated thickly with boulder clay are in striking contrast to the grass and heather tablelands which divide them. Geological causes have produced the main difference between the Galloway valleys and the "dales."

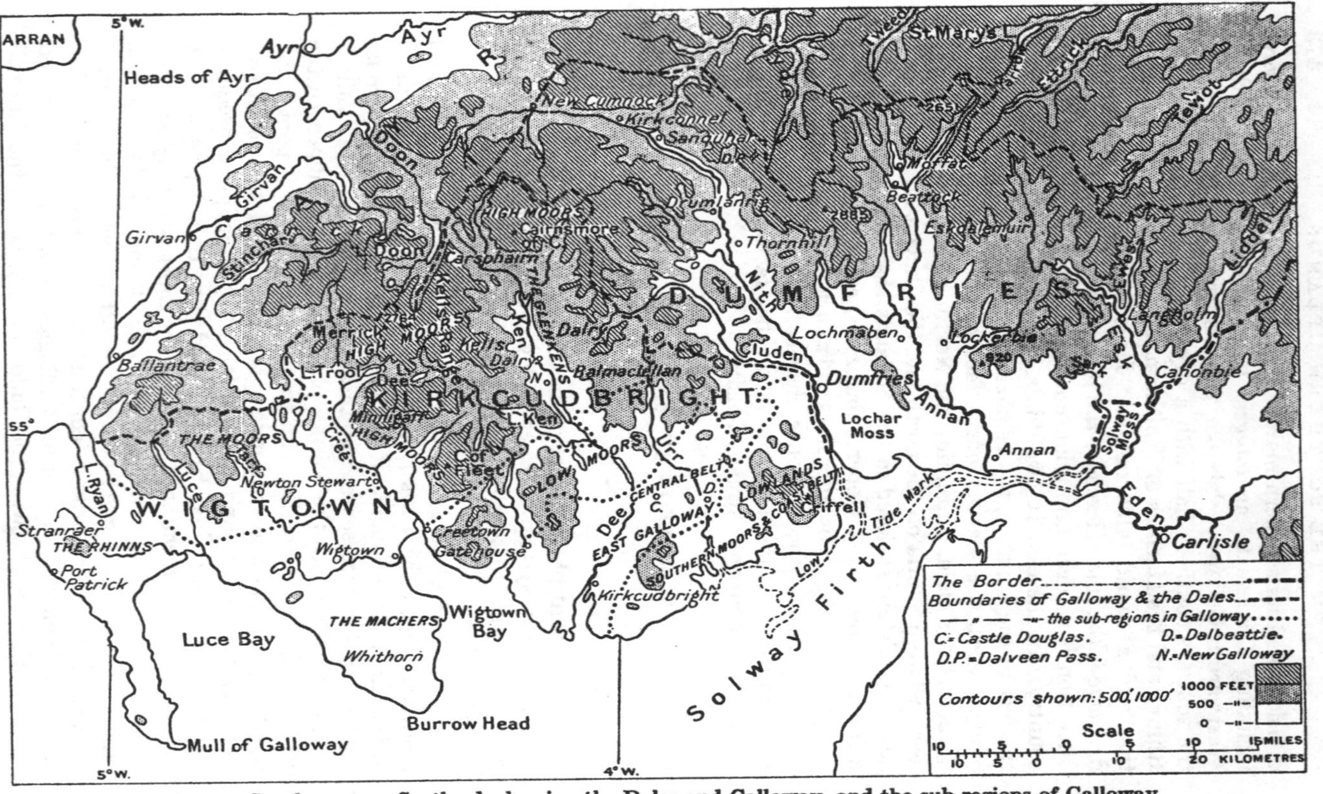

Fig. 57. South-western Scotland, showing the Dales and Galloway, and the sub-regions of Galloway.

The dales occupy inliers of younger and softer rocks set in the older massifs where river evolution was more rapid*. The grits and shales of Silurian age form large plateaus covered with peat, heather and grass while the valleys have mixed soils derived from New Red Sandstone, limestone and boulder clays. The intrusion of granite bosses into the Silurian floor accounts for the more rugged outlines of the Galloway highlands, but the smoother contours of the grassy uplands beyond the Nith may perhaps be partly due to the difference in climatic exposure of the two areas. It is not difficult to see why Galloway became the country for dairy farming and the eastern uplands for wool†.

Glaciation has superimposed its own topography in a region which was already reduced by peneplanation. The boulder clay formation which covers much of the plains and valleys has a hummocky surface with many of the hollows occupied by lochans. The moraines of the hill areas represent a later phase of glaciation and are limited in general to areas above 1500 ft. (460 m.). The lowest moraine on the mainland of southern Scotland is at the foot of the western spurs of the Criffel massif near Southwick station at a height of 175 ft. (53 m.)‡. The wide spread of marsh flats and lochans in Wigtown and lower Kirkcudbright and the ridges of fluvio-glacial sands and gravels which rise from them, are the main results of this phase in the physical evolution of the region.

Galloway

The Representation of the People Act of 1918 restored the ancient province of Galloway to political significance by creating a new constituency out of the administrative areas of Kirkcudbright and Wigtown. The name "Galloway" was derived from the Celtic "Gallgaidhel" the "stranger Gaels," a name applied to the early Gallovidians who were isolated then and for a long period after from the general life of Scotland. The Carrick division of Ayrshire adjoining the modern Galloway was until 1186 part of the province. Galloway had its own native chiefdoms until the 11th century, and down to the 15th the local provincial tradition was preserved in a

* *Vide supra*, p. 469. In the wide depression of western Galloway, now occupied by Loch Ryan and Luce Bay, are similar rocks alien to the fundamental topography.

† For a more detailed treatment of the physical features see an article on "Galloway" by the author, *Scot. Geog. Mag.* vol. XXXVIII, 1922, pp. 22–39.

‡ J. W. Gregory, "The Moraines, Boulder Clay and Glacial Sequence of South-Western Scotland," *Geol. Soc. of Glasgow*, vol. XVII, 1925–26, pp. 354–376. According to this authority the lower boulder clay is a marine deposit laid down in a still sea when the mainland was 1000–1200 ft. lower than at present.

code of civil and criminal law, while the Gaelic language persisted until the close of the 16th century.

The two main divisions, the "Shire" of Wigtown and the "Stewartry" of Kirkcudbright, are believed to have originated in the period of reorganisation which followed the War of Independence. In 1369 Galloway east of the Cree was granted to Archibald the Grim, an early Marcher earl, who appointed a Steward to collect his revenues and to administer justice in his "Stewartry." Galloway west of the Cree (Gaelic, *Crioch*, a boundary) was outside the control of the Marcher lords, but it had its own native nobility among whom the Kennedy family was most prominent. Geographically it was more closely bound to Ayrshire, and it was in fact under the jurisdiction of the royal sheriff who probably governed the southern "shire" by proxy.

The western peninsula of Galloway is known as the "Rhinns" (Celtic, headlands) the Wigtown peninsula, terminating in Burrow Head, is called the "Machers" (Celtic, plains) while the "Moors" of the north form the third division of the "Shire." This last is not used for administrative purposes to-day but it appears as a local name in Timothy Pont's maps in Blaue's Atlas of 1655.

In Kirkcudbright the divisions are less well defined, and there are no regional place-names except in the Dee valley. Here in the valley of its tributary the Ken, above the loch of that name, are four riparian parishes, Kells, Dalry, Balmaclellan and Carsphairn, which are locally grouped together as the "Glenkens"; they are noted for sheep and cattle rearing.

The men of Galloway are remarkable for their abnormal stature, and they are perhaps the tallest men in Europe. This stature appears to be a Nordic rather than a Mediterranean trait, and a considerable distribution of red hair and light eyes points to an association with the Scandinavian elements in the Lothians. Culturally also, Galloway seems to have been more influenced by the Anglian church of Northumbria than by the Celtic church of Ireland.

Galloway is exposed on the south and west to sea influences, and the effect of these is seen in the milder conditions of the coastal margins, while climatic variations inland are determined by the orographical nature of the country. The province shares in the general cyclonic weather control. The main belt of travelling low pressures lies permanently off the north-west shores of Britain but sends branches across the British Isles. In the three months, October to December, such a branch crosses Galloway, while in spring it may be said to lie between two parallel storm paths, one along the Solway and the other across Central Scotland. The main direction of the isotherms over the province is from north to south, and in general the temperatures diminish from west to east. The cooling effect of elevation is the most marked feature. The normal Scottish type of rainfall is characterised by an even distribution during the year with a tendency towards a spring minimum and an autumn maximum. This type is seen in Galloway, although west of the Cree the maximum period is in winter rather than autumn. The area with under 40 inches is confined to the Rhinns and Machers and the southern margins of the Kirkcudbright table-land, the minimum being registered at the Mull of Galloway (26·8 in. or 661 mm.). The zone having from 40 to 50 in. (1000–1250 mm.) is generally limited to the eastern county but in wet years it extends farther west. In Kirkcudbright the rainfall increases from south to north, in Wigtown from west to east. The very mild climate which permits cattle to remain in the fields the year round is due to the prevailing warm sea winds and the latent heat released by frequent precipitation. The heavy rainfall and cloudiness are deterrents to cereal growing and wheat is restricted to the coastal areas. The province is eminently suited to sheep and cattle rearing, and the proximity of good markets for dairy produce has stimulated the latter at the expense of the former. Fig. 57 illustrates the principal natural units in Galloway, and these may be briefly considered. In the western county the three units of Rhinns, Machers and Moors are well defined. The Rhinns peninsula including also the plain of Stranraer is the richest farming land in the province. There is no observable contrast in the Silurian and more recent formations as all are covered with glacial drifts. The urban nuclei of Stranraer and Port Patrick communicating with Ireland and Central Scotland are the markets for this area. The Machers have a variety of soils. The estuarine belt along Wigtown bay is largely reclaimed land. The centre is "drum and esker" country, while at Burrow Head, Wenlock shales outcrop and cultivation is confined to the valleys, while in the west is a hill pasture zone of glaciated Silurian rocks. Early maps of the Machers show a wide distribution of "lochans," or lake-filled hollows in the boulder clay, and the numerous grassy flats in the peninsula often occupy the drained sites of these glacial lakes. "Flow peat" still occupies the undrained hollows, and the farms are situate as a rule on rock outcrops above the general level. It was Chalmers in his *Caledonia* (1824) who first noted that "on some of the heights the soil is more fruitful than it is in the vales below." To-day, turnip and oats cultivation occurs on the rounded gravel eskers or the clay drumlins while the level stretches between the mounds may be occupied by undrained peat bog. The boulder clay deposits undoubtedly aided peat growth, and the chilling effect they have in arable lands has proved a great hindrance to agriculture. This widespread distribution of glacial drift may account for the universal use of dry dykes of round stones in the province, this being the most economical method of getting rid of the stones when new fields were being opened up.

The "Moors" of Wigtown are really continuous with the higher uplands of Minnigaff and Kells in the adjoining county, but are neither so elevated nor so extensive. The mature river valleys carry the agrarian belt of the Machers right up into this hill pasture zone. In the Stewartry of Kirkcudbright the natural units are less well defined. We may distinguish the high moors of Merrick-Kells, the Glenkens valley and the low moors all in the north and west, from the east Galloway lowlands which comprise the southern moors and coastal belt, and the central plain with its three large towns, Castle Douglas, Dalbeattie and Kirkcudbright. In both counties there is a natural and economic gravitation towards the junction of hill and plain, and the railway to Stranraer roughly indicates the line of densest population.

In Galloway there are four main types of husbandry of special importance: (*a*) dairy farming, (*b*) stock raising, (*c*) sheep rearing, and (*d*) arable farming. Of these the first and third are the chief. The Ayrshire cow, introduced 1850–70, is the typical dairy animal, and the dairy country *par excellence* is Wigtown. The establishment of creameries in the main dairy-cattle areas is an important recent development. These are located along the main lines of communication, at Stranraer, Bladnoch, Sorbie and Colfin in Wigtown, and in Kirkcudbright at Tarf, Gatehouse and Dalbeattie. They are run by the farmers who hold shares in the different companies. One of these is controlled by the Scottish Wholesale Co-operative Society at Glasgow. A large proportion of the milk is collected by motor lorries and dispatched as whole milk to various English towns such as Newcastle, Birmingham, Liverpool and even London, while much of it is manufactured into cheese, butter and cream, and sold in similar markets. One company* alone disposed of 2½ million gallons (*c.* 9½ million li.) of milk in 1926. A pig-rearing industry has been created as a result of the butter and margarine manufacture of the creameries, and the waste products are thus utilised with substantial profit. Areas situated far from the centres where milk is required, *e.g.* the southern

* The Galloway Creamery Coy.

Rhinns and Machers, specialise in the cheese industry on their own account, and here again the "whey" residue is used as pig feed. In the Stewartry dairy farming is pre-eminent in the central and coastal belt but nowhere is it so important as in the neighbouring shires.

The Galloway sheep are noted as a mutton rather than a wool stock. The Black-faced breed is probably the oldest. It is now mainly confined to the moors, and southern breeds like Leicesters and Cheviots are more common in the lower hills and the agrarian zone. The Stewartry has about three times the number in Wigtown. The rearing of the native black "Galloway" cattle is an old industry, probably at its height in the "droving" days of the 18th century, when the cattle were driven by road to Norfolk and other places near London. Mixed farming is fairly widespread in both counties and owes its high development to the activities of certain local squire farmers of the Agrarian Revolution period (1750). Arable farms are from 60 to 600 acres (24–240 ha.) and are rented from £80 to £700 a year. Five- or six-year crop rotations are usual. The hill sheep farms are of course much larger. In the Glenkens some arable farms in the main valley have their attached hill farms on the adjoining moors.

In farm buildings the local "whinstone" or greywacke and the Queensberry grits of the Silurian country are used. Granite may be used in the dwelling houses, and the fine New Red building sandstone of Locharbriggs and elsewhere in Nithsdale has also been introduced. The market towns of mid-Galloway, Stranraer, Newton Stewart, Wigtown, Castle Douglas, Dalbeattie and Dumfries are the natural meeting place of sheep and cattle farmers.

Apart from farming there are no occupations of a conspicuous nature. Fishing is much less important than might be expected in a country with such a long coast-line. A writer in 1684 said: "our sea is better stored with good fish than our shoare is furnished with good fishers." And these words are still true to-day. Again there are few good harbours, and coastwise commerce is very small. The province has good building stones and the granite of Creetown and Dalbeattie is known in almost every quarter of the globe. The Liverpool docks are built largely of Creetown granite. The dykes of felsite, which run with the strike of the rocks from south-west to north-east throughout Galloway, provide the best road metal to be obtained. Macadam, a native of Carsphairn, owed his early successes in part to these dykes.

The population of the province has declined steadily since 1851. In 1921 the census returns give a total of 37,155 for Kirkcudbright and 30,783 for Wigtown. Of the urban centres the following may be noticed:

Maxwelltown ...	6094	Stranraer	6138
Dalbeattie ...	2998	Newton Stewart ...	1831
Castle Douglas	2801	Wigtown	1299
Kirkcudbright	2101	Whithorn	1033
Gatehouse ...	893		

The relative proportion in the different natural units is indicated by the following figures of population per 1000 acres (405 ha.):

Rhinns ...	144	Low moors	55
Moors	59	East Galloway	
Machers ...	76	(central belt) ...	143
Glenkens ...	16	East Galloway	
Minnigaff ...	12	(coast belt) ...	187

Historically, the proportion has remained fairly constant, but the decline is most marked in the upland parishes, notably since 1861. In Wigtown 14 out of a total of 17 parishes had their maximum population in that year, as had 10 out of 26 parishes in Kirkcudbright. This general decline was in part due to the increased facilities for rural exodus provided by the improved communications, and also to the attraction of the industrial towns of both England and Scotland. The change from stock rearing to dairy farming appears to have been coincident with the fall in population. The historic interest of the province and its attractive scenery has stimulated an annual tourist invasion which is not without its material compensations. Its economic function would appear to be to supply food materials for the industrial towns of Central Scotland and possibly Northern Ireland.

The Dales

In Nithsdale, Galloway is joined to the "march" country of the dales. There is no natural boundary between England and Scotland, and the historic "debatable" land between Esk and Sark serves to emphasise this. The Plain of the Solway on the Scottish side is a continuation of the Plain of Carlisle beyond the border, and the New Red rocks underlie the mosses which hamper the egress of the Annan and Nith, and form a wide band up to their junction with the Carboniferous Limestone which flanks the Silurian hill country to the north. The topographical forms are everywhere more mature than they are in Galloway. The softer rocks and their greater diversity together with the greater erosive power of the streams have produced the essential differences. East of Annandale, a line joining Lockerbie and Langholm roughly marks the northern limit of the newer rocks, and Silurian forms predominate to the north. The basins of Thornhill in Nithsdale and Lochmaben in Annandale are filled with rocks of Permian and Triassic age, while Coal Measures occur at Sanquhar and Kirkconnell in Nithsdale and at Canonbie in the junction of Esk and Liddel. Coal-mining, quarrying and brick and tile manufacture are all associated with these basins contained by the dales. Whole districts in Glasgow have been built of the red "freestone" quarried at Locharbriggs and other parts of the dales. The Silurian shales were formerly used for roofing, but the lighter Welsh slates are now more popular. The blue clays of the lower river basins on the 50-foot beaches are used in brick and tile and similar industries. Along the Solway shore is a succession of extensive peat mosses such as Lochar Moss which are in part cultivated but largely waste land. An attempt to work the peat chemically at Ironhurst proved a commercial failure. The local residents use the peat as fuel.

With the exception of a portion of the Liddel valley and the upper Nith, the "dales" lie in the county of Dumfries. About half of its total area is under grass and heath, and one-fifth of it is arable land. The highest points are in the north. Queensberry Hill (2285 ft.) is central and commands Nithsdale and Annandale, but is surpassed by Hartfell (2651 ft. or 808 m.) which divides Moffatdale from Tweeddale. Eskdale is dominated by Whita Hill (1162 ft. or 354 m.). Volcanic rocks of Old Red Sandstone age add to the variety of hill contours. Birrenswark Hill (920 ft. or 280 m.), famed for its Roman camp, is a lava flow of this kind. Lower Annandale is divided from Nithsdale by a spur which separates the Plain of Dumfries and the Plain of Lochmaben, and sinks at its southern end under the Solway Moss*. Both dales are notable for their rich cultivated soils especially in the holms of river alluvium. Important passes such as Dalveen, Enterkin and Mennock lead from the dales into the hill country where sheep-rearing is the main occupation. The Esk flows for 26 miles (42 km.) to join the Liddel water, and in that distance crosses four geological formations—the Silurian, the Old Red Sandstone, the Coal Measures and the New Red rocks. The latter portion is the most fertile part of Eskdale. In the hill country there are no large areas of hill peat, as in Galloway, and the typical round grass-covered hills of the Southern Uplands prevail. The advance of agriculture was

* For an excellent summary account of the Dumfries region, see J. D. Ballantyne in *Observation*, Part 7, 1926, and Part 10, 1927.

accompanied by a decline in natural forests. Most of the timber of to-day therefore consists of plantations along the river courses, beech and ash, and the hardier spruce and larch.

In the region between Nith and Esk the climate is mild (Jan. 38°·5 = 3°·6 C., July 59°·5 = 15°·3 C.), but the rainfall is heavy (53 in. or 1325 mm.). At present cattle-rearing appears to be declining and giving place to sheep farming. Cheviots are bred on the higher grass walks and half-bred lambs—a cross of Cheviot and Leicester—in the lowlands. Ewes, Eskdalemuir and Langholm are the chief sheep-rearing districts. The "dales" are pre-eminent for arable mixed farming, and towns such as Dumfries, Thornhill, Lockerbie, Moffat and Langholm are market towns. The arable farms vary in size between 100 and 300 acres (40–120 ha.), while the pastoral hill farms vary from 300 to 3000 acres.

The total population of the county of Dumfries in 1921 was 75,370. Dumfries (15,728), the county town, has woollen, linen and silk mills. Annan (3928) is the most important fishing port of the Solway on the Scottish side, and is noted for salmon and shrimps. Langholm has tweeds and tanning industries and finds its markets in England rather than in Scotland, as it cannot compete with the Tweedside mills. The general coal depression has resulted in the Canonbie mines being closed.

About 24 per cent. of the population is engaged in agriculture, by far the most important occupation. There is a big contrast between the "county districts," *e.g.* Thornhill (12,699), and the half-empty moorland parishes such as Westerkirk (368) where the population has declined by 47 per cent. since 1811. The density in the Langholm area is 17·3 per 100 acres, while in Eskdalemuir it is under 1 per 100 acres. The Esk-Liddesdale region may be said to be transitional between the agrarian "dale" country and the true hill grass moors which form the background to the staple industry of the Tweed valley.

From the dales four main roads radiate and lead to the several dales of the Tweed basin. Of these two pass from Moffat in Annandale to the upper Tweed and to the Yarrow respectively. The other two lead to Teviotdale, the one by Liddesdale, the other from Langholm in Eskdale by Ewes Water to Teviothead. Between these two pairs of routes lies upper Eskdale, with no main road. The isolation of this district led to the selection in it of a site for a magnetic observatory, opened in 1909. Eskdalemuir Observatory, standing some three miles from the head of the dale, and at an altitude of 800 ft. (244 m.), is thus likely always to be free from artificial magnetic disturbance due to electric traction and power circuits. It is now one of the principal magnetic observatories in the world, and an important meteorological station, while observations are made also in other branches of geophysics*.

C. THE TWEED BASIN

Alan G. Ogilvie, with a contribution by the late William G. Smith

The eastern part of the Southern Uplands, drained by the Tweed and its tributaries, forms a compact geographical unit. It has a lowland nucleus for human development which lies open to the sea eastward; but elsewhere the basin is wellnigh shut off by the encircling bare and bleak hills that form the watershed. One important tributary, the Till flowing north-west, lies in England; and one head-stream, the Lyne, comes from the Midland Valley. But essentially the natural region coincides with the drainage basin.

Within the basin there is a striking contrast between the Merse lowland—bordering the lower Tweed—and the uplands through which the upper and middle Tweed and all of its greater affluents flow in deep and generally narrow valleys. The people of these—Tweeddale, Ettrickdale, Teviotdale, and Lauderdale—have a strong local patriotism arising from the secluded nature of the valleys, a patriotism which had full play throughout centuries of Border warfare and which is still evident in athletics and other ways. The contrast is fundamentally geological, as is explained above (p. 468). The topographical contrast is sharpened by the occurrence of a zone of andesitic lavas and of other igneous rocks extending from the abrupt isolated boss of the Eildon Hills north-north-eastward to meet the greywackes of the Lammermuirs and, south of the Tweed, eastward to the spurs of Cheviot. This hard edge has so retarded erosion of the Old Red Sandstone to the west of it that a considerable expanse of a relatively smooth surface, apparently continuous with that of the Midland Valley described as the higher Lowland peneplane†, has been preserved, and most of the valley floors of the "dales" may also be referred to this higher surface.

The Basin as it was

The Tweed basin owes to its geographical position athwart the easiest routes between England and mid-Scotland a war-troubled history during many centuries. And, if we may judge from the nature of prehistoric remains in the region, vigorous warfare must

* Information kindly supplied by the late Director, A. Crichton Mitchell, D.Sc.

† *Vide supra*, p. 426 and Fig. 53, p. 427.

have here also characterised the periods for which written records are wanting. Few evidences of the peaceful life of the prehistoric inhabitants are extant. On the other hand the numerous hill forts as well as cairns and tumuli seem to bespeak a considerable warlike population perhaps in Roman times or earlier but certainly during the dark ages that followed the withdrawal of the Legions. On Fig. 58, 217 forts have been plotted, of which about 90 are in two groups that appear to indicate contests for command of entrances

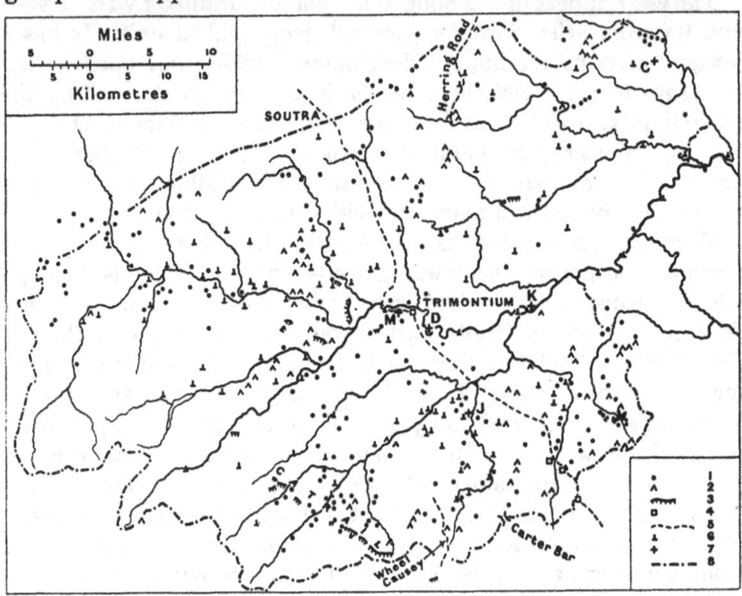

Fig. 58. The Tweed Basin: prehistoric and historical sites; 1, hill forts; 2, cairns and tumuli; 3, trenches, etc.; 4, Roman camps; 5, ancient roads; 6, mediaeval castles and towers; 7, monasteries; 8, boundary of the region. For river-names, see Fig. 59, p. 493.

to or exits from the basin west and south respectively. These "duns," capping hilltops and spurs and occasionally valley crags, encircle the concentration of valley routes upon the Tweed above Innerleithen, and again are spread upon the slopes of the Cheviot Hills from their north-eastern spurs westward to the middle Teviot. The distribution on the Lammermuirs also indicates defence of routes up the valleys, over the summit and round the coast*. The small number of such structures about the lower Tweed where suitable sites exist may perhaps be taken as evidence that the fort builders were hillmen, which is likely in view of the swampy nature

* *Cf.* p. 435.

of the lowland in ancient times and of the undoubtedly greater luxuriance of the forest there.

No forts crown the hills in the south-west. In Teviotdale remains are practically limited in this direction by a double-banked ditch running across the hill and valley known as the "Catrail," the purpose of which—whether defence, road or boundary—is unknown*. Similar works remain on the hills between the Tweed and the valleys of Gala and Ettrick on either hand.

The hills are scored in many directions by broad drove roads, some of which are probably routes of great antiquity. Two tracks are traditionally ancient: the "Herring Road" on the eastern Lammermuirs and the "Wheel Causey" at the head of the Jed valley†. The Roman road, "Dere Street," leading from the Tyne at Corbridge near Hadrian's Wall to the mouth of the Lothian Esk, traverses the Tweed basin nearly in a straight line. Crossing the Cheviot divide just east of the Carter Bar pass, where a great camp still remains to be excavated, its course is discernible, and still partly used, north-westward to the eastern foot of the three Eildon Hills; whence the name of the large medial Roman station here—Trimontium (now Newstead). From here the route (now uncertain in detail) ascended the plateau west of the Leader valley to cross the watershed at Soutra.

The nearest modern counterpart of mediaeval life in Tweeddale is probably to be found in Albania; and the Albanian *kula* is the equivalent of the Border peel tower‡. The distribution of all of these keeps, in so far as vestiges remain, as well as of larger castles—about 100 in all—is shown on Fig. 58. They stand for the most part at lower altitudes than the prehistoric forts, but otherwise their general arrangement is similar. They were the houses of the lairds, and round them clustered for protection the humble dwellings of the people. Thus, the map gives some idea of the distribution of population throughout the long period in which raids from England were to be anticipated. The concentration of peels in the dales of Jed and Teviot reminds us that this was a vulnerable sector and that the bands of *comitadjis* in Balkan countries had their parallel on both sides of the Border in the maintenance of bands of outlaws whose business was pillage.

* This is shown where certain on Fig. 58; but there are more doubtful traces north-westward across Ettrick, and thence north-east to link with the other works referred to.

† On Spartleton Edge near the point where the "Herring Road" descends to the Whitadder there are groups of hut circles surrounded by enclosures.

‡ These are quadrangular stone towers, but originally they were merely enclosures of palisades; hence the derivation: Fr. *pel*, Lat. *palus*, a stake.

To turn now to evidences of peaceful development. The first rifts in the cultural gloom of the region may be indicated in the sites of Celtic monasteries that flourished from the 7th century at Coldingham in the north-east and Old Melrose on the high bluff (*meall ros* in Gaelic) in the Tweed meander below Newstead. From here the torch was passed on, for the latter House produced St Cuthbert who converted Northumbria. The Tweed valley was wrested from Northumbria for Scotland in the 11th century; and in the 12th century the region definitely entered the Middle Ages under enlightened Scottish kings. Among the real pioneers in the Arts, in agriculture and in commerce were the monks of the great abbeys: Dryburgh, Kelso, Melrose, Jedburgh and Coldingham, all but the last founded or confirmed by King David I.

The abbeys became the nuclei of small towns; and the same century saw the foundation in Tweeddale of two of the first four royal burghs of Scotland, Roxburgh and Berwick. Roxburgh grew around the king's castle at the confluence of Tweed and Teviot—now completely disappeared and replaced by Kelso across the Tweed. Berwick, doubtless an important haven from earliest times, was soon to become the leading seaport of Scotland, in spite of the exposed situation of the estuary, for the same wise king encouraged Flemings to settle and foster trade with the Low Countries in the wool and hides of Tweeddale. The monks and the burghers, then, in co-operation may be said to have welded Tweeddale into an economic unit with the same products as it has to-day save for the modern manufactures of the middle basin. Agriculture of mediaeval type spread on the lower ground which was gradually cleared and fenced. Wheat, bere, oats, pease, beans and flax were grown, and with the plentiful supply of river salmon, food was sufficient. The bridges of Berwick and Peebles and sundry fords as well as numerous wind- and water-mills were maintained; several roads were made—well enough for a heavy eastward traffic of ox-carts—to the natural outlet, Berwick; and the monasteries developed the breeding of sheep and the systematic organisation of flocks and of hill pastures, and so founded the economic mainstay of the region. Thus Tweeddale began to take advantage of its natural aptitudes. It was the wealthiest district of Scotland till the end of the 13th century. Trade was centred in Berwick, Roxburgh, Jedburgh, Peebles and Selkirk, all royal burghs of the 12th century*, and in the abbey towns of Melrose and Kelso. But material progress soon became greatly hindered by recurrent raids and more serious English invasions. For purposes of defence the Scottish Border was divided

* Lauder became a royal burgh in 1494, and Hawick in 1537.

into three sectors or marches, of which Tweeddale formed the Middle and East Marches. The latter and most vulnerable coincided with the eastern lowland, which is still "the Merse," *i.e.* March.

Berwick unfortunately possessed an importance other than that of Tweeddale's port. It guarded the main road from England to Central Scotland. After changing hands several times it was finally lost to Scotland in 1482, and thenceforward till the Union the Tweed basin was without its port. Moreover the 16th century saw the decline and destruction of the monasteries. Life was peaceful but economically stagnant in the 17th century, for the frontier, though no longer a military line, was still a fiscal barrier till 1707. Not until the economic Revolutions, Agrarian in the 18th century, Industrial in the 19th, could Tweedside attain to the full prosperity for which it was naturally adapted. Until communications by road and railway were improved to meet the needs of industry, life in the dales remained backward. It will suffice to mention the extreme difficulty which the inhabitants experienced in obtaining fuel and lime even at the end of the 18th century. Both had to be carried on horseback for long distances; the coal came either from Northumberland or Midlothian and was consequently a great luxury, often unobtainable on account of the fords being impassable. Wood was almost as scarce, since the forests had been denuded, even whin (*Ulex*) was difficult to get, and peat was frequently far from the consumer. The lack of lime hampered domestic building, and thatched houses of "clat-and-clay" were still common. The mediaeval monks, however, had played their part in promoting agriculture and fostering spinning and weaving of flax and wool for home use. The glorious ruins of their convents to-day form an important asset to the region. These, and the local associations with Sir Walter Scott who first gave them wide publicity, attract a great throng of tourists to the middle Tweed.

The Basin as it is

The economy of the region to-day resembles that prevailing in the Middle Ages in the nature and general distribution of its agricultural production; but the methods of each are of course now vastly more efficient. In addition, a highly specialised textile industry is concentrated in the towns of the lower dales; and trade, no longer using its natural sea gate, now thrives upon an excellent road and rail system crossing the divides by several routes north and south.

As indicated above, the main topographic division is that between the Merse and the plateaus with their dales. The region as a whole lies mostly in the four counties: Berwickshire, composed of three districts—the Merse, Lammermuir and Lauderdale; Roxburghshire—mainly Teviotdale, but crossing the Tweed on the low ground and also overlapping the Solway drainage to include Liddesdale; Selkirkshire, anciently called Ettrick Forest; and Peeblesshire, formerly known as Upper Tweeddale; while the upper Gala valley lies in Midlothian.

The Merse, whose upper limit (see Fig. 59) may be taken as 400 ft. (120 m.), forms one of the largest expanses of drumlins in Britain; and the pattern of streams, roads, plantations, hedges and houses conforms almost exclusively to the trends of the drumlin axes and to the perpendicular directions. This area alone in the Tweed basin has a mean annual rainfall of less than 30 inches (750 mm.), fairly evenly distributed by months, but with maxima in August and October. The climate is in other ways immensely superior to that of the upper basin from the point of view of the "arable farmer." Indeed it may be said that for the growing of crops the Merse is physically one of the favoured spots of Scotland. It is all potential farm land, and Fig. 54 shows that over a large area less than 20 per cent. of it is under permanent pasture even in present economic conditions (1926) which are highly unfavourable to arable farming. Nevertheless agricultural conditions here cannot be understood without reference to the stock-raising industry, to which we must return.

Meanwhile we may note (from Fig. 54, p. 439) that while west of the lowland the land rises and the relief becomes rapidly greater, yet the area enclosed for agriculture as opposed to hill pasture broadens out and penetrates the Lammermuirs by the valleys of Leader and Gala where some crops are reaped above the 1000 ft. (300 m.) level. Thus there is a zone of hilly country extending nearly southward from the Gala across the Tweed at Melrose and penetrating the southern "dales" in which a notable proportion of the land is under crops. This zone abuts everywhere upon the great sheep runs and grouse moors of the Uplands, where only the narrowest strips of valley land are tilled. Altitude, relief, soil and climate—with rainfall varying from 35 to 60 inches (875–1500 mm.)—combine in the upper Tweed basin virtually to exclude the growing of crops on a large scale. Here we are emphatically in the heart of the hill sheep country. The Tweed basin as a whole provides pasture for a large proportion of Scottish sheep; and Fig. 56 shows approximately how they are distributed*. It will be seen that they

* The number represented is the maximum for the year, *i.e.* after the lambing season.

occupy upland and lowland alike in large numbers, though the density is considerably reduced on the best of the arable land mentioned above. But if Figs. 54, 55 and 56 are considered together it is clear that there is a contrast in the utilisation of land, east and west. In the lowland a high type of mixed farming prevails, with varied crops, most of which are used in feeding sheep and store cattle for the market and dairy cattle sufficient for the local needs. Thus oats is the chief cereal; much land is under turnips; and beans are grown on heavy soils. Barley for malting—occupying about one-quarter the area of oats—and wheat are also important crops. It is a district of large and usually prosperous farms, the average size for Berwickshire, 211·9 acres (85·7 ha.), exceeding the figure for all other Scottish counties. The "dales," on the other hand, are practically given up to sheep farms.

Hill Pastures and Sheep Rearing

Contributed by the late William G. Smith

The hill pastures of Tweeddale form a part of a large area of Scotland (from 50 to 60 per cent.) which is land never cultivated, or in part once cultivated, but now so long unploughed that it may be regarded as almost natural grassland. Most of the lower hills were in past times forest or scrub of oak, birch or pine, but from various causes, such as deforestation, followed by sheep-grazing and burning of heather and moor bent, tree growth has been replaced by grassy or heathery herbage. Some idea of the former extent of forest may be gleaned from observation of the present woods and plantations from Broughton to Melrose, rising in places to the watershed.

In upper Tweed, as far as Broughton, the farms are large and almost entirely pastoral, generally extending from the watershed to the streams. Such farms are rarely less than 1000 acres (400 ha.), several exceed 5000 acres, and they are almost purely sheep farms with few cattle. Where the haughs broaden out, from Broughton downwards, the arable land is increased, though the same type of farm predominates, extending from watershed to stream, but with more cattle. Lower down the valley, from Peebles, the holdings tend to be segregated into lowland farms with arable and grass, sheep and cattle, while the higher slopes and summits are pastoral sheep farms. From St Boswells downwards through the Merse, the lowland farm predominates, more or less arable, but the type of husbandry throughout tends to be modified by the requirements of sheep-grazing.

The Tweed area includes both green pasturage and heather, and this has its influence on the sheep stocks. Green pasturage is generally grazed by the white-faced, hornless Cheviot sheep, whereas the coarser herbage and heather are better utilised by the hardier horned Black-face sheep. The low-ground farms with rotation grass and turnips carry mainly cross-bred sheep, Cheviots or Black-faces crossed with larger earlier-maturing breeds such as Border Leicester or South-Down.

Just as the Tweed has its sources in the upland streams, so the sheep industry of the lowlands of the Tweed and far south into England has its foundations in the upland sheep farm. The ewes are born there and each year ewe lambs are added to the permanent flock to replace old ones sold away to become the mothers of cross-bred lambs in the low country. In this way the ewe stocks of the uplands become acclimatised and are fitted to utilise the herbage of their own hills better than introduced sheep. The flocks are graded up by the introduction of rams, often bought at high prices. The limiting factor of a hill farm is the number of ewes that can be maintained over winter. The stock is more or less doubled during the lambing season, but before the winter all surplus ewes and male sheep are removed to bring the hill back to its winter carry. There is thus an annual stream of hill sheep to the lowlands, and the hill farm is the seat of a great basic industry which influences the economics of the greater part of Scotland.

The natural herbage determines the carrying capacity of the hill farm, since ploughland is relatively insignificant. In the south of Scotland during winter each ewe requires about two acres (0·8 ha.) of grazing on a good hill farm, but where the natural herbage is less nourishing four acres may be needful. Hence the rental per acre is always small, a little above or below two shillings, but varying from farm to farm. In a country of hills and valleys like Scotland, the hill farm generally extends from one of the larger streams upwards to the watershed, so that topography has a considerable influence on the area of each farm and of its utilisation by stock during the winter. Where the watersheds are high, as amongst the larger Bens of the Highlands, large areas can only be used for summer grazing, hence very low rentals and the conversion of these into the more profitable deer forest or grouse moor.

Sheep farming has introduced many changes in the primitive vegetation of Scotland. The most significant is the destruction of forest, so evident in the south. This is due directly to the destruction of tree seedlings by grazing sheep or rabbits, hence afforestation must be preceded by fencing. Indirectly in laying out the sheep walks there has been suppression of forest, since the better forest land coincides with the better grazing land. The absence of trees has favoured other types of herbage: heather has replaced pine forest, and bracken, originally a plant of the open spaces of oak and birch woods, has increased so much as to become a pest, since it is not grazed by sheep. Man has also influenced the sheep farm in other ways so that what may appear to be the natural vegetation of the hills is actually the result of a process of farm management. Each spring large areas are burned to clear off old dry herbage and to promote new growth. Old heather when burned returns from seedlings or from new shoots from the old stumps. Experience indicates that heather makes better grazing if it is burned before it becomes fifteen years old, but over large areas the return has been so slow that the heather is being replaced by inferior herbage, including Blaeberry (*Vaccinium*), White Moor Bent (*Nardus stricta*), and Blowgrass (*Molinia*). This leads to marked depreciation, since young heather, next to alluvial grassland, is the most important food of the hill sheep. Moor-burning of the grasses named above is also practised, and considerable improvement of the grazing follows the burning of upland peat moors. The open sheep drain of the hill farm, said to have been first used in Eskdalemuir about 1770, has played an important part in the improvement of the herbage. These conduct superfluous water to the streams, thereby making wet ground accessible to stock, and at the same time favouring the growth of more useful grasses, etc. An adaptation of the sheep drain is also used to conduct water from wells or springs to drier parts where the irrigation improves the herbage; for example, heather can be converted into grass in a year or two*.

Immediately below the hill pasture there is in many parts a zone of enclosed fields which were once ploughland, but have now reverted to a mixed grassland that is not far removed from the naturally wild. Much of this can be considerably improved by manuring with lime and phosphates and by draining.

The wider valleys of the Tweed have more extensive alluvial areas, hence more grassland, and they are used for sheep farms of a different type. Ewes from the hill farms are crossed with rams of larger breeds, such as Border Leicester and South-Down, and

* For an account of the vegetation of the Moorfoot Hills and a detailed map of a typical upland area, see William G. Smith, "The Distribution of *Nardus stricta* in Relation to Peat," *Jour. of Ecology*, vol. VI, No. 1, 1918.

lambs are obtained earlier in spring. This type of farming involves the provision of turnips for winter and young grass for spring, both of which require ploughland. Increased arable land opens the way for cattle feeding, so that the lower farms of the Merse are of the mixed lowland type.

The Textile Industry

Alan G. Ogilvie

Prior to the late 18th century spinning and weaving were mainly house industries, though in some of the burghs the websters were an incorporated body at least by the 16th century. Flax and wool were both manufactured; lint mills existed in Melrose, Ancrum and Earlston, and towards 1700 their products were known beyond Scotland; but in the same period woollen-weaving was of only local importance. As linen declined, mainly owing to competition, cotton was tried—about 1790—and the goods attained some fame. But about the same time the weavers of Galashiels and Hawick were turning their attention to machinery, with much more important results.

Given machinery, the local advantages of the towns in the dales for the development of spinning and weaving on a modern basis were quickly utilised. These, as has often been pointed out, were abundance of wool and plenty of water-power; while benefit was now derived from the absence of limestone in the hills in that the water for wool-washing and dyeing was soft. As the industry grew and water-wheels gave insufficient power, the distance from coalfields emerged as a disadvantage. In view of this and of the greater distance from markets and from the sources of imported wool, the Tweeddale towns have had to arm themselves specially against the obvious competition of the more favoured Yorkshire mills. They have therefore specialised in goods of fine quality. The original rough cloths made—"Galashiels grey" and "Hawick blue"—gave way to checks, based on the old shepherd plaiding. These "tweels" became fashionable as "tweeds" owing to the "appropriate blunder" of some English clerk; and so the product of Tweeddale mills maintains its place in a world-wide market where high class cloths of this character are in demand. Hawick in increasing degree has specialised in hosiery that has a reputation, at least in Britain, rivalling that of the cloths. But for fine goods fine wools are required. This has led first to the increase of Cheviot sheep in the district, since the coarse wool of the Black-face serves only for rough cloths and for carpets, and secondly to vast importation of foreign wools. Thus the industry to-day is an example of geographical momentum; for the initial advantages have greatly dwindled. The industry has even outgrown the space it was founded in. The valleys of Gala and Teviot are so narrow as to cramp the growth of Galashiels and Hawick, especially the former, so that the mills of Selkirk were largely built to supplement those of Galashiels.

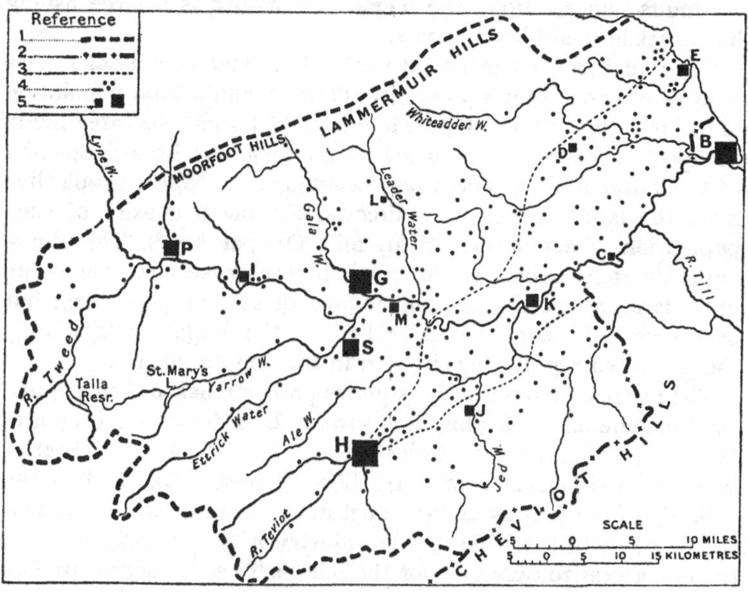

Fig. 59. The Tweed Basin: distribution of the population (1921); 1, boundary of the region; 2, Anglo-Scottish Border; 3, limit of the Merse; 4, rural population (dots = 200); 5, towns (proportional to population).

Distribution of Population

From what has been said it will be evident that the two divisions of the region will present a contrast in density of population. Fig. 59 is an attempt to show the distribution (in 1921) by the dot method. The small dots, each representing 200 persons, have in many instances been placed at villages, and otherwise near the centre of a group of farms*. The Merse as there delimited contains about 265 sq. miles (686 km.^{2}). Its population is about 28,000, with a density, therefore, of 104 per sq. mile (41 per km.^{2}). This

* A numerical value smaller than 200 would have given a better representation; but this value is adopted to make the map comparable with that of Mr Crowe (see p. 438, footnote), the scale of which, however, is smaller.

population is almost entirely rural and evenly spread. In Berwickshire the percentage (40) of the occupied population engaged in agriculture is higher than in any other county south of the Highlands. Of the four burghs, with some 9000 people in all, Kelso and Coldstream are typical "country towns" and Duns merely the largest of a string of villages placed where the rivers, gnawing their way through the edge of the smooth plateau fronting the Lammermuirs, emerge upon the Merse. Eyemouth is a large fishing haven set in a cliff-bound coast.

The remainder of the Tweed basin—the uplands and dales—has an area of about 1435 sq. miles (3717 km.^{2}) and a total population of 82,000. But of these 48,000 live in eight burghs that are mainly occupied in the woollen industry. Deducting, at a hazard, one half of the burghal total, and thus assuming that 58,000 people live from the land directly or indirectly, the mean density of such population is about 40 per sq. mile (16 per km.^{2}). This figure approximately expresses the truth in the sense that the entire uplands represent the sustenance area of the sheep farmers; but of course the people actually live in the larger valleys, only shepherds' cottages being situated in the remoter glens*.

The burghs, represented by squares proportional to their population, include all the leading mill towns. In order of size they are: Hawick (16·4 thousand), Galashiels (12·9), Selkirk (5·8), Peebles (5·5), Innerleithen (2·4), Jedburgh (2·4), Melrose (2·2), Lauder (0·8).

Study of the railway and road net in the Tweed basin shows that the continued life of the textile industry is due in large measure to good access to railways; for the middle Tweed is served by two routes from the north and two from England, one of each constituting the "Waverley" through route. This route brings out the strong nodality of St Boswells at the east foot of the Eildons, where it meets the line from Berwick. This place, at the geographical centre of the basin, is on the brink of the plateau where the Tweed has carved two meanders between Leaderfoot and its entrance upon the Merse. In one of these was the original Melrose Abbey, and Dryburgh stands in the haugh of the lower. Nearby on the plateau was Trimontium, and Melrose is just beyond; Newtown St Boswells† is a small market town, for long the site of the leading summer fair in the region. To-day, while Hawick and Peebles are also great sheep markets, St Boswells, drawing stock from all the

* In comparing this map (Fig. 59) with that of sheep distribution (Fig. 56) where one dot represents 500 sheep, it should be remembered that one shepherd with his dogs can normally tend from 600 to 750 sheep.

† St Boisel was the abbot of St Cuthbert's time.

basin east of Innerleithen, attracts most of the English buyers of lambs; and this on account of the rail facilities. Kelso has a weekly grain market and it shares the sale of fat stock with St Boswells. This quiet rural abbey town stands apart from the clatter of the up-stream mill towns yet it is a link in the industrial chain, for it takes first place in Scotland for the sale of rams of the valuable wool-giving breeds and sends its stock to Australia, New Zealand and South Africa. The harvest returns anon to feed the Tweedside looms*.

* The Editor wishes to thank Miss C. P. Snodgrass, M.A., and Miss F. C. Welch, B.A., for access to unpublished essays by them on parts of the region.

been called 'a Lincolnshire ... [illegible] ... of the English buyers of lambs' and this on account of the tall lambs. Kelso has a weekly grain market and it shares the sale of fat stock with St Boswells. [illegible] ... stands ... [illegible] ... in the [illegible] ... on the side of ... [illegible] ... [illegible] ... and [illegible] ... the lowest returns ... [illegible] ... the Tweedside [illegible].

[illegible] ... and [illegible] ... [illegible] ... on [illegible].

INDEX

[It has been thought that the systematic subdivision of the book would help readers to find references to particular physical and human features and names, and the compiler of the index has therefore avoided lengthening it by the inclusion of a large number of names, but has rather specialised on references to geographical subjects.]

Printed in the United States
By Bookmasters